T0311677

## Sky and Ocean Joined

As one of the oldest scientific institutions in the United States, the U. S. Naval Observatory has a rich and colorful history. It was initially founded as the Depot of Charts and Instruments in 1830; in 1844 it became the first national observatory of the United States, analogous to the famous observatories at Greenwich and Paris. It remained the only U. S. national observatory until the 1950s.

This volume is, first and foremost, a story of the relations among space, time, and navigation, from the rise of the chronometer in the United States to the Global Positioning System of satellites, for which the Naval Observatory provides the time to a billionth of a second per day. It is the story of the development of the Master Clock of the United States, of star catalogs and almanacs, and of numerous discoveries and adventures along the way. Among the latter, this history encompasses the discoveries of the moons of Mars and Pluto, the sixteen American Transit-of-Venus expeditions dispatched by the Observatory in the nineteenth century, and colorful figures like Matthew Maury, Simon Newcomb, and Asaph Hall. It is also a story of the history of technology, in the form of telescopes, lenses, detectors, calculators, clocks, and computers, over 170 years. It is a story of how one scientific institution under government and military patronage has contributed, through all the vagaries of history, to almost two centuries of unparalleled progress in astronomy.

Sky and Ocean Joined will appeal to historians of science, technology, scientific institutions, and American science, as well as astronomers, meteorologists, and physicists. It provides a fascinating insight for scholars and general readers interested in Naval history and navigation, timekeeping, positional astronomy, and celestial mechanics, and also charts the relationship between astronomy and astrophysics, the rise of national observatories, and the development of science under government and military patronage.

STEVEN J. DICK has worked as an astronomer and historian of science at the U. S. Naval Observatory since 1979. He obtained his B.S. in astrophysics (1971), and M.A. and Ph.D. (1977) in history and philosophy of science from Indiana University, and is well known as an expert in the field of astrobiology and its cultural implications. Among other publications he is author of Plurality of Worlds: The Origins of the Extraterrestrial Life Debate from Democritus to Kant (1982), The Biological Universe: The Twentieth Century Extraterrestrial Life Debate and the Limits of Science (Cambridge University Press, 1996), and Life on Other Worlds (1998). He was also editor of Many Worlds: The New Universe, Extraterrestrial Life and the Theological Implications (2000). Dr Dick served on Vice President Gore's panel to examine the societal implications of possible fossil remains of life in the Mars rock, and is the recipient of the NASA Group Achievement Award for helping to initiate NASA's multidisciplinary program in astrobiology. He has served as Chairman of the Historical Astronomy Division of the American Astronomical Society (1993–94), and as President of the History of Astronomy Commission of the International Astronomical Union (1997–2000).

Frontispiece: The seal of the U. S. Naval Observatory depicts Urania reaching toward the stars, with the terrestrial globe in her outstretched hands. The motto, taken from the Fourth Book of the *Astronomicon* of Manilius, may be translated

> Then, too, the pilot's care: the stars are scaled
> And sky with ocean joined.

Both the seal and the motto emphasize the Observatory's origin in the need for improved navigation. See C. H. Davis, "Explanation of the Seal of the U. S. Naval Observatory," *Astronomical and Meteorological Observations Made at the U. S. Naval Observatory During the Year* 1865 (Washington, 1867).

# Sky and Ocean Joined

## THE U.S. NAVAL OBSERVATORY 1830–2000

STEVEN J. DICK

CAMBRIDGE
UNIVERSITY PRESS

CAMBRIDGE UNIVERSITY PRESS
Cambridge, New York, Melbourne, Madrid, Cape Town, Singapore, São Paulo

Cambridge University Press
The Edinburgh Building, Cambridge CB2 8RU, UK

Published in the United States of America by Cambridge University Press, New York

www.cambridge.org
Information on this title: www.cambridge.org/9780521815994

First published 2003
This digitally printed version 2007

*A catalogue record for this publication is available from the British Library*

*Library of Congress Cataloguing in Publication data*

Dick, Steven J.
Sky and ocean joined : the U.S. Naval Observatory, 1830–2000 / Steven J. Dick
   p.   cm.
Includes bibliographical references and index.
ISBN 0 521 81599 1
1. United States Naval Observatory – History – 19th century.   2. United States Naval
Observatory – History – 20th century.   3. Astronomy – United States – History.   I. Title.

QB82.U62 U553 2002    522′.19753–dc21    2002017386

ISBN 978-0-521-81599-4 hardback
ISBN 978-0-521-03750-1 paperback

# Contents

Contents

# Acknowledgments

Already upon my arrival at the Naval Observatory in 1979, the need for a detailed history of the institution was clear. Here was one of the oldest scientific institutions in the U. S. government, analogous to the famous Greenwich, Paris, and Pulkovo Observatories, with a history that had been almost entirely ignored for a half century. The brief studies by J. E. Nourse, *Memoir of the Founding and Progress of the United States Naval Observatory* (1873), and Gustavus A. Weber, *The Naval Observatory: Its History, Activities and Organization* (1926), while they are useful sources, were outdated and not intended as comprehensive histories. Jan Herman's *A Hilltop in Foggy Bottom: Home of the Old Naval Observatory and the Navy Medical Department* provided a skillful, illustrated overview of the Foggy Bottom site, and Gail S. Cleere's *The House on Observatory Hill: Home of the Vice President of the United States* did the same for the present site from the point of view of the Vice President's residence (originally the residence of the Observatory Superintendents). Two sesquicentennial publications (*Sky with Ocean Joined*, an internal publication edited by Dick and Doggett in 1983, and *Proceedings, Nautical Almanac Office Sesquicentennial*, edited by Fiala and Dick in 1999) contain historical sections but are by no means comprehensive. In the most fundamental sense, then, this book fills a gap both in naval history and in the history of science in the United States.

This book was written over 15 years at the Naval Observatory, during which my other duties in astronomy, history, and public affairs gave me the enormous advantage of learning about the institution first-hand. My thanks go first to the Observatory's Superintendent, Captain Richard Anawalt, who made it possible for me to begin work on this volume in 1987, and to a series of Superintendents (see Appendix 2) who have supported it since then, ending with Captain Ben Jaramillo and Deputy Superintendents Commander Doug Groters and Commander Susan Greer.

On the scientific staff of the Observatory, my thanks go first to Scientific Directors Gart Westerhout and Kenneth Johnston. Robert W. Rhynsburger, an astronomer in the 6-inch Transit Circle Division, first stimulated my interest in the Observatory's history upon my arrival into that Division in 1979. Even though we overlapped on the staff for little more than a year, he did much to lay the groundwork for such a history by identifying photographs and freely offering his substantial knowledge of the Observatory's history. My thanks go to him for also reading the entire manuscript, and to Dennis W. McCarthy, Thomas C. Corbin, Theodore Rafferty, Dan Pascu, Brian Mason, P. Kenneth Seidelmann, Demetrios Matsakis, George Kaplan,

John Bangert, Sethanne Howard, and Gernot Winkler for reading chapters or sections relevant to their expertise.

Other members of the staff to whom I am indebted include Theodore Rafferty of the Astrometry Department, whose intimate knowledge of instrumentation greatly benefited me; LeRoy Doggett of the Nautical Almanac Office, who provided advice and moral support before his untimely death in 1996; Brent Archinal for his inventory of historic instrumentation and other discussions; Merri Sue Carter for her work on women employees at the Observatory; James DeYoung for his compilation of references to Observatory clocks and timekeeping apparatus; and a whole series of scientists who willingly submitted to detailed oral-history interviews (see Appendix 1). The longevity record in the latter respect goes to Paul Sollenberger, who came to the Observatory in 1914 – and remembered every detail when I interviewed him at the age of 96.

My sincere thanks go to Brenda Corbin and Gregory Shelton, Librarians of the U. S. Naval Observatory, who, as keepers of one of the finest astronomy libraries in the United States, have been tremendous resources throughout this study. Thanks also go to Brenda for help in identifying Nautical Almanac Office staff in Figure 8.7 (one of many challenges in photo identification in this study), and unstinting moral support. Unless indicated otherwise, the photographs are from the Naval Observatory photograph collection organized by Mabel Sterns as a labor of love. In this regard thanks are due also to my colleague Geoff Chester for help in photograph preparation, and for taking on public-affairs duties that allowed me to finish this history. My appreciation also goes to a series of Observatory summer students over the years who provided support on biographical databases, historical research, and archives inventory: Michael Liu, Beth Johnson, Whitney Treseder, and Shawnette Adams. Finally, my thanks go to the Naval Observatory History Committee, which is charged with the preservation of the Observatory's instruments and history.

Outside of the Naval Observatory I wish to thank my colleagues at the Navy Historical Center for unfailing provision of resources and support through two of its Directors, Dean Allard and William S. Dudley. The Navy Department Library has also been an invaluable resource. Outside of the Navy, my thanks go to Ian R. Bartky, a long-time student of timekeeping and former scientist and administrator at the National Bureau of Standards (now the National Institute of Standards and Technology). He read the entire manuscript and gave detailed commentary, and his own book *Selling the True Time: Nineteenth-Century Timekeeping in America* has been an indispensable source. Craig B. Waff also read and commented on my chapter on the early history of the Nautical Almanac Office, for which his own research has been indispensable. David DeVorkin, Marc Rothenberg, Deborah Warner and Robert Smith, fellow historians of astronomy in Washington, provided professional support and collegiality.

Although this might be considered an "official history" in the sense that it was written while I was employed by the Naval Observatory, the terminology should not be considered pejorative. I have been given free rein to "tell it like it was." Parts of chapter 1 appeared in S. J. Dick, "Centralizing Navigational Technology in America: The U. S. Navy's Depot of Charts and Instruments, 1830–42" *Technology and Culture*, **33** (July, 1992), 467–509. Parts of chapters 1 and 7 appeared in S. J. Dick, "John Quincy Adams, the Smithsonian Bequest, and the Origins of the U. S. Naval Observatory," *Journal for the History of Astronomy*, **xxii** (February, 1991), 31–44, and S. J. Dick, W. Orchiston and T. Love, "Simon Newcomb, William Harkness and the Nineteenth-Century American Transit of Venus Expeditions," *Journal for the History of Astronomy*, **xxix** (1998), 221–255. Parts of chapters 3, 8 and 12 appeared in S. J. Dick, "History of the American Nautical Almanac Office," in A. Fiala and S. J. Dick, *Proceedings, Nautical Almanac Office Sesquicentennial Symposia*, 11–53. I thank the publishers for permission to use those parts in this history.

Finally, at Cambridge University Press, my thanks go to Simon Mitton, Susan Francis, Carol Miller, Jillian Culnane, Alison Litherland, and Steven Holt.

# Abbreviations used in references and tables

| | |
|---|---|
| AAAS | American Association for the Advancement of Science |
| ACRS | Astrographic Catalog Reference Stars |
| AENA | *American Ephemeris and Nautical Almanac* |
| AGK | Astronomische Gesellschaft Katalog |
| AHP | Asaph Hall Papers, Manuscripts Division, Library of Congress |
| AJ | *Astronomical Journal* |
| AJSA | *American Journal of Science and Arts* |
| AMS | *American Men of Science* |
| AN | *Astronomische Nachrichten* |
| APAE | *Astronomical Papers Prepared for the Use of the American Ephemeris* |
| ApJ | *Astrophysical Journal* |
| APM | American Prime Meridian papers |
| APS | American Philosophical Society |
| AR | *Annual Report of the Superintendent of the U. S. Naval Observatory* |
| BAAS | *Bulletin of the American Astronomical Society* |
| BAC | British Association Catalogue |
| BAN | *Bulletin of the Astronomical Institutes of the Netherlands* |
| BDAS | *Biographical Dictionary of American Scientists* |
| BMNAS | *Biographical Memoirs of the National Academy of Sciences* |
| BONC | Board of Navy Commissioners |
| CAB | *Appleton's Cyclopedia of American Biography* |
| CCD | Charge-coupled device |
| CGPM | Conférence Général des Poids et Mesures (General Conference of Weights and Measures) |
| CIPM | Comité Internationale des Poids et Mesures (International Committee of Weights and Measures) |
| CNA | Committee on Naval Affairs |
| CPC | Cape Photographic Catalog |
| CR | *Compte Rendus de séances de l'Académie des Sciences* |
| CWG | Charles W. Goldsborough |
| DAB | *Dictionary of American Biography* |
| DE | Development Ephemeris |
| DSB | *Dictionary of Scientific Biography* |

| | |
|---|---|
| E | Entry. Used in conjunction with RG to specify location in National Archives |
| ET | Ephemeris time |
| FAME | Full-sky Astrometric Mapping Explorer |
| FC | Fundamental Catalog |
| FK | Fundamental Katalog |
| GC | General Catalogue |
| GMT | Greenwich Mean Time |
| GPS | Global Positioning System |
| IAU | International Astronomical Union |
| ICRF | International Celestial Reference Frame |
| IRS | International Reference Stars |
| JHA | *Journal for the History of Astronomy* |
| JMG | James Melville Gilliss |
| JPL | Jet Propulsion Laboratory |
| JRASC | *Journal of the Royal Astronomical Society of Canada* |
| LC | Library of Congress |
| LMG | Louis M. Goldsborough |
| MNRAS | *Monthly Notices of the Royal Astronomical Society* |
| NA | National Archives and Records Administration, Washington |
| NAO | Nautical Almanac Office |
| NAOSS | *Proceedings of the Nautical Almanac Office Sesquicentennial Symposium, U. S. Naval Observatory, March 3–4, 1999* |
| NAS | National Academy of Sciences, Washington |
| NCAB | *National Cyclopedia of American Biography* |
| NFK | Neuer Fundamental Katalog |
| NHC | Navy Historical Center, Washington Navy Yard, Washington, D. C. |
| NHF | Navy Historical Foundation |
| OHI | Oral-history interview, deposited at USNO Library |
| PASP | *Publications of the Astronomical Society of the Pacific* |
| PGC | *Preliminary General Catalogue* |
| Proc. | *Proceedings* |
| PUSNO | *Publications of the U. S. Naval Observatory* |
| PZT | Photographic zenith tube |
| QJRAS | *Quarterly Journal of the Royal Astronomical Society* |
| RG 45 | Record Group 45, "Naval Records Collection of the Office of Naval Records and Library," National Archives and Records Administration, Washington, D. C. |
| RG 78 | Record Group 78 "Records of the Naval Observatory," National Archives and Records Administration, Washington, D. C. |
| RSC | Royal Society (London) *Catalogue of Scientific Papers* |

| | |
|---|---|
| RSN | Annual Report of the Secretary of the Navy |
| SAOC | Smithsonian Astrophysical Observatory Catalog |
| SecNav | Secretary of the Navy |
| SI | Système Internationale |
| SNP | Simon Newcomb Papers, Manuscripts Division, Library of Congress |
| SOJ | *Sky with Ocean Joined* |
| SRS | Southern Reference Stars |
| TAC | Twin Astrographic Catalog |
| *Trans.* | *Transactions* |
| UCAC | USNO CCD Astrograph Catalog |
| *USNIP* | *U. S. Naval Institute Proceedings* |
| USNO | U. S. Naval Observatory |
| USNOA | U. S. Naval Observatory Archives, located on the grounds of the Naval Observatory in Washington |
| USNOA, AF | U. S. Naval Observatory Archives, Administrative files |
| USNOA, BF | U. S. Naval Observatory Archives, Biographical files |
| USNOA, SF | U. S. Naval Observatory Archives, Subject files |
| UT | Universal Time |
| UTC | Coordinated Universal Time |
| VLBI | Very-long-baseline interferometry |
| WO | *Washington Observations*, the first series of Naval Observatory publications, also known as *Astronomical Observations* |
| WWS | *Who's Who in Science* |

To Watchers of the Skies

Everywhere

# Introduction

The omniscient scholar would analyze climates of opinion and all their component ideas and idea-clusters. He would reveal their development in time, their action upon one another, and their relationship to the events of social and political history. He would psychoanalyze individual thinkers and study their thinking as expressions of group and class behavior. He would range easily through all fields of thought, from theories of physics to theories of art. He would study the function of language in the formation of ideas. He would expound the culture of elites and the culture of masses, the minds of nations and of whole civilizations, ancient and modern, Eastern and Western. . . . In the absence of such polymaths, intellectual history reduces in real life to a profusion of specialized studies employing many methodologies, fulfilling a variety of purposes, and based on conflicting definitions of the discipline.

W. Warren Wagar[1]

The statement above gives a good indication of the theoretical scope of intellectual history, and why in practice all who write it fail. For intellectual history in its institutional setting, which is the subject of this volume, the task is at once more and less daunting. We have for our subject a single institution and a well-defined span of 170 years. We need not worry about the rise and fall of civilizations. On the other hand, we must take into account not only individuals and their interactions in an institutional context, but also the development of science, the roles of the federal government and the military, the history of technology, and some measure of political and social history, in addition to a host of other components.

Histories of scientific institutions are sparse among the literature of history of science, especially in the United States. In the case of astronomical institutions, only in recent years has this sparsity begun to be remedied; this study may be seen as part of that remedy. As one of the oldest scientific institutions in the U. S. government, the Naval Observatory is in many ways unique. It maintains the Master Clock of the United States, and hence provides time for the nation and is the chief contributor to world time; it is one of the premier institutions for positional astronomy, a field whose applications are essential but largely unappreciated; it determines Earth-orientation parameters needed for a variety of practical purposes; it compiles a variety of almanacs

---

[1] W. Warren Wagar, *World Views: A Study in Comparative History* (Dryden Press: Hinsdale, Illinois, 1977), p. 4.

for navigation, geodesy and astronomy; and it continues to make signal contributions to astronomy even outside of its rather circumscribed mission. Each of these aspects has its history. With its unique time span as a Federal astronomical institution in the United States, it is hoped this history of the U. S. Naval Observatory will serve as a foundation toward a better understanding of the history of astronomy, and the history of science in general, especially in its American context. As the Prelude makes clear, this study is also an important part of the broader history of national observatories, and forms a seldom-recognized part of the history of the U. S. Navy and navigation. It is the American story of the long-standing relationship of astronomy in the service of navigation; thus the title Sky and Ocean Joined.

I have divided this study into three parts: the founding era, the golden era, and the twentieth century. While the twentieth century might have been broken into several eras, the overriding organizing principle is a continuous story in three main areas: positional astronomy, time and navigation. Of course these areas are interrelated, and thus, while documenting their unique aspects, I have also tried to make clear their connections, as well as to place the sum total in its institutional context.

The founding era of the Naval Observatory encompasses three foundings: the Depot of Charts and Instruments in 1830, the permanent Depot and de facto Naval Observatory in 1842, and the Nautical Almanac Office in 1849. The development of the permanent Depot and its supposedly "small observatory" into the first National Observatory of the United States is the subject of the first chapter; the origins of the Nautical Almanac Office (closely related to the Naval Observatory but not officially incorporated into it until the turn of the century) are the subject of the third chapter. In addition to the three founding events, the founding era is given coherence by the four men who successively headed the Depot and the Observatory: Louis M. Goldsborough, Charles W. Wilkes, James Melville Gilliss, and Matthew Fontaine Maury; two other chapters concentrate on the administrations of two Superintendents, Matthew Maury and James M. Gilliss. The resignation of Maury to join the Southern cause at the beginning of the Civil War, and Gilliss's death at its end, mark the termination of the founding era of the Naval Observatory.

The golden era is marked by the tenure of Asaph Hall, Simon Newcomb, William Harkness, and other Naval Observatory astronomers, by the discovery of the moons of Mars, the far-flung transit-of-Venus and solar-eclipse expeditions, and by progress in the more routine work of positional astronomy and celestial mechanics. During this era the Naval Observatory acquired the largest telescope in the world, excelled in most of its activities, and achieved worldwide fame as a result. However, the realm of astronomy was rapidly growing, and this was the last era in which world-wide prominence in classical astronomy could be equated with worldwide prominence in astronomy.

The twentieth century is actually grounded in events that began in 1893, when the Observatory moved to its present site, with new buildings and instruments, a new lease on life, and new opportunities in a rapidly growing discipline now fully engaged in incorporating astrophysics and photography. We will find here that, despite its continuing excellence and innovation in specific areas such as radio time dissemination, the Observatory did not immediately take advantage of astrophysics, and did not make full use of new techniques in photography. Sometimes this tardiness resulted from issues of mission and targeted government patronage beyond its control. The issue of military versus civilian control of the Observatory, and whether government astronomy would be better undertaken in a different branch of government, also came to a head early in the century, sapping the energies of the institution. The promise of a new lease on life expected from new buildings and instruments was overtaken by external changes to the field, to which the Observatory did not respond rapidly.

Yet, during the course of the century, in the areas of time, positional astronomy, and navigation, the Observatory gradually not only carried out its duties, but also once again became preeminent in its field. Throughout the century, the Observatory both affected, and was affected by, many other major landmarks of American political and scientific history. Its time and navigation functions proved essential through both World Wars. The first operational Ritchey–Chrétien telescope (the optical design now used for many telescopes including the Hubble Space Telescope) was built for the Observatory in the midst of the Great Depression. Responding to the inexorable growth of light pollution in the Washington area, during the 1950s a major field station was established in Flagstaff, Arizona; a smaller station was built in Florida, and far-flung stations operated temporarily at the Southern-Hemisphere sites of Samoa, Argentina, New Zealand, and Chile. The Space Age and the computer revolution also shaped events at the Observatory, as at other scientific institutions. The 61-inch astrometric telescope dedicated in 1964 would never have seen the light of day had the space race not freed funds for science. Although the problems of navigation were now expanded from the oceanic to the interplanetary regime, at first the Observatory made a conscious decision to limit its connections to NASA; newer institutions, such as the Jet Propulsion Laboratory, filled the space navigation gap. Nevertheless, the technological necessities of war and peace in the Space Age became drivers for new accuracies in the traditional functions of positional astronomy, time determination, and celestial mechanics.

Lest the reader lose sight of the forest for the trees, it is important to emphasize that only a finite number of problems dominates Naval Observatory history; they reappear again and again in ever more subtle and expressive forms. The determination and dissemination of time; the definition of an increasingly accurate and dense stellar reference frame; the determination of the astronomical constants; the closing of the gap between observation and gravitational theory as applied to the motions of the planets, satellites, and other solar-system objects; and the uses of all these aspects for navigation, all set in the American context, are the main subjects of this book. That is not to say that there were no unexpected developments. The discoveries of the moons of Mars and Pluto, of interstellar polarization of starlight, of the first asteroids found by a U. S. institution, and cutting-edge work in charge-coupled devices, infrared astronomy, and optical interferometry, are all part of Naval Observatory history.

Six generations of scientists carried out this work at the Naval Observatory. For many of them, their efforts are documented here for the first time, and bringing these scientists and their work back to life has been one of the pleasures of writing this history. A few, including Asaph Hall, Simon Newcomb, Wallace Eckert, and Gerald Clemence, have entered the pantheon of American astronomy because of their contributions. Others, such as Lt James Melville Gilliss, George W. Hill, William Harkness, William Eichelberger, H. R. Morgan, C. B. Watts, and Edgar Woolard, among many, deserve to be better known. Some of them find a place in this history; others remain for future research.

For the reader who wishes to follow developments thematically, sections 2.2 and 5.3 as well as chapter 11 cover developments in time determination, timekeeping, and time dissemination over 170 years. Nautical Almanac Office history is found in chapters 3, 8, and 12. Positional astronomy is dealt with primarily in sections 2.2 and 5.2 and in chapter 10. The Observatory's administrative history after the founding era is covered in section 5.1 and chapter 9. For purposes of perspective I have felt it necessary to cover the last two decades of impressive progress in positional astronomy, time, and navigation at the Observatory. While this provides a context for the previous 150 years, it is more difficult to see how the future will judge the last 20 years; the reader will therefore find that the more recent years are covered in less detail than the rest of the history.

No history covering a subject expansive in space and time can claim to be exhaustive in detail. While clearly many avenues could be explored using the approaches mentioned at the head of this chapter, I have sought to capture in this volume the major institutional themes spanning more than a century and a half of unparalleled progress in astronomy.

# Prelude. Perspectives and problems: The nation, the Navy, the stars

It is with no feeling of pride, as an American, that the remark be made that, on the comparatively small territorial surface of Europe, there are existing upward of one hundred and thirty of these light-houses of the skies; while throughout the whole American hemisphere there is not one.

President John Quincy Adams, 1825[1]

Before we plunge into the history of one of the oldest scientific institutions in the United States, it is essential to gain a perspective from at least three points of view: the history of astronomy, the history of science in America, and the history of navigation in the context of the U. S. Navy. These perspectives are not mutually exclusive; indeed the links between astronomy and navigation, particularly in the U. S. Navy, and their place in the history of science in America, are major themes of this study. They serve not only to provide the essential context of our story, but also to highlight some of the problems with which we will be concerned.

## 1. History of astronomy

The U. S. Naval Observatory must ultimately be placed in an international context. Specifically it falls squarely within that genre of astronomical institutions known as "national observatories." What happened in the United States in the 1830s and 1840s had already happened in other countries: a need was perceived for an astronomical observatory, not so much to advance the pure science of astronomy, but to serve practical purposes essential to national interests. Indeed, although the details differ in each country, a global perspective on the history of astronomy reveals that the Naval Observatory is part of what we may term a "national observatory movement," a 300-year pattern of government support for astronomical observatories. This movement gave rise to some of the longest-lived scientific institutions in the world. Throughout those three centuries the number of national observatories has increased, and the idea of "the national interest" has broadened to include pure research and national prestige, in addition to more practical concerns. National observatories are not immortal, however, as witnessed by one of its most venerable members, the Royal Greenwich Observatory, which ceased to exist as a scientific institution in 1998.

Table P.1 shows some of the more important national observatories, and

---

[1] John Quincy Adams's first annual address to Congress, December 6, 1825, in *A Compilation of the Messages and Papers of the Presidents, 1789–1902*, ed. James D. Richardson (New York, 1904), p. 56.

Table P.1. *Some important national observatories and their patrons*

|  | Institution | Founded | Patron |
|---|---|---|---|
| First era | Uraniborg (Tycho Brahe) | 1576 | Frederick II |
|  | Paris Observatory | 1667 | Louis XIV |
|  | Royal Greenwich Observatory | 1675 | Charles II |
|  | Berlin Observatory | 1701 | Frederick I |
|  | St Petersburg Observatory | 1725 | Peter the Great |
| Second era | Royal Observatory, Cape | 1820 | Great Britain |
|  | U. S. Naval Observatory | 1830 | U. S. Navy |
|  | Pulkovo Observatory | 1839 | Nicholas I |
|  | Chilean National Observatory | 1852 | Chile |
|  | Argentine National Observatory | 1870 | Argentina |
|  | Potsdam Astrophysical Observatory | 1874 | German Academy of Science |
|  | Smithsonian Astrophysical Observatory | 1891 | Smithsonian/United States |
|  | Dominion Observatory | 1903 | Canada |
|  | Dominion Astrophysical | 1918 | Canada |
| Third era | National Radio Astronomy Observatory | 1956 | NSF/AUI, United States |
|  | Kitt Peak National Observatory | 1957 | NSF/AURA, United States |
|  | National Radio Astronomy Observatory | 1959 | CSIRO, Australia |
|  | Cerro-Tololo Inter-American | 1963 | NSF/AURA/Chile |
|  | European Southern Observatory | 1964 | Five countries, more later |
|  | Anglo-Australian (Siding Spring) | 1967 | Great Britain/Australia |
|  | Space Telescope Science Institute | 1981 | NASA/AURA |

*Abbreviations:*
NSF     National Science Foundation
AUI     Associated Universities, Inc.
AURA    Associated Universities for Research in Astronomy
NASA    National Aeronautics and Space Administration
CSIRO   Commonwealth Scientific, Industrial and Research Organization

suggests three eras in this movement: the first era, in which the prototype Paris, Greenwich, Berlin, and St Petersburg observatories were founded; the second era, characterized by offshoots from previous national observatories, by new observatories of younger nations, and by the rise of astrophysical observatories; and the third, post-World-War-II era, characterized by national or international consortia, large budgets relative to those of previous years, and increasingly sophisticated telescopes, detectors, and spacecraft observing an expanded region of wavelengths. Observatories in the latest era heavily emphasize astrophysics. The founding of the Naval Observatory falls clearly in the second era, and suggests comparison with the Royal Greenwich Observatory, the Cape Observatory in South Africa, and the Pulkovo Observatory in Russia, in terms of origins, programs, and evolution. The comparative history of such similar institutions is an important and unrealized task that we can only touch on in this volume.[2]

[2] A more detailed description of the national observatory movement, and a brief comparison of the U. S. Naval Observatory and Pulkovo Observatory as exemplars of that movement founded within a decade of each other under very different political conditions, may be found in Steven J. Dick, "Pulkovo Observatory and the National Observatory Movement," *Inertial Coordinate System*

How this venerable national observatory theme was first played out in the context of American history is the subject of this book. The United States had its own internal national observatory movement, that is, its series of false starts before the U. S. Navy's Depot of Charts and Instruments evolved into the first national observatory of the United States in the mid-1840s. Four attempts have generally been recognized, including those related to the U. S. Coast Survey, the establishment of a prime meridian of the United States, and the Smithsonian bequest, before the successful attempt of the U. S. Navy.[3] We briefly describe each of these efforts because they were in some ways related, and events surrounding the Smithsonian bequest would have a significant effect on the founding of a permanent Naval Observatory.

When the Survey of the American Coast was authorized in 1807, President Thomas Jefferson and Treasury Secretary Albert Gallatin selected Ferdinand R. Hassler to undertake the work. Among Hassler's plans were two observatories, to serve as fixed points in the Survey for determining base latitude and longitude. One of the observatories, Hassler noted, might be placed in the city of Washington "as observatories are placed in the principal capitals of Europe, as a national object, a scientific ornament, and a means of nourishing an interest for science in general." Hassler even drew up plans for such an observatory, to be located on a hill north of the Capitol.[4] Though President Madison and Secretary of the Treasury Dallas approved Hassler's astronomical plans in 1816, the repeal of part of the Survey act in 1818 brought these astronomical plans, and the work of the Coast Survey itself, to a halt. When the Survey was revived in 1832, the new law specified that nothing in the act should be construed "to authorize the construction or maintenance of a permanent astronomical observatory." This prohibition is often attributed to Congressional dislike of John Quincy Adams, who favored the idea of an observatory, but it was undoubtedly also a cost-cutting move.

on the Sky, Proceedings of International Astronomical Union Symposium 141, Leningrad, October 17–21, 1989, Jay H. Lieske and Victor K. Abalakin, eds. (Kluwer: Dordrecht, 1990). Also very useful is Kevin Krisciunas, *Astronomical Centers of the World* (Academic Press: New York, 1988). Substantive histories exist for only a few national observatories; see for example the Greenwich Tercentenary volumes *Greenwich Observatory: The Royal Observatory at Greenwich and Herstmonceux, 1675–1975*; volume 1, Eric G. Forbes, *Origins and Early History (1675–1835)*, (Taylor and Francis: London, 1975); volume 2, A. J. Meadows, *Recent History (1836–1975)*, (Taylor and Francis: London, 1975); and volume 3, Derek Howse, *The Buildings and Instruments* (Taylor and Francis: London, 1975). See also C. Wolf, *Histoire de l'Observatoire de Paris de sa fondation à 1793* (Gauthier-Villars: Paris, 1902), and the broader and briefer history in S. Debarbat, S. Grillot, and J. Levy, *L'Observatoire de Paris: Son Histoire (1667–1963)* (Observatoire de Paris: Paris, 1984); A. N. Dadaev, *Pulkovo Observatory: An Essay on its History and Scientific Activity*, trans. by Kevin Krisciunas (Washington, 1978); and Richard A. Jarrell, *The Cold Light of Dawn: A History of Canadian Astronomy* (University of Toronto Press: Toronto, 1988).

[3] Charles O. Paullin, "Early Movements for a National Observatory, 1802–42," *Records of the Columbia Historical Society*, **25** (1923), 36–56. See also David F. Musto, "A Survey of the American Observatory Movement, 1800–1850," *Vistas in Astronomy*, **9** (1967), 87–92.

[4] Paullin (ref. 3), 38–39. This is the same site where the Depot of Charts and Instruments and its small observatory would be located, 1834–42. For Hassler's observatory plans and its instruments see "Papers on Various Subjects Connected with the Survey of the Coast of the United States," *Transactions of the American Philosophical Society*, **2**, new series (Philadelphia, 1825), 357–370. The quotation is from page 242. See also Florian Cajori, *The Chequered Career of Ferdinand Rudolph Hassler* (Christopher Publishing House: Boston, 1929; facsimile reprint, Arno Press: New York, 1980).

The second attempt for an astronomical observatory was connected to William Lambert's proposal for determining a prime meridian for the country. It was first proposed to Congress in December 1809, and this and similar proposals of Lambert were renewed until 1824. Though some observations were made in 1821, a permanent astronomical observatory was never established. In light of later events, it is interesting to note that the climate of opinion was generally favorable, as evidenced in the remarks of Secretary of State James Madison in July, 1812: "An observatory would be of essential utility. It is only in such an institution, to be founded by the public, that all the necessary implements are likely to be collected together, that systematic observations can be made for any great length of time, and that the public can be made secure of the result of scientific men. In favor of such an institution it is sufficient to remark, that every nation which has established a first meridian within its own limits has established also an observatory. We know that there is one at London, at Paris, Cadiz, and elsewhere."[5]

The third attempt to found a national observatory is associated with the name of John Quincy Adams, whose earliest published efforts in this regard began with his Presidential message of December 6, 1825. A select committee of Congress responded to this proposal with an elaborate report of its own, composed by Major General Alexander Macomb, Chief of Engineers of the Army. No action was taken, and Adams renewed his proposal for an observatory in 1840 in connection with the bequest of James Smithson and the founding of the Smithsonian Institution. This proposal, which was repeated until 1846, is directly associated with the evolution of the Navy's Depot of Charts and Instruments into an observatory, and will be discussed in that context in chapter 1.[6]

If the American government found it difficult to support a national observatory, astronomy in general in the United States did not fare much better prior to 1840. At that time almost nothing qualifying as an observatory existed in the country. However, by 1856 Elias Loomis, in *The Recent Progress of Astronomy, Especially in the United States*, could discuss some 25 observatories, mostly private or attached to educational insti-

[5] Paullin (ref. 3), 41. According to Paullin, Lambert was a resident of Virginia and then the City of Washington, and a clerk in the War Department until 1821, when he resigned to make longitude observations, using a temporary observatory. For a history of Washington meridians see Silvio Bedini, *The Jefferson Stone: Demarcation of the First Meridian of the United States* (Professional Surveyors: Frederick, Maryland, 1999), and Matthew H. Edney, "Cartographic Confusion and Nationalism: The Washington Meridian in the Early Nineteenth Century," *Mapline* (Spring/Summer, 1993, 4–8). On American prime meridians in general, see Joseph Hyde Pratt, "American Prime Meridians," *Geographical Review*, **32** (1942), 233–244. On the broader history of prime meridians see *Longitude Zero, 1884–1984, Vistas in Astronomy*, **28** (1985), 7–407, and USNOA, SF, "Prime Meridians" folder.

[6] Steven J. Dick, "John Quincy Adams, The Smithsonian Bequest, and the Founding of the U. S. Naval Observatory," *JHA*, **22** (1991), 31–44; Marlana Portolano, "John Quincy Adams's Rhetorical Crusade for Astronomy," *Isis*, **91** (2000), 480–503; and Paullin (reference 3), 44–48. On Adams see also Samuel Flagg Bemis, *John Quincy Adams and the Union* (Academic Press: New York, 1970), especially Chapter 23, "Lighthouses of the Skies, 1825–46." The 23-page Congressional document is "National Observatory," 19th Congress, first session, March 18, 1826, House Report No. 124, copy in USNO, SF, Legislation File.

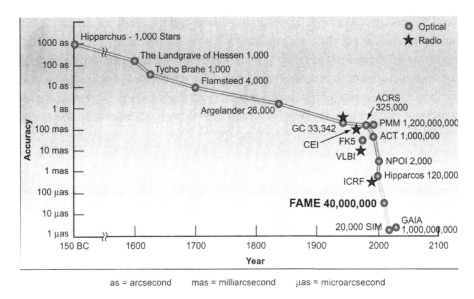

Figure P.1. Astrometric accuracies, showing the "Golden Age of Astrometry" beginning in the 1980s, as accuracies surpassed the milliarcsecond range. Items after ICRF are projected. Source: U. S. Naval Observatory.

tutions.[7] Aside from the observatory at West Point, which was used for teaching and not research, the Navy's observatory was the only government-supported astronomical institution of its era. Moreover, though astronomy experienced even more rapid growth in the United States following the Civil War, for more than a century after its birth the Naval Observatory remained the only government-supported observatory in the country.[8]

The work of the Naval Observatory, like that of other national observatories during much of their history, has been largely devoted to positional astronomy, timekeeping, and celestial mechanics. These have been the principal concerns of astronomers for most of astronomy's history, from the Babylonians and Greeks through the nineteenth century. The likes of Hipparchus and Ptolemy, Brahe, Kepler, and Newton, Laplace and Lagrange embraced this "classical astronomy" as their life work, and their contributions have been examined well.[9] However, what of their nineteenth- and

[7] Elias Loomis, *The Recent Progress of Astronomy, Especially in the United States* (third edition, Harper and Bros.: New York, 1856), pp. 202–292.

[8] With the exception of the Smithsonian Astrophysical Observatory, which has always been considered as "quasi-governmental" because of its dual funding sources, not until the founding of Kitt Peak National Observatory and the National Radio Astronomy Observatory in the mid-1950s did this situation change. On the Smithsonian Astrophysical Observatory see Bessie Zaban Jones, *Lighthouse of the Skies: The Smithsonian Astrophysical Observatory: Background and History, 1846–1955* (Smithsonian Institution: Washington, 1965), and the "Smithsonian Astrophysical Observatory Centennial" issue of the JHA, **21** (February, 1990). On the later American national observatories see Frank Edmondson, *AURA and its U. S. National Observatories* (New York: Cambridge University Press, 1997).

[9] See, for example, Robert Grant, *History of Physical Astronomy* (London, 1852), A. Pannekoek, *A History of Astronomy* (George Allen and Unwin: London, 1961), and a host of more recent studies.

twentieth-century successors? Figure P.1 places those centuries in the context of earlier astrometric accuracies, which are expressed in terms of "arcseconds," 1/3,600th of a degree (the full Moon subtends about one half degree, or 1,800 arcseconds). Increasing technological sophistication brought observational capabilities for position determination from 60 arcseconds (or one arcminute) for Tycho Brahe in 1600, to a few tenths of an arcsecond at the time the Naval Observatory was founded, more than 100 times better.[10] The improvement over the next century and a half to 1980 was only by about a factor of ten, despite the use of devices like the traveling-wire micrometer, ingenious methods for reading the graduated circle, and improvements to refraction theory. Most of our history deals with this period of slow improvement. During the last two decades of the century, however, new technologies routinely achieved milliarcsecond accuracies (an improvement by a factor of a million over the ancient Greeks), and technologies for microarcsecond accuracies were being developed. As Figure P.1 certainly indicates, developments over the last two decades have led to a "golden age of astrometry."

These developments over 170 years give rise to important historical questions: What problems did new levels of accuracy enable astronomers to tackle and resolve? Conversely, what have been the drivers behind their quest for greater accuracy? How have practical problems of navigation and more purely scientific problems such as stellar motion and galactic structure intermingled? As one of the few American institutions to carry on this work, and the only one over such an extended period, the Naval Observatory is a natural choice for an analysis of this kind.

The improvements in the capabilities of positional astronomy affected strongly another task of the Naval Observatory – the accurate determination of time. This necessarily follows because for centuries astronomy held a monopoly on time determination, since no more accurate periodic phenomenon was known to exist than the rotation of the Earth on its axis, measured with respect to a celestial body. However, in the 1920s and 1930s, quartz-crystal clocks, and then in the 1950s atomic clocks, showed the irregularities in the Earth's rotation. The man-made clock now outdid Nature itself, but only by exploiting another part of Nature, the natural properties of the atom. "Atomic time" now joined the "Earth time" that had reigned for centuries, and made physicists rather than astronomers the keepers of the most accurate time. Taking a broad view, Figure P.2 shows that the accuracy of clocks improved by a factor of a billion over three centuries, which is even more impressive than the improvements in positional astronomy.[11] This was possible because of the development of

[10] The technology behind improvements in astrometric accuracy before the founding of the Naval Observatory is described in Allan Chapman, *Dividing the Circle: Development and Critical Measurement of Celestial Angles, 1500–1850* (E. Horwood: New York; 1990); "The Accuracy of Angular Measuring Instruments used in Astronomy between 1500 and 1850," *JHA*, **14** (1983), 133–137. See also David W. Hughes, "Astronomical Angular Accuracy," *Nature*, 307 (January 5, 1984), 15–16.

[11] From Derek Howse, *Greenwich Time and the Discovery of the Longitude* (Oxford University Press: Oxford, 1980), and the new edition *Greenwich Time and the Longitude* (Philip Wilson: London, 1998); and Eric Forbes, *The Birth of Scientific Navigation: The Solving in the 18th Century of the Problem of Finding Longitude at Sea*, National Maritime Museum monograph 10 (1974), p. 181.

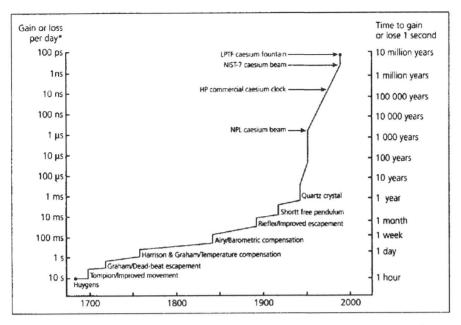

Figure P.2. The increasing accuracy of precision clocks. NPL is the National Physical Laboratory (U. K.), NIST is the National Institute of Standards and Technology (United States), and LPTF is the Laboratoire Primaire du Temps et des Fréquences (France). Source: Derek Howse (ref. 11), p. 170, used with permission.

fundamentally new technologies, such as the quartz-crystal oscillator and the atomic clock. Along with novel questions generated by the increase in accuracy of clocks, we shall examine astronomers' reactions to these developments, again at the one American astronomical institution that eventually held sole responsibility for time determination, maintenance, and dissemination. How the Naval Observatory came to have an monopoly in the timing function among astronomical observatories, and the Observatory's interaction with the American standards laboratory also undertaking work related to time, is another part of our story. The changing technologies of time dissemination, from the transport of timekeepers to time balls, to telegraph, radio, and satellite, is another thread of our story.

These developments in positional astronomy and timekeeping closely relate to celestial mechanics, the third major function of the Naval Observatory. It is positional astronomy that provides the data against which theories of planetary motion must stand or fall. Time is the measure of all change, or, to put it another way, the argument or independent variable of all theories of planetary motion. Changing measures and definitions of time are therefore bound to be of essential interest to this enterprise, as are new levels of accuracy in positional astronomy, which, via gravitational perturbations, might reveal the effects of new planets or necessitate the revision of old theories. Celestial mechanics is the business of the Nautical Almanac Office, and some of the giants of American science, including Simon Newcomb and G. W. Hill, devoted their lives to it in conjunction with their work in that office. It is work that requires substantial international cooperation, so we shall discuss not only the work itself, but also the efforts focused on standardization and cooperation.

Taking a broader view, this history of the Naval Observatory may be seen as a case study of the interactions between the "old astronomy" of position and the "new astronomy" of astrophysics. The rise of astrophysics in the latter half of the nineteenth century was a momentous event that represented a sharp break with old techniques and traditions.[12] This was in many ways a crisis for the old astronomy, since new talent was increasingly siphoned off to new, more exciting areas of research. At the time of transition what was the institutional reaction to these developments, particularly among Naval Observatory astronomers of international stature in the field of positional astronomy at the time of transition? Could a naval institution, which also claimed to be a national institution, justify funding for research in the new astronomy, when such research had nothing to do with the Navy or navigation? Or could substantial links be forged between the two areas of work? Conversely, what was the attitude of those practitioners of the new astronomy toward the work of the Naval Observatory, the chief representative in the United States of the old tradition? Did this attitude change over time? These are questions of broad interest to the history of astronomy.

No more basic concepts exist in astronomy than position and motion in "space" and the passing of "time." Although the term "old astronomy," which encompasses these concepts, was at one time purposely meant to be pejorative, today astrophysicists widely recognize the links of their subject with positional astronomy.[13] For much of the twentieth century the Naval Observatory, while gradually entering the field of astrophysics, has been the sole guardian of that older tradition in the United States. Its history is therefore a good opportunity to consider the interactions between the two.

## 2. History of science in America

The United States in 1830, the year of the founding of the Depot of Charts and Instruments, was a country still very much in its youth. The American Revolution lay

---

[12] On the rise of astrophysics, see Agnes Clerke, *A Popular History of Astronomy through the Nineteenth Century* (A. & C. Black: Edinburgh, 1885) and subsequent editions, and her *Problems of Astrophysics* (A. & C. Black: London, 1903); also Owen Gingerich, ed., *Astrophysics and Twentieth Century Astronomy to 1950*, Part A (Cambridge University Press: Cambridge, 1984). Throughout the history of astronomy there have been several "new astronomies," from Kepler's *Astronomia Nova* with its novel theory of planetary motion to Langley's *New Astronomy* (1888) and the era of new wavelength regions observable from the ground and outer space.

[13] See, for example, Martin Schwartzschild, "New Impetus to Astrometry," in *Aspects of Stellar Evolution*, Proceedings of a Conference held in honor of Ejnar Hertzsprung at Flagstaff, Arizona, on June 22–24, 1964, *Vistas in Astronomy*, **8**, 3–6. (The Naval Observatory's 61-inch astrometric reflector was also dedicated at this time.) Also Ivan Mueller and Barbara Kolaczek, eds., *Developments in Astrometry and Their Impact on Astrophysics and Geodynamics* (Kluwer: Dordrecht, 1993).

barely 50 years in the past, the Constitution only 40 years, and the burning of the nation's capital during the War of 1812 still stirred the memory of the inhabitants of the City of Washington.

Science in the United States in the 1830s was not yet a substantial pursuit, at least in part because it was not sanctioned by Congress and therefore not the object of government patronage. Even before the Revolution, an interest in science was apparent in the existence of the American Philosophical Society, the contents of its *Transactions* (first officially published in 1771), and David Rittenhouse's observation of the 1769 transit of Venus, among other sporadic activities.[14] However, two momentous decades later, the Constitution still did not specifically give Congress the power to promote science, except in the issuing of patents. The federal role in promoting science was a matter of interpretation, and, caught up in power struggles between the states and the Federal government as well as in the daily survival of the nation, Congress was not inclined to take a liberal view of the Constitution.

The three decades 1800–1829 have been called the "Jeffersonian era" of science, after the man who not only assumed the Presidency for the first eight years of that period, but who was also one of the country's leading men of science and culture throughout that period. Under Jefferson's administration Lewis and Clark made their famous explorations of the West, the U. S. Coast Survey was founded in 1807 to perform a specific task only, and serious attempts were made to found a National University where science might play some role. The founding of the West Point Military Academy in 1802 bore fruit for science beginning in 1817, when Sylvanus Thayer made it the country's chief source of engineers, who were essential in the Army's Corps of Topographical Engineers and for the public works of an expanding nation. The Jeffersonian period ended after the Presidency of John Quincy Adams, another lover of science, whose futile efforts at establishing a national observatory beginning in 1825 have been described above, and whose frustration at the state of astronomy in America in comparison with that in Europe is evident in the passage at the head of this chapter.[15]

The accession of Andrew Jackson to the Presidency in 1829 inaugurated the age of Jackson. Although President Jackson had little interest in science, exploration and surveying were "fundamentally in tune with the main task of the American people in this period – the conquest of a continent." It was in this context that the Coast Survey was revived in 1832 under Hassler, and that the U. S. Exploring Expedition, first proposed under Adams's administration in 1828, was finally dispatched in 1838 under Lt Charles Wilkes. The success of the U. S. Navy's Depot of Charts and Instruments was related to this American exploration imperative by virtue of the need for chronometers

---

[14] The best treatment of science during this era of American history is Brooke Hindle, *The Pursuit of Science in Revolutionary America, 1735–1789* (Academic Press: New York, 1956).

[15] On the Jeffersonian era in science, see John C. Greene, *American Science in the Age of Jefferson* (Ames: Iowa, 1984), and A. Hunter Dupree, *Science in the Federal Government* (Harvard University Press: Cambridge, Massachusetts, 1957), chapter 2.

and instruments aboard ships (including Wilkes's) to determine longitude, as well as the need for astronomical observations for comparison with those made on the Wilkes expedition. Such practical beginnings laid the foundation for scientific curiosity and more theoretical work.[16]

The 1840s, during which the Navy's Depot was transformed into an Observatory and Hydrographic Office, saw the real foundations of a permanent American science. The Coast Survey was placed under the strong leadership of A. D. Bache, and the Smithsonian Institution, which had been born after protracted debate, was placed in the able hands of Joseph Henry. Although the National Institute for the Promotion of Science struggled and failed, the American Association for the Advancement of Science was born in 1848.[17] The birth of the U. S. Naval Observatory was closely connected to many of these events; its leaders and staff functioned in the milieu of this growing scientific establishment. It shared with some of them problems such as military-versus-civilian control of scientific activities, an issue debated repeatedly before Congress.[18]

Slowly the needs of the country gave rise to other government scientific institutions, including the Geological Survey (1879), the Weather Bureau (1890), the National Bureau of Standards (1901), and the National Institutes of Health (1930). There is, however, a need for perspective of scale: the U. S. Naval Observatory has always been small by comparison with other Federal scientific institutions and most Navy agencies. Its staff reached a peak of just over 200 in the 1970s, and today its staff numbers about 100, while agencies such as the Geological Survey, the National Bureau of Standards (now the National Institute for Standards and Technology), and the Naval Research Laboratory employ thousands. The Observatory's research, and, development budget is in the millions, while other scientific agencies (including naval oceanography) may have budgets in the hundreds of millions or even billions. In proportion to the importance of scale in such institutions, we must be careful not to over-

---

[16] On science in the age of Jackson see Dupree (ref. 15), chapter 3, and George H. Daniels, *American Science in the Age of Jackson* (Columbia University Press: New York, 1968). The quotation is from Dupree, p. 51. For the context of exploration, including some of the people discussed in this volume, see William Goetzman, *New Lands, New Men* (Columbia University Press: New York, 1986), and the same author's *Exploration and Empire* (Columbia University Press: New York, 1966).

[17] On this period see Robert V. Bruce, *The Launching of Modern American Science, 1846–1876* (Knopf: New York, 1987). On the AAAS see Sally Gregory Kohlstedt, *The Formation of the American Scientific Community: The American Association for the Advancement of Science, 1848–1860* (University of Illinois Press: Urbana, 1976), and Sally Gregory Kohlstedt, Michael M. Sokal, and Bruce V. Lewenstein, *The Establishment of Science in America: 150 years of the American Association for the Advancement of Science* (Rutgers University Press: New Brunswick, New Jersey, 1999). Nathan Reingold, *Science in Nineteenth Century America: A Documentary History* (Columbia University Press: New York, 1964) gives a broad view of nineteenth-century science centered around some of its most important correspondence and documents.

[18] Military–civilian control was one of the issues discussed in *Testimony before Joint Commission to Consider the Present Organization of the Signal Service, Geological Survey, Coast and Geodetic Survey, and the Hydrographic Office of the Navy Department* (49th Congress, first session), Senate Miscellaneous Document 82, March 16, 1886, p. 179. See Clarence G. Lasby, "Science and the Military," in David D. Van Tassel and Michael G. Hall, *Science and Society in the U. S.* (Homewood: Illinois, 1966), pp. 251–282.

generalize our conclusions.[19] It is perhaps best to see the Naval Observatory in terms of a case study of the activities of a long-lived but relatively small government agency, embedded in one of the largest government agencies, the Department of Defense. It is an organization driven on the one hand by the mission-oriented needs of an enormous bureaucracy from above, and on the other hand by the advancement of astronomy. These driving forces are not necessarily mutually exclusive; their interaction is a major theme of this study.

### 3.     Navigation and the U. S. Navy

The most immediate context for the origin of the Depot of Charts and Instruments was the U. S. Navy's need to improve navigation, in particular the determination of longitude. It is this practical connection that would by degrees lead to an astronomical observatory, and, when that observatory grew beyond practical needs, navigation still provided the core *raison d'être*. It is important, then, that we understand the state of navigation at this time, and in particular the relation of the marine chronometer to astronomical observation.

Astronomy and navigation have existed in symbiotic relationship since oceanic navigation began. The early Polynesians used celestial bodies for determining latitude during their long ocean journeys. By the fifteenth and sixteenth centuries, Columbus and the other great Western voyagers were attempting to determine latitude by more refined celestial observations, measuring the altitude of the Pole star or of the Sun, the latter in conjunction with declination tables for the Sun. On the rolling deck of a ship, and with only crude instruments, this method was not accurate even to one degree (60 nautical miles at the equator), but it was eventually perfected with the improvement of the quadrant. By contrast, no method existed for finding longitude other than estimating it from dead reckoning.[20]

To the "discovery of the longitude," the other coordinate necessary for ocean navigation, some of the greatest minds in Western civilization applied themselves, in vain until the eighteenth century.[21] Astronomy was of utmost importance in this endeavor, for all methods of longitude determination at sea depend on knowing simultaneously the local time at two places – one whose longitude is known, the other

[19]  The history of American scientific institutions is gradually receiving increasing attention. On the National Bureau of Standards, which figures peripherally in this history, see Rexmond C. Cochrane, *Measures for Progress: A History of the National Bureau of Standards* (National Bureau of Standards: Washington, 1966); and Elio Passaglia, *A Unique Institution* (National Bureau of Standards: Washington, 1999).

[20]  On the early history of navigation see E. G. R. Taylor, *The Haven-Finding Art: A History of Navigation from Odysseus to Captain Cook* (London, 1956). Full histories are given in J. B. Hewson, *A History of the Practice of Navigation* (Brown, Son & Ferguson: Glasgow, 1951), and W. E. May, *A History of Marine Navigation* (Norton: Henley on Thames, 1973). In his chapter "How Columbus Navigated, 1492–1504," Samuel Eliot Morison concludes that dead reckoning using charts and compass served Columbus better than did celestial navigation during his voyages to the New World, in *Admiral of the Ocean Sea: A Life of Christopher Columbus* (Boston, 1942), pp. 183–196.

[21]  Howse (ref. 11). Newton, Halley and others became involved in longitude proposals, and Newton even designed a quadrant about 1700. See H. C. Freiseleben, "Newton's Quadrant for Navigation," *Vistas in Astronomy*, **22**, (1980) 515–522.

on the ship itself. For this the observation of predicable events in the heavens is well suited, whether eclipses of Jupiter's satellites, as proposed by Galileo in 1616, or lunar occultations of stars, or lunar eclipses. However, the first two are not easily accomplished on a moving and pitching ship, and the latter are not very frequent occurrences.

It is remarkable that what would become the two chief methods for finding longitude at sea – the method of lunar distances, and the chronometer – were conceived theoretically less than two decades apart: the method of lunar distances in Johann Werner's commentary on Ptolemy's *Geography*, published in 1514, and the chronometer method in a work of Gemma Frisius published in 1530. In order to determine the difference between the local time at the ship's position and the time at the meridian of Greenwich, lunar distances used the motion of the Moon against the background of the starry sky, much like the moving hands of a clock; the observed angular distance between the Moon and a particular star, compared with the calculated distance at the Greenwich meridian (provided by an almanac), gives the Greenwich time. An altitude observation of the Sun or star gives the observer's local time, and the difference between this local time and Greenwich time is the longitude. The chronometer method is even better, for in this case one can carry the Greenwich time aboard ship, and determine the local time from a single celestial observation, or (in practice) the mean of several such observations.[22]

It is even more remarkable that, after two centuries of sustained attempts, both the chronometer and lunar-distance methods became practical in the same decade: the 1760s. It was only after a century of observational and theoretical work, largely at the Royal Observatory at Greenwich, which had been founded specifically for the purpose, that Neville Maskelyne and the British *Nautical Almanac* (1767) brought the method of lunar distances to fruition through a better knowledge of stellar positions and the motion of the Moon.[23] It was another British subject, John Harrison, who, after long attempts, finally built a chronometer accurate enough to pass sea trials in 1761–62 and 1764, according to specifications set by the Board of Longitude.[24] Thanks to John Hadley's invention of the reflecting octant in 1731, celestial observations could be made accurately enough to employ the two methods.

---

[22] On the method of lunar distances see Howse (1998, ref. 11), pp. 183–185, and D. H. Sadler, "Lunar Distances and the Nautical Almanac," *Vistas in Astronomy*, **28** (1985), 113–122. In these calculations one revolution of the Earth (360 degrees) is equivalent to 24 hours of time.

[23] On the British Nautical Almanac see Howse, *Nevil Maskelyne: The Seaman's Astronomer* (Cambridge University Press: Cambridge, 1989).

[24] This is a story so full of personality and drama that it has been told many times, most recently in the best-selling book by Dava Sobel, *Longitude: The True Story of a Lone Genius Who Solved the Greatest Scientific Problem of his Time* (Walker and Company: New York, 1995), and Dava Sobel and William J. H. Andrewes, *The Illustrated Longitude* (Walker and Company: New York, 1998). These books were based on a Harvard conference whose proceedings are to be found in W. J. H. Andrewes, *The Quest for Longitude* (Collection of Historical Scientific Instruments, Harvard University: Cambridge, Massachusetts, 1996). On the marine chronometer see Rupert T. Gould, *The Marine Chronometer: Its History and Development* (J. D. Potter: London, 1923, reprinted 1960), and Marvin Whitney, *The Ship's Chronometer* (Cincinnati University Press: Cincinnati, 1988).

(Thomas Godfrey, a self-taught astronomer in Philadelphia and associate of Benjamin Franklin, made a near-simultaneous invention of the instrument.) The further eighteenth-century inventions of the repeating circle by Tobias Mayer and the nautical sextant by Captain John Campbell, together with Jesse Ramsden's revolutionary advances in the art of graduating scales in the 1770s, gave to navigation a retinue of tools worthy of the task.[25]

Once they had been invented, the methods of lunar distances and chronometers were for more than a generation rival methods. Britain's Astronomer Royal Nevil Maskelyne can be forgiven if, having spent great care on the tables of lunar distances in his *Nautical Almanac*, he was less than enthusiastic about chronometers. In retrospect he had little cause for immediate worry that "lunars" would be superseded; due to their expense and scarcity, chronometers did not come into widespread use immediately, even in Britain, except for important voyages of exploration. Cook carried them with good results on his second (1772–75) and third (1776–79) voyages to the Pacific, as did Bligh (1787) and Vancouver (1791–95).[26] Around the beginning of the nineteenth century navigators found that the chronometer method was easier, and that, when it was applied carefully, it yielded better results than did the use of lunar distances. With cheaper prices the demand for them gradually increased. In England chronometers were not issued to all Royal Navy ships until about 1825, and the worldwide demand exceeded the supply until 1840.[27]

In the United States, to the extent that scientific navigation was practiced at all in the late eighteenth century, Yankee sailors made use of the method of lunar distances and the British *Nautical Almanac* during long trading voyages.[28] The first American name widely associated with improvements in navigation is Nathaniel Bowditch, a Salem, Massachusetts seaman who participated in five voyages between 1795 and 1803 in the China and East India trade. In 1802 Bowditch published his *New American Practical Navigator*, a corrected version of an English text by J. H. Moore, with Bowditch's new and simpler method for computing lunar distances. This landmark

---

[25] J. A. Bennett, *The Divided Circle: A History of Instruments for Astronomy, Navigation and Surveying* (Phaidon Christie's: Oxford, 1987). On the invention of the octant and sextant, and the relative roles of Hadley and Godfrey, see Howse (ref. 11, 1980), pp. 57–60.

[26] Hewson (ref. 20), pp. 246–249.

[27] Gould (ref. 24), p. 131; Forbes (ref. 11), p. 16. According to Samuel Eliot Morison, in 1833 the Royal French Navy, with 250 ships (including 52 ships of the line), had only 44 chronometers, *The Great Explorers: The European Discovery of America* (Columbia University Press: New York, 1978), p. 31. See also G. W. Nockolds, "Early Timekeepers at Sea: The Story of the General Adoption of the Chronometer between 1770 and 1820," *Antiquarian Horology*, **iv** (1963), 110–113, 148–152, and Alun C. Davies, "The Life and Death of a Scientific Instrument: The Marine Chronometer, 1770–1920," *Annals of Science*, **35** (1978), 509–527. Also, W. E. May, "How the Chronometer went to Sea," *Antiquarian Horology* (March, 1976), summarized in *Vistas in Astronomy*, **20** (1976), 135–137.

[28] Not much has been written on the history of navigation in America, but see, for example, Dirk Struik, "The Practical Navigators," *Yankee Science in the Making* (Scribner: New York, 1962), pp. 93–133, and Silvio Bedini, "From His Known Shore," *Thinkers and Tinkers: Early American Men of Science* (Scribner: New York, 1975), pp. 346–363. In his *Maritime History of Massachusetts* (Boston, 1921), Samuel Eliot Morison spends less than five pages on American navigation prior to 1830, pp. 113–117, but does give a good feel for the extent of colonial American sea trade.

publication would go through many editions and, along with the Nautical Almanac, became the bible for American seaman.[29]

Editions of Bowditch, still published even today under the name American Practical Navigator, came out only at irregular intervals and are therefore not a particularly sensitive barometer of navigational changes in America.[30] It is notable, however, that already in his first edition of 1802 Bowditch gave brief instructions, with examples, concerning how "to find the longitude by a perfect time-keeper." This by no means signifies the widespread use of this method in America, and it is clear from this volume that the use of lunar distances was still the method of choice. The same was true for land expeditions; although Lewis and Clark carried a foreign-made chronometer with them on their famous expedition (1804–06), they also used lunar distances for longitude. However, as early as 1804, young William C. Bond (who later became the first Director of Harvard College Observatory) constructed a chronometer in Boston, which for several years he used as the standard by which other chronometers, imported from England, were measured, at least for some ships sailing out of Boston.[31] Bond began building his first seagoing chronometer during the war of 1812, and completed it in 1815. When it went to sea in 1818 on the U. S. Navy ship Cyrus, the ship's Captain reported that it functioned well by comparison with lunar distances, but warned that he was not experienced in the use of the chronometer.[32] Between 1825 and 1830 Bond investigated the rates of marine chronometers with various effects, and the table appended to his results is also indicative of their increase in worldwide usage

[29] Dirk J. Struik (ref. 28), p. 108. On Bowditch see, for example, R. E. Berry, Yankee Stargazer: The Life of Nathaniel Bowditch (McGraw-Hill: New York, 1941), and Nathan Reingold, Dictionary of Scientific Biography.

[30] On editions of Bowditch, see John F. Campbell, History and Bibliography of the New American Practical Navigator and the American Coast Pilot (Peabody Museum: Salem, Massachusetts, 1964). See pp. 87–88 for chronometers. Bowditch's son Jonathan Ingersoll Bowditch carried on his father's tradition until the 35th edition in 1867. The U. S. Navy Hydrographic Office was assigned responsibility for the publication beginning in 1868, and in 1972 the Defense Mapping Agency Hydrographic Center (now NIMA, the National Imagery and Mapping Agency) took over publication. The current title is The American Practical Navigator: An Epitome of Navigation.

[31] On the Lewis and Clark chronometer see Silvio Bedini, "The Scientific Instruments of the Lewis and Clark Expedition," Great Plains Quarterly, 4 (Winter, 1984), 54–69, and on their method for determining longitude, Richard Preston, "The Accuracy of the Astronomical Observations of Lewis and Clark," Proceedings of the American Philosophical Society, 144 (June, 2000), 168–191. Regarding the 1804 chronometer, Bond's elder brother Thomas wrote "In an old French book he found the description of the chronometer used by the celebrated navigator, La Perouse, and determined to try his skill in making one on the same plan, to be kept in motion by weights instead of springs. Notwithstanding his constant attention in the shop, he carried out his plan, and it still remains as a memento of his skill and patience when but fifteen years old." Edward S. Holden, Memorials of William Cranch Bond . . . and of his Son George Phillips Bond (C. A. Murdoch & Co.: San Francisco, 1897), p. 7. See also Struik (ref. 28), p. 118. For more on the introduction of the marine chronometer into the United States see Silvio Bedini, Thinkers and Tinkers: Early American Men of Science (Scribner: New York, 1975), 352–355.

[32] Carlene E. Stephens, "Partners in Time: William Bond & Son of Boston and the Harvard College Observatory," Harvard Library Bulletin, 35, no. 4, 359–360. This article also places Bond in the broader context. Regarding the 1815 chronometer, "the severe test of a voyage to India and return proved it to be an excellent timekeeper," according to Holden, (ref. 31), p. 7.

by the mid-1820s.[33] By 1826 Bowditch had expanded his discussion of the "perfect time-keeper" method, because "the moderate price of good chronometers now, in comparison with their values many years since, together with various improvements in their construction, have caused this method of determining the longitude to be much more used within a few years, than it was when the first editions of this work were published."[34]

While Bond calibrated chronometers for merchant ships sailing out of Boston, the question of central importance for us is the growth in use of chronometers in the U. S. Navy. The American Navy, first established on a permanent basis in 1798, had been active in distant seas as early as the Tripolitan war with the pirates in the Mediterranean in 1801.[35] However, the scarcity and expense of chronometers at that time undoubtedly meant that few, if any, were available in the American Navy. The comment of the Captain of the *Cyrus* is consistent with the view that they were still scarce in the American Navy in 1818. Only with the increasing presence of U. S. Navy ships in distant seas such as the Pacific, and the increasing availability of chronometers in the 1820s, did the American Navy begin to acquire them in large numbers. The Navy in 1830 consisted of almost 50 ships, of which 19 were active, 19 others inactive and in need of repair, and 12 under construction.[36] In comparison with Great Britain's 139 ships and France's 52 ships of the line at this time, the demand for chronometers was obviously not as great in the U. S. Navy.

It is not only the rising number of Navy-owned chronometers but also their delicate nature that is essential to this history. Marvelous as the new technology was, the chronometer was not perfect. Like every timepiece ever constructed, each one had a "rate," that is, how fast or slow it ran, and the chronometer could not be used to full

[33] "Observations on the Comparative Rates of Marine Chronometers," *Memoirs of the American Academy of Arts and Sciences*, volume 1 (Cambridge University Press: Cambridge, 1833), pp. 84–90. Bond's table shows that he rated two chronometers per year in 1821–23, nine in 1824, ten in 1825, 26 in 1826, and 78 in each of the years 1828 and 1829. These were for ships sailing out of Boston, and though the nationality of the ships is not indicated, they were undoubtedly predominantly, British, with some American, so one cannot use it as an indicator of the growth of chronometer usage in America. Moreover, it is impossible to disentangle the growth of Bond's personal business from the growth of chronometer usage. Carlene Stephens, private communication, December 8, 1989.

[34] Bowditch (1826), p. 182. In a footnote the publisher and seller of navigational instruments, Edmund Blunt, states that "the chronometers most celebrated for correctness are those made by Mr. French, London, and for sale by James Ladd, no. 30 Wall Street, New York, who mechanically understands that valuable instrument."

[35] Much has been written on the U. S. Navy, but general histories of the Navy include little on its scientific or navigational aspects. There are a few specialized studies, notably Harvey M. Sapolsky, *Science and the Navy: The History of the Office of Naval Research* (Princeton University Press: Princeton, 1990), and Gary E. Weir, *Ocean in Common: American Naval Officers, Scientists, and the Ocean Environment* (Texas A & M University: Houston, Texas, 2001).

[36] *Report of the Secretary of the Navy*, December 6, 1830, p. 185. Of the active ships five were frigates, ten sloops of war, and four schooners. Ten years later there would be 60 ships, of which 37 were active. Four foreign squadrons were active in 1830: the Pacific, Mediterannean, Brazilian, and West Indian. The Annual Reports of the Secretary are by far the best source for Navy issues, activities, and budget requests for any particular year, and thus for the amount of growth or decline over a period of years.

effect unless its characteristic rate was known. This rate could then be taken into account in the longitude calculation. The determination of chronometer rates for official naval purposes was already being undertaken at the Greenwich Observatory as early as 1766, when Harrison's timepieces were under trial. During the years 1823–35 the Admiralty instituted trials at Greenwich in order to determine the best chronometers for the purchase of the Royal Navy, and, beginning in 1844, the Greenwich trials became routine annual events.[37] The same activity became prominent at other observatories of maritime nations around the world.

It is here that we come to the direct connection with our story. For, just as the Royal Observatory at Greenwich was founded principally to both acquire and improve astronomical data for the navigational method of lunar distances, so 150 years later was the Depot of Charts and Instruments founded in order to improve the second major navigational method – determination of longitude by use of chronometers – by making astronomical observations to determine precisely the characteristic rate of each timepiece. This was absolutely crucial to the successful use of the method, for the longer the voyage, the more crucial an accurately determined rate. It is in this context that we shall begin our story in chapter 1.

Beyond the *origins* of the institution in navigation, the Naval Observatory's further development is closely connected with the continuous drive for improvement in navigational accuracy. Improvements in star positions and timekeeping translated directly into improvements in navigation. Eventually electronic navigation would outstrip celestial navigation; just how much by the early 1980s is shown in Figure P.3.[38] We shall consider the shifting role of the Naval Observatory in the development and implementation of navigation techniques in the U. S. Navy, from the 1830s when it was largely responsible for the capability of fixing accuracies on the ocean of tens of miles via chronometer, to the end of the twentieth century, when it was only one of many agencies aiding in the maintenance of electronic means for determining ship positions to within a few meters. The response of the Observatory to changing navigational techniques, developments that go to the heart of its mission, is a primary question of this study.

Naval administration is also important for our story, for the chain of command has surely affected the day-to-day operations of the institution. When it was founded in 1798 the U. S. Naval administration consisted of little more than the Secretary of the Navy and his clerks. In 1815, in the wake of the war of 1812, the Board of Navy Commissioners was established, consisting of three prominent officers to help in the civil policy of the Navy. This arrangement lasted until 1842, when the Bureau system was inaugurated, initially with five bureaus, then eight in 1862.[39] Thus we will find the

---

[37] Gould (ref. 24), "Chronometer Trials at the Royal Greenwich Observatory," pp. 253–266.
[38] Wolf Kuebler and Sharon Sommers, "A Critical Review of the Fix Accuracy and Reliability of Electronic Marine Navigation Systems," *Navigation: Journal of the Institute of Navigation,* **29** (Summer, 1982), 137–151.
[39] Charles O. Paullin, *Paullin's History of Naval Administration, 1775–1911* (1968), chapter 5, "The Navy Commissioners, 1816–42," and chapter 6, "The Naval Bureaus, 1842–61."

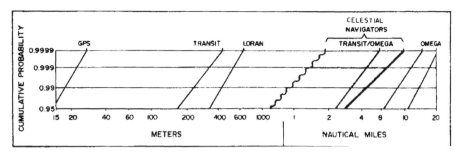

Figure P.3. Fix accuracies for marine navigation by the early 1980s, at the 95% to 99.99% probability range, showing where highest-precision celestial navigation fits among electronic navigation systems. Source: Kuebler and Sommers (ref. 38), courtesy Institute of Navigation. By the turn of the century GPS accuracy was about 3 meters at the 95% level, in contrast to 15 meters in 1980. In the 1850s longitude was determined routinely only to within about ten nautical miles.

Depot of Charts and Instruments at first under the Board of Navy Commissioners; the Depot and its successor Naval Observatory and Hydrographic Office under the Bureau of Ordnance and Hydrography from 1842–65; and the Naval Observatory then under the newly created Bureau of Navigation until 1889. Table P.2 shows the organizational sponsors of the Observatory throughout its history.

Of particular importance is the fact that, since 1942, the Naval Observatory has come under the Chief of Naval Operations, a recognition of its operational mission grounded in navigation. In this respect the Observatory has never been considered an official "Navy Laboratory," such as the Naval Research Laboratory, which is under the Chief of Naval Research. Rather, it comes under the heading of operations, a circumstance that both circumscribes its mission and affects its funding. Though the Observatory has remained under the Office of the Chief of Naval Operations since 1942, many reorganizations have taken place within that Office, and, since 1978, the Observatory has come under the purview of the Oceanographer of the Navy as its "resource sponsor." The Naval Observatory is a so-called "Echelon-2" Command, similar to the Naval Oceanography Command, but much smaller.

Finally, we should note that the Naval Observatory has intermittently absorbed or given up navigation-related functions, and thus is the parent of several organizations that now completely dwarf it. The Nautical Almanac Office, founded as a separate Navy agency in Cambridge, Massachusetts in 1849, officially became a part of the Naval Observatory near the end of the nineteenth century. The Hydrographic Office, which was split off from the Naval Observatory in 1866, in 1962 became the Naval Oceanographic Office; in the mid-1980s it had a budget of $175 million, employed 500 scientists, and operated 12 research ships. Its mapping functions were in turn split off (in 1972), and they became a part of the Defense Mapping

Table P.2. *Navy organizations with command*
*authority for the U. S. Naval Observatory*

| | |
|---|---|
| Board of Navy Commissioners | 1830–42 |
| Bureau of Ordnance and Hydrography | 1842–65 |
| Bureau of Navigation | 1865–89 |
| Bureau of Equipment | 1889–1910 |
| Bureau of Navigation | 1910–42 |
| Chief of Naval Operations | 1942–present |

Agency (now the National Imagery and Mapping Agency), an Army/Navy/Air Force conglomerate whose major component is the Hydrographic/Topographic Center and whose budget now exceeds that of its parent Naval Oceanographic Office. Internally, the navigational instrument and chronometer rating and repair function, having reached a peak during World War II, left the Observatory for the Norfolk Navy Yard in 1951. Thus our concern with hydrography as an integral part of this institutional history ends in 1866, as does our concern with navigational instruments in 1951. On the other hand, since the Nautical Almanac Office has been, and its function continues to be, an integral part of the Naval Observatory's mission for almost a century, I tell its story from the beginning. Although there have been many reorganizations within the institution, the three major functions related to time, positional astronomy, and the *Nautical Almanac* (in the twentieth century) have remained throughout its history.

In the terminology of Thomas Kuhn's *Structure of Scientific Revolutions*, the science undertaken at the Naval Observatory is undoubtedly "normal science" as opposed to "revolutionary science," and falls largely within that category which he calls "data gathering." Romantic notions and public perception notwithstanding, anyone who has worked in a field of science knows that the vast majority of all scientific work is not revolutionary, but the fleshing out and fulfillment of revolutions long since past. Data gathering, whether its results are put to use inside or outside an institution, is an essential part of science, especially if it is accurate and unique data. Unlike the majority of written history of science, which emphasizes revolutionary events and heroic figures, this history may be taken as a contribution to the history of normal science, which represents the majority of history of science that remains as yet unwritten. As an institutional history, this science is seen from the sometimes chaotic viewpoint of the scientist "as it really happened," rather than in the disembodied form often seen in textbooks or in the history of ideas.

Thus the perspectives on astronomy, science in America, and the U. S. Navy and navigation presented here not only serve to give us our bearings at the outset of our study, but also provide the essential context that must be taken into account during the institution's entire history. We shall relate in this volume not an isolated story of the U. S. Naval Observatory, but its proper place in science, the Navy, and navigation. For it is surely a truism that no institution exists in a vacuum. Each is subject to many forces, changing with time, acting at levels ranging from the individual to the bureaucratic, the sum total of which comprises what we call history.

Part 1

# The founding era,
# 1830–65

# 1    From Depot to National Observatory, 1830–46

> ... first, that a suitable place be designated to serve as a general depot for all the Chronometers, Instruments of reflection, theodolites, circles, telescopes, charts etc., belonging to the Navy ... 2d that to this depot there be attached a competent officer and an artist of known merit and capacity. The former to act under the immediate orders of the Navy Commissioners; to be made personally responsible for all instruments submitted to his charge; and especially required to determine the rates and characters of Chronometers.
>
> L. M. Goldsborough, November 1830[1]

> I should have regarded it as time misspent to labor so earnestly only to establish a *depôt*. My aim was higher. It was to place an institution under the management of *naval officers*, where, in the practical pursuit of the highest known branch of science, they would compel an acknowledgment of abilities hitherto withheld from the service.
>
> J. Melville Gilliss, 1845[2]

If Greenwich Observatory had its Charles II, Paris Observatory its Louis XIV, and Pulkovo Observatory its Tsar Nicholas I, the American counterpart, in good Yankee tradition, began with nothing quite so regal. Like Greenwich, Paris, and Pulkovo, the U. S. Navy's Depot of Charts and Instruments was born of navigational necessity. It did not come about by decree from above; neither was it first conceived as an observatory at all. The idea first surfaced in 1829 in a letter from the Board of Navy Commissioners to the Secretary of the Navy, proposing a reorganization within the Navy, including a new Department of Docks and Navy Yards. Among the new duties of this Department would be to provide for "a special officer to take charge of all the nautical instruments, books, and charts not on board of ship, to keep them in order for use when required." Moreover, this officer would "attend particularly to the timepieces, or chronometers, to ascertain precisely their character, such as their rate of deviation from true time, whether they are affected by changes of weather, etc."[3]

---

[1] LMG to SecNav, November, 1830. RG 45, NA, E 22, Letters from Officers to SecNav, volume 9, microcopy 148, roll 64, 10–41–8, 53.

[2] JMG, in *Report of the Secretary of the Navy, Communicating A Report of the Plan and Construction of the Depot of Charts and Instruments, with a Description of the Instruments*. 28th Congress, second session, February 18, 1845, Senate Executive Document 114, p. 66.

[3] BONC to SecNav, November 23, 1829, RG 45, E 213, Letters to SecNav. John Rodgers served as President of the Board from 1827 until 1837, so that this letter and others cited later are under his signature on behalf of the Board.

Though the larger reorganization plan did not come to pass until 1842, a depot for charts and instruments was an idea whose time had come. As we have seen in the Prelude, navigational technology was becoming ever more important in the U. S. Navy, and in particular the chronometer for determining longitude was rapidly assuming the role of centerpiece in that technology. The surprise is therefore not the founding of a practical establishment for rating chronometers and storing charts and instruments, but that this institution grew, within the span of a dozen years, into a formal astronomical observatory well beyond the needs of the Navy. Even more, it did so in the face of active hostility to such an observatory, stemming in part from Congressional dislike of John Quincy Adams' proposal for a "lighthouse of the skies" and other political considerations as well. As one historian has put it, the founding of the Naval Observatory in 1842 was "the classic example of the surreptitious creation of a scientific institution by underlings in the executive branch of the government in the very shadow of congressional disapproval. No more hated proposal existed, and nowhere had more pains been taken to prevent the creation of a new agency. Yet despite this vigilance the forces that required an observatory gained their ends."[4] We shall see that this significantly overstates the case. While it is true that Congress never passed legislation for a national observatory, it did authorize a Depot and small observatory, which was converted into a national observatory only with the crucial approval of a series of Secretaries of the Navy, hardly "underlings" in the government. Moreover, by 1842 the "hated proposal" of Adams for a national observatory had given way to more practical political jockeying involving the bequest of James Smithson.

How the U. S. Navy's Depot of Charts and Instruments grew into both a naval observatory and a national observatory will be the major theme of this chapter. In order to examine this theme we must study the origins of the Depot in 1830, the growth of its activities in subsequent years, and the forces that brought about a new Depot and associated naval observatory, the springboard from which the status of "national observatory" could be claimed.

## 1.1 Origins: Goldsborough and the G Street Depot

In November 1830 Lt Louis M. Goldsborough (Figure 1.1) wrote the Secretary of the Navy suggesting that just such a Depot be established as the Board of Navy Commissioners had recommended the previous year. Indeed, Goldsborough's proposal mentions the earlier reorganization plan, and it is likely that his father, Charles W. Goldsborough, the influential Secretary of the Board, had played a role in drafting that plan.[5] Indeed, one can plausibly surmise that father and son conspired to implement that part of the proposal related to the Depot, which would just happen to employ Louis at a time when responsible duty for Navy officers was scarce. The younger

---

[4] A. Hunter Dupree, *Science in the Federal Government* (Harvard University Press: Cambridge, Massachusetts, 1957), p. 62.

[5] Charles Washington Goldsborough (April 18, 1779–December 14, 1843) was chief clerk of the Navy Department, Secretary of the Board of Navy Commissioners 1823–42, and author of *The United States Naval Chronicle* (1824).

Figure 1.1. Louis M. Goldsborough
(1805–77), founder of the Depot of Charts
and Instruments in 1830, and first
Officer-in-Charge, 1830–33. He was
promoted to Commander (as seen here)
in 1841, Captain (1855) and Rear Admiral
(1862).

Goldsborough (1805–1877), who was part of a prominent Maryland family with much political influence, had received a midshipman's warrant at the age of seven, was ordered to Boston on duty at the age of 11, cruised the Mediterranean and Pacific during the years 1817–24 aboard the *Franklin 74*, and received a lieutenant's commission in 1825. In the five years just prior to 1830 he had studied 16 months in Paris, and, in 1827, while cruising in the Mediterranean on the schooner *Porpoise*, had commanded four boats that recaptured the English brig *Comet* from Greek pirates, a foray during which 90 pirates were killed or wounded.[6] Returning from Mediterranean duty in 1830, a young man of 25 undoubtedly eager to make his mark, Goldsborough wrote his "paper upon the subject of establishing a deposit for the nautical instruments and charts belonging to the Navy."[7]

[6]  By far the most comprehensive source on Goldsborough is Jean Ponton, *Rear Admiral Louis M. Goldsborough: The Formation of a Nineteenth Century Naval Officer*," Ph.D. Dissertation, Catholic University (Washington, 1996). See also *DAB*, volume 7, pp. 365–366; *A Naval Encyclopedia, Comprising a Dictionary of Nautical Words and Phrases, Biographical Notices, and Records of Naval Officers* (Philadelphia, 1881), pp. 314–316, and J. T. Headley, *Farragut and our Naval Commanders, Comprising the Early Life and Public Services of the Prominent Naval Commanders who . . . Brought to a Triumphant Close the Great Rebellion of 1861–1865* (New York, 1867), pp. 196–208. The bulk of Goldsborough's papers are to be found in 23 volumes in the Library of Congress, but, since they have no bearing on Goldsborough's duties during this period, the Navy records in the National Archives (see Appendix 1) are most crucial for our story. It is notable that Goldsborough went to Paris at his own expense, where he studied French and "the scientific branches of his profession, – mathematics, astronomy, etc.," *A Naval Encyclopedia*, p. 314.
[7]  LMG to SecNav, November, 1830 (ref. 1 above). The ten-page document was undated and untitled; the title here is that given by the Secretary of the Navy in his correspondence to BONC, November 30, 1830 (ref. 12 below). The date is determined from subsequent responses to the proposal. The full paper appears in Steven J. Dick, "Louis M. Goldsborough's Proposal to Establish a Depot of Charts and Instruments in the U. S. Navy: Text and Commentary," *Rittenhouse: Journal of the American Scientific Instrument Enterprise*, **4** (May, 1990), 79–86.

The ten-page paper made a persuasive argument. Goldsborough pointed to the progress of navigation in the last three decades, particularly in three areas: the "grand and beautiful" invention of the chronometer, improvements in the sextant of reflection used for celestial observations, and the improvement of lunar tables necessary for the method of lunar distances. Use of the chronometer and the motion of the Moon were independent methods of determining the longitude of a ship at sea. Normally both methods were used as a check on each other; according to Goldsborough, the chronometer method was by now considered preferable since it was less liable to error. However, care in the purchase, maintenance, and rating of the chronometers was crucial if this method were to achieve full promise: "An error of four minutes in the time shown by a chronometer, creates an error of one degree of longitude in the reckoning of a ship; and a mistake or neglect in ascertaining its daily rate of a few seconds, would, in the course of an ordinary voyage, accumulate not only to this number, but to one even greater and fraught with far more serious consequences . . .".[8]

In order to bring home his point, Goldsborough cited from personal experience the case of the *Franklin* 74, which sailed from New York to the Pacific in 1821 with a chronometer and sextant as its chief method of determining longitude. (Even today this example is important in giving us a feel for the accuracies of the best marine timepieces of the period, and for the links among time, longitude, and geographic position on which the lives of sailors and the merchant marine alike depended.) After the ship was 25 or 30 days out of port, Goldsborough wrote,

> . . . one of the Cape Verd [sic] Islands was suddenly and unexpectedly descried – the Chronometer not pleasing us within sixty miles of our actual position. The first impression was that the instrument was radically defected, and by its differing so far from the truth no further confidence could be placed in its correctness. Thus were we deprived of the mainstay in our navigation. Upon our arrival at Rio, Commodore Stewart thought proper to have ascertained by observations, whether its correct rate had been given us on leaving N. York. So far from such proving to be the case, instead of *three* seconds and a fraction (the daily rate given us on our departure) it was found to be *eleven* seconds and a fraction. Had the precautionary measure of making one of the Cape Verd Islands not have been taken, this glaring error of eight seconds per day would have caused, in our passage of 45 or 50 days, a discrepancy between the calculated and actual position of our ship, of not less than 85 or 100 miles – a distance sufficient to jeopardize the safety of half a million of public property and the lives of 750 souls.[9]

All of this, Goldsborough noted, not because the chronometer was faulty – any timepiece was bound to have imperfections – but because it had not been rated accu-

---

[8] Ibid., 5.   [9] Ibid., 5–6.

rately so that those imperfections could be taken into account. Goldsborough went on to state, on the basis of his experience and inquiries, that not two of ten U. S. ships have their chronometers accurately rated, that not more than three of five chronometers have been procured with proper care, and that repairs to these instruments are improper and overcharged. The person now rating chronometers for the Navy, he noted, also rated them for merchant ships, and the method for rating was designed to save trouble rather than to ensure accuracy: "In a word, the process is mechanical when it should be scientific; the results blundering when they should be precise."[10]

Thus from the crucible of direct experience, and especially from experience with the chronometer, the need for a central Depot grew. Mindful of economy, the young lieutenant was quick to add that only one additional instrument would be needed for rating chronometers at such a Depot, an instrument "of reflection and great radius, with tripod, vertical and oblique place screws," estimated to cost $200. The method of rating was to be astronomical observation, for no more accurate natural clock was known than the Earth's rotation as measured by the passing of an astronomical body over the meridian. The inherent imperfection of the chronometer, and the need to determine this imperfection correctly, thus strengthened the ancient navigational link between the sea and the stars. However limited at first, this proposal was the entering wedge for a greater astronomical observatory.

On November 20, Secretary of the Navy John Branch forwarded Goldsborough's proposal to the Board of Navy Commissioners for action, asking for a report of the actual condition of the instruments, and a recommendation as to how the situation could be corrected. Less than a week later John Rodgers, President of the Board, reported that they had no information on the state of the chronometers and nautical instruments. The Board recommended that a Depot be established along the lines of Goldsborough's proposal, adding only that it be should be located at the seat of Government so that the Navy Commissioners could more easily and frequently inspect it.[11] Four days later Secretary Branch requested the Board to make estimates of the costs of carrying out the plan. The Board replied that the plan could be implemented immediately by sending an officer to collect the instruments, and by renting a house for storage until a suitable building could be constructed. The costs would be a lieutenant's salary, a house rental of $250 per annum, and about $80 for stationery and fuels. The proposed "artist" could be engaged for each job. No mention was made of the proposed rating instrument. Thus the Depot of Charts and

---

[10] Ibid., 7. Chronometer raters with small observatories gradually sprang up in most U. S. ports; see Ian Bartky, *Selling the True Time: Nineteenth-Century Timekeeping in America* (Stanford University Press: Palo Alto, California, 2000), pp. 12, and 216 note 12.

[11] SecNav to BONC, November 20, 1830, attached to Goldsborough's report, and BONC to SecNav, November 26, 1830, RG 45, E 213. Rodgers's son would later become a Superintendent of the Naval Observatory. On Branch see W. Patrick Strauss, "John Branch" in *American Secretaries of the Navy*, volume 1, 1775–1913, Paolo E. Coletta, ed., pp. 143–150. Branch's brief tenure (9 March, 1829–12 May, 1831) came to an inglorious end when his wife insulted the wife of the Secretary of War, who also happened to be a personal friend of President Jackson.

Instruments had its first budget – a total of $330, in addition to personnel, comprising a single person.[12]

The Secretary's reply the very next day established the Depot, and is testimony again to the influence of Goldsborough's father and a personal connection with the Secretary: "Your letter of the 3d inst. has been received and the mode therein suggested for giving efficiency to the measure brought to your notice by this Dept. on the 12th ult approved." He ordered Lt Goldsborough to report to the Board for instructions, to be given immediately so that the young lieutenant could proceed to collect the nautical instruments before the winter set in.[13]

The action of the Board was equally swift. On December 6 they wrote to Goldsborough "You will proceed forthwith to Philadelphia, New York, Boston, Portsmouth, N. H., and Norfolk, Virginia, where you will receive from the respective commandants, all the chronometers, sextants, theodolites, circles, and nautical instruments of value, and not in use – all which are to be transported to this place." Cautioning Goldsborough to handle the chronometers especially with great care, the Board also forwarded letters instructing the commandants of the Navy yards to box and hand over their instruments to Goldsborough.[14]

Goldsborough was now faced with the practical matter of carrying out his self-imposed task. At once he appealed to the Secretary for $250 to carry out his instructions, and the funds were granted. Then, after the Christmas holidays, he set off for New York and Norfolk. He did not avoid severe winter weather as hoped, and instead found himself forced into "involuntary exile" for more than a month in Norfolk due to ice. Meanwhile in Washington his father attempted to arrange suitable quarters for the Depot. "I have been much worried about a house for the reception of your chronometers and nautical instruments," he wrote to his son on February 18. Not finding anything permanent, he had decided to rent a two-story brick house at 17th and G Streets, near the President's House, the Navy Department, and the Goldsborough residence, that would be ready to receive the instruments upon Goldsborough's return (see the map in Figure 1.2). In addition to two rooms about 16 feet square, it had a dry cellar.[15]

---

[12] SecNav to BONC, November 30, 1830, RG 45, E 222; BONC to SecNav, December 3, 1830, RG 45, E 213.

[13] SecNav to BONC, December 4, 1830, RG 45, E 222. No letter from SecNav to BONC dated November 12 has been found; this may refer to the letter of November 20 mentioned above.

[14] BONC to Goldsborough, December 6, 1830; RG 45, E 214, Letters to Officers. On the same day a circular was issued to the Commandants of the Navy Yards apprising them of this action. RG 45, E 212, December 6, 1830, Circulars Issued.

[15] On the $250 see LMG to SecNav, December 7, 1830, and the SecNav reply December 8, 1830, RG 45, E 22, Letters from Officers to SecNav, microcopy 148, roll 64, 10–41–8. Our knowledge of the trip to New York and Norfolk is from the correspondence between Goldsborough and his father during the trip, which has been preserved in the NHC archives, ZB file, CWG to LMG, January 3, and February 2 and 18, 1831. The house, described in the elder Goldsborough's letter of February 18, was rented from Mr Nourse and "stands directly opposite Mr Wirt's" (Wirt was the Attorney General). On November 1, 1831, Goldsborough would marry Wirt's daughter Elizabeth.

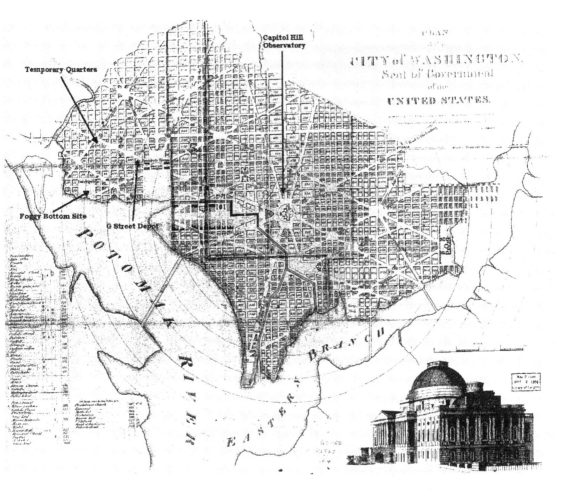

Figure 1.2. Early sites of the Depot of Charts and Instruments are shown on a map of the City of Washington dated 1835. Goldsborough first located the Depot near the President's House. In 1834 Wilkes moved it near the Capitol Building, it was in temporary quarters on Pennsylvania Avenue under Maury from 1842 until 1844, and finally from 1844 to 1893 the new Depot grew into the U. S. Naval Observatory at the "Foggy Bottom" site.

Even while exiled in Norfolk, Goldsborough was not idle. He seems to have actually begun rating chronometers while in Norfolk, and was also observing occultations. Obviously impatient to begin his duties in Washington and full of ideas, his father counseled patience: the full value and importance of his new duties must be developed by degrees, not all at once. Still in Norfolk on February 22, Goldsborough concurred in his father's decision to rent the house in question, asked for a bed to be placed in the Depot so he could sleep near the instruments, and was eagerly awaiting the departure of the steamboat *Potomac*, with its treasure of instruments.[16]

Also during his weeks in Norfolk, Goldsborough proposed the purchase of a transit instrument that he had seen while in New York, for sale by Richard Patten at a price of $350. Though Board President Rodgers requested approval for this purchase, it was apparently denied, and Goldsborough was asked about using one of Ferdinand Hassler's transit instruments from the moribund Coast Survey, which was now in storage at the War Department. But Louis objected in a letter to his father: "Of all the transits I examined in the War Dept (and I was told that all were shown me) there was not one at all calculated, or indeed even susceptible of use, for my purposes. Those of the W. Dept are all portable transits, or attached to theodolites; and have by no means the fixtures *essential* to a *permanent* one." There was no doubt they were the best of their kind, Goldsborough wrote, but he did not want a portable instrument. He would apply to the Navy Secretary himself for a proper instrument and would state, if necessary, "that the amusing decision of the Commissioners in favor of taking one from the War Dept is based upon a total ignorance of what is absolutely necessary to the efficient and creditable discharge of my duties." With the impatience of youth, the lieutenant declared that, if he had thought any of the Coast Survey instruments stored at the War Department would have sufficed, he never would have applied for the purchase of a new one. He saw this "stubborn imbecility" of the Board of Navy Commissioners as a personal threat: "My reputation in the science is now too deeply involved for me to say that I failed in the discharge of my duties because I had not yet proper instruments to make accurately my observations!" But he also drew a larger lesson: such an episode so high in the ranks of the Navy goes far to establish "that all scientific efforts will ever be discouraged . . . so long as our Navy remains headed as it is!"[17]

Such frustrations aside, Goldsborough returned to Washington, safely stored the instruments in the house his father had rented, and, still without his permanent transit instrument, commenced his duties.[18] For the first few months chronometer ratings were carried out via sextant and "repeating-circle" observations, both day and night. In addition, Goldsborough was responsible for delivering all navigational equipment to the various ships before they set sail from the several Navy Yards. This

---

[16] CWG to LWG, February 2 and 18, 1831; LWG to CWG February 22, 1831. Goldsborough also asked that six or eight "trusty negroes" be waiting when he arrived in Washington to hand carry the instruments to the house.

[17] LMG to CWG, February 18 and 22, 1831, NHC archives. The latter letter also contains Goldsborough's observations of an eclipse on February 12. The reference to problems in Navy management probably refers to the reorganization plan of 1829, which was still under lively discussion at the time, under which the Bureau system would replace the Board of Navy Commissioners. The plan was not passed until 1842.

[18] Our knowledge of the Depot's early years is derived from correspondence between Goldsborough and his superiors at the Board of Navy Commissioners, as found in NA, RG 45, BONC Records, especially Entry 228, "Letters from the Depot of Charts and Instruments, 1831–1842," seven volumes, and Entry 214, "Letters to Officers, 1815–1842," two volumes.

in itself was a considerable task; even though only 19 of 50 ships were now active in the American Navy, Goldsborough's deliveries would have been particularly crucial to the overseas squadrons in the Pacific, Brazil, the West Indies, and the Mediterranean.[19] Depending on how distant their destination, the ships would receive one or two chronometers, along with appropriate charts, sextants or quadrants, artificial horizons, compasses, and spyglasses, all now under Goldsborough's care.[20] It is little wonder that by June 1831 Goldsborough was complaining to the Board that his frequent absences required an assistant, preferably a passed midshipman, a request that was soon approved.[21]

Meanwhile, the Secretary of the Navy was unsuccessful in persuading the War Department to loan one of the stored transit instruments.[22] So the Patten transit instrument of 30-inch focal length was purchased, and was used to rate the chronometers from the summer of 1831 to the summer of 1833. According to a later account "the transit was mounted within a small circular building, upon a brick pier, having a base about 20 feet below the surface. To Lieutenant Goldsborough, therefore, is due the erection of the first astronomical instrument for the Navy at Washington."[23] This meager description of the G Street Observatory and Patten transit instrument is all that remains.

The duties at the Depot quickly grew, and represent an early example of centralization of technology and what we would now call "quality control" in the U. S. Government. In July 1831 the commandants of the navy yards were ordered to ship all their charts and nautical books to the new Depot, and, in yet another example of the kind of work needed, Goldsborough proposed to translate a survey of the coast and harbors of Brazil, and modify the Brazilian charts to the meridian of Greenwich.[24]

Most importantly, Goldsborough was also given duties of purchasing new chronometers and charts, and disposing of those now obsolete, a function that allows us to see the state of navigational technology in the U. S. Navy, and the sources of that

19  See Prelude, *infra*, ref. 36; also RG 45, E 213, BONC to SecNav, January 14, 1830, 296. This number of ships did not include those on the Great Lakes, which did not need chronometers.
20  Even frigates and sloops bound for distant shores were receiving two chronometers by 1833; RG 45, E 213, January 12, 1833.
21  LWG to BONC June 10, 1831; BONC to SecNav, June 10, 1831. The assistant approved upon Goldsborough's recommendation was passed midshipman Robert B. Hitchcock.
22  SecNav to BONC, June 15 and 17, 1831. The Secretary of War declined on the grounds that the instruments were too valuable to loan!
23  Gilliss, *Report* (ref. 2), 64. BONC to SecNav, June 27, 1831 describes the Patten instrument as "believed to be the only one in the United States for sale." SecNav to BONC, June 29, 1831 indicates that the Navy still wanted to obtain a transit instrument from the War Department, but Gilliss states that the Patten instrument was the one purchased. The Patten instrument might not have been purchased until summer of 1832, for RG 45, E 214, June 19, 1832 contains the following letter from BONC to Goldsborough in New York: "Capt. Wadsworth having required the repeating theodolite for the survey of Narragansett Bay, you are authorized to purchase a transit instrument at not exceeding 40 guineas and to bring it with you to the Depot at Washington to be used in place of the theodolite in ascertaining the rates of the Chronometers." On Patten see Deborah Jean Warner, "Richard Patten (1792–1865)," *Rittenhouse: Journal of the American Scientific Instrument Enterprise*, **6**, no. 2 (1991), 57–63.
24  Circular from BONC to Commanders of Navy Yards, July 2, 1831; BONC to SecNav, July 6, 1831. The survey was that of Raufsin; see LWG to BONC, November 29, 1831.

technology. Discussing replacements for seven chronometers he proposed to dispose of in July 1831, Goldsborough noted that the Parkinson and Frodsham firm and Mr French were regarded as the best chronometer makers in England. Parkinson and Frodsham had been most often used by the U. S. Navy in the past, and, considering their chronometers by far the best in the U. S. Navy, Goldsborough proposed that the Navy continue this practice. He suggested that the Navy purchase its chronometers directly from London via the American consul, transport them by the regular line of packets from London to Philadelphia, and place them in the hands of a chronometer vendor there until they could be picked up by Goldsborough.[25]

With regard to the interesting question of the number of chronometers now in use in the U. S. Navy, Goldsborough surprisingly indicates that the Navy had enough for its use in 1831, but there was a problem with quality: "Should all the chronometers now belonging to the service be of the most approved construction and of the best quality, it would not seem that more than a sufficient number to meet the contingencies of the Navy would be on hand."[26] By the end of the year, when a private citizen offered four chronometers for sale to the Navy, he declined, writing that there were "a sufficient number of chronometers and instruments to meet the present demands of the service."[27] Goldsborough nowhere states the exact number of chronometers currently in use in the U. S. Navy, but, considering the number of its ships, the rates of purchase and disposal, and the upper limit of 54 chronometers known to be owned by the Navy in 1835, we may estimate the number at between 30 and 40 when the Depot was founded in 1830.[28] It is therefore certain that already in the 1820s the U. S. Navy was using a considerable number of chronometers, a fact supported by Goldsborough's story of the *Franklin* voyage of 1821.

Aside from the all-important chronometers, Goldsborough also purchased and disposed of other instruments (Figure 1.3). Old sextants, quadrants, and spyglasses were sold in 1831 to Blunt and Patten in New York; but when early the next year eight sextants, four quadrants, and 12 spyglass telescopes were required, all had to be purchased from the U. K.[29] By the end of the same year Goldsborough reported the need for "12 double-framed sextants, graduated to 10 [arcminutes] by Troughton and

[25] LMG to BONC, RG 45, E 228, July 18, 1831. Goldsborough proposed that the seven old chronometers be placed for sale in the hands of the chronometer vendors Demilt of New York and Lukens of Philadelphia, and that the new ones be delivered to the latter. On Demilt and Lukens see Marvin Whitney, *The Ship's Chronometer* (Cincinnati University Press: Cincinnati, 1985), pp. 360–361 and 380.

[26] LMG to BONC, RG 45, E 228, 18 July, 1831.

[27] LMG to BONC, ibid., December 13, 1831. On the offer to the Navy and details of the four chronometers see Fuller to CWG, December 2, 1831.

[28] On the number and quality of Navy chronometers in 1835 see ref. 49 of this chapter. On purchases of chronometers after March, 1832 see "Letter from the Secretary of the Navy, transmitting a Report on the Subject of Chronometers, & c.," July 1, 1840, House of Representatives Document number 249, 26th Congress, first session, 1–17.

[29] The receipt of the instruments purchased was acknowledged in LWG to BONC, March 12, 1832. The instruments sold to Blunt and Patten included four sextants at $25 each, six quadrants at $5 each, and 31 spyglasses at $2 each, for a total of $192. LWG to BONC, September 14, 1831.

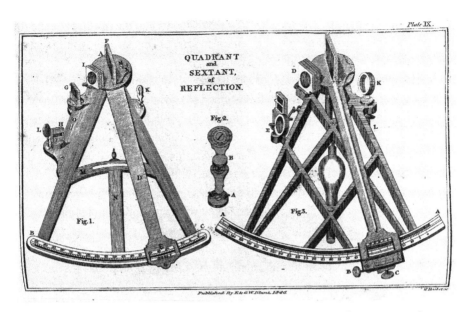

Figure 1.3. Navigational instruments of the early nineteenth century are shown in these plates from Bowditch's *New American Practical Navigator*. The quadrant and sextant were used for determining angular distances between the Moon and stars, or for determining local time by a variety of methods. These plates first appeared in the 1807 edition, and continued appearing in editions through the first half of the century.

Simms, 24 Quadrants of Parkinson and Frodsham, and 36 spyglasses of Dolland."[30] This pattern of the purchase of chronometers from Parkinson and Frodsham, spyglasses from Dolland, and sextants and quadrants from Troughton and Simms or Parkinson and Frodsham, was a steady one until American makers could compete.[31]

The duties of the purchase, repair, and distribution of navigational instruments, the observation of Sun or stars necessary for rating the chronometers, and the care and distribution of charts kept Goldsborough and his assistant fully occupied for two years, until he resigned his duties on February 11, 1833. A four-page inventory of instruments on hand at the Depot at that time reveals the extent of the Depot's duties and the central role it had quickly assumed in relation to the navigational technology of the U. S. Navy. Among the items on hand were 11 box chronometers, an assortment of sextants and quadrants, artificial horizons, 669 sheets of charts, observation logs for chronometer rating, and an assortment of books that included Pearson's *Astronomy*, Vince's *Astronomy*, Bowditch, the British *Nautical Almanac*, and Laplace's *Mécanique Céleste*.[32] As these instruments and books indicate, current navigational

---

[30] LMG to BONC, RG 45, E 228, December 18, 1832.
[31] The one volume comprising entry 229 of RG 45 contains "Letters from the American Consul in London," 1831–1835, including invoices from Parkinson and Frodsham, Dolland and Troughton, and Simms.     [32] Hitchcock to BONC, February 13, 1833.

methods compelled the U. S. Navy, like all other navies around the world, to attend to the heavens. This practical need had given much impetus to astronomy in the past, and it did so now in the American context.

Though no immediate cause for Goldsborough's resignation has been found, his subsequent career suggests that he resigned from the Depot and took leave of absence from the Navy because he was a man of action and disliked being confined to an office. Over the previous two years Goldsborough had not only founded the Depot of Charts and Instruments, but also married Elizabeth Wirt, the daughter of the famous Attorney General, who lived across the street from the Depot. Taking a two-year leave from the Navy, Goldsborough now led a group of German immigrants to his father-in-law's estate in Florida, and became involved in the Seminole Indian War. Returning to the Navy in 1835, he eventually went on to a distinguished career, including becoming Superintendent of the Naval Academy (1853–57) and Commander of the North Atlantic Squadron during much of the Civil War, attaining the rank of Admiral.[33]

Goldsborough himself probably viewed his brief years at the Depot as little more than a prelude to an active and colorful life. However, the growth of the Depot during those years, and its continuance after his departure, are ample testimony to the utility of his idea, and Goldsborough would live to see it become much more than a Depot.

## 1.2   Struggle: Wilkes and the Capitol Hill Observatory

On March 12, 1833, Lt Charles Wilkes (Figure 1.4) was ordered to take charge of the Depot.[34] At 34, Wilkes (1798–1877) was also a veteran of sea duty in the Mediterranean and the South Pacific, sailing the latter in the *Franklin* on the same cruise as Goldsborough, who was seven years his junior. During the 1820s Wilkes spent long periods on furlough or awaiting orders, and received training in surveying under Ferdinand R. Hassler, first Superintendent of the Coast Survey. The knowledge thus gained was put to use in his duty of surveying Narragansett Bay during the years 1832–33. Fresh from this experience, with a second son less than a month old, he took on his new duties in the City of Washington.

Though Wilkes firmly believed that the Depot served an important purpose, decades later he recalled the trepidation with which he accepted the position. ". . . I at

---

[33] A personal reminiscence of Goldsborough's later life is given by Rear Admiral Caspar F. Goodrich in "Memorabilia of the Old Navy," *Proceedings of the U. S. Naval Institute*, **30**, 823–830; Goldsborough is reported to have been "a very large man, reputed to be six feet four inches tall and to weigh nearly four hundred pounds."

[34] The Board recommended to SecNav on February 18 that Wilkes and two other officers were qualified to replace Goldsborough, BONC to SecNav, February 18, 1833. The order from the Board to take charge is BONC to Wilkes, March 12, 1833. The best (though biased) biographical source for Wilkes is his own autobiography, published as *Autobiography of Rear Admiral Charles Wilkes, U. S. Navy, 1798–1877*, ed. William James Morgan *et al.* (Department of the Navy: Washington, 1978). See also DAB, volume 20, 216–218, and Daniel M. Henderson, *Hidden Coasts: A Biography of Admiral Charles Wilkes* (Columbia University Press: New York, 1953).

Figure 1.4. Charles Wilkes (1798–1877), Officer-in-Charge of the Depot from 1833 until given command of the Exploring Expedition in 1836. Engraving from the painting by Thomas Sully.

once Saw there was a great feeling that all that Goldsborough had done in the way of arranging the duties and manner of conducting them would involve me in difficulties with the Board through their Secretary, who was Goldsborough's father and had a most excellent opinion of his Son's acquirements & talents, and situated as he was, actually having Control though knowing very little about the necessary duties and observations to be made." His concern was not entirely misplaced, for Wilkes reported that "Mr. Goldsborough always showed me great kindness and was inclined to assist me except when it came to supersede what His Son had established."[35]

It was with some boldness, therefore, that, little more than a month later, Wilkes asked the permission of the Board to move the Depot to a new site on Capitol Hill. Thinking such permission had been given, Wilkes was surprised on June 18 to have the elder Goldsborough deliver a message from a Board member ordering him to suspend the removal of the Depot. Wilkes replied that most of the items had been removed a month before, except for the contents of two small buildings removed the previous week. This incident led him to detail his reasons for making the move: first to be near his home, second because the wet soil around the original observatory caused "vapor" to form, which affected his observations.[36] A new location not being found in the vicinity of the Navy Department, Wilkes had moved it in May to Capitol Hill, a site, he noted, that Hassler had selected for an observatory in connection with

---

[35] Wilkes, *Autobiography*, p. 294.
[36] Wilkes's June 19 reply is in the Wilkes papers, Library of Congress, Box 1, Wilkes to BONC. This lengthy letter asked Board President Rodgers to recall a meeting with Wilkes on April 22 at which Rodgers agreed to removal of the Depot to Capitol Hill, if it could be undertaken at Wilkes's expense.

the Coast Survey in 1816. The issue dragged on, with the Commissioners complaining of the distance from the Department, and even when Wilkes offered to build at his own cost a Depot at the site, the Board declined.[37] However, his persuasive powers won out, and Wilkes said of the elder Goldsborough that "after I had been successful in removing the temporary obsy [observatory] to Capt Hill [Capitol Hill] and thus swept all vestige of his small arrangement away, he forgave me, and cordially as far as I was concerned, and the duties appertaining to the position I had were offered all the facilities which lay in his power."

It was no coincidence that the site Wilkes had chosen was the site where he had relocated his family from New York. Having left in civilized New York "a most delightful house and every Comfort and pleasure we desired," they moved into a large three-story brick house that Wilkes had rented on North Capitol Street. Adjoined by a second house, both structures had been built by George Washington only about 1000 feet from the Capitol (Figure 1.5). Since he was bothered by noisy neighbors, Wilkes also rented the second house at the first opportunity. When both were offered for sale shortly afterward, Wilkes bought them for $4000, along with adjoining lots. It was here, in one of the stories of the North house, that Wilkes moved the Depot, and, in the center of the lot, he built an observatory.[38] From here, through June 1842, the Depot and its Observatory would function.

Most of our knowledge of the new observatory comes from the description of it by Wilkes's successor, published some 13 years later. "A small frame building, to be used as an astronomical observatory for the Depot of Charts and Instruments, was erected by Lieutenant Wilkes on Capitol hill, some time during the years 1833–34. Its location was 1,200 feet (nearly) N. 5 degrees W from the centre of the Capitol, on the brow of an eminence sloping rapidly to the north and west, and commanding unobstructed views of the range of hills which bounded the horizon in those directions." To the south and east vision was obstructed to an altitude of 7 degrees by the Capitol building itself![39] As for the structure of the building:

---

[37] The proposal to build a Depot is in Wilkes to BONC, September 16, 1833, RG 45. A proposed Coast Survey observatory at this same site in Washington is described in Hassler, *Transactions of the American Philosophical Society*, volume II, new series (1825), pp. 365–370. It is possible that the site, or one nearby, was used as an observatory even earlier by Mr William Elliot, assistant to William Lambert, for determining the longitude of the Capitol. According to an article in the *Records of the Columbia Historical Society*, 2 (1899), 67, Wilkes moved the Depot "to a small frame building on a high elevation that was located at the rear of Mr William Elliot's residence, No. 222 North Capitol street, situated on the west side between B and C streets, and N. 5 degrees 0 minutes W., 1200 feet (nearly) from the center of the Capitol, being the same observatory as was built by Mr. Ellicott [sic] in March, 1824." Wilkes makes no mention of a previous observatory, but Silvio Bedini accepts this identification in *The Jefferson Stone* (Frederick: Maryland, 1999), p. 54.

[38] Wilkes, *Autobiography*, pp. 299–302, a source that also gives the Washington ambience and social customs of the times.

[39] Gilliss, *Astronomical Observations made at the Naval Observatory, Washington, under Orders of the Honorable Secretary of the Navy, Dated August 13, 1838* (Washington, 1846), p. viii. Gilliss gave its latitude and longitude as 38 degrees 53 min 32.8 s N, 5 h 08 m 08.0 s west of Greenwich. Gilliss's copy of this volume, autographed "Naval Observatory from Lieut. Gilliss," is in the Naval Observatory Library.

Figure 1.5. A guidebook view of the U. S. Capitol Building and environs, about 1820. The Navy Depot, Wilkes's house, and the observatory built by Wilkes were located about 1,200 feet north of the Capitol, and may be among the buildings sketched to the left of the Capitol in this view.

The length of the observatory, east and west, was fourteen, its breadth thirteen, and height from the floor to the eaves, inside the plastering, ten feet – its roof sloping to the north and south at the usual inclination in buildings covered with wood.

Meridian doors, each nearly two feet wide, and making, together, an aperture more than three and a half feet, reached from the floor to the eaves; they opened outwardly, and were lined inside with baize. The roof doors were of the same width, but extended only to within three feet of the ridge pole. These were raised by pulleys leading over uprights on the roof, through sheaves in the east and west sides of the building. There was a door of ordinary dimensions in each of these sides; and to prevent the transmission of terrestrial vibrations, the building was surrounded by a ditch five feet wide and deep.

Curiously, this arrangement left a zone 13 degrees wide on either side of the zenith invisible. In this building was placed a transit instrument, designed by Hassler and constructed by Troughton for the Coast Survey in 1815, which was loaned to the Navy Department on the application of Wilkes (Figure 1.6). The lens had a clear aperture of 3.75 inches, and a focal length of 63 inches. Aside from this instrument, which was permanently mounted on granite piers protruding six feet above the floor, the only other instruments during Wilkes's tenure at the observatory were a sidereal clock

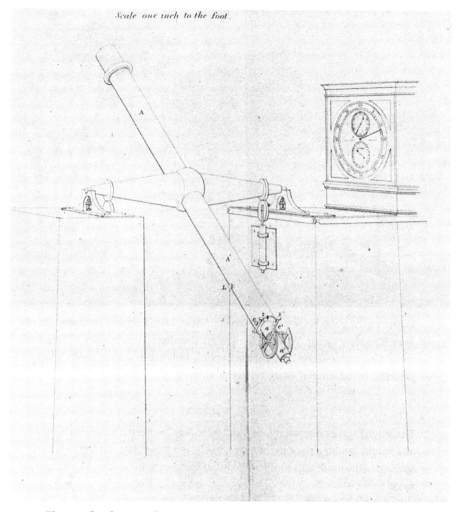

Figure 1.6. The Troughton transit instrument, constructed for the Coast Survey in 1815, was loaned to the Navy Department in 1834. It was used for rating chronometers, and later for Gilliss's star catalogs at the Capitol Hill Observatory. From J. M. Gilliss, *Astronomical Observations* (Government Printing Office: Washington, 1846). The clock is a Parkinson and Frodsham sidereal.

and a number of portable instruments. The latter included a 3.5-foot-focal-length achromatic telescope built by Jones, and the Patten transit instrument.[40]

Though occasional observations were made of eclipses and other phenomena, this observatory was used prior to 1836 primarily for chronometer rating; Wilkes's observing log shows that this was done by observation of the Sun, using the

---

[40] Gilliss, ibid., p. xiii. Gilliss later mounted the Patten instrument "for the use of the assistants on the pier near the south door."

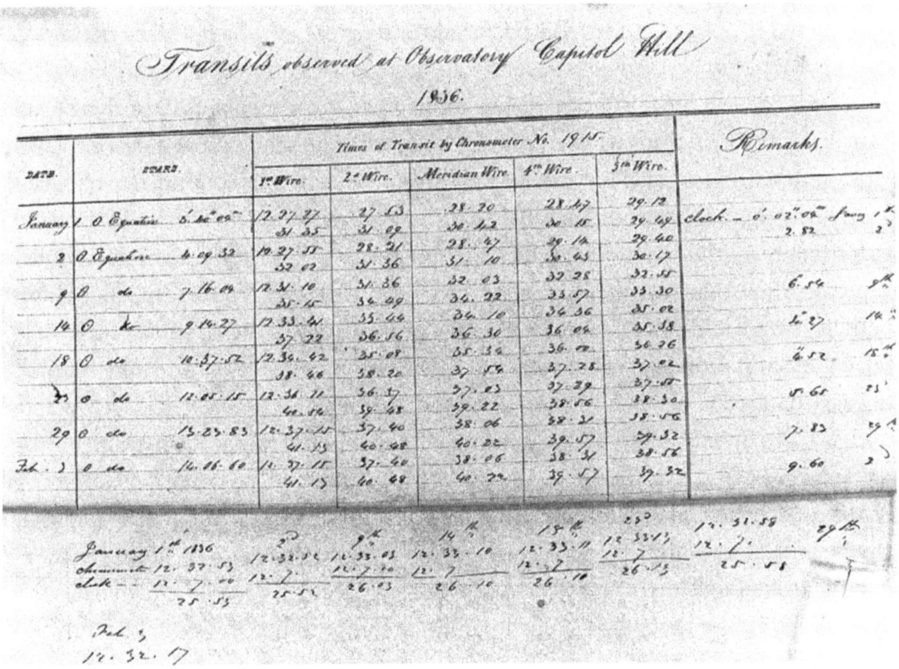

Figure 1.7. A page from Wilkes's observing book shows eight observations of the Sun over the meridian in early 1836; taking into account the equation of time, the data could be used for rating chronometers. U. S. Naval Observatory Library.

Troughton transit instrument (Figure 1.7).[41] The manner of rating chronometers was adopted from the method used at the Royal Observatory, Greenwich, and a trial number of eight seconds was given to suppliers as the minimum acceptable rate for a chronometer.[42]

In addition to such rating, much still needed to be done in the purchase and repair of chronometers and other navigational instruments. "We are now much in want of good Chronometers," Wilkes wrote in mid-1833, taking a tougher stand than Goldsborough toward the necessity for purchasing quality chronometers. "Those now at the Depot were never good ones and from age and usage have become very defective. According to my calculations, taking into consideration the state of the Chronometers now belonging to the Navy and the number that would probably be rejected, thirty Chronometers would be the number to put on trial to meet the wants

[41] Ibid., pp. xi–xiv. Wilkes's observing log for chronometer rating, (transits) at Capitol Hill Observatory, is at the Naval Observatory Library.

[42] Wilkes to BONC, December 16, 1833. The manner of finding the trial number "is the same as that made use of at the Royal Observatory from Dr. Young's formula, viz, by taking the difference between the greater and lesser mean monthly rates, multiplying by 2 and adding the mean monthly variation for the trial number." Wilkes noted that the trial number accepted at the Royal Observatory, Greenwich was 12 seconds for the prize reward when trials were begun in 1822, whereas the third-prize reward was now 4.5 seconds. The mean of 8 seconds was adopted for chronometers used by the U. S. Navy. A paper accompanying this letter describes the exact method of rating chronometers at the Depot.

of the service."[43] By 1835 the Navy had 54 chronometers, of which Wilkes character-ized 12 as the very best, eight as second rate, and 34 as indifferent and unreliable.[44]

Along with procuring and disposing of other instruments, Wilkes also began to make use of local firms whenever he could, as when he sent 34 spyglasses to a local Washington firm for repair, at a cost of $9 apiece.[45] He not only continued to circulate charts on demand, but also examined all the charts for accuracy, and requested early in his tenure a lithographic press for chart production.[46] His monthly reports to his superiors show that the work during his three-year tenure was routine, but increasing both in volume and in scope. As an example of the latter, Wilkes reviewed proposals related to navigation, including one from Matthew Fontaine Maury for an instrument to be used in reducing observations by lunar distances.[47]

When, in July, 1836, Wilkes was ordered to Europe to purchase instruments for the U. S. Exploring Expedition, Lt Hitchcock took charge of the Depot. In November Lt James Melville Gilliss was ordered to report as his assistant, and then was put in charge of the Depot the following spring. The Exploring Expedition, another indica-tion of awakening American science, was thus indirectly responsible for giving the Depot a key figure in its early history. Of no less importance, it would also provide impetus for converting the Depot into a true astronomical observatory.

### 1.3  Success: Gilliss and a permanent observatory

It was under James Melville Gilliss (1811–65) (Figure 1.8) that the Depot began to blossom into something beyond its humble beginnings. Gilliss was a local boy, born in nearby Georgetown, whose father had helped defend the City of Washington in the War of 1812, was appointed by President Monroe to an accounting position in the Treasury Department, and served in that Department through six Presidents until 1846. Gilliss became a midshipman at the age of 15, performed the usual sea duty, spent a year beginning in 1833 at the University of Virginia, and, like Goldsborough, studied for 6 months in Paris before returning to duty at Philadelphia and finally the Depot in Washington in 1836. According to his biographer an early formative experi-ence, in which members of Congress were told in his presence that "there is not an officer of the Navy capable to conduct a scientific enterprise," inspired him to prove them wrong: "from that hour no effort has been spared by which the standard of intel-ligence in the service might be increased and its reputation enhanced."[48]

[43] Wilkes to BONC, RG 45, E 228, July 29, 1833.
[44] Wilkes to BONC, July 7, 1833 and November 23, 1835.
[45] The job went to Messrs Joseph & Deuchman [?], "opticians of this city . . . the only opticians who grind their own lenses in this country." Wilkes to BONC, RG 45, July 7, 1833.
[46] Wilkes to BONC, August 23, 1833 gives details and reasons for wanting a press.
[47] Wilkes to BONC, October 5, 1833, E 228. Recalling a similar proposal by the Washington surveyor William Elliot, Wilkes noted the imperfections to which all such instruments were liable, but praised the ingenuity of Maury's plan.
[48] The chief source of Gilliss's life is B. A. Gould, BMNAS, 1 (1867), 1–57. The main source of his genealogy and family is Frances Howard Ford Greenidge, *Ancestors of Raymond Oakley Ford and Frances Howard Ford* (privately published, 1994), volume 1, Naval Observatory Library. Gilliss's father George is discussed on pp. 148 ff, and Gilliss himself on pp. 153 ff. The incident referred to occurred in 1831 and is described in Gould, BMNAS, 1 (1867), 4.

Figure 1.8. Lt James Melville Gilliss, in charge of the Depot 1836–42, responsible for founding a permanent Depot and the Naval Observatory, and its second Superintendent, 1861–65. He was promoted to Commander and Captain in 1861, and saw the Observatory through the Civil War.

Gilliss inherited the full range of activities begun by Goldsborough and Wilkes, including the care and rating of chronometers, which continued to be a central concern, now commanding the attention of American manufacturers. An indication of growing American commercial interests in chronometers, and of the Navy's role as a user in this endeavor, is a resolution of the House of Representatives in 1840, which called upon the Navy to report on the procedure used for the purchase of chronometers. A statement reporting data beginning in 1832 showed a stream of more than 100 new purchases during the 1830s, including chronometers for the U. S. Exploring Expedition. The makers were largely from London, and included Parkinson and Frodsham, Molyneux & Sons, J. M. French, and Thomas Cotterell. As early as 1833 the Navy purchased a chronometer from Little & Elmer of Bridgeton, New Jersey, followed by one from Eggert in New York (1838) and two from Bliss and Creighton in New York (1840). By 1836 the Navy procedure was to have chronometers placed on trial for one year, after which they would either be purchased or returned to the maker's agent in the United States. The issue which brought the House resolution was that these agents were selling the rejected foreign chronometers without duty charge, to the detriment of American makers, a problem the Navy now promised to resolve.[49]

---

[49] "Letter from the Secretary of the Navy, transmitting a Report on the Subject of Chronometers," as cited in ref. 28. This report, with its accompanying table of chronometer purchases compiled by Gilliss, gives detailed information on the makers, their American agents, types, and prices of chronometers, and the Navy's trial procedure. In 1840 the Navy was paying $300 for a chronometer with a trial number of 8 seconds, and up to $400 for one with a trial number of 4 seconds or less. The table also shows that, after 1834, William Bond & Son no longer acted as an agent dealing with the Navy, Arthur Stewart and B. and S. Demilt in New York and S. Willard in Boston taking over that role for the various makers. It is notable that 29 chronometers accompanied the U. S. Exploring Expedition in 1838 according to Wilkes, who led it; see *Narrative of the United States Exploring Expedition during the years 1838, 1839, 1840, 1841, 1842* (Philadelphia, 1845), volume 1, p. xxi.

It was not in his routine work with chronometers but in astronomy that Gilliss would show his capacity for hard work and precise observation, and act on his desire to advance science in the Navy. Like those of his predecessors, Gilliss's telescopic observations were at first limited to those necessary for chronometer rating. During the winter of 1837–38 he also observed transits of the Moon and reference stars for determining longitudes connected to Wilkes's survey of the Savannah River. Finally, on August 13, 1838 came a crucial order from Secretary of the Navy J. K. Paulding to make similar observations for determining longitude differences between Washington and locations visited by the Exploring Expedition. Detailed instructions were drawn up by Wilkes, and the Navy also contracted W. C. Bond in Boston for the same task.[50]

The order and the accompanying instructions were considerably broader in scope than the primary purpose of determining longitude differences between Washington and specific sites along the route of the Exploring Expedition, and they were the decisive factor that gave impetus to astronomical observations. Wilkes's instructions to observe the Moon and its reference stars, as well as "falling stars" and eclipses of the Sun, Moon, and Jupiter's satellites, were all methods of various accuracy for determining longitude by the simultaneous observation of astronomical phenomena. Wilkes's further instruction to "lose no opportunity of making observations on any astronomical phenomenon, describing the place in the heavens of its occurrence," was undoubtedly also inspired by practical reasons, but can also be seen as a broad mandate to observe almost anything in the heavens, a mandate Gilliss was quick to seize. Moreover, Bond and Gilliss were to make magnetic and meteorological observations "useful to the solution of the great physical problem under investigation by order of European Governments," as Gilliss later put it, referring to the attempt to determine the nature of the Earth's magnetic field. From these orders sprang the first volume of astronomical observations to be published in America, as well as a substantial volume of magnetic and meteorological observations.[51]

Taking advantage of the new orders, Gilliss reported that the Depot had neither suitable instruments nor the staff to undertake the desired observations. The

[50] The method of "moon culminating stars" was a supplementary means of determining longitude by observing the right ascension of the moon, with respect to nearby reference stars, from two different stations. On the method see Elias Loomis, *An Introduction to Practical Astronomy* (New York, 1855), p. 312. Allied to the method of lunar distances, it was normally practiced from fixed observatories. For example, during the Exploring Expedition, Wilkes recounts landing in Tahiti in 1839, and setting up a temporary fixed observatory: "I had been in hopes of obtaining a full series of moon culminating stars on Point Venus; but I was disappointed, for it rained almost every night. I was, therefore, compelled to rely for the longitude on the chronometers alone, and restricted even in that method to observations of the sun." Wilkes (ref. 49), volume 2, p. 5.

[51] The full orders are reprinted in Gilliss, *Magnetic and Meteorological Observations Made at Washington under Orders of The Hon. Secretary of the Navy, Dated August 13, 1838* (Washington, 1845), pp. vi–vii. The volume autographed "U.S. Naval Observatory from Lieut. J. M. Gilliss" is also in the Naval Observatory Library. See also Gilliss, *Report*, p. 65. The first volume of astronomical observations is cited in ref. 39.

3.75-inch Troughton transit would be used for the longitude determinations, but, for the observation of phenomena such as Jupiter's satellites, Gilliss persuaded the Secretary of the Navy to authorize purchase of an achromatic telescope with a focal length of 42 inches. For some time Gilliss had reported meteorological observations quarterly to the Navy Department, using a float-gauge barometer, thermometers of the same kind as those supplied to ships of war, and a Daniel hygrometer. For magnetic observations, the Navy had a variation transit and a dip circle, both constructed in 1828, but these were in such disrepair that Gilliss was now able to obtain a declinometer (Figure 1.9), an 8-inch dip circle and a sidereal chronometer. An instrument maker could not be found in the United States to construct the magnetic instruments, so Gilliss ordered them from Gambey in Paris, the source of some of the Exploring Expedition's instruments; when Gambey declined Troughton and Simms were contracted and in June 1840 the instruments arrived.[52] Table 1.1 summarizes the instruments on hand at the Depot as of about 1840.

For the magnetic observations a wooden-frame building was erected 50 feet south of the astronomical observatory and 40 feet northwest of the house occupied as a Depot. Here on two sandstone piers instruments were placed, the declinometer in July of 1840 and the magnetometer about a year later. The dip circle was placed on a pier 40 feet to the northeast of the declinometer, completing a modest cluster of scientific instruments within clear sight of the Capitol. Here for two years, until June 1842, Gilliss and his assistants faithfully made their magnetic observations, the American contribution to a worldwide system of magnetic observations.[53] All the while they continued their primary duties of making astronomical observations with the transit telescope, but with a greatly expanded observing program again motivated by a desire that the Navy contribute to science. Although the list was at first limited to stars of the *Nautical Almanac* and those listed in the astronomy volumes of Pearson and Vince, on Gilliss's own initiative the list grew to include the stars in the *Catalogue of the Royal Astronomical Society*. Between September, 1838 and the return of the exploring expedition in June, 1842, Gilliss observed more than 10,000 transits of the Moon, planets, and stars, including a yearly average of about 110 transits of the Moon and 20 occultations.[54]

⋆    ⋆    ⋆

It is not even for this expanded series of accurate observations that Gilliss is chiefly remembered, but for his key role in bringing about a permanent Depot that could also

---

[52] The magnetic and meteorological instruments are described in detail, and the results published, in Gilliss, *Magnetic and Meteorological Observations*, op. cit., pp. x–xxi. The assistants and their dates of duty are listed on p. xxi.

[53] For the context of the magnetic work see Robert P. Multhauf and Gregory Good, *A Brief History of Geomagnetism and a Catalogue of the Collections of the National Museum of American History* (Smithsonian Institution Press: Washington, 1987). A description of various magnetic terms and instruments is also found in this volume.

[54] Gilliss, *Report* (ref. 2), p. 65. The results were published in *Astronomical Observations*, as cited in ref. 39.

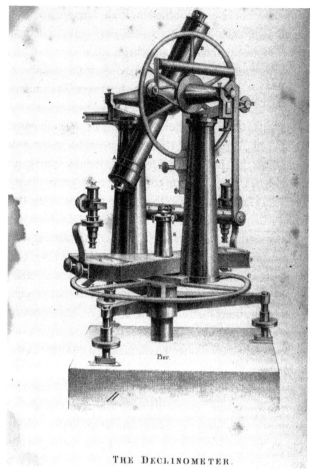

THE DECLINOMETER.

Figure 1.9. A declinometer for measuring magnetic "variation," that is, the angle between the magnetic compass needle and geographic north. It was based on a design by Gambey, was built by Troughton and Simms for the Navy, and arrived at the Depot of Charts and Instruments in June, 1840.

function as a substantial naval observatory. By 1841 "the little box on Capitol Hill," as B. A. Gould later called the makeshift observatory adjacent to the Navy's Depot, was clearly inadequate to its task. As Gilliss recalled, several factors caused him to press for a new Depot and associated observatory. Among these were the unsuitability of the building, defects in the transit instrument, and especially the lack of space for a permanent meridian circle that he was finally authorized to purchase for measuring declinations.[55] He

<hr />

[55] Gilliss, ibid., p. 65. Late in his tenure Gilliss was authorized to purchase a sidereal and mean time clock from Parkinson and Frodsham, and a meridian circle from Ertel & Son. Gilliss, *Astronomical Observations*, p. xiii. An Ertel circle was ordered January 20, 1842, and on April 6, 1842 Ertel & Son company reported they had begun construction, to be completed by the following October. RG 78, letters received. On the Ertel circle, see Gould (ref. 48), p. 26.

Table 1.1. *Navigational and astronomical instruments of the Depot of Charts and Instruments, 1830–42*

**Navigational instruments**

Chronometers: about 35 in 1830; 54 in 1835; 131 in 1846, 80 more on trial

Sextants and Compasses (early 1840s):

>100 sextants (for 40 ships active) from Troughton and Simms

600–800 compasses

"numerous" thermometers, barometers, timepieces, spyglasses (Dolland), and hour glasses

Charts: 15,000–20,000 in early 1840s

**Astronomical and magnetic instruments**

| | | | |
|---|---|---|---|
| Transit instrument | 1831 | Patten | Only one for sale in United States |
| 3.75-Inch transit instrument | 1834 | Troughton | On loan from Coast Survey; purchased by Hassler in 1815 |
| 3.2-Inch refractor | 1838 | Simms | 42-Inch focal length |
| Declinometer | 1840 | Troughton & Simms | Replaced older 1828 variation transit |
| Dip circle (inclinometer) | 1840 | Troughton & Simms | Replaced older 1828 dip circle |

thus urged the Board to recommend an appropriation for a permanent building, which they enthusiastically did in November 1841. The utility of such a Depot, the Board wrote to the Secretary of the Navy, "as a matter of economy only, is fully manifested in the careful preservation, and distribution to sea-going vessels, of the valuable instruments and charts which have been carefully selected and procured for the service, at considerable expense. To this may be added the facilities which such an establishment presents to officers of the navy, for obtaining useful, valuable, and, indeed, necessary knowledge, in some of the higher branches of their profession." Secretary of the Navy Abel Upshur, whom Gilliss characterized as interested in advancing science, expressed his "entire approval" of the proposal. The charts and nautical instruments of the Navy, he wrote, "have been procured at great labor and expense, and are indispensable in the naval service. The small expenditure which will be necessary to preserve them in a condition always ready for use, is not worthy a moment's consideration when compared with the great purposes which they are designed to answer. They are a necessary part of a naval establishment worthy of the present and growing greatness of our country."[56]

The path of the proposal in Congress, where it first had to survive both the House and Senate Naval Committees, may be seen from Figure 1.10. In the House Naval Committee one Congressman, the 35-year-old Francis Mallory of Virginia, favored it

---

[56] Gilliss to BONC, September 24, 1841, E 228 argues briefly for "the propriety of erecting a suitable building for the Depot of Charts and Instruments." BONC to SecNav, November 30, 1841, in *Report of the Secretary of the Navy*, December 4, 1841, p. 373, endorses the proposal. The Secretary of the Navy's recommendation is in ibid., p. 367.

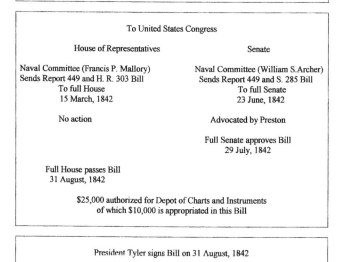

Figure 1.10. Steps in founding a permanent Depot of Charts and Instruments, 1841–42.

openly, but the majority expressed no opinion.[57] On March 15, 1842 Gilliss persuaded the one skeptical member of the Committee to visit the Depot, so close at hand to the Capitol. On the same day, Mallory "reported" to the full House a bill "to authorize the construction of a depot for charts and instruments of the navy of the United States."[58]

The Bill was accompanied by a report, which Gilliss had drafted for Mallory. It is a revealing document that stresses the utility of the Depot to the Navy, the inadequacy of the present Depot and its accompanying observatory, the past accomplishments of the Depot in making magnetic, meteorological, and astronomical observations, the necessity of new quarters in order to protect the valuable instruments in both the Depot and the Observatory, and the unsettling fact that the property now occupied was privately owned. Most of the four page document dealt with the

---

[57] Francis Mallory, born December 12, 1807, was a Congressman from Virginia in the 25th Congress, 26th Congress, and 27th Congress (December 28, 1840 to March 3, 1843). His interest in naval matters probably derived from his having served as a midshipman in the U. S. Navy from 1822 until 1828. He graduated from medical school at the University of Pennsylvania in 1831, and, after practicing in Norfolk pursued agricultural interests. He died March 26, 1860. *Biographical Directory of the American Congress, 1774–1971*, pp. 1330–1331.

[58] House Committee on Naval Affairs, *Depot of Charts, &c*, Report number 449 to accompany bill H. R. 303, 27th Congress, second session, March 15, 1842. On the visit of the skeptical Congressman see Gilliss, *Report* (ref. 2), p. 65.

various duties of the Depot, and a few sentences addressed the need for an observatory: "Observatories, though not expensive, cannot prosper in our country until we can obtain rest from the pursuit of mercantile affairs, or their charge is undertaken by the Government. The duties are confining; if properly executed, arduous; and but few are qualified by experience or habits to undertake them. If officers can be found with taste for such duties, an observatory will give more information to the world, under a military organization, in one year, than under any other direction in two. A small observatory is absolutely essential to the depot: without it, the duties cannot be performed." Aside from this clear statement, and a few paragraphs on the utility of astronomy for navigation, no stress was laid on an astronomical observatory, certainly not one that might be considered a national observatory.[59]

On March 15, the day Mallory reported the bill to the House, it "was read a first and second time, and committed to the Committee of the Whole House on the State of the Union," where no action was taken.[60] Mallory suggested that the bill would have a greater chance of success in the House if the Senate would first pass a bill. Gilliss then tried to drum up support in the Senate Naval Committee by asking members to inspect the poor condition of the Capitol Hill Observatory, but his requests for such inspections "were put off from time to time."[61]

Gilliss later wrote that he believed the Naval Committee of the Senate was finally persuaded by an astronomical event. In April Gilliss had observed comet Encke with the 3.5 foot telescope. He subsequently reported this at a meeting of the National Institute for the Promotion of Science, and, as it happened, in the audience was a certain Senator Preston.[62] When a few days later Gilliss paid a visit to the chairman of the Senate Naval Committee, he found Preston with him. What happened next is one of those fortuitous events that occur in Washington, then as now: "As soon as I began the conversation about the little observatory, Mr. Preston inquired whether I had not given the notice of the comet at the institute, and immediately volunteered, 'I will do all I can to help you'."[63] In fact on June 23, Senator William S. Archer (from Virginia,

---

[59] Report number 449 (ref. 58), 3. The first draft of this report, in Gilliss's hand, is in RG 45, E 228, March 1, 1842.

[60] *Journal of the House of Representatives*, 27th Congress, second session, March 15, 1842, 542. The Journal indicates that John Quincy Adams was on the House floor at the time.

[61] Gilliss, *Report* (ref. 2), p. 65.

[62] William Campbell Preston, born December 27, 1794 served as Senator from South Carolina November 26, 1833 to his resignation on November 29, 1842. He studied law at the University of Edinburgh, practiced in Virginia and South Carolina, and was president of South Carolina College from 1845 to 1851. He died in 1860, the same year as Congressman Mallory.

[63] Gilliss *Report* (ref. 2), p. 66. Comet Encke has the shortest known period of any comet, 3.3 years. It was discovered in 1786 by Pierre Mechain, and is named after J. F. Encke, who first established its periodic nature in 1819. On the 1842 return it reached perihelion on April 13. Since it was of magnitude 5.5 and had no tail, it is not a case of the Senator having been impressed with the sight of the comet, but with the fact that Gilliss had seen it in the nation's capital. I have found no observations published by Gilliss. However, Sears Cook Walker and E. O. Kendall observed it in great detail at the Philadelphia High School between March 27 and April 11, "Observations of Encke's Comet, at the High School Observatory, Philadelphia, March and April, 1842, with the Fraunhofer Equatorial, by Sears C. Walker and E. Otis Kendall. Read May 20, 1842," *Transactions of the American Philosophical Society*, **8** (1843), 311–314. Their results were also reported in three other journals.

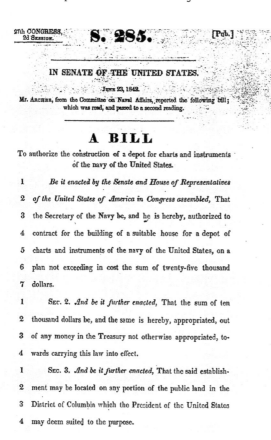

Figure 1.11. The Senate bill for the Depot of Charts and Instruments.

like Mallory) reported the bill identical to that of the House, from the Naval Committee to the Senate (Figure 1.11).[64]

Meanwhile ominous events were coming to a head on another front, and a circumstance that Gilliss had perhaps foreseen came to pass. Upon his return from the Exploring Expedition, on June 15, 1842 Wilkes had informed the Board that, as of the end of the month, the premises occupied as the Depot would no longer be available to the Government for that purpose.[65] On June 22 (one day before the Naval Committee reported the bill to the full Senate) Gilliss was ordered to find another house for the Depot, and by June 30 the contents of the Depot had been removed to a house on Pennsylvania Avenue.[66] On July 2, Gilliss turned the key of the old observatory over to Wilkes. Then, in a curious move, Gilliss resigned his duties, and on July 12 reported to the Board that he had turned over charge of the temporary Depot to one Matthew Fontaine Maury, who had received orders to report on June 29.[67]

[64] *U. S. Senate Journal*, 27th Congress, second session, June 23, 1842, 417.
[65] Wilkes to BONC, June 15, 1842, Wilkes papers, LC, box 1, folder 1.
[66] BONC to Gilliss, June 22, 1842, NA, RG 45, and Gilliss to BONC, June 25, 1842. Gilliss further describes the house on Pennsylvania Avenue as belonging to the late Hon. John Forsyth. Further evidence indicates that it was at 2222–2224 Pennsylvania Avenue, as shown in Figure 1.2.
[67] Gilliss to Wilkes, July 2, 1842; Gilliss to BONC July 12, 1842. On Maury's orders see Frances Leigh Williams, *Matthew Fontaine Maury, Scientist of the Sea* (Rutgers University Press: New Brunswick, 1963), pp. 142–143 and 518.

The abrupt removal of the Depot must have been a disappointment to Gilliss, made more so because of charges from Wilkes that Gilliss had not taken care of the property. This perhaps had something to do with Gilliss's resignation, but the reasons are not clear. One day later the Board simply expressed its regret that Gilliss had found it necessary to retire from his duties, and praised him for duties well performed.[68]

Despite his resignation Gilliss was not about to let plans for the new Depot drop. On July 29, the bill, designated S. 285, was passed in the Senate.[69] The next day the bill went to the House, and Gilliss later described the situation there: "A majority was known to be favorable, but its number on the calendar, and the opposition of one or two members, were likely to prevent action upon it; and that it did received the sanction of the House of Representatives at the last hour of the session of 1841–42, the navy is indebted to the untiring exertions of Dr. Mallory."[70] As Gould reported, the Senate bill "went to the House on the 30th of July; was referred to the same committee as before; but as a Senate bill was treated with courtesy. It was reported back without discussion, passed by the House without debate, and on the 31st of August, 1842, became a law."[71] So, with the signature of President Tyler on that date, the last day of the second session of the 27th Congress, the Secretary of the Navy was authorized to contract for a building and instruments for a depot, not to exceed $25,000 (just over a half-million dollars in 2001 dollars). The bill actually appropriated only $10,000 of that amount, leaving the source of the remainder still uncertain, but an internal Navy matter rather than a Congressional one.

Paradoxically, on the same day, a bill was passed and signed into law to reorganize the Navy Department, along the lines suggested in 1829, and out of which the idea of a Depot of Charts and Instruments had first surfaced. As if to wipe the slate clean, the Board of Navy Commissioners, which had been created in 1815, was abolished, and replaced with a new system of five Bureaus. The new Depot, whatever shape it would take, was placed under the new Bureau of Ordnance and Hydrography.[72]

With knowledge of these details, it is crucial to ask, given a Congress that had ridiculed every attempt by John Quincy Adams to found a national observatory, why it now voted to fund even a "small observatory." Those members of the House and Senate Naval Committees who were familiar with the observatory specified in the report accompanying the bill may have been persuaded by two factors: it was not

---

[68] BONC to Gilliss, July 13, 1842.

[69] U. S. Senate Journal, 27th Congress, second session, July 29, 1842, 514.

[70] Gilliss, Report (ref. 2), p. 65.

[71] Gould (ref. 48), p. 24. House Journal, 27th Congress, second session, July 29, 1842, 1194, 1201, and 1475 indicates that, on July 29, the Secretary of the Senate informed the House that S. 285 had passed the Senate; on August 2 it was read a first and second time in the House; and on August 31 it was read a third time "and passed in the affirmative." The bill was signed by President John Tyler on the same day, ibid., 1478 and 1482. Gould's statement notwithstanding, debate did take place on the floor of the House. See Congressional Globe August 27, 1842, 960 and 978, indicating (unspecified) initial opposition to the bill in the House from Representatives Cave, Johnson, Spring, and others. The final law is found in The Public Statutes at Large of the United States of America (hereafter U. S. Statutes at Large), volume 5 (Boston, 1846), p. 576, chapter 277.

[72] U. S. Statutes at Large, ibid. (ref. 71), "An Act to reorganize the Navy Department of the United States," volume 5 (Boston, 1846), 579–581. This act repealed the Act of February 7, 1815.

Adams who was proposing it, and Gilliss had made a good case that the Navy needed a "small observatory" for practical needs. According to Gould, the motive in allowing an observatory for the Navy under a different name was the same motive as the proviso against an observatory stipulated when the Coast Survey was revived in 1832: to avoid at all costs giving Adams any credit for the idea. "When," he wrote, "at last Congress did appropriate the means for erecting an Astronomical Observatory, and subsequently for its support, it was under a fictitious name; the authors of the laws intending an Astronomical Observatory, and being well aware that the funds would be so applied, but causing the insertion of the proviso in the one case, and of the feigned name in the other, for the purpose of preventing the institution from being attributed to the influence of Mr. Adams."[73]

The story, however, is considerably more complicated than Gould's conjecture – taken from Adams's cousin Josiah Quincy and written as part of a eulogy to Gilliss – of lingering hatred toward Adams's proposal for a national observatory. In fact we must look to broader and more subtle events in order to understand how the authorization for a naval observatory came about in Congress. In particular it was the events surrounding the bequest of James Smithson, and the struggle over the fate of the fledgling National Institute for the Promotion of Science, that secured the crucial support of Senator W. C. Preston for Gilliss's proposal in the Senate.[74] We recall what Gilliss tells us about his serendipitous meeting with Preston: "As soon as I began the conversation about the *little observatory* [my emphasis] . . . Mr. Preston immediately volunteered 'I will do all I can to help you'." Why was Preston so interested? Preston, the Senate colleague of John Calhoun of South Carolina, had argued, with Calhoun in 1839, against the U. S. acceptance of the Smithsonian bequest because it increased national power at the expense of the states. After that failed he was an ardent opponent of Adams's desire to use the Smithsonian bequest for an astronomical observatory. As chairman on the Senate side of the Joint Committee on the Smithsonian bequest (as Adams was on the House side), in 1840 Preston presented resolutions counter to Adams's proposal for an astronomical observatory.[75]

His reasons for doing so became clearer the following year. In 1841 Preston took up the idea of Joel Poinsett – a guiding light of the National Institute, and

[73] Gould (ref. 48), p. 20, who cites Josiah Quincy's *Memoir of the Life of John Quincy Adams*. Because Josiah Quincy was Adams's cousin, family bias needs to be taken into account in assessing the validity of this statement exalting John Quincy Adams. For more on Adams's proposals for a national observatory see Samuel Flagg Bemis, *John Quincy Adams and the Union* (Columbia University Press: New York, 1970), chapter 23, "Lighthouses of the Skies, 1825–1846."

[74] Steven J. Dick, "John Quincy Adams, the Smithsonian Bequest, and the Founding of the U. S. Naval Observatory," *Journal for the History of Astronomy*, **22** (1991), 31–44. On the origins of the Smithsonian see, for example, A. Hunter Dupree, *Science in the Federal Government* (Harvard University Press: Cambridge, Massachusetts, 1957), chapter 4, "The Fulfillment of Smithson's Will, 1829–1861." On the National Institute see Sally Gregory Kohlstedt, "A Step toward Scientific Self-Identity in the United States: The Failure of the National Institute, 1844." *Isis*, **62** (Fall, 1971), 339–362.

[75] Marlana Portolano, "John Quincy Adams's Rhetorical Crusade for Astronomy," *Isis*, **91** (2000), 480–503; Bemis (ref. 73), p. 505.

President van Buren's Secretary of War before retiring to South Carolina after van Buren's defeat – to place the Smithsonian bequest under the control of the National Institute.[76] It was Preston who in 1841 introduced one bill to incorporate and give Congressional blessing to the National Institute for the Promotion of Science, and another bill to put the Smithsonian fund under the control of that Institute.[77] No action was taken in 1841, so April of 1842, the same month that Gilliss was pushing the Depot of Charts and Instruments bill through the Senate, found Preston again introducing in the Senate the bill to dispose of the Smithsonian Fund through the National Institute.[78] In the House Adams simultaneously introduced his bill pushing the use of the Smithsonian bequest for an astronomical observatory.[79] Why did Preston latch onto Gilliss's proposal in June and personally push it through the Senate? Surely it had occurred to him that, by establishing an observatory within the Navy Department, he would undermine Adams's position for an observatory based on the Smithsonian bequest, and, not incidentally, put those monies instead under the control of the National Institute, as his own bill proposed.

Thus the creation of a permanent Depot of Charts and Instruments with its observatory was in great part a byproduct of the tug-of-war over the Smithsonian bequest. Though Preston did not get his way with that bequest for the National Institute, neither did Adams realize his life-long dream, and Adams's failure can be directly linked to the founding of the Depot with its observatory. For the "small observatory" of the House report materialized as much more. Gilliss later wrote "Taking the report of the Naval Committee . . . as the exponent and will of Congress" the Secretary of the Navy ordered him to "visit the principal Northern cities, for the purpose of obtaining information respecting a plan, which, whilst it combined essentials, should not exceed in cost the appropriated sum."[80]

The nature of the building and the excellence of the instruments considered for purchase from the outset betray Gilliss's purpose that this would be no mere Depot. Though he kept the rooms for the charts and navigational instruments large, and those for the telescopes small "lest the limits of the law should be exceeded," Gilliss ended up with a sizeable observatory. This raises an important question: at what point

[76] Bemis (ref. 73), p. 511. Poinsett was thus an influential constituent both of Senator Preston and of Senator Calhoun.

[77] William J. Rhees, The Smithsonian Institution. Documents Relative to its Origin and History (Washington, 1879), and Twenty-Sixth Congress, Proceedings in the Senate, 238–242. The bills were introduced February 17, 1841.

[78] Ibid., 27th Congress, 1841–43, Proceedings in the Senate, 247. The bill (S. 224) was introduced April 11, 1842, and was identical to the bill (S. 259) of 1841.

[79] Ibid., Proceedings of the House, April 12, 1842, 249–260. The bill (H. R. 386) specifies "That the sum of thirty thousand dollars, part of the accruing interest on the same Smithsonian fund, be, and the same is hereby, appropriated towards the erection and establishment, at the city of Washington, of an astronomical observatory, adapted to the most effective and continual observations of the phenomena of the heavens; to be provided with the necessary, best, and most perfect instruments and books, for the periodical publication of the said observations, and for the annual composition and publication of a nautical almanac."

[80] Gilliss, Report (ref. 2), p. 2. We shall discuss Gilliss's consultations and the resulting building and instruments in chapter 2.

did Gilliss decide that a substantial observatory should be built, rather than the "small observatory" provided in the Congressional action? Looking back from 1845 Gilliss himself was unequivocal that he had worked with more than a Depot in mind: "I should have regarded it as time misspent to labor so earnestly only to establish a *depot*. My aim was higher. It was to place an institution under the management of *naval officers*, where, in the practical pursuit of the highest known branch of science, they would compel an acknowledgement of abilities hitherto withheld from the service." This was by no means equivalent to a national observatory, and in fact the evidence shows that Gilliss himself never had a national observatory in mind. In September 1841, just at the time Gilliss was proposing a new Depot, he had encouraged Hassler to continue agitating for a national observatory, because "We propose a Depot of Charts and Instruments for the use of the Navy, and not a national observatory. If a national observatory is to be erected, I think from your abilities and experience, you have a right to originate it, and I would gladly add my mite to carry it through whenever you may prepare it." In 1845 Gilliss still advocated keeping his creation a naval observatory, with rotating naval personnel who could disseminate the knowledge within the navy. "The personnel of a National Observatory would be governed by different motives and objects, and permanence should certainly be a sine qua non; but, regarding this only as a naval observatory, it is of the utmost importance that we give to the service the greatest possible benefit from it."[81]

Gilliss clearly had a naval observatory in mind in 1845. However, upon its completion in 1844 Gilliss was not made head of the institution, and so would have no control over its destiny. Abel Upshur had gone on to become Secretary of State, and both he and his replacement as Secretary of the Navy had been killed in a shipboard cannon accident. March 1844 thus found a new Secretary of the Navy, John Y. Mason, in office. To Gilliss's astonishment, Mason named fellow Virginian Matthew Fontaine Maury to head the new institution. With Maury's known interest in hydrography, this may have been a conscious decision to emphasize another of the Depot's duties rather than astronomy.[82]

If so, the resulting national observatory is even more surprising. That even after its construction had been authorized no one yet considered the naval observatory a national observatory is made evident by Adams's repeated attempt in 1844, two years after the President had signed the legislation for the Depot, to have the Smithsonian

[81] Gilliss, *Report* (ref. 2), p. 66. The letter to Hassler, dated September 29, 1841, is quoted in Florian Cajori, *The Chequered Career of Ferdinand Rudolph Hassler* (reprint edition, Arno Press: New York, 1980), p. 175. In 1843 Gilliss described a staffing plan he had sent to the Secretary of the Navy "In the plan which I submitted to him; I proposed to have a Director who should be a Captain in the Navy, and 4 Lieutenants; (one to each principal inst:) with 8 Passed Midshipmen to attend to the Magnetic and Meteorological Observations and details. There would be no difficulty, however in having any number of Assistants which the Director may desire and my only anxiety is, that an energetic officer may be placed at its head." Gilliss to Loomis, July 4, 1843, in Nathan Reingold, *Science in Nineteenth Century America* (Hill and Wang: New York, 1964), p. 137.

[82] On Maury's appointment see Frances Leigh Williams, *Matthew Fontaine Maury, Scientist of the Sea* (Rutgers University Press: New Brunswick, 1963).

funds used for a national observatory.[83] The final irony is that it seems it was Matthew Maury who co-opted Adams. He failed in a bid to hire William Bond from Harvard, but did hire in rapid succession John C. Coffin, Joseph S. Hubbard, and Sears Cook Walker. With their help and that of naval personnel, Maury was able to publish in 1846 what he called "the first volume of Astronomical Observations that has ever been issued from an institution properly entitled to the name of Observatory on this side of the Atlantic."[84] Maury's co-option of Adams took shape not only in the form of the impressive contents of this volume, but also by its dual title pages, one of which called the institution a "Naval Observatory" and the other a "National Observatory" (Figure 1.12).

Maury could not have achieved this alone, and in fact it is clear that he had the support of the new Secretary of the Navy George Bancroft, the famous historian and founder of the U. S. Naval Academy. Bancroft later recalled "When I became Secretary of the Navy the Observatory was already in existence and under the superintendence of Maury. It was then known officially as the Depot for Charts, but Congress had not expressly sanctioned the Observatory by name. Mr. J. Q. Adams still cherished the hope of being the founder of a National observatory. In conjunction with Lt. Maury and taking counsel also of the best scientific men, I got large appropriations for the Institution, introduced under Mr. Maury, Scientific men, for example Sears Walker, and in a word did all I could to carry forward and perfect what I found begun. I have no right to be called in any sense the Originator of the Observatory. But I contributed my part while in office, to procure for it so complete instruments and observers, as superseded Mr. Adams' scheme, as he himself once said to me."[85] Though Maury is usually seen as having given short shrift to astronomy, he did take the crucial step, with the decisive support of Bancroft, of claiming it not only as a naval observatory, but also as a national observatory.

Thus, whether or not it was Senator Preston's intention to undermine Adams, in the end this is the effect it had. During the final debate in 1846 on the Smithsonian bequest, Adams stated "I am delighted that an astronomical observatory – not perhaps so great as it should have been – has been smuggled into the number of institutions of the country, under the mask of a small depot for charts . . .". "There is not one word about it in the law," he noted, but he concluded regarding the Smithsonian bequest that "I no longer wish any portion of this fund to be applied to an astronomical observatory."[86] Though Adams expressed delight, his disappointment at not having his own name associated with a dream of 20 years must have been very real. Less than a year before he had made the difficult journey westward to dedicate the Cincinnati

---

[83] Rhees (ref. 77), 28th Congress, House of Representatives, June 7, 1844, pp. 293–302. The bill (H..R. 418) is found in full on pp. 299–301.

[84] *Astronomical Observations made under the direction of M. F. Maury . . . During the Year 1845 at the U. S. Naval Observatory, Washington* (Washington, 1846).

[85] M. A. DeWolfe Howe, *The Life and Letters of George Bancroft* (Scribners: Port Washington, New York, 1908), pp. 277–278.     [86] Rhees (ref. 77), 29th Congress, pp. 442–443.

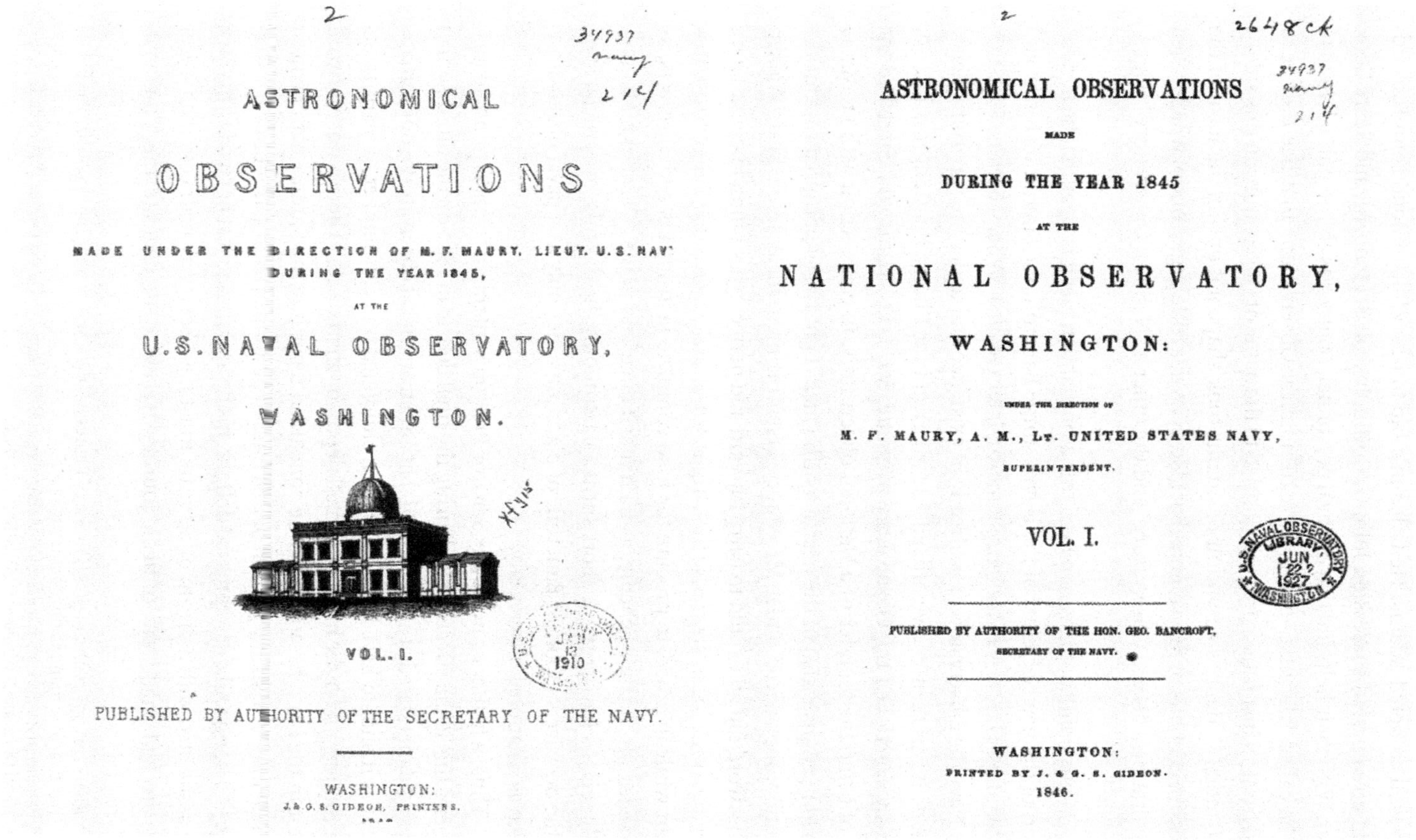

Figure 1.12. Dual title pages for the first volume of *Washington Observations*, 1846.

Observatory; political realities now prevented him from dedicating a national observatory in Washington.

The early 1840s thus form a crucial juncture for three institutions in the history of American science. The rapid development of the Depot of Charts and Instruments into a Naval Observatory denied Adams his national observatory. Smithson's bequest was left free for the development of the Smithsonian Institution guided by Joseph Henry. And with Preston's failure to tie the National Institute to the Smithsonian bequest, it was left to fade away into history, making way for the American Association for the Advancement of Science.[87]

From modest and purely practical origins in navigation 16 years before, the Depot of Charts and Instruments, startlingly, now assumed the role of the first national observatory of the United States. It did so, without concerted plan, via a number of crucial steps. Goldsborough and Wilkes brought centralization, scientific method, and an increase in efficiency to the Navy's growing navigational technology. Gilliss, inspired by a deeply rooted personal desire to increase the scientific reputation of the Navy, not only perfected the work of his predecessors, but also seized on the needs of the U. S. Exploring Expedition as a means of expanding astronomical, magnetic, and meteorological observations beyond the immediate needs of navigation. Aided by partisan politics centering around John Quincy Adams and the Smithson bequest, Gilliss furthermore pushed through Congress an authorization for a permanent Depot, which he converted into a naval observatory. In a surprise move again indicative of a broad interest in science and an increase in the Navy's reputation, Matthew Fontaine Maury seized for this naval institution the title of National Observatory. All of this happened with the direct approval of a series of Secretaries of the Navy, who saw the relevance of science and technology to the naval service, appreciated the increased prestige that a naval observatory lent to the naval establishment, and even perceived that an expanded national role might contribute to a growing American science, to the international credit of the U. S. Navy and the nation.

Having captured the prized title of National Observatory, Matthew Fontaine Maury would lead it through a crucial era in American science in the years prior to the Civil War. How Maury balanced the needs of the Navy with the broader hopes and burdens inherent in the title of National Observatory, we examine in the next chapter.

---

[87] On the relation of the National Institute to the AAAS, and the founding of the latter, see Sally Gregory Kohlstedt, *The Formation of the American Scientific Community: The American Association for the Advancement of Science, 1848–1860* (Univeristy of Illinois Press: Urbana, 1976), especially pp. 87–88, and Sally Gregory Kohlstedt, Michael M. Sokal, and Bruce V. Lewenstein, *The Establishment of Science in America: 150 years of the American Association for the Advancement of Science* (Rutgers University Press: New Brunswick, 1999).

## 2   A choice of roles: The Maury years, 1844–61

> However superbly equipped with instruments an observatory may be, it is
> needless to expect from it magnificent contributions to astronomy unless there
> be not only eyes to observe, but power to treat observations after the instruments
> have performed their office . . . I do most respectfully and earnestly invoke such
> action by the department, and such legislation by Congress, as may be necessary
> to give full employment to the instruments and a becoming expression to the
> powers of this observatory.
>
> Matthew Fontaine Maury 1860[1]

From his appointment on October 1, 1844, until his departure on April 20, 1861 to join
the Confederate cause at the beginning of the Civil War, Matthew Fontaine Maury
(Figure 2.1) served as the first Superintendent of the U. S. Naval Observatory.[2] With
one exception, no other Superintendent would approach Maury's length of service of
16 years, 6 months and 20 days.[3] No other Superintendent would face the enormous
challenges, and wield the considerable power, inevitably brought by being placed in
charge of a new institution. No other Superintendent would achieve the international
renown – and the national notoriety – of Matthew Maury.

Perhaps for all of these reasons, no period in the history of the institution is
better known than the Maury years. Several biographies, from an uncritical treatment
by his daughter in 1888 to an exhaustive scholarly study 75 years later, have examined
Maury the man, emerging largely with a sympathetic view.[4] Historians examining
Maury's key role in early hydrography have often conferred on him the title "The father

---

[1] Maury to Captain D. N. Ingraham, Chief of the Bureau of Ordnance and Hydrography, August 7,
1860, in RSN, December 1, 1860, 215–216.

[2] We should be clear at the outset that it was not formally designated "U. S. Naval Observatory"
when Maury took charge. As we have seen in the first chapter, technically Congress had funded
only a Depot of Charts and Instruments. Functionally, it was also an observatory and
hydrographic office, a fact only officially recognized in 1854 when the institution was designated
"U. S. Naval Observatory and Hydrographical Office."

[3] This period does not include Maury's more than two years (July 11, 1842–September 31, 1844) as
the last "Officer in Charge" of the old Depot of Charts and Instruments. The only
Superintendent who served a similar length of time was J. F. Hellweg (June 9, 1930–February 18,
1946).

[4] Diana Fontaine Maury Corbin, A Life of Matthew Fontaine Maury (S. Low, Marston & Searle:
London, 1888), and Frances Leigh Williams, Matthew Fontaine Maury, Scientist of the Sea (Rutgers
University Press: New Brunswick, 1963). Other biographies include Charles L. Lewis, Matthew
Fontaine Maury: The Pathfinder of the Seas (United States Naval Institute: Annapolis, 1927), and John
W. Wayland, The Pathfinder of the Seas: A Life of Matthew Fontaine Maury (Garrett & Massie:
Richmond, 1930).

Figure 2.1. Lt Matthew Fontaine Maury (1806–73), first Superintendent of the Naval Observatory (1844–61), proclaimed it a National Observatory. From a daguerreotype taken in the early 1850s, courtesy of Jan K. Herman.

of oceanography," a title that practitioners in the field today – especially U. S. Navy oceanographers – widely accept.[5] Historians of science in the American context, on the other hand, have largely agreed in placing him as an outsider, meddling in the activities of civilians more qualified to play a leadership role in the launching of American science just underway in the 1840s.[6] All of these studies, of course, shed light on Maury's work at the U. S. Naval Observatory. However, they have left unanswered a variety of questions raised by a more direct approach to the history of the institution where Maury spent his most productive years.

In this chapter, while making full use of past historical scholarship, our primary focus must be on Maury in the institutional context: how did Maury fulfill the conflicting promise of both a naval observatory, based on practical work, and a national observatory, which implies a broader role in astronomy? What work program did Maury devise, and to what extent did navigation, as opposed to pure astronomy, drive this program? How were Maury's choices constrained by his superiors and Navy policy? What links were forged with other national and international scientific institutions, and for what purposes? In order to answer these questions we shall first examine the resources at Maury's command – the site, its buildings, and the instruments available for the work. We then move on to Maury himself, the staff he assembled, and their early activities. Finally we delve into the problem of astronomy and hydrography as scientific research programs in competition for limited resources at the Naval Observatory during the pre-Civil-War years. We shall find that Maury was not the enemy of astronomy he was often made out to be, and that his hydrographic work was just as important as his astronomical work to the Navy and to fledgling science in the United States.

## 2.1    The setting: Site, building and instruments

The legislation passed by Congress and signed by the President in August of 1842 authorizing a Depot of Charts and Instruments specified "That the said establishment may be located on any portion of the public land in the District of Columbia

---

[5] Williams (ref. 4), pp. vii–viii; Susan Schlee, *The Edge of an Unfamiliar World: A History of Oceanography* (Academic Press: New York, 1973), Margaret Deacon, *Scientists and the Sea 1650–1900* (London, 1970), and John Leighly, introduction to reprint of Maury's *The Physical Geography of the Sea and its Meteorology* (Cambridge University Press: Cambridge, 1963). Also, Marc Pinsel, *150 Years of Service on the Seas: A History of the U. S. Naval Oceanographic Office, 1830–1980* (Department of the Navy: Washington, 1981).

[6] Nathan Reingold, *Science in Nineteenth Century America: A Documentary History* (Hill and Wang: New York, 1964), especially pp. 145–146, depicts him as "an enemy of the circle" of Bache and Joseph Henry and their associates collectively known as the "Lazzaroni," emphasizing that an important segment of the American scientific community of the time viewed Maury as "an outsider, a rival, and a fake." Robert Bruce, *The Launching of Modern American Science, 1846–1876* (Cornell University Press: Ithaca, 1987), pp. 171–186, especially 183 ff, discusses the differences between Bache and Maury. Harold Burstyn, "Maury" entry in DSB, pp. 195–197 concludes that the hostility toward Maury was justified. A. Hunter Dupree, *Science in the Federal Government*, pp. 105–109 agrees with the view of Maury as an outsider.

which the President of the United States may deem suited to the purpose."[7] The task of choosing a site therefore fell to President Tyler, who decided on 19 acres on the north bank of the Potomac River in the southwestern part of the city (see Figure 1.2). On the Federal Plan of the City the site was known simply as "Reservation No. 4," but, because George Washington had recommended in 1796 that a National University be built, in 1842 the site was also known as "University Square," even though the plan was never carried out. Here, on a hill 95 feet above the waters of the Potomac, a mile and a half west of the White House, the new Depot and the observatory building would rise.[8]

Choosing the site was easy compared with deciding exactly what was to be placed on it. The nature of the building to be proposed was of course closely connected to its perceived function, including the instruments it was to house. As a guide, two general considerations were paramount in preparing the first drawings for the structure. The $25,000 authorized for both building and instruments naturally placed a constraint on the size of the building. Gilliss later wrote that "lest the limits of the law should be exceeded," the rooms for the charts and instruments were kept "of suitable dimensions," while those for the astronomical instruments were "of the smallest size possible." It was with these broad constraints in mind that, on September 9, 1842, only nine days after the Bill became law, Gilliss received orders from Secretary of the Navy Upshur to visit the "principal Northern cities" to gather information for carrying the plan into effect.[9]

The names of those with whom Gilliss consulted make an interesting commentary on the state of astronomy in America at the time. In Washington, he consulted with F. R. Hassler, in his last years as Superintendent of the Coast Survey, and a veteran of astronomical observation for survey purposes. In Philadelphia Gilliss visited A. D. Bache, who in this year had resumed a professorship at the University of Pennsylvania after working on the reorganization of public schools in Philadelphia, and would soon succeed Hassler in 1844. In Philadelphia he also found Sears Cook Walker at the Philadelphia High School Observatory, and R. M. Patterson, Director of the U. S. Mint,

[7] "An Act to authorize the construction of a depot for charts and instruments of the Navy of the United States," *The Public Statutes at Large of the United States of America*, volume 5 (Boston, 1846), p. 572, chapter 277.

[8] The site has a colorful history that has been traced back as far as 1755, when General Braddock landed troops and encamped on the hill during his march against Fort Duquesne in the Colonial Wars. Here, in 1813 and 1814 the American Army had also camped, before advancing to Bladensburg for the defense of Washington against General Ross and Admiral Cockburn. This use had earned the site the nickname "Camp Hill." For background on the site see J. E. Nourse, *Memoir of the Founding and Progress of the United States Naval Observatory* (Government Printing Office: Washington, 1873), pp. 26–29, and Jan K. Herman, *A Hilltop in Foggy Bottom: Home of the Old Naval Observatory and the Navy Medical Department* (reprinted from *U. S. Navy Medicine*, 1984).

[9] JMG, in *Report of the Secretary of the Navy, Communicating A Report of the Plan and Construction of the Depot of Charts and Instruments, with a Description of the Instruments*. 28th Congress, second session, February 18, 1845, Senate Document 114, p. 2 (Hereafter cited as *Report*); reprinted in I. B. Cohen, *Aspects of Astronomy in America in the Nineteenth Century* (Arno Press: New York, 1980).

a founder of the Franklin Institute and active member of the American Philosophical Society. In New York he visited at West Point Military Academy W. H. C. Bartlett, Professor of Natural and Experimental Philosophy, who had made a trip to European Observatories in 1840 to consult about plans for the West Point Observatory.[10] In some ways this observatory could be seen as the Army analog to the proposed Naval Observatory, and Gilliss undoubtedly listened with special interest to Bartlett, who may also have helped Gilliss plan for his own trip to Europe. In Boston Gilliss found William C. Bond, Director of the Harvard College Observatory since 1839, and his 17-year-old son George Phillips Bond, a future director of that Observatory. Here also was R. T. Paine, who prepared the astronomical contents for the *American Almanac*, and was the Chief Engineer for the trigonometric survey of Massachusetts, a task that included much astronomical work. The net result of these consultations was to recommend a single cruciform building, as opposed to the original plan of a central building for the offices and residence, and two detached buildings for the instruments. On Gilliss's return G. F. De la Roche drafted the plans under his direction; in November Mr William Bird received the building contract to carry out the plans.[11]

These consultations in the United States undoubtedly set Gilliss upon a course that he himself had already begun with his order of a meridian circle from Munich for the old Depot in the spring of 1842: as in the case of the Philadelphia High School and, to a more limited extent, West Point, the instruments were to be constructed largely in Germany, rather than in the U. K. These American consultations, especially with Sears Cook Walker at Philadelphia, most probably predisposed Gilliss toward the German instrumentation. However, when it came to the detailed plans and purchase of astronomical instruments in the 1840s, there was no substitute for a trip to Europe such as had previously been made by Elias Loomis (1837), W. H. C. Bartlett (1840), and O. M. Mitchel (1842) in purchasing instruments for the Hudson, West Point, and Cincinnati Observatories. Gilliss had to be careful not to exceed the $25,000 authorized, and recalled in 1843 that he "pointed out to the Secretary of the Navy how the instruments might be obtained without a special appropriation, the discussion of which might not aid the whole plan: and he directed me to obtain them."[12] The list of instruments sought included an achromatic refractor, meridian transit, prime vertical transit, mural circle, comet seeker, and magnetic and meteorological instruments, along with books.[13] It was now a matter of deciding from which manufacturers the instruments should be purchased.

[10] E. S. Holden, "Biographical Memoir of William H. C. Bartlett, 1804–1893," BMNAS, 7 (June 1911), 173–193: 176. From July 1 to November 20, 1840 Bartlett visited the observatories at Greenwich, Oxford, Cambridge, Dublin, Armagh, Edinburgh, Paris, Munich, and Brussels; and the workshops of Troughton and Simms, Dolland, Jones, Grubb, Gambey, Ertel, Merz, and others. He submitted his "Report on the Observatories, etc. of Europe" on February 16, 1841; the 52-page report exists only in manuscript form in the U. S. Military Academy library. Many of the instruments Bartlett obtained were similar to those that Gilliss would purchase.

[11] Gilliss, *Report* (ref. 9), p. 2.

[12] Gilliss to Loomis, July 4, 1843, Loomis Papers, Yale University, cited in Reingold (ref. 6), p. 137.

[13] Gilliss, *Report* (ref. 9), p. 9.

For the achromatic refractor Gilliss at least had some guidance from the direct experience of his American colleagues. At Philadelphia he would have seen the 6-inch Merz and Mahler refractor, and at West Point Bartlett undoubtedly told him about the similarly sized instrument he had ordered from Lerebours of Paris, mounted by Thomas Grubb of Dublin.[14] These were the largest refractors in the United States at the time of Gilliss's trip (though Cincinnati would mount a 12-inch refractor in 1844 and Harvard a 15-inch refractor in 1847). The meridian instruments were another matter, for no one in the United States as yet had them of any size. But Gilliss knew where to go: the London firm of Troughton and Simms would be first on his list, followed by Pistor and Martins, Merz and Mahler, and Ertel and Son in Germany.[15]

On November 19, 1842 Gilliss received instructions from Secretary of the Navy Upshur to go to Europe. He traveled for three purposes: to make refinements to the building plan, to consult regarding the instruments and to make their purchase, and to purchase books. On December 2, 1842, with the preliminary plans in hand, Gilliss embarked from Boston to Liverpool and thence to continental Europe to obtain the suggestions of astronomers and place the orders.[16] On December 16, after a stormy passage of 14 days, Gilliss arrived in Liverpool and made his way to London by railroad. He first called on Troughton and Simms, who informed him of the way to the Royal Observatory at Greenwich, four miles away by rail, where Gilliss found Airy absent, but Sheepshanks showed him through the Observatory. On Monday he returned to Greenwich, where Airy "expressed great satisfaction at the location and general arrangement of the instruments." Airy had specific recommendations about the mural circle, not all of which Gilliss accepted. Two days later, on the basis of his discussions with Airy, Gilliss had Simms draw up plans for a mural circle. After showing the plans to Airy, on Christmas Eve Gilliss ordered the mural circle from Troughton and Simms, the first order of a major instrument for the new observatory.

It was in Germany that Gilliss would place most of his orders for major instruments, in particular in the cities of Berlin and Munich. At Altona near Hamburg he met Schumacher, Director of the Observatory and editor of the *Astronomische Nachrichten*. Having been impressed with the Pulkova Observatory's prime-vertical instrument and its observations recently published by Struve, Schumacher urged Gilliss to purchase a similar instrument, calling it "infinitely superior to the zenith

---

[14] On the Philadelphia and West Point instruments see William H. C. Bartlett, "Account of the Observatory and Instruments of the United States Military Academy at West Point, with Observations of the Comet of 1843," *Trans. APS*, new series (1846), 191–203.

[15] For background on these instrument makers see Henry C. King, *The History of the Telescope* (Dover: New York, 1979). On Troughton and Simms see Anita McConnell, *Instrument Makers to the World: A History of Cooke, Troughton and Simms* (William Sessions: York, c. 1992).

[16] Gilliss reported the details of his trip to Europe, including his diary entries, in a letter to the Secretary of the Navy dated March 23, 1843, NA, RG 45, Letters Received by Secretary of the Navy from Officers below the rank of Commander, Microcopy 148, roll, 151, 177 ff. Included in this report are alterations proposed to the plan for the Depot, reports on the instruments and books purchased, diary and memoranda, and letters to instrument makers in Europe.

tube."[17] Rümchen, the director of the Hamburg Observatory, also advised Gilliss to purchase a prime-vertical transit and meridian circle instead of the mural circle, having had previous bad experience with a mural circle. Apparently convinced, Gilliss visited the famous Hamburg firm of Repsold, where he was informed that, due to heavy orders (they were then working on the Oxford heliometer), they could not make such an instrument within four years.

It was only when he reached Berlin that Gilliss could purchase his first German instrument. Here the famous Encke supported the recommendation for a prime vertical, and furthermore advised a Fraunhofer mounting (on a stone pier rather than one of wood) for the great telescope, which was to have the same dimensions as his. Fully convinced, two days later Gilliss placed the order for the prime vertical with Pistor and Martins. The instrument was to be precisely like Struve's, and Encke promised to keep an eye on its progress.[18]

After stops in Leipzig and elsewhere, Gilliss arrived in Munich on January 16. Here he would place orders for three more instruments: the transit instrument, the refractor, and the comet seeker. His first stop was at the firm of Merz and Mahler, successors to the legendary Fraunhofer. The extent to which they were still keepers of Fraunhofer's optical secrets is apparent in Gilliss's description of their glass casting process: "The furnaces are in the Tyrol, and surrounded with high walls; and so jealously is the art guarded, that Mr. Merz does not permit even his workmen to be present at the casting, but turns them all outside the walls. His son is probably the only person living capable to manufacture the glass, Mahler being the mechanician."[19] Here Gilliss found completed and fitted in their cells one lens of 14, two of 12, two of 10.5, and one of 9 French inches. For budgetary reasons, he chose the latter. Then he went to Ertel and Son, from whom in 1842 he had already ordered a 4.5-inch meridian instrument, later known as the "German circle." Here he requested they prepare drawings for a transit instrument of 7-foot focal length, and on January 20 he ordered the 5.3-inch optics for the instrument from Merz and Mahler, and the instrument itself from Ertel. The following day he gave Merz and Mahler the order for the refractor and comet seeker. Fraunhofer in 1824 had first designed the Dorpat 9.6-inch instrument, and Gilliss ordered a duplicate of that instrument.[20]

Before Gilliss's return to Washington in March he had visited the observatories at Greenwich, Dublin, Oxford, Cambridge, Hamburg, Berlin, Leipzig, Munich, and

---

[17] Ibid., entry for January 3, 1843.

[18] Ibid., January 7 entry. Later in Gilliss's visit to Greenwich, Airy objected to the prime-vertical instrument, especially one like Struve's supported on only one axis; February 28 entry.

[19] Gilliss, *Report* (ref. 9), p. 9.

[20] Ibid., p. 9. A "French inch" was not exactly equivalent to an English inch, so that the 9-French-inch lens cited here is more commonly known as the 9.6-inch refractor once it had been installed in the United States. An indication of where the optical frontiers were in 1843 is Gilliss's remark that "Mr. Merz offered to construct one of *eighteen inches*, if I would allow five years for the completion of the telescope." According to the Observatory's *Astronomical Observations* (ref. 36 below), volume 2, p. lxiv, the 9.6-inch instrument is the "counterpart" of the Dorpat and Berlin instruments.

Table 2.1. *Astronomical instruments of the Naval Observatory purchased 1842–45*[a]

| Instrument (aperture) | Date of purchase | Maker | Maker's location | Cost | Focal length |
|---|---|---|---|---|---|
| 9.6-Inch achromatic refractor | 1843 | Fraunhofer/ Merz & Mahler | Munich | $6,000 ($3,000 lens) | 14 ft |
| 5.3-Inch transit | 1843 | Ertel & Son Merz & Mahler | Munich | $1,800 ($320 lens) | 7 ft |
| 4-Inch mural circle | 1843 | Troughton & Simms | London | $3,550 | 5 ft |
| 5-Inch prime vertical | 1843 | Pistor & Martins | Berlin | $1,750 | 6.5 ft |
| 3.9-Inch comet seeker | 1843 | Merz & Mahler | Munich | $320 | 2 ft 10 in |
| 4.5-Inch German circle (30-inch circles) | 1842 | Ertel & Son | Munich | – | 5 ft |
| 6.6-Inch refraction circle | 1845 | Ertel & Son | Munich | 10,000 Bavarian florins | 8 ft |

Notes:
[a] Data from Maury, *Astronomical Observations* (1846), volume 1, Appendix, p. 2. Many of the instruments are shown in detailed plates in this volume.

Altona, along with some private observatories near London. He had made the acquaintance of many of the most prominent astronomers in Europe during his ten weeks abroad, establishing a strong "European connection" with the first national observatory of the United States. The hopes of these astronomers must have left Europe along with Gilliss. On March 4, 1843, having laid the foundations for the observatory's instruments for the next half century, Gilliss left Liverpool for Boston. His accomplishments during that time are shown in Table 2.1, which summarizes the instruments purchased during the trip to Europe, for a total cost of $13,100, more than half of the entire $25,000 authorization – which was to cover the building also.

Meanwhile, back in Washington preparations for the building were proceeding, under the direction of contractor William Bird. Even before Gilliss left for Europe, the excavations for foundations and piers at the site had begun before the ground became too frozen to proceed. The ground was excavated to a depth of 8 feet for the foundations of the walls and piers, except for the great equatorial pier, which had a base 9 feet below the surface. Luckily, the building itself had not yet been begun by the time Gilliss returned from Europe, for the position for the building was then shifted 25 feet to the west, upon the recommendation of the Astronomer Royal at Greenwich.[21] By July 1843 the granite piers for the instruments had all been erected,

---

[21] Gilliss, *Report* (ref. 9), pp. 2–3. Gilliss had intended to align the meridian transit with 24th Street, but the Astronomer Royal thought that there was no advantage to this, and so it was moved 25 feet for architectural effect to the intersection of 24th and D Streets. See Figure 5.5.

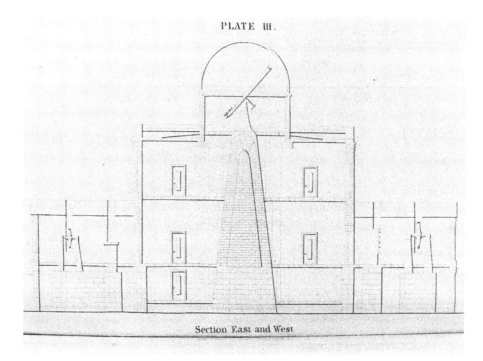

PLATE III.

Section East and West

Figure 2.2. An east–west cross section of the new Observatory, showing the location of the 9.6-inch refractor and transit instruments. From Gilliss's 1845 report on the construction of the "Depot" (ref. 9).

and by October the brickwork around the walls had been completed. By late November Gilliss reported to the Secretary of the Navy "The roofs have been sheathed, and the coppering will be completed in a few days. A part of the floors have been laid, the furnace for heating the building put up, the cast iron work for the revolving dome fitted, and plastering commenced. I hope, therefore, the work will be finished during the month of May next."[22] Several plates in Gilliss's report show cutaway drawings of the edifice (Figure 2.2) and the floor plan (Figure 2.3), and a year later the title page to the first volume of observations included an artist's sketch of the completed building (Figure 2.4). The magnetic observatory, placed underground as at Munich, was finished by September, 1843.[23] In his report, Gilliss also submitted an estimate for grading and enclosing University Square, and for construction of a residence for the superintendent, a post that he clearly believed he would most likely hold. His estimate of $20,000 was almost equal to the original appropriation for the construction of the Depot! Clearly, more funds would be required to supplement the original $25,000 authorization.

In 1844 the instruments began to arrive. The Pistor and Martins prime-vertical

---

[22] JMG to SecNav David Henshaw, November 23, 1843, in RSN, 25 November, 1843, 577.
[23] Gilliss, Report (ref. 9), p. 8, and JMG to SecNav, November 23, 1843, 577 (ref. 22).

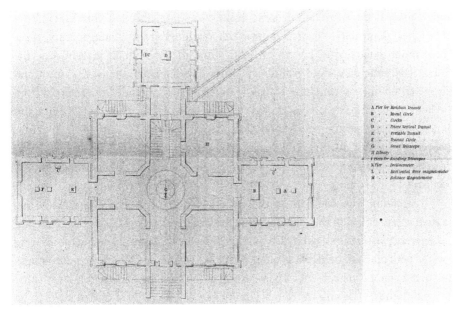

Figure 2.3. The layout of the observatory, with the 9.6-inch refractor located in the dome at the center, flanked by wings to the east, west, and south housing the transit instruments. The tunnel to the magnetic observatory angles off toward the southwest. From Gilliss's 1845 report.

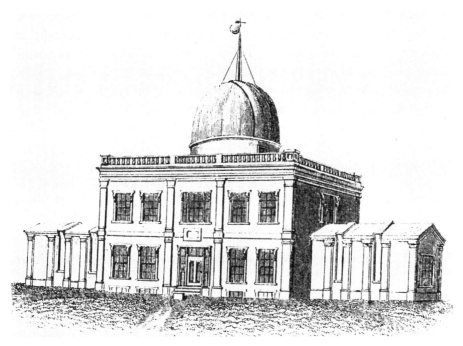

Figure 2.4. The Naval Observatory with a time ball, from the first volume of *Washington Observations*, 1846.

transit, made under the watchful eye of Encke, was shipped from Berlin via Hamburg in October, 1843; due to ice it did not reach Washington until late February, 1844. Three months later, on May 28, 1844, the first of the Munich instruments, the Equatorial Refractor, arrived via Bremen. Three days later, on June 1, 1844, the Mural Circle arrived in New York, and Gilliss "personally superintended the removal from the hold of the ship; its transportation to and re-stowage on board the Washington packet, and exercised the same supervision on its arrival here."[24] With the help of four laborers and J. H. C. Coffin, Gilliss mounted these instruments, and was adjusting them and the magnetic instruments even after he announced the building ready for occupancy.

Attention was also paid to the esthetic details of the building, which was painted cream color on the outside, while the walls and ceilings of all the observing rooms were plastered and painted sky blue. The floors of all the observing rooms were covered with oil cloth, which was never swept but wet mopped to prevent the circulation of dust on the instruments. The west wall of the building had a window with blue Venetian blinds.[25] During the antebellum era several significant changes would be made to the building: in 1846 the south wing was extended 20 feet for a second "circle in the prime vertical", in 1847 the Superintendent's dwelling was erected east of the Observatory (Figure 2.5), and in 1848 the East wing was extended to connect to the Superintendent's house.[26]

Another result of Gilliss's trip to Europe was that it resulted in the beginnings of a library for the Observatory. Although a small number of books, among them Bowditch, Vince's *Astronomy* and Pearson's *Astronomy*, were transferred from the old Depot, the books presented to Gilliss and purchased by him formed the real core. In addition the Naval Observatory was placed on the exchange list of publications for many observatories, making the beginnings of one of the best astronomical libraries in the world.[27] In short, in instruments, building, and books, Gilliss provided a magnificent launching of the U. S. Naval Observatory.

## 2.2 Settling in: Maury and his staff in the formative years, 1844–49

The early years of any institution are crucial to its subsequent history, for during this time plans are laid, goals formulated, and an institutional identity established that shape the future well beyond the initial investment. For this reason it is essential, before assessing the achievements of the Maury years, to scrutinize the early years of Maury's tenure regarding decisions on staff, the relative roles of hydrography

---

[24] On the prime vertical see JMG to SecNav, November 23, 1843, 577 (ref. 22); on the refractor, Gilliss, *Report* (ref. 9), p. 12; on the mural circle, Gilliss, *Report* (ref. 9), p. 29.

[25] *Astronomical Observations* (ref. 36), volume 1, p. 1, also known as "*Washington Observations*" and cited hereafter as WO.

[26] WO (1862) reviews the buildings and grounds up to 1862. The second prime-vertical circle was damaged, returned to the makers in 1847, and remounted only in June, 1861. However, its design was radically defective, and it was discarded.

[27] The books obtained as a result of Gilliss's trip are listed in Gilliss, *Report* (ref. 9), pp. 56–63.

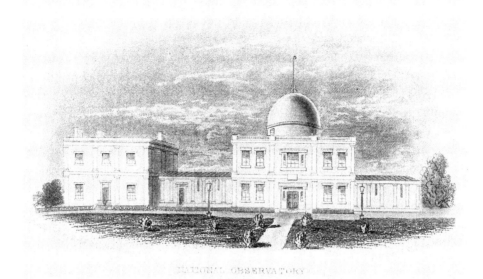

Figure 2.5. The observatory with the Superintendent's residence, which was added in 1847–1848, on the left. Note small time ball on dome.

and astronomy, the published results of scientific observations, and their reception at home and abroad.

The first decision of utmost importance had been made by Secretary of the Navy John Y. Mason, for it was he who would choose the head of the new institution. One can forgive Gilliss, if, having led the effort which resulted in the approval of the new Depot and observatory by Congress, having traveled up and down the Eastern United States and much of Europe drawing up plans for the building and ordering instruments, and having mounted and tested the instruments himself, he was bitterly disappointed when not he, but Matthew Maury, was appointed Superintendent of the new Observatory. Writing to Elias Loomis about the Observatory in the fall of 1844 Gilliss confirmed "My instructions from the Department terminated with its completion, and although there is no duty so agreeable to me as astronomical labours, yet, I have too much pride to solicit a connection with an establishment whose existence is owing solely to myself. You know, the law recognizes it only as a Depot for the Charts etc of the navy; and Lieut. Maury has moved into it, with his Instruments, charts etc. What the intention is, I know not, for I have also too much pride to enquire, although, I believe my head would be brought with sorrow to the grave, if I thought I had laboured to found a mere Depot."[28] Gilliss's connection with the Observatory was

---

[28] JMG to Loomis, October 18, 1844, Loomis Papers, Yale University, cited in Reingold (ref. 6), p. 138. Gilliss expressed hope that he would have his report on the construction of the new building and instruments by December 1, and said he would resume work on the observations made in connection with the Exploring Expedition, which he hoped to have finished by the summer. The analysis of the Exploring Expedition observations was not published.

severed and his future in astronomy uncertain, though, as we shall see in chapter 4, he went on to make important contributions to the field and at the end of his life would return to the Observatory once again.

There were, of course, logical arguments in Maury's favor. Maury was born in 1806 near Fredericksburg, Virginia, became a midshipman in 1825 and had three periods of sea duty between then and 1834. He was the author of two articles on navigation in 1834, and of a textbook on navigation that would become widely used.[29] Maury was promoted to Lieutenant in 1836, and in September 1837 was appointed astronomer of the U. S. Exploring Expedition. He resigned this position, along with other officers, a few months later after disagreements with Secretary of the Navy Mahlon Dickerson, but not before he had gone to Philadelphia for several months of instruction in astronomy at a small observatory in Rittenhouse Square.[30] A leg injury in 1839 confined him to shore duty, a circumstance that allowed him to write on naval reform, and to push for the replacement of the Board of Navy Commissioners with the Bureau system. As we have seen, Congress approved that system on the same day that Gilliss had procured an appropriation for the new Depot.

Of more immediate significance, since July of 1842, Maury had taken Gilliss's place as Officer-in-Charge of the temporary Depot of Charts and Instruments, which the new building was designed permanently to replace. Maury's activities during his two years as Officer-in-Charge included all those duties of charts, instruments, and chronometer rating carried out by his predecessors, though now under difficult circumstances in temporary quarters on Pennsylvania Avenue. Early in his tenure at the temporary Depot Maury realized the need for hydrographic work, and it was in this field that he had begun to make his mark. Here he used old ships' logs to extract information, and had his first thoughts about the wind and current charts that would soon make him famous.[31] By late 1842 he had won approval from his superiors that all crucial information from ships be sent to the Depot. In the fall of 1843 a new edition of his *Navigation* was published, and the following April Maury, as one of the directors of the National Institute for the Promotion of Science, gave a lecture on the Gulf Stream. Thus, as Gilliss worked to ready the new Depot and Observatory, Maury was becoming widely known for his hydrographic work.

On August 4, 1842, even before the final passage of the bill for the Depot by the House, Maury was naturally thinking about its effect on his future. He wrote to his cousin Ann Maury that "A Bill has passed the Senate for building a depot, as 'tis called

[29] Matthew F. Maury, "On the Navigation of Cape Horn," *American Journal of Science and Arts*, **26** (July 1834), 54–63; "Plan of an Instrument for Finding the True Lunar Distance," ibid., 63–65; *A New Theoretical and Practical Treatise on Navigation* (Philadelphia, 1836).

[30] Williams (ref. 4), p. 115. The "planked observatory" had "telescopes, transit instruments, chronometers, theodolites . . . clocks, sextants etc. and all sorts of magnetic apparatus." The instructor was Walter R. Johnson, who may have been a member of the Corps of Professors of Mathematics of the Navy.

[31] Williams (ref. 4), pp. 148–149. On Maury's two-year tenure see ibid., pp. 144–157.

... What Part I am to have in it, or what bearing it is to have upon me in my present situation, I am not able even to guess."[32] Maury continued to speculate on his future in letters to his friends and relatives, writing of the Superintendency of the new institution in late 1843 "Who is to have it I do not know – I suppose the competitors will be multitudinous and I shall not swell the list. I am trying to leave my mark before they wipe me out." As late as September 12, 1844: "I have been officially informed that the new observatory will be ready on 1st October ... I have not been told whether I would be continued in charge ... I shall await events."[33] However, Maury was not entirely idle in the matter. As he later recalled to a friend "You know I did not want the place, and only decided to keep it when I heard it had been promised to a civilian, under the plea that no one in the Navy was fit for it. I then went to [Secretary of the Navy] Mason, pronounced *that* the repetition of a practical libel, and told him he must stand by me. He did so, and though I had never seen an instrument of the kind before, and had no one with me who had, I was determined to ask no advice or instruction from the *savans*, but to let it be out and out a navy work."[34] The sentiment that the Navy should control the observatory is remarkably reminiscent of Gilliss's words two years earlier, but Gilliss hardly had Maury's aversion to learned men.

On October 1, Secretary of the Navy Mason, a fellow Virginian, appointed Maury the new Superintendent. With a minimum of astronomical credentials, and thanks to political connections and no small measure of fate, the present belonged – for better or worse – to Matthew Maury. On October 8, having been absent from Washington delivering instruments to ships, Maury first strode into his new Depot and observatory. The ironies of his position were abundant and the ambiguities of the institution must have been clear to knowledgeable outsiders, as well as to the staff. As Adams's call as late as 1845 for a national observatory indicates, it was still completely unclear what the future of the naval institution would be.[35] Seldom had a Director entered an institution with more power to shape its future.

We may infer Maury's immediate goals from his duties at the old Depot, for whatever the future of the institution, the necessity of carrying out the old duties continued unabated at the new site. Maury described these duties as consisting "in a regular series of meteorological observations, in a partial series of magnetic observations, in collecting Hydrographical information, and in constructing charts, in trying and rating Chronometers, and in the purchase and supply of all nautical books, maps, charts and instruments required for the use of the Navy." In fact all of these duties fully occupied the staff of three lieutenants, six passed midshipmen and a machinist

---

[32] Maury to Ann Maury, August 4, 1842, in Corbin (ref. 4), p. 45.

[33] Maury to William M. Blackford, November 19, 1843, Maury to Ann Herndon, September 12, 1844, both cited in Williams (ref. 4), p. 523, note 83.

[34] Maury to William Blackford, January 1, 1847, in Corbin (ref. 4), pp. 49–50.

[35] See chapter 1.

brought from the old Depot.[36] At the old Depot one of the lieutenants was responsible for the chronometers, another for the charts and other instruments, and a third for collecting and collating hydrographic information. In order to free one officer for astronomical duty at the new Depot, the care of chronometers, charts, and other instruments was given to one lieutenant. Given that the Navy now had more than 100 chronometers and sextants, some six or eight hundred compasses, numerous thermometers, barometers, timepieces, spyglasses, hour glasses, and mathematical instruments, and fifteen to twenty thousand charts, one could see that this lieutenant was kept well-occupied. A second lieutenant continued the hydrographic work, which consisted mainly in the production of the "Wind and Current Charts," of which the first "Chart of the Atlantic" was about to be issued as the second volume of *Washington Observations* was published in 1846. The passed midshipmen were kept busy with meteorological observations, correspondence, accounts, and reducing observations. However, their ongoing task of making magnetic observations was soon over. First attempts to get the magnetic observatory at the new location in working order ended in failure, as preparations dragged on and then rains filled the underground chamber. The magnetic observatory was eventually abandoned.[37]

However, in addition to all of its past duties, the House and Senate Committees on Naval Affairs, the Navy, and the nascent scientific community obviously expected more from the institution, now stocked with the finest-quality astronomical instruments and prominently situated on University Square. The first problem was staff. The Navy could supply Maury with lieutenants and midshipmen, but the real problem, if good use were to be made of the astronomical instruments, was expertise in astronomy. This problem was solved largely through the Navy's established tradition of Professors of Mathematics, a position that had grown out of the post of Schoolmaster created in the Navy in 1813. The Professors were expected to instruct midshipmen, aboard ship, on all techniques needed for navigation. In 1834 the first three Professors of Mathematics were named; their usefulness is attested by their expanded number (22) by 1845.[38] The founding of the U. S. Naval Academy in that year by Secretary of

---

[36] *Astronomical Observations made under the Direction of M. F. Maury, Lieut. U. S. Navy, During the Year 1845, at the U. S. Naval Observatory, Washington* (Washington, 1846), Appendix, p. 1. This rambling 119-page appendix to the first volume, also known as "*Washington Observations*" is the best printed source for the first two years of Maury's administration at the new Observatory. It is in the form of a report to Secretary of the Navy Bancroft, and is dated July 1, 1846 at its beginning, and September 1, 1846 at the end. The three lieutenants mentioned in the text were Benjamin F. Sands (later a Superintendent), Gustavus H. Scott, and William L. Herndon. The latter was Maury's brother-in-law, later to achieve fame for his exploration of the Amazon, and for his bravery when, under his command, the U. S. Mail steam packet *Central America* went down during a hurricane in 1857 with some $450 million in gold from the California Gold Rush. (The wreckage was located in 1986 in 8,000 feet of water 200 miles off the coast of South Carolina, and artifacts were auctioned at Sotheby's in 2000). The disaster, the nineteenth-century equivalent of the Titanic, and the recovery of the ship are vividly described in Gary Kinder's *Ship of Gold in the Deep Blue Sea* (Vintage: New York, 1998).    [37] *Ibid.*, p. 2.

[38] On the Professors of Mathematics see Charles J. Peterson, "The United States Corps of Professors of Mathematics," *Griffith Observer* (February, 1990), 2–14. In 1848, following a wrangle at the Naval Academy involving civilian stature in a military chain of command, the Professors

Table 2.2. *Permanent staff of the Naval Observatory during the Maury years*

| Name | Dates at USNO | Education | Age at appointment |
|---|---|---|---|
| John H. C. Coffin | 1845 (January)–1853 | Bowdoin (1834) | 29 |
| Joseph S. Hubbard | 1845 (May 8)–1863 | Yale (1843) | 21 |
| Reuel Keith | 1845 (August 1)–1854 | Middlebury (1845) | 19 |
| Sears Cook Walker | 1846 (February)–1847 (March) | Harvard (1825) | 40 |
| James Major | 1847–1859 (September 3) | Belfast, Ireland | 34 |
| James Ferguson | 1848–1867 (September 26) | Unknown | 50 |
| Mordecai Yarnall | 1852 (October)–1878 | Bacon (c. 1837) | 36 |
| Thomas J. Robinson | 1860 (May)–1861 (July) | Unknown | Unknown |

the Navy George Bancroft abolished the need for Professors on board ship, and gave the opportunity to assign them for duty elsewhere. Even though the use of midshipmen for short periods was not calculated to breed continuity in the work or methods of the office, Maury claimed the Depot itself "was an excellent school, affording them, in the course of a year, much highly valuable information in an important branch of their profession . . . It is the means of disseminating much valuable information among Navy officers, and, more than that, of exciting in them a thirst for knowledge in the higher branches of those departments of science, which apply, in an eminent degree, to the sailor at his calling."[39]

During the Maury years, eight men with advanced training or experience in navigation or astronomy joined the staff (Table 2.2), six of them in the Navy's Corps of Professors of Mathematics. By the end of 1844 Maury had the astronomical instruments in working order, and was "waiting the pleasure of the Department to place at my disposal an additional force which I proposed to employ as Astronomical observers."[40] Secretary Bancroft soon obliged. The first to arrive, in January 1845, was Professor of Mathematics John H. C. Coffin (1815–90), who had helped Gilliss set up and adjust the Observatory's instruments.[41] Coffin was a graduate of Bowdoin College in 1834, and had gone with his uncle in that year on a lengthy sea voyage, during which he first learned navigation and seamanship. He became a Professor of Mathematics in 1836, whereupon he taught midshipmen at sea and ashore at Norfolk, served at times as the navigation officer on ship, and participated in surveys of the Florida coast. He

were made commissioned officers appointed by the President (*Statutes at Large*, volume 9, p. 272). They were, however, staff officers rather than line officers, and thus not able to exercise military command. (For more on this distinction see chapter 5, ref. 11). See also Marc Rothenberg, "The Educational and Intellectual Background of American Astronomers, 1825–1875," Ph.D. dissertation (Bryn Mawr, 1974), p. 205.

[39] Maury, *WO*, volume 1, Appendix, p. 5.    [40] Ibid., Appendix, p. 3.

[41] A. N. Skinner, *Science*, **9** (January 6, 1899), 5 says in an article on the Naval Observatory that Coffin assisted Gilliss in 1843 in "fitting up the new Observatory." He was not permanently attached to the new observatory until January, 1845. The best source on Coffin's life is George C. Comstock, "Biographical Memoir of John Huntington Crane Coffin, 1815–1890," BMNAS, **8** (1919), 3–7; first published separately by the Academy in 1913. See also Clark Elliott (ed.), BDAS, p. 58. The January date is given in Elias Loomis, *The Recent Progress of Astronomy* (third edition, Harper & Brothers: New York, 1856), p. 235.

Figure 2.6. Joseph Stillman Hubbard, observer, theoretician, and member of the National Academy of Sciences.

would stay at the Observatory for nine years, and would go on to a distinguished career at the Naval Academy and become Director of the Nautical Almanac Office (Figure 8.4).

In May 1845, Joseph Stillman Hubbard (1823–63), arrived at the Observatory. Hubbard was a recent graduate of Yale (1843), who (Figure 2.6) had gained valuable experience as an assistant of Sears Cook Walker at the Philadelphia High School Observatory in 1844, before coming to Washington to reduce the latitude and longitude observations of Captain Frémont's western expeditions. Here, upon completion of his duties, A. D. Bache at the Coast Survey and Frémont himself interceded with Secretary Bancroft on Hubbard's behalf to obtain for him a position as Professor of Mathematics.[42] Hubbard would stay at the Observatory until his early death in 1863. The last Professor of Mathematics to arrive at the Observatory in 1845 was Reuel Keith (1826–1908), who came in late summer straight from Middlebury College in Vermont. Almost simultaneously with his graduation, Keith was appointed a Professor of Mathematics in July, and would spend most of his decade-long naval career at the Naval Observatory.[43]

[42] B. A. Gould, "Eulogy on Joseph S. Hubbard," *Annual of the National Academy of Sciences for 1863–64* (Cambridge University Press: Cambridge, 1865), pp. 71–112; reprinted BMNAS, **1** (1865), 1–34. See also DAB, **9**, 329–330.

[43] Keith was appointed July 8, but remained in classes at Middlebury to finish his degree, and arrived at the Observatory on August 1. Little is known of Keith; he has no entries in BDAS, DAB, or NCAB. His publications include only two brief letters in AJ. The last of these, dated September

Figure 2.7. Sears Cook Walker, the first civilian employee at the Naval Observatory, arrived in 1846 from the Philadelphia High School Observatory.

The Professors of Mathematics had good educations but little experience in astronomy. In order to fill this void, the Navy tried and failed to hire William C. Bond from Harvard, but in February, 1846 succeeded in persuading Sears Cook Walker (1805–53) to come from Philadelphia; he was the first civilian to join the work force at the Observatory.[44] A graduate of Harvard in 1825, Walker (Figure 2.7) taught in Boston and Philadelphia before becoming an insurance actuary in 1836. This job left some leisure time, and Walker became heavily involved with the Philadelphia High School Observatory. He was responsible for recommending the German instruments for this observatory, and with his half brother E. O. Kendall observed with them and published results. A series of "unfortunate investments and commercial operations" left Walker

2, 1854, gives a lower limit on his dates at the USNO. After leaving the Observatory in 1853, he worked as a surveyor, and resigned from the Navy July 11, 1856 (General Navy Register, 308). See Rothenberg (ref. 38), p. 167, and Rothenberg, "Observers and Theoreticians: Astronomy at the Naval Observatory, 1845–61," in *Sky with Ocean Joined: Proceedings of the Sesquicentennial Symposia of the U. S. Naval Observatory*, Steven J. Dick and Leroy Doggett, eds. (U. S. Naval Observatory: Washington, 1983), 29–43: 33 and notes 10–12.

[44] On the attempt to hire Bond see draft letter, Bond to SecNav George Bancroft, May 14, 1845, Bond Papers, Harvard University Archives, which was brought to my attention by Carlene Stephens. While declining the offer to be a civilian astronomer at the Observatory, Bond consulted with Peirce and wrote that they both thought it "a point of the first importance that you should have the assistance of a practical astronomer;" they recommended William Mitchell of Nantucket for the position. On Walker's date of appointment see Maury, *WO*, volume 1, Appendix, p. 13. On Walker see B. A. Gould, *An Address in Commemoration of Sears Cook Walker, Delivered before the American Association for the Advancement of Science, April 29, 1854* (1854); also DAB, 20, 359–360.

penniless in 1845; even so Bache had to convince him to accept Bancroft's offer to come to Washington.

By July, 1846, as Maury wrote the Explanatory Appendix for his first volume of observations, the facility had a total staff of 16, seven Lieutenants, five passed midshipmen, three Professors of Mathematics, and one civilian. Of these, Maury himself points out, ten were employed at the Depot prior to the Observatory's construction and now handled chronometer and hydrographic duties, leaving three lieutenants, three Professors and Mr Walker to undertake the astronomical work. Maury, himself only 39 years of age when he assumed the Superintendency, thus presided on the astronomical side over a youthful corps of three lieutenants and three Professors: Hubbard was 21, Keith 19, and Coffin 29. Walker, 40 at the time of his appointment and with a reputation in practical astronomy, was more Maury's equal. On the Depot side of the institution the midshipmen and lieutenants must have also been a youthful corps.

One more Professor of Mathematics would be added during Maury's early period. James Major (1813–98) had entered the naval service in 1838, joined the Corps of Professors in 1844, served briefly on the sloop *Falmouth*, and was assigned to the Naval Observatory in 1847. Perhaps because of the neighboring influence of Georgetown University, he became a Jesuit in 1858, and resigned his navy commission in 1859.[45] Civilian James Ferguson would replace Walker in 1848 and Professor Mordecai Yarnall would replace Coffin in 1852; both had lengthy careers at the Observatory, and, as we shall see, their impact began to be felt in Maury's later years. A final Professor of Mathematics, Thomas J. Robinson, came in the last year of Maury's tenure and had little impact.[46]

Tensions were present from the outset among Maury and his young staff, not least in the matter of salary. The pay of lieutenants was $1,500, while the more highly trained Professors of Mathematics, and Walker, received only $1,200, which was the pay of a lieutenant when he was on leave of duty. Even midshipmen received $750 per year. On the other hand the lieutenants could not complain about the pay of their boss; the Superintendent received the same $1,500 salary as they did. Maury calculated that the extra cost of maintaining an astronomical observatory for the Navy, over and above its depot functions, was only $5,700, "a sum which would hardly pay the principal Astronomer alone at one of the European Observatories." This not-so-subtle request for a salary increase, not only for himself but also on behalf of the Professors of Mathematics, was only "reluctantly made" in the first volume of observations, Maury told the Secretary of the Navy, and "through you to the public."[47]

[45] On Major see USNOA, BF.
[46] Most of these figures are discussed in the context of the education of early American astronomers in Rothenberg, ref. 38 (Hubbard, pp. 64–65; Walker, pp. 81–85; Major, p. 155; Keith, p. 167; and Yarnall, pp. 182–184), and in Rothenberg (ref. 43).
[47] WO, Appendix, p. 15. By act of Congress, the pay of the Superintendent was increased to $3,000 in 1848 (retroactive one year) in honor of Maury's success with the wind and current charts. Williams (ref. 4), p. 184.

Tensions went deeper than money, however. Perhaps because of his Navy training, Maury kept tight control over his subordinates. As Coffin's biographer, the astronomer George Comstock, wrote seven decades later "Coffin's official relation to this work, however, was that of a subordinate executing the orders of an omniscient superior, who assumed full responsibility and credit for everything done, even precluding the young professor from so much initiative as is implied in altering the adjustments of his instrument. This restriction was subsequently withdrawn perhaps in recognition of the young man's quality, but it is sufficiently evident that during his entire connection with the Observatory Coffin chafed under this official impediment cast about his scientific work."[48] Comstock noted that this characteristic was not peculiar to Maury, or even to the Navy, but "in some measure even the civilian science of that day put into the relation between senior and junior a feudal character of lord and vassal that would hardly be tolerated now." Such tensions were perhaps in part the cause for the departure of Walker in 1847, and may have also hastened Coffin's departure (usually attributed to eyesight problems) in 1853. Tensions notwithstanding, a sense of excitement, anticipation, and promise pervaded the new institution as it laid out its program of work, took advantage of new technology, played its role in astronomical and hydrographic developments, and began contacts with the outside world.

### The early observing program

Was there a research program that Maury had in mind when assigning his staff to astronomical duties? If so, was it driven by navigational needs or pure astronomy? At first Maury clearly concentrated on navigational needs: "The principal aim and object of the Observatory is to assist in perfecting and procuring the requisite data for the American Nautical Almanac. This has been considered its first duty. With this aim in view, the instruments have been directed chiefly to the sun, moon, planets, and a few of the principal fixed stars," Maury wrote in July, 1846.[49] It was with this aim in view that Maury assigned his astronomical staff to their first duties. Upon Coffin's arrival, Maury placed him in charge of the mural circle (Figure 2.8), to be assisted by Lt Thomas Page.[50] Although Coffin had no experience with astronomical observing instruments (indeed almost no one in the United States did in 1845), this staffing inaugurated the pattern of having a Professor teamed with a lieutenant, whenever possible. Observations on the 5.3-inch transit instrument (also to be seen in Figure 2.8) began on April 22, 1845 with Lt Maynard. He was teamed with Professor Hubbard

---

[48] Comstock (ref. 41), p. 6.

[49] Maury, WO, Appendix to volume 1, pp. 31–32. An American Nautical Almanac had, however, not yet been approved, and, when it was, would not be placed under Maury, a story told in chapter 3.

[50] Thomas Jefferson Page (1808–99) at the age of 87, wrote an autobiographical sketch from his home in Rome, "Autobiographical Sketch of Thomas Jefferson Page," USNIP, volume 49. Regarding his three years (1845–48) at the Observatory he wrote (p. 1,672) "It was not an easy berth, for day and night in good weather, I was at work with the Mural Circle. The Observatory was in its infancy and its character before the world was to be established. Placed thus upon our mettle, we all labored diligently in consequence."

Figure 2.8. The Mural Circle and the West Transit Instrument. The 4.1-inch-aperture mural circle was mounted as shown in 1845; the 5.33-inch transit instrument was originally mounted in the "west room" in 1845, and mounted as shown here in 1864. Plate I from *Instruments and Publications of the U. S. Naval Observatory* (Washington, 1845–1876).

until Professor Keith arrived, when Hubbard moved to the German Circle, assisted by Lt Porter. Only for the prime vertical was no Professor available; Maury placed his brother-in-law Lt William L. Herndon in charge of this instrument, and assisted himself, though over the longer term Hubbard would become the expert in its operation. During the next few years the Professors would stay with their instruments, while the lieutenants came and went.[51]

These arrangements left the equatorial telescope (Figure 2.9) with no permanent observer. Maury originally had intended to put Walker in charge of the equatorial, giving him a detailed letter of duties.[52] However, the instrument was not fully ready until April, 1846, so Maury put him to work as a computer rather than an observer, especially in discussing the latitude, and admittedly his duties were "laborious and tedious." Walker never did get to observe on a regular observing program before his abrupt departure a year later, but, as we shall see, his achievements during that year were important to the Observatory's budding reputation.

The meridian observations during 1845 were directed more toward immediate needs rather than observing programs. On the mural circle Coffin and Page observed standard navigational stars, those bright stars traditionally used by navigators. Hubbard, Keith and Maynard on the transit instrument did the same, and the prime vertical observed alpha Lyrae and 61 Cygni, located near the zenith and therefore important for determining the latitude of the Observatory without refraction effects. Beyond that Maury proposed to construct a catalog of every star brighter than sixth magnitude from the zenith to 20 degrees north.

The final result for 1845 was, in Maury's words, "the first volume of Astronomical Observations that has ever been issued from an institution properly entitled to the name of Observatory on this side of the Atlantic." Of its 538 pages, 146 described in detail the instruments and their adjustments, 271 gave the printed observations of each star, and 119 described the duties of the office for its first year. Only two pages gave the final result: for all the meridian instruments a catalog of 96 stars, many observed 15 or more times (with 63 observations of alpha Lyrae the record high), for a total of almost 1,000 observations.[53] Considering that it was the first year of the Observatory, these results were promising for astronomy, and the volume was praised both in the United States and in Europe.

The limited program of 1845 soon gave rise to a much more ambitious program that had nothing to do with navigation. Since the limited program left spare time for the observers, who had to wait at their telescopes for the selected objects to pass over the meridian, at the end of 1845 Maury conceived in addition the idea of a "systematic review and exploration of the whole heavens, in ascertaining Right

---

[51] WO, volume 1, Appendix, pp. 3–4. Keith started observing August 17, and Almy September 22.

[52] WO, volume 1, pp. 33–36.

[53] "Catalogue of all the Stars observed both in Right ascension and Declination at the U. S. Observatory during the year 1845, reduced to their mean places for January 1, 1845," WO, volume 1, pp. 272–273; for the Maury quotation see p. 118.

Figure 2.9. The 9.6-inch Equatorial Telescope, which was mounted in 1844. Plate II from *Instruments and Publications*.

Ascension, Declination, and assigning position to every star, cluster and nebula" within the reach of the instruments. Exemplifying the same spirit that led Gilliss to begin an expanded catalog of stars in connection with the U. S. Exploring Expedition, Maury's proposal was immediately approved by Secretary of the Navy Bancroft, who noted in his formal authorization letter that, in addition to results of importance to maritime science, "the country expects, also, that the Observatory will make adequate contributions to Astronomical Science." On January 5, 1846 Maury gave the order to his observers to begin this expanded program, which embraced stars down to tenth magnitude.[54] This system of "zone observations," devised by Coffin and Hubbard, made use of the transit instrument for right ascension, the mural circle for declination, and the meridian circle observing the same stars as a check on both instruments. Observations commenced in April, 1846. Reaching further south than the major observatories of Europe, and realizing the degrading effects of refraction at low altitudes, the program began at declination 45 degrees south, only six degrees from the horizon. By the end of July five or six thousand stars had been observed in the 15-degree band from 45 to 30 degrees south declination, the first of more than 15,000 that would be observed repeatedly over the next five years.

The observations themselves, ambitious as they were, were only the beginning of the new program. The analysis, or "reduction," of the raw observations was no less important and even more time-consuming, as is apparent from the selected meridian observations published in the first volume. Some 45 pages of Maury's 119-page Appendix dealt with various factors and tables for reducing the observations. Of special importance was the refraction correction, which depended both on temperature and on zenith distance. Standard tables existed for this, but Maury noted that they were inconsistent, differing by as much as an arcsecond even at the relatively high altitude of about 51 degrees. There was little Maury could do about improving the refraction tables – he decided to adopt Bessel's tables, and set Professor Coffin to work putting them into a more convenient arrangement. However, temperature was important for applying the refraction correction, and he did take great pains to calibrate his standard thermometers, of which he had purchased six from Troughton and Simms of London, at least one of which had been compared with the standard of the Royal Astronomical Society. These and other tables serve to show the formidable task of observational analysis that European observatories had grappled with for centuries, which was now being faced by Americans for the first time in a comprehensive and systematic way. As we shall see, the reduction of the Naval Observatory zone observations became a contentious issue later in the Maury era.

In addition to the ambitious program of meridian observations, other astronomical work was not neglected. The first published equatorial observations were of the comet of June 4, 1845, but the split of Biela's comet in January, 1846 drew Maury's special attention.[55] Hubbard's work with the prime vertical on the parallax of Vega began in 1848, as well as his extended work on the orbits of the asteroids and the comet of 1843 that would bear fruit later in the Maury era. In short, like the catalog of 96 stars discussed earlier, the broader astronomical observing program was off to an auspicious start.

---

[54] Bancroft's immediate approval is discussed in WO, volume 1, Appendix, p. 32. Maury's order to his staff "Observations for 1846" dated January 5, 1846 is to be found on p. 33, and Bancroft's formal authorizing letter to Maury of March 6, 1846 on p. 38.

[55] WO, volume 2, p. 2; observations on pp. 341–356, results pp. 421–425.

### Of time and longitude

As an institution that owed its founding to the necessity for accurate timekeeping in the form of the marine chronometer, time as determined by astronomical observation remained at the very heart of the institution under Maury. Time was important at the Naval Observatory – as at other national observatories around the world – for at least three related reasons that were by now classic: to rate chronometers, to determine star positions, and to determine longitude; a fourth objective, the dissemination of time to the public, was an innovation undoubtedly influenced by recent European example. Whether routine or innovative, the pursuance of these tasks required meticulous attention to precision methods, and constant vigilance to any potential improvements in such methods. In fact the Naval Observatory during the early Maury era not only acquired new precision clocks, but also soon took advantage of a new American technology for registering time more precisely.

Certainly among the first imperatives of the new Observatory was the continuation of stellar observations for the purpose of rating chronometers. The instrument used for this purpose was no longer the old Coast-Survey Troughton transit, but rather the 5.3-inch Ertel transit telescope, which was also engaged in the star-catalog work. By 1846 there were 80 chronometers on shore ready for service or on trial, and 120 more onboard ships. Their rating was a routine but meticulous process, much the same as that carried out at the Capitol Hill Observatory, and attended by much record-keeping. This function amounted to a major task; indeed, as the Observatory grew in stature, some wondered whether rating chronometers should remain at a scientific institution. It did, until 1951, when the function moved to the Norfolk Navy Yard. The same issue faced the Astronomer Royal at Greenwich, also under the aegis of a navy.

Very early in Maury's administration he began to disseminate time to the citizens of Washington via a visual signal; suggestions that navigators on the Potomac used it to check their chronometers have not been verified. Curiously, it was Secretary of the Navy Mason, rather than Maury himself, who initiated action for the plan. It is certainly far too much to expect that any memory remained of the suggestion made to the American Navy in 1830 that they adopt a time-ball method about to be placed in operation in England at its major naval base. Nevertheless, the Secretary wrote Maury on December 10, 1844 ordering a time signal, emphasizing its usefulness to the public rather than to navigators: "You will be pleased to devise some signal by which the mean time may be made known every day to the inhabitants of the city of Washington. When you are prepared to put your signal into operation, you will give notice of the kind you have adopted in the city newspapers, & at the same time inform the Department." This was the first Federal Government order for time dissemination in the United States, and, although it applied initially only to those who were within sight of the ball, gradually it would be expanded with improvements in technology.[56]

---

[56] NA, RG 78, Letters Received, 1838–84, box 2, John Y. Mason to M. Maury. See Ian R. Bartky and Steven J. Dick, "The First North American Time Ball," JHA, **13** (1982), 50–54, reprinted with

It was the method of using a falling ball, which had been tested at Portsmouth, U. K. as early as 1829 and in use since 1833 both on the important Indian-Ocean island of Mauritius and at the Royal Observatory at Greenwich, that Maury adopted for his signal. He probably learned of it from the Royal Observatory's annual volumes for 1836 or 1840, which mention the "signal ball." Three days after the Secretary's letter, Maury ordered from the Navy agent in Washington "one flagstaf 20 feet long and three stays for do [ditto?] each 40 feet long two pair of Sig [signal?] halliards. One black ball 3 feet diameter . . .". In the *Daily National Intelligencer* for January 2, 1845 one finds a notice that "A ball will be hoisted above the dome . . . every day (Sundays excepted) . . .", followed by another report on April 15, 1845 that "On top of the dome there is a pole, up which a ball is hoisted and let fall every day precisely at 12M . . .". The first volume of *Washington Observations* (1846) makes no mention of the time signal, but the engraving on its title page shows the ball mounted on a flagstaff atop the dome of the 9.6-inch telescope (Figure 2.4).[57]

The idea was simply that the ball would drop at a precisely known instant, 12 noon on the Observatory's meridian. The early experimental stage of this operation is evidenced in Maury's correspondence; in February 1845 he wrote the New York agent for Charles Goodyear, the inventor of vulcanized rubber whose patent had been awarded in 1844, "Be pleased to make and send . . . four air tight balls of Gumelastic Composition capable of being inflated into spheres. They are wanted for signal balls. They should therefore have a bracket at opposite poles firmly secured so that the halliards for hoisting up may be bent on to one bracket and a rope for hauling down to the other. Let two of the balls be at least 4 feet in diameter when inflated one of 3 feet, and one of 18 inches . . .". Coffin followed up on this order, writing to Maury in late February that the "India-rubber balls" are "in progress and will be completed as soon as the weather will permit." Maury wrote in May to Messrs Blunt, chart suppliers and instrument makers in New York, inquiring "What will you charge for that copper ball? I may want it for a time ball . . .".[58] There were also various methods for dropping, including controlled lowering, dropping onto the dome, and throwing by hand, all of which were probably used in the early period; eventually an electromagnetic relay caused the ball to fall.[59]

addenda and a list of early time balls in NAWCC [National Association of Watch and Clock Collectors] Bulletin, **41** (December, 1999), 741–744. For the 1830 suggestion and the world's first time balls, Ian R. Bartky and Steven J. Dick, "The First Time Balls," JHA, **12** (1981), 155–164. On the expansion of the time system in the United States, see Ian Bartky, *Selling the True Time: Nineteenth Century Timekeeping in America* (Stanford University Press: Palo Alto, California, 2000), chapters 4–9. Hereafter *Selling the True Time*.

[57] Bartky and Dick (ref. 56, 1982), especially references in notes 13 and 19 of that article. The two *Daily National Intelligencer* citations are in ibid. (ref. 56, 1999). Maury to Blunt, NA, RG 78, May 27, 1845, Letters sent, 396.

[58] RG 78, Letters Sent, July, 1842–November 1862, volume 1, 317; Maury to Goodyear February 7, 1845, ibid., 336; J. H. C. Coffin to Maury, February 20, 1845, ibid., Letters Received, box 2.

[59] Bartky and Dick (ref. 56, 1982), 51–52. The magnetic relay was in use by 1862, perhaps even earlier. As Gilliss wrote "For the purpose of giving correct time to the city, a staff has been placed on top of the dome, and a large, but light, ball is hoisted ten minutes before 12 o'clock of each

Whatever the material or the method, the result was the same: anyone within sight of the ball could set his timepiece exactly to noon. "Ball time" became synonymous with accurate time, and a popular and practical attraction; it was reported that "John Quincy Adams, who was a devoted friend of the Observatory, and who used to visit it frequently in the last days of his life [he died February 23, 1848], has been known to walk all the way up to the Observatory from his lodgings, to see the ball fall."[60] It was a watershed of sorts, for accurate time now left the Observatory and was available to a restricted public. Either directly or via telegraph signal, the Naval Observatory would continue to drop a time ball in Washington every day except Sunday until 1936, long after it had been superseded by telegraphic and radio technologies.

A preoccupation with time pervaded the work of the observatory. The determination of time for chronometer rating or for the public depended on observations of the Sun or of stars ("clock stars") whose positions were well known. Their time of crossing the meridian *defined* the correct time. Once time had been determined by these well-known objects, the chronometers could be rated or the clocks set. More than that, these data, and the timing operation in general, were essential to a second goal of the Naval Observatory, namely determining star positions. The right ascension of any star can be determined by timing exactly when that star crosses the meridian, and applying the "clock correction" determined by a set of clock stars observed with it. The entire program of positional astronomy that Maury laid out – whether in its core form for the *Nautical Almanac* stars or for the immense program of zone observations – therefore depended on knowledge of the correct time. For this purpose the Observatory as of January 1849 had three clocks and a chronometer for use among the five instruments.[61]

In addition to chronometer rating and star positions, the determination of longitude was also related to time, for, as we have seen in chapter 1, differences in longitude are merely differences in time. The longitude of a place could be determined by

(footnote 59 cont.)
day, except Sunday. The pulley is connected with an electro-magnetic battery after the ball is up, and the circuit is broken by the assistant in the chronometer room at the instant of noon." Gilliss, WO (1862), p. x. By order of Secretary of the Navy Richard Danzig, the time-ball tradition was recreated on New Year's Eve between 2000 and 2001, when a new time ball was dropped at the Observatory's site on Massachusetts Avenue. The Observatory also coordinated an around-the-world time-ball drop on this occasion, as reported in *NAWCC Bulletin*, 41 (December, 1999), 737.

[60] *Bohn's Hand-book of Washington* (Bohn: Washington, 1852), pp. 51–52. On the incorrect 1850s attributions see Bartky and Dick (ref. 56, 1982), 52, footnote 6. The first description of the Washington time ball did not appear in the *Washington Observations* until 1862.

[61] Maury, WO, volume 2, Appendix, p. 6. Gilliss, *Report* (ref. 9), p. 9 notes that two clocks from the Depot were passed on to the new Observatory, as well as two clocks purchased by Wilkes for the U. S. Exploring Expedition. For a compilation of information about clocks at the Naval Observatory see James A. DeYoung, "United States Naval Observatory Clocks and other Time-Keeping Apparatus found in the Washington Observations and the Publications of the USNO, 2nd series," 35-page typescript (c. 1983), USNO Library, and the same author's "United States Naval Observatory Clocks Mentioned in the Annual Reports, 1842–1930" (1982), 18-page typescript, USNO Library.

means of occultations or Moon culminations, as Gilliss had done at the Capitol Hill Observatory, or differentially by determining the difference in longitude from some other place whose longitude was known. Maury tells us in his first volume that observations for determining longitude of the Observatory were under way, but had not been analyzed "for want of computors and other requisites." He adopted a provisional value of 5 h 8 min 14.64 s west of Greenwich.

A novel "differential" approach to determining longitudes became the subject of considerable controversy involving the embryonic Naval Observatory, the introduction of electric telegraphy, turf battles with the Coast Survey, and ultimately a landmark invention for measuring time in astronomy. In June, 1844, even as the new Naval Observatory was preparing to open its doors, Commander Charles Wilkes made the first telegraphic determination of longitude in the United States, the difference between the meridian bisecting the Capitol dome in Washington and that running through Baltimore's Battle Monument Square. Only three weeks after American telegraphy began with the words "What hath God wrought!," Wilkes (with Samuel F. B. Morse's permission) used the telegraph line to compare the chronometers at Baltimore and Washington, finding the difference between the two cities to be 1 min 34.868 s.

Clearly the telegraphic method held great promise for an age-old problem. The determination of longitude differences was already a part of the effort of the Coast Survey under Hassler and Bache. In his first Report as Superintendent of the Survey dated December, 1844, Bache wrote that the determination of such differences between points on the American East Coast and European observatories was under way, using traditional methods. William Bond was making the observations at Cambridge and E. O. Kendall at Philadelphia, with the calculations in charge of Sears Walker, soon to depart Philadelphia for his position at the Naval Observatory in Washington. In addition to occultations and Moon culminations observed by this group, Bond was also preparing a report on the exchange of chronometers between Boston and Liverpool.[62]

In light of the electric telegraph the goal was to supplement or replace these traditional methods. In late 1845, a few months before Walker left for Washington, Bache requested that Walker make arrangements with the telegraph companies for the use of their lines in determining longitude differences. In their discussions on the subject they arrived at the conclusion that transits of stars should be substituted for Wilkes's method, which involved only the exchange of signals over the telegraph line.

---

[62] "A Report of the Superintendent of the Survey of the Coast, showing the Progress of the Work during the Year ending November, 1844," in *Report of the Secretary of the Treasury*, December 23, 1844, Senate Document 16, 28th Congress, second session, 15. Bartky, *Selling the True Time* (ref. 56), chapter 3, describes these events in more detail. For the context of Bond's work see Carlene E. Stephens, "Partners in Time: William Bond & son of Boston and the Harvard College Observatory," *Harvard Library Bulletin*, **35** (1987), 351–384; however, see the cautionary notes in Bartky (ref. 56), p. 219, note 12 and p. 228, note 43.

After some delay occasioned by his removal to Washington and assumption of new duties in February 1846, Walker first carried out this approach at the Naval Observatory. Because the Observatory had the premier transit instruments in Washington, Maury and Bache had agreed to cooperate in this endeavor.[63] The method of star transits was first proven in part by an experiment undertaken by Walker and Lt Almy between the Naval Observatory and Philadelphia. Walker recalled the events of October 10, 1846: "the transit of the star 2838 Bailey, over the seven wires of the west transit instrument of the Washington Observatory, was signalized by Lieut. Almy, U. S. N., the officer having charge of that instrument. This transit was noted on the Washington clock, by Lieut. Almy, and also by myself, comparing together, by the ear, the seven key beats, with the clock beats. The same key beats were also noted by Prof. Kendall, at Philadelphia. This was the first practical application of the method of star signals, which is sooner or later to perfect the geography of the globe."[64] The same experiment was repeated later in October, and in July and August of 1847, the differences between Washington, Jersey City and Philadelphia were determined. Again the Naval Observatory was used as the Washington station, this time with Professor Keith at the instrument. In the end, however, Walker's hopes for widespread use of the method were frustrated; for technical reasons the Coast Survey's standard method would be telegraphic exchange of clock signals, not stellar transits.[65]

At the Naval Observatory the determination of longitudes was not yet an official part of the duties. Despite their cooperation in early telegraphic longitude determinations, Bache's Coast Survey and Maury's Naval Observatory competed for the task. Nationwide longitude determination was not the mission of either agency, and, despite Maury's repeated arguments to the Secretary of the Navy that a program of longitude determination via telegraph "comes eminently within the province of the Observatory," such a program was never authorized. By March, 1847, Walker, no longer able to work under Maury, had left the Naval Observatory for the Coast Survey. By 1849 the Coast Survey had its own station, known as Seaton Station, in Washington for determining longitude. By 1853 Maury had clearly lost the battle for a longitude program at the Observatory. Bache and Walker at the Coast Survey had stolen his thunder. Only in the 1860s – after Maury's departure and the Civil War – would the Observatory once again become involved in determining longitude differences. Such

---

[63] On Naval Observatory–Coast Survey cooperation in this instance see Bartky (ref. 56), chapter 3.

[64] Walker, "Report on the Experience of the Coast Survey in Regard to Telegraph Operations, for Determination of Longitude, & c.", *American Journal of Science and Arts*, **10** (November, 1850), 153, extracted from Walker, *Proceedings of the AAAS* (1849), p. 182. For the report on the work see Walker, "Account of the telegraphic operations for fixing the longitude of Washington, Jersey City and Philadelphia," AN, **27** (1848), 121–126. Almy gave his recollections of the experiment in *Record of the United States Naval Institute*, **5** (May 29, 1879), 456–457.

[65] For a first hand (but biased) history of these events see Sears Cook Walker to A. D. Bache, November 10, 1846 as published in AN, no. 632 (1848), 121–126, and Walker (ref. 64). Gould also describes Walker's role in his eulogy, pp. 14–20, but a much more balanced view is given in Bartky, *Selling the True Time*, chapter 3. Bartky (p. 34) terms the first Washington–Philadelphia experiments "promising but unsuccessful," with the first real success in August, 1847. Bartky (p. 35) describes the technical weaknesses of the method.

turf battles were not unusual in the young American republic where scientific activity was growing.[66]

Although the Naval Observatory was out of the longitude-determination business for the rest of the Maury era, it would be greatly affected by an invention stemming from the method of star transits for longitude – an invention on which reputations were staked and that soon became the center of great controversy. The issue was not the method of star transits itself, but rather the method of registering the time more precisely. After all, the temporal tasks of an observatory – the determination of time, of stellar positions in right ascension, and of longitude – could greatly profit by any ability to register the time more precisely than the standard "eye-and-ear" method of observing and listening to a clock. At stake was the ability to measure precisely one of the fundamental properties of nature – time.

The essence of the new method of time registration – which would come to be known as "the American method," consisted in an electrical switch attached to a clock. Although the idea had been broached before at the Coast Survey, Bache and Walker rejected it, feeling that it would affect the clock's performance. William Bond had similar ideas at Harvard, but they were not advanced enough for Walker's use in an upcoming longitude campaign in Cincinnati. Two scientists – O. M. Mitchel at the Cincinnati Observatory, and Cincinnati physician John Locke – responded to Walker's pleas for help. It was Locke who followed through, working closely with Walker. Locke fitted an existing astronomical clock with a circuit-breaking device, and in late November, 1848 found the optimal method for registering both time and an observed event on the same record. Time was registered by connecting a closed electrical circuit to the recording pen of a moving Morse telegraph register, and observed events were registered on the same moving record by breaking a circuit with a telegraph key.[67]

The Naval Observatory's direct involvement began on December 30, 1848 when Locke informed Maury of a discovery that Maury characterized as "one of the most important inventions of the age." With it time could be determined more accurately, making it, in Maury's words, "as easy and as practicable to divide seconds into hundredths, as, before, it was to divide minutes into seconds."[68] Writing to Secretary

---

[66] Maury to SecNav George Bancroft, July 1, 1846–September 1, 1846, WO, volume 1, Appendix, p. 87. Maury, WO, volume 2, Appendix, p. 3, and WO, volume 3 (1853), Introduction, p. i. Maury refers to Gilliss's discussion in *Proc. APS* (20 April, 1849), and the Coast Survey in *AJ*, **3**, 23. See Holden's 1879 subject index for later USNO involvement in longitude determination.

[67] See Bartky, *Selling the True Time*, chapter 3, pp. 35 ff, for an expert untangling of the sequence of events in this controversy. On Locke see Adolph E. Waller, "Dr. John Locke, Early Ohio Scientist (1792–1856)," *The Ohio State Archaeological and Historical Quarterly*, **55** (1946), 346–73, and BDAS. It is notable that Locke had met Maury, Bache and others in 1844 at the National Institute meeting, at which he chaired a session, and that Locke and Bache had become friends. For the reaction of the British Astronomer Royal to the American method, see G. B. Airy, "On the Method of Observing and Recording Transits, lately Introduced in America; and on Some other Connected Subjects," *MNRAS*, **x** (1850), 27–34 and 93.

[68] Maury to John Locke, January 5, 1849, in WO, volume 2, Appendix, p. 5. Maury describes the events surrounding this invention and his reaction to it, on pp. 4–40, where some of the letters have been altered from those in the National Archives.

of the Navy John Mason the same day, Maury explained the benefits of the invention in more detail: "The discovery consists in the invention of a magnetic clock, by means of which seconds of time may be divided into hundredths with as much accuracy and precision as the machinist, with rule and compass, can sub-divide an inch of space. Nor do its powers end here. They are such that the astronomer in New Orleans, St. Louis, Boston, and in every other place to which the magnetic telegraph reaches, may make his observations, and, at the same moment, cause this clock, here in Washington, to record the instant with wonderful precision. Thus the astronomer in Boston observes the transit of a star as it flits through the field of his instrument, and crosses the meridian of that place. Instead of looking at a clock before him, and noting the time in the usual way, he touches a key; and the clock here subdivides his seconds to the minutest fraction, and records the time with unerring accuracy. The astronomer in Washington waits for the same star to cross his meridian, and, as it does, Dr. Locke's magnetic clock is again touched; it divides the seconds, and records the time for him with equal precision. The difference between these two times is the longitude of Boston from the meridian of Washington."[69]

Maury's opinion of this invention was unambiguous: "Thus this problem, which has vexed astronomers and navigators, and perplexed the world for ages, is practically reduced at once, by American ingenuity, to a form and method the most simple and accurate. While the process is so much simplified the results are greatly refined. In one night the longitude may now be determined with far more accuracy by means of the magnetic telegraph and clock, than it can by years of observation, according to any other method that has ever been tried." It was perhaps a bit much to call this invention a "national triumph," but it was certainly a triumph for astronomy. Recalling Wilkes's pioneering longitude experiment, and Locke's former role as a member of the Navy Medical Corps, Maury used the opportunity to request purchase of Locke's "magnetic clock" and to plead with the Secretary for a continued role for the Navy in longitude work. In particular, such a contrivance would "enable the National Observatory to perform a most important part of its appropriate duties, and a most acceptable service to the world in perfecting the geography of the country, and in affording so many well-determined points of departure for the traveler, the surveyor, and the navigator." Maury suggested that Locke change the name of the "magnetic clock" to "electro-chronograph," for the "whole apparatus consisting of clock, battery, registering apparatus, and all the machinery and implements used to record, *electro-chronographically*, an astronomical observation . . .". This name was engraved on the clock itself (Figure 2.10), and the term "chronograph" (Figure 2.11) for the registering apparatus was in widespread use over the next century.[70]

[69] Ibid., p. 7.
[70] Ibid., pp. 4 and 8. The clock itself is now in the National Museum of American History, at the Smithsonian Institution. On the history of the chronograph see Rand B. Evans, "Chronograph," in R. Bud and D. J. Warner, eds., *Instruments of Science: An Historical Encyclopedia* (Garland: New York and London, 1998), pp. 110–112.

Figure 2.10. The clock portion of Locke's electro-chronograph, photographed at the new Naval Observatory on Massachusetts Avenue. It is now located at the Smithsonian Institution.

With the success of the new method, Locke, Bache, Walker, Mitchel, and even Bond became involved in a bitter priority dispute. It was Bond who would market the first commercially successful chronograph in the early 1850s. However, by act of Congress on March 3, 1849 – pushed by Maury as part of the Navy Department's appropriations bill – it was Locke who was authorized to receive $10,000 and to erect his clock for the Observatory in return for unrestricted use of the invention. Thus in the end, although Maury did not get his longitude program, he did acquire the signal clock and chronograph, and institute the new method for astronomical observation that spread throughout the world.[71] The "American method" revolutionized the

---

[71] John Locke, *Report of Professor John Locke on the Invention and Construction of His Electro-Chronograph for the National Observatory, in Pursuance of the Act of Congress, Approved March 3, 1849* (Wright, Ferris and Co.: Cinncinati, 1850). Simon Newcomb later ridiculed the clock in *Reminiscences of an Astronomer* (Houghton, Mifflin & Company: Boston and New York, 1903), p. 118. Other sources, including Naval Observatory reports, indicate that it served a useful purpose. In any case, certainly the method proved itself. A variety of chronographs would be used well into the twentieth century, including one 24-inch cylinder originally made by William Bond, with Howard and Davis, "furnished with the Locke clock," which was used until 1872 with the 8.5-inch transit circle. At that time a new chronograph by Alvan Clark and Sons was used, one of several devised for the transit of Venus. Both revolved once each minute. *WO* for 1892 (1899), pp. viii–ix. These were followed by Bausch, Lomb, and Saegmuller chronographs.

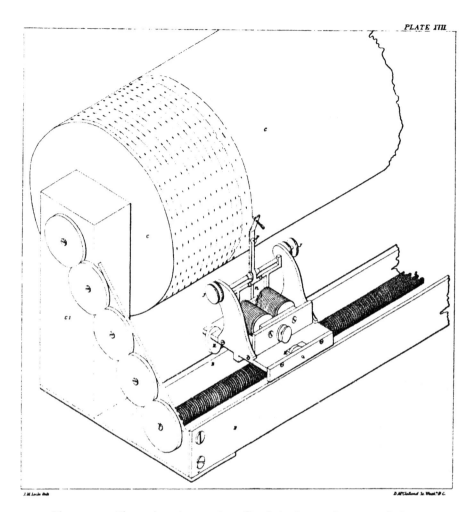

PLATE XIII

Figure 2.11. The registration portion of Locke's electro-chronograph, later known simply as the chronograph. It consisted in a uniformly rotating cylinder with paper and pen, which allowed the time of an event to be registered to hundredths of a second, ten times better than previous methods. From *Report of Professor John Locke* (ref. 71).

observing procedures at the Observatory, and eventually in all of practical astronomy when the registration of time was involved.

### Sears Cook Walker and Neptune

One of the greatest discoveries of nineteenth-century astronomy occurred less than two years after the Naval Observatory had opened its doors: the discovery of the planet Neptune, the first to be found since William Herschel's discovery of Uranus in 1781. The existence of Neptune had been predicted by Leverrier and it was first spotted at the Berlin Observatory by J. G. Galle on September 23, 1846; on October 20 the news reached the United States when the steamship Caledonia landed in Boston.[72] The Bonds were the first to spot it in the New World with the Harvard 4.5-inch refractor on October 21 and 22; during the next two days it was observed in Washington by Maury with the equatorial, by Almy and Keith with the West Transit Instrument, and by Hubbard with the Meridian Circle.[73]

Early in November Maury placed Sears Cook Walker in charge of investigations of the new planet. To Maury's credit he put the right man to work on the right problem, with results that played an important role in gaining an early reputation for the Observatory. Walker immediately set to work not on observations of Neptune (Maury was the sole observer of Neptune on the Equatorial until November 28, when Walker took over) but on the theory of its motion. The idea was that, on the basis of the few observations that had become available since its discovery, an orbit could be computed for Neptune and the planet's position traced backward in order to determine whether it had been observed mistakenly before as a fixed star. If it had been, those earlier observations could in turn be used to refine the orbit.

Assuming a constant distance from the Sun for the period of three months since its discovery, Walker used European observations from September 24, and Washington observations for October 24 and November 21 to determine that the orbit did appear to be nearly circular. By late December he had additional observations of his own with the equatorial, and began a rigorous computation of the circular elements of the orbit, resulting in the first Ephemeris of Neptune, covering the period from August 1, 1846 to February 1, 1847. Additional observations, including a large number from Washington, were used to refine the orbit, dropping the assumption of circularity. It was at this point that Walker decided to examine old star catalogs for the purpose of detecting Neptune as a missing star. By a process of elimination only Lalande's *Histoire Céleste* was found to be adequate for a star of magnitude 7.8 in that part of the sky. Examining the stars that Lalande observed in the year 1795 in the region he calculated Neptune might have been, on February 2, 1847, Walker reported to Maury the results of his labor, and singled out one star in particular that best fit the position and magnitude of Neptune.[74] Indeed, to have found a star of magnitude 7.8

---

[72] On the story of the contentious discovery of Neptune see Morton Grosser, *The Discovery of Neptune* (Harvard University Press: Cambridge, Massachusetts, 1962); Allan Chapman, "Public Research and Private Duty: George Biddell Airy and the Search for Neptune," *JHA*, **19** (1988), 121–139, Robert W. Smith, "The Cambridge Network in Action: The Discovery of Neptune," *Isis*, **80** (1989), 395–422, and the references in all these works on the numerous literature on the subject. On the reaction to the discovery in America see John G. Hubbell and Robert W. Smith, "Neptune in America: Negotiating a Discovery," *JHA*, **23** (1992), 260–291.

[73] The observations at Washington were announced in "The New Planet," *Washington Union*, October 26, 1846 and the *National Intelligencer*, October 27, 1846. Subsequent observations of Neptune with the equatorial telescope at the Naval Observatory during 1846 were published in *WO for 1846*, pp. 341–356.

[74] Maury, *WO for 1846* (1851), Appendix, p. 1, gives the date as Feburary 1 and says that 14 stars from Lalande "answered to the description of the new planet," but these were the 14 stars in the field; Walker, "Investigations which led to the Detection of the Coincidence between the Computed Place of the Planet Leverrier, and the Observed Place of a Star recorded by Lalande, in May, 1795," *Trans. APS*, **10** (1853), 141–155: substantially the same account is given as section 2 of Walker's "Researches Relative to the Planet Neptune," *Smithsonian Contributions to Knowledge*, **II** (1851), 10–17. The latter contains a list of 1,156 observations of Neptune through January 1848.

in the field Walker considered an "astonishing coincidence" if it did not indeed turn out to be Neptune.

The weather was stormy on the night of February 2, and on the third Maury, Coffin, and Hubbard huddled together to examine Walker's list of Lalande's stars in the area calculated for Neptune in 1795. So sure were they that it was really Neptune, and that it would now show up missing there, that on the same day they sent an account of Walker's work, and his prediction, to Gilliss, Bache at the Coast Survey, and Joseph Henry at the Smithsonian. Walker drew up a list of 14 field stars for Hubbard to observe, one of which was the suspected planet. What happened next is excitedly described by Walker: "The first clear night after detecting this coincidence was the fourth of February. Prof. Hubbard examined the region with the Equatorial, and reported to Lieut. Maury the next morning, February 5th, that he had found all the guide-stars in this list; but that the star that was designated as *expected to be missing*, was indeed *missing*. Prof. Hubbard reviewed the region several times. The star which should have preceded the missing star by 1 [degree] was in place. It was brought to the middle transit wire and to the bottom of the field. The Lalande star should have been in the upper portion of the field. It was not there. Nor was there in the vicinity any star that could be reasonably supposed to have been erroneously recorded as in the place of the missing star. I may add, that this region has since been examined by Lieut. Maury and Prof. Hubbard. The star is certainly *missing*." Lalande had unsuspectingly observed Neptune, on May 10, 1795, giving astronomers the benefit of observations more than 50 years before its discovery as a planet.[75] Walker lost no time in using Lalande's observations to improve the orbit of Neptune, beginning his calculations the following day, and continuing throughout the month of February.[76]

This pre-discovery observation of Neptune was announced to the American public on February 9. The news reached Leverrier in Paris on the same day (March 29) from two quarters, for astronomers at Altona had also made the discovery. Leverrier made the announcement of the double discovery to the French Academy that very day.[77] Even so, the hypothesis of identity occasioned a number of doubts until the arrival in Washington on May 19 of the news that M. Mauvais had discovered another Lalande observation of Neptune on May 8, 1795.[78] Even then the short period compared with the hypothetical periods of Leverrier and John Couch Adams in England occasioned skepticism and controversy. The claim of Benjamin Peirce at Harvard that the planet discovered by Galle was not the one Leverrier predicted also caused great controversy, but here our account of the Naval Observatory's involvement must end.[79]

Shortly afterward Walker left the Observatory, tired of Maury's overbearing

[75] Walker, "Investigations . . .," 18–19, and "Researches . . .," 148.
[76] The elliptic elements were published in *Proc. Am. Acad*, 1, 67; *Proc. APS*, 4, 319; and AN, 25, 383.
[77] CR, March 29, 1847. Peterson's discovery had been made seven days after Walker's.
[78] CR, April 19, 1847.
[79] B. A. Gould, *Report on the History of the Discovery of Neptune* (Smithsonian Institution: Washington, 1850). Gould's presentation copy to Hubbard is in the Naval Observatory Library.

attitude and Navy rules. Writing to Loomis, Maury explained "Mr. Walker was unwilling to comply with the rules of the office, as the officers do, and it was better therefore that he should quit. He wanted to be excused from attending the office entirely and occupy himself upon such subjects only as he should fancy. Mr. W., moreover, was a much better computer than observer; he could compute day in and day out but our night observations would knock him up."[80] Maury perhaps did not realize what he had lost – one of the pioneers of practical astronomy in America – but Maury's loss was the Coast Survey's gain. Walker immediately went to work with his old friend A. D. Bache and carried out his pioneering longitude work at the Coast Survey, which must have soured Maury's relations with that institution. (Maury reciprocated by hiring James Ferguson a year after he had been fired by Bache at the Coast Survey.) Moreover, Walker soon made arrangements for the full publication of his Neptune researches in Joseph Henry's *Smithsonian Contributions*. This, and Henry's relay of Walker's abstract of these researches to the European Journal *Astronomische Nachrichten*, caused a major rift between Maury and Henry at the Smithsonian. Maury later used this episode to remind the Secretary of the Navy that the work of the Observatory's staff was the property of the Observatory, not their own: "The grounds upon which this rule is based, are obvious; assistants are employed by the Observatory to perform certain labors for it; the fruits of these labors belong to it. They are its property, its capital – which can only be made public through its regularly constituted organs."[81] Although the loss of Walker was in many ways a blow to the Observatory, on this issue Maury upheld the standard view of government agencies.

### Wind and current charts

Interesting as were the astronomical activities of these early Maury years, the hydrographic duties were a major and entirely separate dimension to the new institution. As we have seen, Maury gave close attention to them, beginning with his duties at the temporary Depot in 1842. From these early duties, the hydrographic work grew into a major activity, and resulted in the wind and current charts with which the name Matthew Fontaine Maury is most closely associated. As for other scientists of the time, the driving force behind Maury's work was in part nationalistic: "The second nation in the world in maritime importance, it is remarkable how little we have contributed

---

[80] Maury to Loomis, April 20, 1847, cited in Williams (ref. 4), p. 168.

[81] Maury, *WO*, volume 2, Appendix, p. 2. For the details of the publication controversy, see Williams (ref. 4), pp. 168–172. Patricia Jahns, *Matthew Fontaine Maury and Joseph Henry: Scientists of the Civil War* (Hastings House: New York, 1961) examines in a popular vein the stormy relationship of these two men. Albert Moyer, *Joseph Henry: The Rise of an American Scientist* (Smithsonian Institution Press: Washington, 1997) gives a more nuanced and detailed view. James R. Fleming describes the Maury–Henry relationship in the context of meteorology, and comments on the Walker episode, in *Meteorology in America, 1800–1870* (Baltimore and London, 1990), pp. 106–110. *The Papers of Joseph Henry*, edited by Nathan Reingold and Marc Rothenberg (Smithsonian Institution Press: Washington, 1972– ), is an invaluable resource on American science during this period. Thus far eight volumes have been published, covering the years 1797–1853.

to the general stock of that kind of nautical information without which our vessels could not cross the seas – without which our commerce could not exist," Maury wrote in 1845 in requesting funds for his work. "Always borrowing heretofore, it is time we should become lenders at least of a proportional part of this information." The United States, Maury emphasized, was indebted to the English and French Admiralties not only for charts of the oceans, but also for charts of its own waters: "As yet, an American man-of-war cannot enter the capes of Virginia, or approach this city, the capital of the Union, without applying to the hydrographical office of England for the chart on which to shape her course . . . Charts are now being compiled from materials in this office, which, it is believed, will add something to the general stock of nautical information."[82]

Writing to John Quincy Adams in 1847, Maury described the purpose of the charts as "to generalize the experience of navigators in such a manner that each may have before him, at a glance, the experience of all. The track of each showing the time of the year, the prevailing winds and currents encountered, with all other information obtained, is projected on the charts."[83] In that year, after five years of studying the old ships' log books, the first of eight sheets of the North Atlantic were published, with seven more in the hands of the engraver by the end of the year. By 1849 all eight sheets were in their third edition. Similar charts for the South Atlantic and the Pacific and Indian Oceans were also begun during this period and issued, mostly in the early 1850s. These charts, known as series A "Track Charts," inaugurated a whole series of related charts (Figure 2.12). During the same period another series of charts began to appear, including the more specialized trade-wind charts (series B, beginning 1851), pilot charts (series C, beginning 1849), thermal charts (series D, beginning 1850), storm and rain charts (series E, beginning 1853), and whale charts (series F, beginning 1852). Maury announced the latter in a circular in the spring of 1851; six months later Herman Melville published his masterpiece *The Whale* (later titled *Moby Dick*), and referred to Maury's forthcoming whale charts.[84] It was the wind and current charts, however, that were useful to the largest number of sailors. Once it had been shown that they saved sailing time, they were in great demand – and for their distribution Maury exacted the cooperation of ships at sea.

---

[82] Maury to W. M. Crane, Chief of Bureau of Ordnance and Hydrography, in RSN, December 1, 1845, p. 689.    [83] Williams (ref. 4), p. 178. Maury to Adams, November 14, 1847.

[84] Melville's early literary allusion to Maury and the "National Observatory" is found in chapter 44 of *Moby Dick* (Library of America: New York, 1983), p. 1,004. Referring to Ahab's use of charts in his quest for the whale, Melville writes "Since the above was written, the statement is happily borne out by an official circular, issued by Lieutenant Maury, of the National Observatory, Washington, April 16th, 1851. By that circular, it appears that precisely such a chart is in course of completion; and portions of it are presented in the circular." For a complete bibliography showing the timing of the publication of the various series of Maury's charts, see William D. Horigan, *List of Publications Issued by the United States Naval Observatory, 1845–1908* (United States Naval Observatory: Washington, 1908), also in Appendix 3 to volume 6 (new series) of USNO Publications. See also Williams's bibliography of Maury's published works in ref. 4, pp. 693–710.

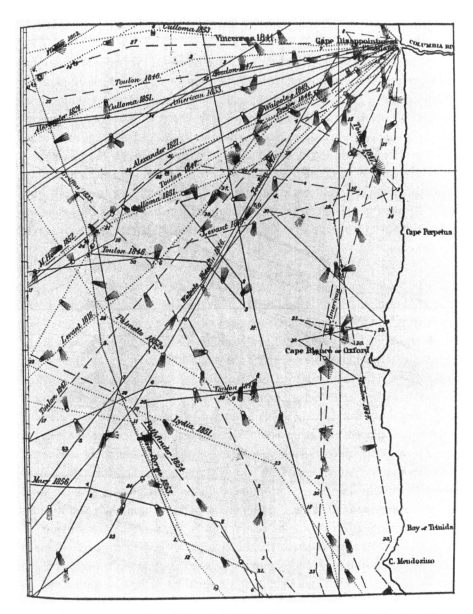

Figure 2.12. A section of Maury's Wind and Current Chart of the North Pacific, No. 5, series A.

By 1850 Maury had launched a variety of programs, made impressive headway both in astronomy and in hydrography, and received international accolades. Yet trouble was on the horizon. As the second half of the century began, the first volume of observations had not yet been followed by a second, and Maury's last ten years assumed a character quite different from the heady first five.

## 2.3 Astronomy versus hydrography: Science, politics, and the Navy in the last decade of the Maury era

The dual astronomical and hydrographic charge of the Depot of Charts and Instruments was officially recognized in 1854 when the name of the institution was changed to the U. S. Naval Observatory and Hydrographical Office. It has often been asserted that astronomy at the Naval Observatory gave way to hydrography, especially in Maury's later years. In order to examine this thesis, we need to look in more detail at how the astronomical and hydrographic activities developed over the full course of Maury's tenure, in particular the little-known but significant activities of James Ferguson, Joseph Hubbard, and Mordecai Yarnall, and Maury's own work in hydrography. We may then examine how Maury balanced the twin duties of astronomy and hydrography in light of national events, Navy policy, and the allocation of resources to the institution.

*The heavens above: Comets, the great asteroid hunt, and meridian work*

Following the departure of Coffin in 1853 three men were the mainstay of the astronomical work at the Naval Observatory during the remainder of the 1850s: Joseph Hubbard, James Ferguson, and Mordecai Yarnall. All were present at the Observatory throughout the 1850s, and in fact would outlast Maury himself. Ferguson would almost single-handedly man the Equatorial Telescope, Yarnall would carry on the meridian work, and Hubbard excelled not only in observational astronomy but also in theoretical work, and in addition played an important role in the increasing professionalization of astronomy and science in America. When Simon Newcomb arrived at the Observatory in 1861 he first called on Hubbard, "the leading astronomer of the observatory," while Gould later wrote that the observations of Ferguson and Yarnall during this time "saved the honor of a national institution" when its head turned to hydrography.[85] It is therefore important that we examine their contributions to astronomy with an eye toward the theme of astronomy's role in the institution.

Of the three astronomers, Hubbard (Figure 2.6) had been the first to come to the Naval Observatory, only a few months after the inception of the institution. We have already seen that he made significant contributions to the Observatory's observational program in its early years, both in the meridian and in the equatorial observations, as is evident in the *Washington Observations*. The scope of his work changed considerably over the next ten years, as he became involved in the founding of the *Astronomical Journal* in 1849, and was forced by ill-health to concentrate on theory during most of the 1850s. Hubbard was a charter member of the National Academy of Sciences in 1863.

It was perhaps the increasing delay in the publication of those observations that led Hubbard in 1849 to play a significant role in the founding of the *Astronomical*

---

[85] Gould, Eulogy on Hubbard (ref. 42), 86; Newcomb, *Reminiscences* (ref. 71), p. 98.

Journal. As part of his project of publishing the "zodiacs," or paths in the sky, of the known minor planets, in November of 1848 Hubbard had turned over a list of these objects to the Smithsonian for publication in its new *Smithsonian Contributions to Knowledge*. Publication here was also delayed, and the summer of 1849 found Gould and Hubbard laying plans for a new journal. Discussions were held with Bache, Peirce, Henry, Coffin, Walker, and Chauvenet, and, in August, 1849, Hubbard laid before the second meeting of the AAAS the plan for an astronomical journal. In addition to Gould and Hubbard, a committee consisting of Coffin, Walker, Henry, Bache, Maury, Davis, and Peirce was formed. The first volume appeared the following November, and the second article (after one by Harvard's Benjamin Peirce) was Hubbard's "Zodiac of Hygea." Not only would Hubbard contribute more than 210 columns of his own research over the next 13 years, twice during Gould's absence from the country he assumed the role of editor. Gould praised Hubbard highly for his role in the origin and growth of the first astronomical journal in the United States, a service rendered free (from Gould's point of view) as a government employee.[86]

The work for which Hubbard was to become best known was his analysis of the orbits of comets, in particular the great comet of 1843. On February 28 in that year the comet was seen in broad daylight in many parts of the world, and "through the early evenings of March, it trailed like a gorgeous banner of flame across the Western sky, the first visitant of its kind within the memories of a full grown man." The comet was of peculiar interest not only for its brilliant tail, which extended nearly parallel to the horizon for some 40 degrees, but also because it came so close to the Sun as to almost graze it, about 90,000 miles or a fifth of the Sun's radius.[87] As late as early April the comet was still being observed in the United States by Walker and Kendall at Philadelphia and Loomis at Hudson, Ohio. Hubbard, within five months of graduation from Yale, observed the magnificent object from New Haven; it fired his imagination and he determined that a study of its orbit would be his special project. By December, 1849 he had published the first part of his exhaustive discussion of the orbit of the great comet of 1843. That analysis would stretch in eight parts of the *Astronomical Journal* through the middle of 1852. A decade later Gould would say that, because of Hubbard's work, "the orbit of no comet of long period has been more thoroughly and exhaustively treated than this."[88] Indeed, Hubbard made use of all known observations of the comet. Both Schumacher at Altona Observatory and Peirce believed that it had a period of about 175 years, and was perhaps identical to the comet of 1668, but Hubbard determined that the most probable orbit was more than 500 years.

[86]  Gould, Eulogy on Hubbard, (ref. 42), 89 and "On the Establishment of an Astronomical Journal in the United States," *Proceedings of the American Association for the Advancement of Science, Second Meeting, Held at Cambridge, August, 1849* (Boston, 1850), pp. 378–381.

[87]  For descriptions of the comet of 1843 see Loomis (ref. 41), pp. 121–131, and Gould, "Eulogy on Hubbard" (ref. 42), 90–91. Halley's comet had appeared in 1835, but was not so brilliant.

[88]  Gould, "Eulogy on Hubbard" (ref. 42), 92. Hubbard's treatment appeared in "On the Orbit of the Great Comet of 1843," AJ, **1** (1849–51), 10–13, 24–29, 57–60, and 153–154; AJ, **2** (1851–52), 46–48, 57, and 153–156.

Gould, with only slight exaggeration, called Hubbard the Encke of the United States, implying that he might have become as well known as the Berlin astronomer had he only lived longer. Hubbard later gave the same exhaustive treatment to Biela's comet, but did not cite Maury's work on its splitting.[89] Late in life Hubbard resumed his work with the prime-vertical instrument on the parallax of Vega, important work that was finished only by Harkness and Newcomb.

In April, 1863, Hubbard was present at the organization meeting in New York of the National Academy of Sciences, one of 20 incorporators (including James Gilliss) from the scientific agencies of the Federal Government.[90] Four months later he was dead. "None was more hopeful, none more buoyant, none more impressed with the magnitude and import of our new duties, than he. It was the realization of the dream of his maturer years," Gould wrote in his eulogy of the first member of the National Academy to die. At the height of the Civil War, by which time Maury had joined the Southern cause, Gould praised Hubbard, and also lost no opportunity to denigrate the despised Maury. By his accounting, the surprise was that Hubbard and his colleagues accomplished as much as they did under the hand of Maury.

⋆   ⋆   ⋆

To be sure, Hubbard's colleagues were also busy, none more so than James Ferguson, (1797–1867; Figure 2.13). Ferguson, the replacement for Sears Cook Walker, had arrived at the Naval Observatory in 1848, the beginning of a long career in charge of the Equatorial Telescope. Ferguson was a native of Scotland who migrated to the United States in about 1800; his educational background is unknown, but in the years 1817–19 he was already at work as a civil engineer on the Erie Canal. He then participated in various survey projects before becoming the civil engineer for Pennsylvania (1827–32) and the "first assistant" in the Coast Survey (1833–47), second only to F. R. Hassler.[91] Ferguson was clearly devoted to his work, and wrote several eloquent arti-

---

[89] Hubbard, "On the Orbit of Biela's Comet in 1845–46," AJ, **3** (1854), 60–68, 73–77, and 89–94; "Results of Additional Investigations respecting the two Nuclei of Biela's Comet," AJ, **4** (July, 1854), 1–5; "On Biela's Comet," AJ, **6** (1860), 110–118, 121–126, 129–134, 137–142, and 157–160. Hubbard also published an extended discussion of the comet of 1825, "On the Orbit of the Fourth Comet of 1825," AJ, **6** (1859), 17–22, 26–31, and 33–37. In total 23 papers are listed in the Royal Society of London's *Catalogue of Scientific Papers (1800–1863)*, volume 3, pp. 453–454.

[90] Thirty-two of the Academy's 50 original incorporators were present at this meeting. On the Academy's organizational meeting and Hubbard's presence see Rexmond C. Cochrane, *The National Academy of Sciences: The First Hundred Years, 1863–1963* (National Academy of Sciences: Washington, 1978), p. 65.

[91] On Ferguson see BDAS, 91–92; Appleton's CAB, **2**, 433; B. F. Sands, "The Late Mr. James Ferguson, of the Naval Observatory, Washington," AN, **71** (1868), 101–102; and Nathan Reingold, ed., *The Papers of Joseph Henry* (Smithsonian Institution Press: Washington, 1975), volume 2, pp. 15–16. Aside from scattered correspondence relating to Ferguson appearing in the latter, little appears to remain of Ferguson's papers. The Naval Observatory library contains a 40-page manuscript "Astronomical Fragments left by Mr. James Ferguson, Assistant Astronomer at the U. S. Naval Observatory at his Death in 1867." This consists mainly of notes taken from other sources on aberration of light, parallax, refraction (from Brünnow, Delambre, Brühns, and Littrow) and mechanics (Newton), together with calculations on the elements of Euphrosyne and other objects.

Figure 2.13. James Ferguson, first American discoverer of asteroids, known for his work on the 9.6-inch equatorial telescope.

cles in defense of the goals and methods of the embattled Coast Survey, aiming in one 1842 article "to secure a clear and impartial investigation; to correct misrepresentations; to counteract the designs of scheming speculators; to defeat, if necessary, the instigations of ill-will; and to uphold the scientific reputation of the country, which is in some measure dependent upon the successful prosecution of this great national undertaking."[92] He was a likely successor to Hassler to head the Survey, but instead Bache was appointed in 1843. Four years later Ferguson was forced out amid charges of incorrect survey work, but also in part due to rivalry with Bache.[93] A year later, in a sort of trade for Walker, Maury hired Ferguson.

By 1850 Ferguson was still in a somewhat precarious positon as a result of attempts to clear his name at the Coast Survey, with Bache threatening to seek his dismissal from the Naval Observatory unless Ferguson desisted in his attempts to clear his name through the Secretary of the Treasury and Congress. However, Ferguson soon demonstrated the "untiring industry and great intelligence" for which he was

[92] Ferguson, review of "Letter from the Secretary of the Treasury, transmitting a Report of F. N. [sic] Hassler, Superintendent of the Coast Survey, showing the Progress made therein up to the Present Time," *North American Review* (April, 1842), 446–447: 447. This review also refers to a previous review of Hassler's paper and documents on the Coast Survey by the same author in *North American Review* (January, 1836), 75–94. Though neither review is signed, both Eliot, BDAS, and Florian Cajori, *The Chequered Career of Ferdinand Rudolph Hassler* (reprint edition, Arno Press: New York, 1980), assign Ferguson as the author, an attribution made likely by the intimate details evident in these papers.

[93] By instructions from the Treasury Department dated February 19, 1847, Ferguson's work on the survey of the Chesapeake and Delaware Bay was referred to a committee composed of Benjamin Peirce, Andrew Talcott, and Charles H. Davis, resulting in Ferguson's dismissal in March of that year. The details of his dismissal are given in Ferguson's *Memorial of James Ferguson, Late First Assistant in the Survey of the Coast: Presented in the Senate by Hon. Thomas H. Benton, July 19, 1850,* where Ferguson presents his case that he was wrongly treated.

praised after his death, as he began the careful observations that were to be the hall-mark of his 20-year career at the Observatory. He was assigned to the Equatorial Telescope with the title of "assistant astronomer" (a title he would hold through the rest of his career), and began as early as March 1848 observing occultations of stars by the Moon. Soon he became part of the great asteroid hunt that had driven so many astronomers since Piazzi's discovery in 1800 that such "small planets" existed, orbiting in a zone between Mars and Jupiter. By 1848 nine of these objects (now called minor planets or asteroids) had been discovered, and, in addition to the search for new ones, they required tracking to determine their orbits.

Though Ferguson's observations of comets were valuable, it was in the nineteenth-century asteroid hunt that he excelled and forged his reputation. Most of his papers on this subject were short and routine, but they occasionally sparked significant controversy. As one example, early in his tenure at the Naval Observatory, in the midst of his acrimonious battle with Bache and four years after Walker's famous work on the pre-discovery observations of Neptune, Ferguson precipitated a controversy over a "missing star."[94] This episode, which was resolved only a quarter century later, is an interesting contrast to Walker's successful location of pre-discovery observations of Neptune, and illustrates the potentially far-reaching effects that the routine work of positional astronomy could have. As part of his work to track newly discovered asteroids too faint to be seen with the meridian instruments, from May 18 to November 24, 1850 Ferguson carefully measured the position of the asteroid Hygeia that had been discovered by astronomers in Europe the previous year.

This was done by reference to certain "comparison stars" whose positions were in turn determined by reference to well-known catalog stars. In this case, the 12th-magnitude Hygeia was referred to a ninth-magnitude comparison star, in turn referred to perhaps a sixth-magnitude catalog star. The problem was that, upon publication in 1851,[95] the English asteroid hunter John Russel Hind could find no evidence of a ninth-magnitude comparison star where Ferguson claimed it was. Either Ferguson had made a mistake, or the "star" had wandered away, in which case it was no star at all, but a possible new asteroid. Hind, inclined to believe the latter, wrote to Bond at Harvard, who in turn informed Maury, who put Ferguson to work to find his comparison star. Unable to find it again, Maury concurred with Hind and reported to the Secretary of the Navy that the missing comparison star was a suspected newly discovered asteroid, the credit for which must go to Hind.[96]

However, Ferguson failed to locate the errant star on the presumption that it was a moving asteroid, and meanwhile Hind, examining an apparent difference in

[94] This episode has been described in detail in Richard Baum, *The Planets: Some Myths and Realities* (Wiley: New York, 1973), pp. 147–162.
[95] "Observations of Hygea made with the Filar-Micrometer of the Washington Equatoreal," AJ, 1 (January 18, 1851), 164–166.
[96] Maury, "Letter of Lieutenant Maury to Hon. William A. Graham, Secretary of the Navy," September 3, 1851, AJ, 2 (October 22, 1851), 53.

Ferguson's published positions of the comparison star, put forth a more startling hypothesis: the motion indicated could not be an asteroid, but must be a real planet.[97] Ferguson then began a search based on this supposition, extending from late August to December, 1851, but again without success.[98]

Only 26 years later did a plausible explanation emerge. Planet-hunters were still enthusiastic in 1878, and an article in *Nature* recalled the Ferguson–Hind episode of the early 1850s.[99] The interest of C. H. F. Peters, Director of the Litchfield Observatory in Clinton, New York, was piqued, and his subsequent examination of the observational record indicated that Ferguson had written down the wrong micrometer-wire number during his observations. When this error was corrected, the ninth-magnitude object turned out to be an ordinary catalog star, an explanation that the Superintendent of the Naval Observatory acknowledged.[100] It was the case of Walker and Neptune in reverse – instead of a suspected star turning out to be a planet, a suspected planet turned out to be a star! Thus, in the judgment of history, Walker is better remembered (although only slightly) than Ferguson.

Despite this somewhat embarrassing episode, Ferguson persisted in his asteroid researches and, using the 9.6-inch refractor, went on to make the first asteroid discovery from the United States. On the night of September 1, 1854 Ferguson had been observing the minor planet Egeria when he found nearby an unexpected object about equal in brightness to it and proved through its motion to be an asteroid, the 31st found since Piazzi's first discovery. Ferguson named it Euphrosyne, one of the three Graces in Greek mythology.[101] The discovery was announced without fanfare in Gould's *Astronomical Journal* and Schumacher's *Astronomische Nachrichten*, and immediately gained Ferguson and the Observatory some badly needed international acclaim. Gould relayed word to Leverrier in Paris, and, one month after its discovery, the latter made the announcement to the French Academy of Sciences. A set of ephemerides computed by Keith followed in the next month. Ferguson's asteroid work received attention in the *Comptes Rendu* several times in succeeding years, a rare distinction for a staff member of the Naval Observatory, with the exception of Maury for his hydrographic work and Walker's work on Neptune. In awarding Ferguson and five other asteroid discoverers the Lalande prize of the French Academy of Sciences the following year, the Academy noted the accelerating pace of asteroid discoveries. Six discoveries had been made in 1854 in Germany, England, France, and the United States,

---

[97] "Letter from Mr. Hind to the Editor," November 12, 1851, AJ, **2** (January 1, 1852), 78.

[98] "Letter from Lieutenant Maury, Superintendent of the Washington Observatory," January 5, 1852, AJ, **2** (March 13, 1852), 91.     [99] *Nature*, **18** (1878), 696.

[100] C. H. F. Peters, "Investigation of the Evidence of a Supposed Trans-Neptunian Planet in the Washington Observations of 1850," December 2, 1878, AN, **94** (1879), columns 115–116; "Letter from Admiral John Rodgers, Superintendent of the Naval Observatory at Washington," December 6, 1878, AN, **94** (1879), columns 113–114.

[101] Ferguson followed the lead of William Bishop, who in 1852 had named the 23rd asteroid Thalia, another of the mythological Graces, upon its discovery by Hind at Bishop's private observatory. In 1857 the third Grace (Aglaja) would give its name to the 47th asteroid.

bringing the total number to 33.[102] Ferguson discovered two more asteroids (50 Virginia on October 4, 1857, and 60 Echo on September 14, 1860) and again shared the Lalande prize for his discovery.[103] By the time of this discovery, the number of asteroids had risen to 66.

Near the end of his busy life of observing, Ferguson once again became embroiled in a political dispute when he campaigned to become civilian director of the Observatory after the departure of Superintendent C. H. Davis for active duty at sea during the Civil War.[104] Despite this failed attempt to gain civilian control of the Observatory, Superintendent B. F. Sands undoubtedly was sincere when he wrote upon Ferguson's death in 1868 that "He was a man of admirable personal qualities, of extensive reading much diversified, and was versed in the languages of Western Europe. His ready and charming conversational power made his society most grateful to all who knew him." Although Ferguson received the astronomical prize medals from the French Academy of Sciences in 1854 and 1860 for his asteroid discoveries, his contribution to astronomy was not particularly original. His was the kind of routine but solidly precise work for which the Naval Observatory was to become well known. By the end of his career at the Naval Observatory in 1867, Ferguson had produced some 90 papers, more than any other American astronomer before the Civil War. Of these about three quarters dealt with observations of asteroids, a few with their elements and ephemerides, and most of the remainder were observations of comets, of which Ferguson discovered several new ones.[105] His scientific publications surpassed those of any other antebellum astronomer.

*    *    *

[102] Gould made the announcement in AJ, "Thirty-First Asteroid," **4** (September 19, 1854), 23, which contained observations for September 2, 6, and 10, and elements computed by Keith. In Germany a note dated September 4, written by acting Superintendent S. Phillips Lee, was printed in AN, **39** (1855), columns 125–128. The announcement to the French Academy is in *Compte Rendus de Séances de l'Académie des Sciences* (henceforth CR), **39**, 643–644; the ephemerides in ibid., **39**, 1021, and the prize award in ibid., **40**, 37–38. The other five asteroid discoveries of 1854 were by Luther in Germany, Marth and Hind in England, and Goldschmidt and Chacornac in Paris. See Gehrels, *Asteroids* (1979), Part I for history, Part VII for a list of all minor planet discoveries up to number 2,125. Also Lutz Schmadel, *Dictionary of Minor Planet Names* (Springer: New York, 1997).

[103] On the announcement of the discovery and observations in France of 50 Virginia see CR, **45**, 693, 810, and 1,102. Ferguson originally named his last asteroid discovery Titania, not realizing that the name had been used by Sir William Herschel more than 60 years earlier for a moon of Uranus. The discovery of what was subsequently called Echo is announced in AJ, "Fifty-Ninth Asteroid," **6** (1861), 160 (subsequently numbered 60 since another had been discovered in Europe a few days prior), and in France in CR, **51**, 547. On the Lalande prize for the latter, see CR, **52**, 557, and 1,146, where Ferguson thanks the Academy and gives credit to the French observer Chacornac for the role played in the discovery by his excellent charts, executed under the French government. In the United States G. Searle discovered 55 Pandora on September 10, 1858, and 66 Maja was found by H. P. Tuttle in Cambridge, Massachusetts, April 9, 1861. Thereafter the pace quickened with discoveries by C. H. F. Peters, H. P. Tuttle, J. C. Watson, and others in the United States in the 1860s and 1870s. The next asteroids discovered at the Naval Observatory would not be until November 16, 1917 (886 Washingtonia) and November 21, 1921 (Anacostia), both of which were discovered photographically by George Peters.

[104] Marc Rothenberg, "Observers and Theoreticians . . .", (ref. 43), 35; Joseph Henry Papers, volume 2, pp. 15–16; Newcomb, *Reminiscences* (ref. 71), pp. 111–112.

[105] For a listing of these see RSC, **2**, 589–592.

Figure 2.14. Mordecai Yarnall, Professor of Mathematics, U. S. N., quintessential meridian-instrument observer and early compiler of star catalogs.

Despite the successes of Hubbard and Ferguson, the astronomical community had come to expect more from a national institution, in particular in the form of star catalogs made famous by the great national observatories of Europe, but whose issue by the Naval Observatory was increasingly delayed. In this regard Mordecai Yarnall (1816–79), who was brought to the Naval Observatory in 1852 to replace Coffin, is a key figure. Like Ferguson, Yarnall (Figure 2.14) also had civil-engineering training, in his case at Bacon College, Kentucky. Unlike Ferguson, he had been appointed a Professor of Mathematics in 1839. Although he had worked with O. M. Mitchel at the Cincinnati Observatory during the 1840s, little is known of the details of his life and work before his arrival in Washington.[106] At the Naval Observatory Yarnall became the quintessential observer with the meridian instruments, including the 5.3-inch transit instrument and the mural circle. More than that, Yarnall became known for his work compiling star catalogs, requiring very different skills from observing. It is notable, however, that, although the observations were undertaken during the early Maury years, the catalogs were not compiled and printed until more than a decade after Maury's departure. This situation leads us directly to the question of the balance of astronomy versus hydrography during Maury's tenure.

The thesis of the decline of astronomical activity introduced at the start of this section may be quantitatively tested by examining the fate of meridian astronomy during the Maury years in terms of observations published. As shown in Table 2.3, it is clear that, by 1850, a dramatic drop in number of observations had taken place. Not only did the number of core observations decrease (with two years' worth of observations not equaling those of one year before), but also the zone observations had ceased

---

[106] On Yarnall see BDAS, and Rothenberg (ref. 38), pp. 182–184.

Table 2.3. *Astronomical observations made during the Maury years*

| Years observed | Instruments[a] | # Stars | Year published | Observers | Remarks |
|---|---|---|---|---|---|
| *Meridian observations for position* | | | | | |
| 1845 | Tr, Mu, PV | 98 | 1846[1] | | |
| 1846 | Tr, Mu, Me | 583 | 1851[2] | | |
| 1847 | Tr, Mu, Me, PV | 642 | 1853[3] | | |
| 1848 | Tr, Mu, Me, PV | 577 | 1856[4] | | |
| 1849–50 | Tr, Mu, Me, PV | 390 | 1859[5] | | |
| 1851–52 | Tr, Mu, Me | 435 | 1862 (67?)[6] | | Reduced by Gould |
| 1853–60 | Tr, Mu | | 1872[7] | | Yarnall |
| 1845–71 | Tr, Mu, Me, PV | 10,658 | 1873[8] | | Yarnall |
| *Zone observations for position* | | | | | |
| 1846 | Meridian Circle | 4,047 | 1860[9] | Hubbard and Maynard | Reduced by Ferguson |
| 1846–49 | Mural Circle | 14,804 | 1872[10] | Coffin, Page, and Steedman | Reduced by Gould |
| 1846–49 | Transit Instrument | 12,033 | 1872[11] | Keith, Beecher Hubbard, Almy, and Parker | Reduced by Gould |
| 1847–49 | Meridian Circle | 7,390 | 1873[12] | Major, Hubbard, and Muse | Reduced by by Gould |

Notes:

[a] Tr, West Transit Instrument; PV, Prime Vertical; Mu, Mural Circle; and Me, Meridian Circle (4.5-inch Ertel & Son).

[1] *Astronomical Observations Made Under the Direction of M. F. Maury, Lieut. U. S. Navy, During the Year 1845, at the U. S. Naval Observatory, Washington* (Washington, 1846). A second title page reads *Astronomical Observations Made During the Year 1845 at the National Observatory, Washington* (Washington, 1846).

[2] *Astronomical Observations Made Under the Direction of M. F. Maury, Lieut. U. S. Navy, During the Year 1846, at the National Observatory, Washington*, volume 2 (Washington, 1851).

[3] *Astronomical Observations Made During the Year 1847 at the National Observatory, Washington, under the Direction of M. F. Maury, LL.D., Lieut. United States Navy, Superintendent*, volume 3 (Washington, 1853).

[4] *Astronomical Observations Made Under the Direction of M. F. Maury, Lieut. U. S. Navy at the U. S. N. Observatory, Washington*, volume 4 (Washington, 1856).

[5] *Astronomical Observations made at the U. S. Naval Observatory during the Years 1849–1850* (Washington, 1859).

[6] *Astronomical Observations made at the U. S. Naval Observatory during the Years 1851 and 1852* (Washington, 1867). Though the title page is dated 1867, the Preface and Introduction are dated 1862.

[7] *Results of Observations made at the United States Naval Observatory with the Transit Instrument and Mural Circle in the Years 1853 to 1860 Inclusive by Professor M. Yarnall, USN, Professor James Major, USN, Professor T. J. Robinson, USN* (Washington, 1872). Appendix II to *Washington Observations for 1871*.

[8] *Catalogue of Stars Observed at the United States Naval Observatory During the Years 1845 to 1871, and Prepared for Publication by Professor M. Yarnall, USN* (Washington, 1873).

[9] *Zones of Stars Observed at the National Observatory, Washington, Volume 1, part 1, containing the Zones Observed with the Meridian Circle in 1846 . . . by Commander M. F. Maury, Superintendent* (Washington, 1860). These were reduced by James Ferguson.

[10] *Zones of Stars Observed at the United States Naval Observatory with the Mural Circle in the Years 1846, 1847, 1848, and 1849 by Professor J. H. C. Coffin, Lt. T. J. Page, and Lt. Charles Steedman, Appendix II of Washington Observations for 1869* (Washington, 1872).

[11] *Zones of Stars Observed at the United States Naval Observatory with the Meridian Transit Instrument in the Years 1846, 1847, 1848, and 1849 by Professor Reuel Keith, Professor Mark H. Beecher, Professor Joseph Hubbard, Lt. John J. Almy and Lt. William A. Parker, Appendix IV to Washington Observations for 1870* (Washington, 1872).

[12] *Zones of Stars Observed at the United States Naval Observatory with the Meridian Circle in the Years 1847, 1848 and 1849 by Professor James Major, Lt. Lafayette Maynard and Lt. William B. Muse, Appendix I to Washington Observations for 1871* (Washington, 1873).

completely by 1851 (and those beyond July, 1849 were never published). Furthermore, Table 2.3 also clearly indicates the increasing delay in analysis and publication of the results. Whereas the 1845 results had been published the following year, those for 1846 took five years, and those for 1849–50 did not appear until ten years later, near the end of Maury's tenure. The delay in publication of the zone observations was even longer: Ferguson published those of the meridian circle for one year in 1860, but the remainder waited almost a quarter-century after their observation for publication, in the years 1872–73. What is open to interpretation is the reason for the precipitous drop in activity around 1850 and publication delays. In order to address this question, we need to take a final look at Maury's work in hydrography.

### The Earth below: Hydrography, meteorology and navigation

During the 1850s Maury's most conspicuous achievements were in the fields of hydrography and meteorology. Not only did the work of refining the wind and current charts continue, but also, beginning in 1850, their use was enhanced by the *Explanations and Sailing Directions to Accompany the Wind and Current Charts*, which went through many editions and had expanded from 18 to 772 pages by 1853. Maury organized the first of several international conferences on cooperation in meteorology in Brussels in 1853, and largely wrote the final report himself. In 1855 he published *The Physical Geography of the Sea*, a book that enjoyed great public interest and remained in print for 20 years. His knowledge of the oceans gained through U. S. Navy deep-sea soundings under his direction enabled Maury to lend substantial aid to Cyrus Field in deciding where to lay the first transatlantic cable, a feat finally achieved in 1858. Through it all Maury and his staff maintained their primary duty of supplying navigational charts and instruments to ships, including the chronometers rated at the Depot. Fame, however, came with a price. After 30 years in the Navy Maury had a considerable number of enemies who begrudged his attempts at naval reform and his lengthy shore duty; it was only after considerable controversy that he was promoted to Commander in 1858, effective retroactively to 1853.

Both the Brussels conference and the Atlantic cable episode illuminate not only Maury's hydrographic and meteorological achievements, but also the politics of American science of the time, in particular the rivalries that had developed among fledgling American scientific institutions. When in late 1851 Britain suggested the possibility of American cooperation in a system of international meteorological observations on land, Maury proposed through his superiors that an international conference be held to discuss not only land, but also sea meteorology. Henry and Bache, however, opposed the plan, believing that the British and American Associations for the Advancement of Science should handle the matter, and perhaps jealously guarding the Smithsonian's system of land meteorological observation which had already begun. In the end the subject was limited to marine meteorology, and, on August 23, 1853, the first International Maritime Meteorological Conference convened in Belgium, with Maury representing the United States. The result was an international system of meteorological observations.[107]

Hardly were the triumphs of the meteorological conference behind him, when Maury was plunged into another project marked by controversy. It is some measure of Maury's national prominence as a hydrographic pioneer that Cyrus Field had approached Maury in early 1854 regarding the feasibility of laying a telegraphic cable on the floor of the Atlantic, just as he contacted Morse to answer the question of the practicality of transmitting messages over this distance if such a cable could be laid. Thanks to improved soundings carried out the previous year by a deep-sea sounding apparatus designed by passed midshipman John Mercer Brooke at the Observatory, Maury could report the existence of a rather shallow plateau in the 1,600 miles of ocean separating Newfoundland and Ireland. This, he urged, was the ideal location for a submarine cable. Field and his financial backers founded the New York, Newfoundland and London Telegraph Company in April of 1854, and, after several attempts, successfully laid the cable and sent messages in August, 1858. In New York Field, Morse, Maury and other participants were toasted at "the greatest victory celebration that city had ever held," and for such a popularly acclaimed event Maury was justifiably better remembered than for his astronomical work.[108]

The significance of the Atlantic-cable story for the Naval Observatory and Hydrographic Office lies not only in its role in determining where the cable should be placed, but also in the subsequent use made of the cable for the first transcontinental longitude determination. For the first time the longitude difference between the Greenwich meridian and the Washington meridian could be directly compared by telegraph line, using methods matured by Walker a few years earlier. That the Naval Observatory would receive the credit for this work rather than the Coast Survey was assured when Maury extracted this promise of first access for a determination from Field instead of a financial remuneration for his role. Only a month after the cable had been laid, Maury wrote the British Admiralty and Astronomer Royal requesting their cooperation in this longitude expedition. Meanwhile Bache and Henry tried to secure the privilege of this determination for the Coast Survey. It is not clear who would have won, but, when the cable went silent in October, the controversy too became mute.[109] Nonetheless, the controversy points out the rivalry between the Observatory and the Coast Survey – between Maury and the so-called "Lazzaroni" including Bache, Henry, and Gould – that marked the decade of the 1850s and certainly affected both institutions.

Maury's accomplishments were in some ways indisputable – that his wind and

---

[107] For the context and effect of this meeting, see James Fleming (ref. 81), pp. 106–110, and Williams (ref. 4), pp. 205–224. For the early correspondence relating to this conference see Maury, *On the Establishment of a Universal System of Meteorological Observations by Sea and Land* (Washington, 1851). On the conference itself see Maury's *Sailing Directions*, sixth edition (1854), and Adolphe Quetelet, *Notice sur Le Capitaine M. F. Maury, Associé de l'Académie Royale de Belgique* (Brussels, 1874).

[108] Williams (ref. 4), pp. 250–252.     [109] Ibid., pp. 233, and 253–256.

current charts shortened oceanic travel times no one could deny. What critics could question was how much of Maury's work was true science, and it is here that we begin to transcend narrow rivalries and go to the fabric of American science of the time. There is no doubt that Maury's was a Humboldtian, empirical science, as his *Physical Geography of the Seas* shows.[110] As Leighly says in his introduction to a modern edition of that volume, we may read Maury's *Physical Geography* today, not for his insights into the nature of the ocean, but as "the highly personal testament of an energetic and self-assertive man unacquainted with the rigorous methods practiced in the academies, but possessing first-hand knowledge of ships and the sea, boundless self-confidence, and a pen well exercised in persuasive writing." Maury's empirical bent is no less true of his astronomy than it is of his hydrography; but there is equally no doubt that data gathering is the bedrock of science, if not its most glamorous activity, and it is an activity not to be ridiculed.

All of this – the personal triumphs and failures, the scientific career, an institutional era – came to an end on April 20, 1861, the end of an era for the nation also. On that date, one week after Ft Sumter surrendered to the rebel forces initiating the Civil War, Maury left the Observatory for the last time, having made the fateful decision to join his native Virginia in the Southern cause. From a Depot of Charts and Instruments he had built a national institution renowned for its hydrographic and astronomical work, even if some charged that the latter had been neglected in favor of the former. It is clearly true that much was left unfinished, especially in the area of astronomy, and, in order to understand why we must take a final look back on these eventful years.

### An assessment of the Maury era

Maury had hardly headed South before the evaluations of his tenure began, colored now not only by the hostility of the Lazzaroni – the scientific elite including Bache, Henry, and Gould – but also by the passions of the Civil War. Whereas publicly expressed opinion of Maury by astronomers had been positive or neutral during the 1850s,[111] Gould, a junior member of the Lazzaroni, now displayed a quite different attitude that was quickly adopted by others. In the midst of the Civil War Gould addressed the National Academy of Sciences on his departed friend Hubbard: "It would be needless, gentlemen of the Academy, did taste not forbid, to describe to you

---

[110] See John Leighly's Introduction to Maury's *The Physical Geography of the Sea and its Meteorology* (Harvard University Press: Cambridge, Massachusetts, 1963), p. xxix. Editions are listed in Williams (ref. 4), p. 698.

[111] Thus Gould had given neutral notices of the second and third volumes of *Washington Observations* in his *Astronomical Journal* for 1851 and 1854, and in his eulogy of Walker in 1854 had nothing damaging to say about Maury. Loomis, commenting on the same volumes in the mid-1850s, believed that they "have placed our National observatory in the first rank with the oldest and best institutions of the same kind in Europe. But few observatories in Europe produce an equal amount of work in a year, and in point of accuracy the observations compare well with those of foreign institutions." Loomis, *The Recent Progress of Astronomy* (Harper & Brothers: New York, 1856), p. 214.

at any length, the embarrassments of astronomers, stationed at the Washington Observatory, while under the charge of the late Superintendent. Few of you, if any, can have failed to appreciate the painful conflict between self-respect and official proprieties, – between the emotions of the scientist, jealous of his country's reputation, and of the subordinate, whose duty in an establishment under military organization demanded tacit submission and apparent acquiescence, under a mortifying or atrocious policy. The sensitive nature of Walker found it impossible to endure the trial; but his pupil, Hubbard, struggled more successfully." More than personal hardship was involved; according to Gould, it was the honor of a national institution: "Would that I might with propriety express my keen sense of the deep debt of gratitude due from American science to those able and disinterested men . . . who bore the mortifications of their position without flinching, that they might save the national scientific institution, which it was partially within their power to protect, from becoming a source of national disgrace . . . They struggled against obstacles which would have deterred most men, in order that the noble instruments might render some service to science, or at least fail to be made implements of national disgrace. How well they succeeded, their record bears witness . . . ."[112]

Little more than a year later, upon Gilliss's unexpected death at the end of the Civil War, Gould was more explicit about the source of his criticism of Maury: "On the memorable 15th of April, 1861, Commander Maury fled from his post at the Naval Observatory, leaving in his haste unquestionable proofs of treasonable correspondence with the public enemy. A day or two later, orders were issued to Gilliss to assume the charge of the Institution, and poetic justice, though long deferred, was at last fulfilled. The sudden transformation which took place was like the touch of an enchanter's wand. Order sprang from chaos, system from confusion, and the hearts of the faithful few who had struggled on for years, hoping against hope, were filled with sudden joy."[113] Both Maury's selection over Gilliss in 1844, and his action at the outset of the Civil War, clearly rankled Gould, and so his opinion of Maury's work can hardly be seen as objective.

By the time J. E. Nourse wrote his *Memoir on the Founding and Progress of the USNO* in 1873, he commented with less bitterness on the relation between hydrography and astronomy in the Maury era: "with the exceptions of the equatorial, and mural circle observations, the zone observations named above, and the unpublished work of other observers during the latter part of the ten years (1851–'61), astronomical work unhappily ceased to be the definite object placed before the Institution. The Wind and Current Charts absorbed the attention of the Superintendent."[114] Long after the Civil

[112] Gould, "Eulogy on Hubbard," 80–81. Such comments and this attitude toward Maury were picked up by R. S. Dugan, author of the DAB article. On Gould as a junior member of the Lazzaroni in the early 1850s see Bruce (ref. 6), p. 221. Also Loomis, *The Recent Progress of Astronomy* (Harper & Brothers: New York, 1856), pp. 233–234.

[113] Gould, "Eulogy on Gilliss," 54. Other sources, as above, give April 20 as Maury's departure date.

[114] Nourse (ref. 8), p. 38. Nourse noted the resumption of astronomical work with Gilliss.

War passions had died down Simon Newcomb offered a still more generous evaluation. He wrote that, although Maury had little experience in the use of astronomical instruments, upon his arrival at the observatory "he went at his work with great energy and efficiency, so that, for two or three years, the institution bade fair to take a high place in science. Then he branched off into what was, from a practical standpoint, *the vastly more important work of studying the winds and currents of the ocean* [emphasis added]. The epoch-making character of his investigations in this line, and their importance to navigation when ships depended on sails for their motive power, were soon acknowledged by all maritime nations, and the fame which he acquired in pursuing them added greatly to the standing of the institution at which the work was done, though in reality an astronomical outfit was in no way necessary to it. The new work was so absorbing that he seemed to have lost interest in the astronomical side of the establishment, which he left to his assistants."[115] The result, according to Newcomb, was that instruments and methods had fallen into disrepair by the time Newcomb arrived a few months after Maury's departure.

We have here a variety of opinions, ranging from malicious to judicious neglect, generally resulting even among modern historians in the conclusion that astronomy was given short shrift in relation to hydrography, and implying that Maury should have done better: "Considering that Maury was in charge of one of the world's major observatories for almost seventeen years and that he had substantial funds at his disposal, his contributions to astronomy seem small. Between 1844 and 1861 he published fewer than twenty papers – all observational – and seven catalogs of observations."[116] In order to evaluate such conclusions – to determine whether Maury was shamefully neglectful as Gould states in his eulogy for Gilliss, or whether he made a judicious, necessary and practical choice of roles as Newcomb asserted in his *Reminiscences*, we need to consider the Maury era from several points of view, bearing in mind not only the duality of the institution, but also the duality of its critics.

It is clear, first of all, that Maury must be credited with transforming a naval observatory into a national observatory, a feat by no means a foregone conclusion when he assumed the Superintendency. In conjunction with Secretary of the Navy Bancroft, Maury not only hired a good staff, but also began an ambitious program in both astronomy and hydrography. Moreover, there is consensus that, during his first five years, Maury did a remarkable job in both fields. His observing program in astronomy, planned by Hubbard and Coffin and executed by the staff, was ambitious but not impossible. Even the critical Gould spoke of Coffin and Hubbard's "careful and laborious organization" of the zone observations, and characterized them as admirably devised. In Europe, Gould pointed out, Argelander had devised a similar program,

---

[115] Newcomb, *Reminiscences* (ref. 71), pp. 103–104.
[116] Harold L. Burstyn, "Matthew Fontaine Maury," DSB, 195–197. It might be noted that, for much of the twentieth century, the Naval Observatory, among its other activities published about one catalog every ten years.

and published the results in 1852. His complaint was not with the program's goal, but with shortcomings in how it was carried out and published: "had the large sums annually voted by Congress for the support of the Observatory been in part devoted to the reduction of these observations, and to the detection of the errors lurking in the observing books, – they would have conferred high honor upon American science, and indeed formed by far the noblest achievement of practical astronomy in America." As it was, the zone observations made since 1849 were rejected, and the remainder suffered from the lack of system under Maury.[117] At the same time as this astronomical program was under way, Maury during his first five years made significant advances in hydrography, which, although it was questioned by the Lazzaroni as to its status as science, was unquestionably of great practical value to navigation.

The concern, then, is with the last decade of the Maury era, during which the astronomical program was downplayed by comparison with hydrography, and even the publication of the observations made during the first five years was greatly delayed. Regarding the publication delay, it must be noted that Maury's publication of volume 1 of the *Washington Observations* within a year of the observations was remarkable, virtually unheard of among the world's astronomical observatories. However, it set a precedent so that people were understandably disappointed when it could not be continued. Though Maury may have tried to keep up, external events ranging from national to local in character delayed volume 2. The Mexican War in 1846 drew away some of the naval officers assisting with astronomy, and resulted in an even more frequent turnover of naval staff. Nor can the burning of the printing plant, along with many of the 1846 observations, be laid at Maury's feet.[118] Nonetheless, the major point is that delay in publication of astronomical observations was not uncommon, even the ten-year delays of the later volumes. Compared with Harvard, Greenwich, and most other national observatories around the world, Maury comes out quite well in this respect.[119] Even the first volume of observations of the Dudley Observatory, which Gould headed, was delayed 20 years.

The wider issue of Maury's relative neglect of astronomy during the 1850s is more complex. In order to assess this, it is important to examine another factor not within Maury's control – his annual budget. As Figure 2.15 shows, the budget during Maury's first 13 years as Superintendent underwent a rather impressive climb, rising

---

[117] Gould, "Eulogy on Hubbard," 83–85.   [118] *WO*, volume 2, Appendix, pp. 1 and 3.

[119] D. J. Warner notes that observatories published annals of their accumulated observations only as money permitted. Furthermore, "There seems to have been very little urgency to reduce or publish any Greenwich observation beyond those needed for the Nautical Almanac." Warner, "Astronomy in Antebellum America," in Nathan Reingold, ed., *The Sciences in the American Context: New Perspectives* (Smithsonian Institution Press: Washington, 1979), pp. 63 and 69. In 1873 Joseph Winlock, third Director of Harvard College Observatory, found a serious lag between observation and publication, even at well-established institutions such as Greenwich and Oxford, where they lagged by three years. For most observatories, including Pulkovo and Cambridge (U. K.), the lag was 10–20 years. B. Z. Jones and L. G. Boyd, *The Harvard College Observatory: The First Four Directorships, 1839–1919* (Harvard University Press: Cambridge, Massachusetts, 1971), pp. 158–159.

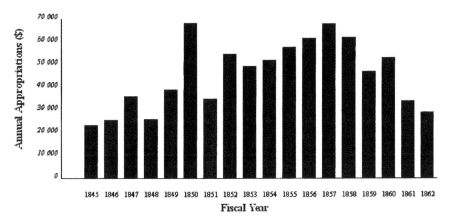

Figure 2.15. Appropriations for Naval Observatory and Hydrographic Office, 1845–62, showing a rise to 1857 and steady decline thereafter. The trend is even more pronounced taking inflation into account; $10,000 in 1845 was about $230,000 in 2001 dollars, while the same amount in 1857 was worth $200,000, and by 1862 only $175,000 in current dollars.

from $23,000 in 1845 to $68,000 by 1857, despite a peak in 1850 primarily due to monies Congress appropriated for Locke's clock and an expanded program of Wind and Current Charts. However, the appropriations then went into a decline so sharp that Maury was practically back to where he had begun at the start of his tenure. Many factors undoubtedly account for these trends, but it is notable that the budgets for the first five years were presided over by Secretaries of the Navy Bancroft and Mason, the former whose explicit policy was to build up the Observatory, and the latter Maury's Virginian compatriot who had personally chosen Maury over Gilliss to head the Observatory.[120] Similarly, the precipitous decline in funding beginning in 1857 was presided over by Secretary of the Navy Isaac Toucey, with D. N. Ingraham as Head of the Bureau of Ordnance and Hydrography, neither of whom took much interest in the Observatory in their annual reports. A more fine-grained analysis would be necessary in order to draw any solid conclusions, but the point is that Maury's superiors had a considerable, near-total, say in determining the resources at his command.

One might well question whether this was any way to run a scientific institution; indeed Maury himself questioned this at the end of his tenure. Yet most other observatories – certainly the few in the United States – looked enviously on the Federal appropriations given Maury. The fact is that Maury's allocation of both manpower and monetary resources within the institution greatly favored hydrography. However, according to Maury it was not money, but expertise and a permanently trained staff that was the real problem behind astronomy's lagging fortunes at the Naval Observatory.

[120] There was of course a time lag in the appropriations process, so that the $68,000 appropriated for the year ending June 30, 1850 had actually been submitted to the President for his budget request in December, 1848, Mason's last year as Secretary of the Navy.

The continual comings and goings of most of the naval staff leave no doubt that this is true. The fraction of Maury's staff assigned to astronomical work – roughly a third at the beginning of his tenure, dwindling to a sixth in later years – certainly mirrors the relative strength of hydrographic versus astronomical research. Since analysis and reduction of the observations took longer than the observations themselves, this lack of trained manpower in the later years was a decisive factor in the demise of astronomy at the institution, and not only by Maury's reckoning. When Ferguson published the first results of the zone observations in 1860, he cited the "many other pressing duties, and the limited computing force allowed" as reasons for delay.[121] Again and again during these years Maury complained to his superiors about the lack of sufficient force kept long enough at the Observatory to be trained for these duties.

These problems were clear for all to see in the late 1850s. What stands out most clearly is that, by the end of Maury's tenure, the Naval Observatory and Hydrographic Office was an institution in crisis. No one recognized it more clearly than Maury himself. His last annual report to his superiors, written less than nine months before his departure for the Southern cause, was devoted almost exclusively to the problems of the office under his charge. Manpower, he continued to stress, was the overriding problem. So limited in staff was Maury at this time – both on the astronomical and on the hydrographic side – that he could not spend the portion of the Observatory's budgeted funds assigned by law to salaries. "The organization of this establishment, both as a hydrographical office and an astronomical observatory, is most defective," wrote the man who 20 years before had played a major role in reorganizing the Navy into the Bureau system. "It should by all means be placed upon a footing that will enable it to vie in usefulness and renown with like establishments in other parts of the world. So to place it requires but little cost of time and money." More specifically, Maury went on, "How defective its organization necessarily is, may be inferred from the fact that though a hydrographical office, it is furnished with neither a chartographer [sic] or a draughtsman. As an observatory, it is, so far as instrumental equipments are concerned, one of the finest in the world; but it lacks force to use these instruments, and is without the power of reduction for publication."[122]

Imploring his superior to take into account his 15 years of experience, Maury noted that the Professors of Mathematics had originally been intended to serve as the computing force, while the Lieutenants and midshipmen were to remain on duty at the Observatory a sufficient amount of time to build up a corps of experienced observers. This plan had worked famously at the beginning, Maury noted, pointing to the favorable reception of the first volume of observations. However, the plan had gradually gone awry, beginning already with the Mexican War in 1846. Now, in 1860, "The

---

[121] *Zones of Stars Observed at the National Observatory, Washington*, volume 1, Part 1, containing the Zones observed with the Meridian Circle in 1846 . . . by Commander M. F. Maury, Superintendent (United States National Observatory: Washington, 1860), reduced by James Ferguson.

[122] Report of Maury to Captain D. N. Ingraham, Chief of the Bureau of Ordnance and Hydrography, August 7, 1860, in RSN, December 1, 1860, p. 215.

superintendent has ceased to have a voice in the selection of officers to assist him. The number allowed has been gradually on the decrease. The time for which, when ordered, they are to remain, is uncertain; it may be a year or more, or it may only be a few months or weeks. But whether the detail is to be for long or short, the superintendent has no information as to its length; and thus by the time an officer has learned how to observe, or to construct charts, even if he chance to have a taste for either, it often happens that he is called away on other duty, and another with as little knowledge of the duties, or aptitude for them, comes in his place. It does not require any lights from experience to show that an observatory so organized has before it the work of Sisyphus." The conclusion must have pained Maury: "For these and other causes, the employment of regular Navy officers as astronomical observers has become simply impracticable, if not impossible."

The unfitness of regular Navy officers for astronomical work left only the Professors of Mathematics, who themselves were rare, and assigned to the Observatory only when they could be spared for other duty: "... so with six fine instruments, requiring each two persons to serve them effectively, the observatory finds itself with but four officers fit as to bodily condition for the duties of an astronomical establishment. The instruments are rusting for the want of force to use them ... However superbly equipped with instruments an observatory may be, it is needless to expect from it magnificent contributions to astronomy unless there be not only eyes to observe, but power to treat observations after the instruments have performed their office."[123] Maury stated the obvious when he said that "in astronomy, an observation that is made and not published, is regarded with as little favor as an idle instrument."

No one can fail to see the sincerity and pleading in Maury's words regarding the failures of the Observatory: "Attention has been frequently called to this subject. It is deserving of consideration. The remedies are simple and of easy application. I do most respectfully and earnestly invoke such action by the department, and such legislation by Congress, as may be necessary to give full employment to the instruments and a becoming expression to the powers of this observatory."[124]

Nor was this situation confined to astronomy. Maury found that the hydrographic office was "even in a worst state as to organization." For the lack of cartographers, series A of the Wind and Current Charts had been suspended for years. The results of the Navy's own surveying expeditions were not analyzed at the Hydrographic Office, but in a "special bureau" set up for each expedition. Maury, who had played such an important role in the reorganization of the Navy to the Bureau system in 1842, now saw that system not working, at least at his level. By the end of the 1850s, not only naval astronomy but also hydrography was suffering as a result.

---

[123] *Ibid.*, 215–216.

[124] *Ibid.*, 216. On Maury's previous calls for action on the matter of personnel see his reports for 1859 (where he points to his recommendations of March 3 and August 5, 1858), etc. The letter of March 3, 1858 is in RSN for 1858, pp. 439–440.

The sentiment of Maury's contemporaries against him was not always misplaced, but was in some ways inevitable. According to the testimony of his staff, Maury had problems in managing civilians in particular. Walker – a pioneer in practical astronomy in America – undoubtedly did not take kindly to being told precisely how to make his observations by one who had done very little of it himself. Hubbard's indictment was more wide-ranging – and representative of astronomical opinion – when he wrote to Gould after Maury's departure that "Your rejoicing . . . cannot exceed mine; for it is a constant gratification to see order quenching chaos, energy overriding the old slowness, and above all our own science raising her triumphant head, and banishing the old humbug." Also the statement of Coffin's biographer that Coffin's relation to Maury was that of a subordinate to an omniscient superior, and that Coffin "chafed under this official impediment cast about his scientific work," could be seen as supporting the view of a military-versus-civilian problem in the Maury era.

However, the problem was not solely, perhaps not even primarily, a military–civilian problem, viewed from either inside or outside the institution. It was, in part, an elite-versus-non-elite conflict, aggravated by the fact that just at this time science was undergoing professionalization in the United States. There is no doubt that Maury was not as educated as Henry, Bache, and Gould, and that he represented a segment of scientists – and a view of science – slowly becoming extinct in the United States.[125] Yet neither is this the entire story, for Gilliss, Maury's successor, would be embraced by the scientific community, despite his being a Navy man with little more formal education than Maury. Unlike Maury, however, Gilliss would cooperate with contemporary scientists, contract to Gould the backlog of observations, and succeed in publishing not only the observations observed under his direction, but also those of the Maury era. Maury could not – or would not – do this, precisely because he was not part of that elite group who could successfully prosecute such work. Neither was Gilliss, but, proud Navy man that he was, he was nevertheless willing to reach out to those who could offer help. One statement early in Maury's tenure about the telescopes under his command would return to haunt him: ". . . though I had never seen an instrument of the kind before, and had no one with me who had, I was determined to ask no advice or instruction from the *savans*, but to let it be out and out a Navy work."[126] For this sort of elitism of his own brand Maury may be strongly faulted, and in this personal sense he himself planted some of the problems of his Federal tenure, quite aside from those placed in his way by the Navy or bureaucratic rivals.

It is important to realize that the balance between astronomy and hydrography was not a product solely of Maury's decision. The U. S. Naval Observatory and Hydrographic Office during the Maury era was part and parcel of the controversy over

[125]  As Mary Ann James has shown in *Elites in Conflict: The Antebellum Clash over the Dudley Observatory* (Rutgers University Press: New Brunswick, 1987), during the Maury era elites, one scientific and the other business class, were in conflict among themselves over the issue of control of the Dudley Observatory.

[126]  Maury to William Blackford, January 1847, in Diane Maury Corbin (ref. 4), p. 50.

a naval versus a national observatory, for, once Maury and Bancroft declared it national in character, they generated high expectations that, in the event, Maury could not fulfill. The institution was also part of the military-versus-civilian controversy that touched many parts of the science of the time; of turf wars with the Coast Survey; of practical versus pure research; of Humboldtian versus theoretical science; and of elite versus non-elite practitioners of science. In short, the institution was a product of the overall status of the young and developing enterprise of science in antebellum America.

Maury's choice of roles, which superficially seems to be a simple conscious choice to emphasize his interests in hydrography, was in actuality a much more complicated affair that mirrored the complexity of the times. How Gilliss extracted the institution from these difficulties – under very different circumstances – we shall see in chapter 4.

# 3 Foundations of the American Nautical Almanac Office, 1849–65

> An opportunity is now offered to the astronomers and men of science in this country, under the patronage of the Navy Department, to promote the cause of sound knowledge and to extend the national usefullness and honor by preparing an ephemeris based upon calculations, which shall be more perfect than those at present employed.
>
> Lt C. H. Davis, March 31, 1849[1]

In 1849, as Maury was deep into his work on the wind and current charts and astronomy was beginning to languish at the Naval Observatory, the Nautical Almanac Office was founded in Cambridge, Massachusetts. Technically, this event had nothing to do with the Naval Observatory; practically speaking it had a great deal to do with it. Recent historical research has uncovered the surprising fact that Congressional approval of the Nautical Almanac Office came about largely through the advocacy of Maury. Even so, not until 1866 would the Office move to Washington, and not until 1893 would it be physically located at the Naval Observatory at its new site. The famous Simon Newcomb, in his final years as Superintendent of the Office, chafed under the change in 1893, which would demote him from the Superintendent of an independent office to the Director of a branch of the Naval Observatory. However, both the Naval Observatory and the Nautical Almanac Office were Navy operations, the former concentrating on astronomical observations, and the latter on theory and predictions making use of those observations, among others. Both agencies had as their goal the improvement of navigation. Thus, even before the co-location of the Naval Observatory and the Nautical Almanac Office at the end of the nineteenth-century, and before the latter unambiguously became an administrative element of the Observatory in 1905, their histories are very much intertwined.

The 16-year period beginning with the founding of the Nautical Almanac Office and extending through the Civil War forms a distinct era in its history. During that time only two men served as Director, Lt C. H. Davis (1849–56 and 1859–61) and Joseph Winlock (1856–59 and 1861–66), the latter in the years before he became Director of Harvard College Observatory. During that time when the office was located in Cambridge, it managed all the problems inherent in any new office: hiring personnel, deciding basic issues on the philosophy and organization of what became an

[1] C. H. Davis to SecNav William Ballard Preston, March 31, 1849, LC Manuscript Division, NHF Collections, Container 218, item 1, folios 1–3.

American Ephemeris and Nautical Almanac, and justifying its very existence to skeptics both in the scientific and in the political world. Under Davis and Winlock the Office tapped some of the best mathematical talent in the United States, and incubated the talents of many more who went on to greater heights in their careers. Benjamin Peirce, Simon Newcomb, Chauncy Wright, and Maria Mitchell were some of the names that graced its early history, forming (in Newcomb's opinion) an "aristocracy of intellect" with the goal not only of improving navigation through the publication of the Almanac, but also of improving the very theories of solar, lunar, and planetary motion on which the Almanac was based. In this chapter we examine the rationale for an American nautical almanac at a time when other nations' almanacs already existed, discuss how the Almanac and the Almanac Office were organized, and analyze how the venture fared in the years before Simon Newcomb brought it worldwide prominence.

### 3.1 Motives for an American nautical almanac

The benefits of an American nautical almanac had been discussed for many years before an Office was actually established to produce such a volume. The benefits were not so clear-cut as one might have thought: by the early nineteenth century there was no lack of almanacs that might be adapted for American use. The French had published their Connaissance des Temps since 1679, the British had been producing a Nautical Almanac and Astronomical Ephemeris since 1766, and, by the late eighteenth century, Spain and Germany were also producing almanacs (Table 3.1).[2] Not only could American vessels use Edmund Blunt's edition of the British Nautical Almanac beginning in 1811, but also since 1830, there was an American Almanac that boasted astronomical tables compiled under the direction of the famous Professor Benjamin Peirce at Harvard.[3]

Given the availability of other nations' almanacs, clearly one driving force for an American almanac was grounded in patriotism. Responding to a proposal for an American almanac in 1831, the Board of Navy Commissioners offered that, although they were not aware of any inconvenience experienced by the U. S. Navy in relying on the British Nautical Almanac, it might be desirable to have one from the point of view of "national pride and independence," and that "an almanac of this kind would be one of the fruits of an observatory, should Congress deem it expedient to establish one." More than a decade later – two months after he had appointed Maury Superintendent of what was still known as the Depot of Charts and Instruments – Secretary of the Navy John Y. Mason noted that the Depot's new astronomical instruments were "well selected, and may be advantageously employed in the necessary observations with a

---

[2] The evolution of national almanacs is discussed in P. K. Seidelmann, P. M. Janiczek and R. F. Haupt, "The Almanacs– Yesterday, Today and Tomorrow," Navigation: Journal of the Institute of Navigation, **24** (1976–1977), 303–312.

[3] Blunt's Edition of the Nautical Almanac and Astronomical Ephemeris for the Year 1818 (Edmund M. Blunt, New York, 1816), advertised its editions beginning in 1811. The American Almanac and Repository of Useful Knowledge, first published in 1829 for the year 1830, contained data of only limited use for navigation.

Table 3.1. *National ephemerides*

| Title | Country | Year first published (for year) | Originator |
|---|---|---|---|
| *Connaissance des Temps* | France | 1679 (1682) | Cassini |
| *Nautical Almanac and Astronomical Ephemeris* | Great Britain | 1766 (1767) | Maskelyne |
| *Berliner Astronomiches Jahrbuch* | Germany | 1776 | Bode |
| *Efemerides Astronómicas* | Spain | 1791 | |
| *American Ephemeris and Nautical Almanac* | United States | 1852 (1855) | Davis/Maury |
| *Annuaire Astronomique* | U. S. S. R. | 1923 | |
| *Japanese Ephemeris* | Japan | 1925 | |
| *Indian Ephemeris and Nautical Almanac* | India | 1958 | |

view to calculate nautical almanacs. For those we are now indebted to foreign nations. This work may be done by our own naval officers, without injury to the service, and at a very small expense."[4]

In his first annual report as Superintendent, Maury himself argued for an American almanac as part of his goals: "If we attempt to compute the 'American Nautical Almanac' – and this we can do at no greater expense than we pay the English for computing theirs for us – from our own data, it is highly desirable that the data should be wholly American." All of the instrumentation for obtaining the requisite data was at hand in his new establishment, Maury noted, except for a meridian circle that he recommended be purchased for determining the refraction correction. Maury's pleas had their effect; not only did he get his new instrument, but also, the following year, now pointedly speaking of the "Observatory" rather than the "Depot," Secretary Mason was pushing for action on the matter: "There can be no doubt that, with the facilities of the Observatory, we might produce our own nautical ephemeris, for which we are now dependent on foreign nations, and without which our ships that are abroad could not find their way home, nor those at home venture out of sight of our own shores. A small appropriation would be sufficient to accomplish the object; and it may well be anticipated that the expenditure would be returned, by supplying our merchant vessels with the American Nautical Almanac at cost."[5] Related to the patriotic feeling was the issue of establishing an American prime meridian, to replace that of Greenwich.

Patriotism, however, was not the only reason for an American almanac. The anonymous author of an article that appeared in January 1849 was critical of the

---

[4] John Rodgers, BONC, to SecNav S. Woodbury, December 10, 1831, RG 78, Entry 213; John Y. Mason, *Report of the Secretary of the Navy*, November 25, 1844, p. 520. The wording for the latter was taken from Maury's letter to his supervisor W. M. Crane, head of the Bureau of Ordnance and Hydrography, dated November 7, 1844. See Waff (ref. 5), p. 88.

[5] Matthew F. Maury, in RSN, October 20, 1845, 690–691; and Mason, RSN, December 5, 1846, 385. The political maneuvering between 1845 and 1849 is described in detail in Craig B. Waff, "Astronomy and Geography vs. Navigation: Defining a Role for an American Nautical Almanac, 1844–1850," in NAOSS, 83–128.

nationalistic reasoning behind an American almanac. The question, the author argued, was not simply "shall the United States have an astronomical ephemeris of their own?," but "does *astronomy* need a new one and a better one than it already possesses?" The author, who may have been either B. A. Gould (a fresh Ph. D. under Gauss at Göttingen looking for a job) or Benjamin Peirce (one of the country's foremost mathematical astronomers), pointed out that the ephemerides of Greenwich, Berlin, and Paris were based on tables from 15 to 40 years old. American astronomers had the opportunity to undertake a full revision of the theories of the Sun, Moon, and planets, and the construction of tables from them. The American nautical almanac should not be an interpolation from European work to the meridian of Washington, the author argued, nor even computations from the same tables. "Nor do we propose an exclusively nautical work. A complete astronomical ephemeris will of course include what is requisite for the purposes of navigation; but if it were confined to this object it would be of little use to the astronomer engaged in improving the science itself. He requires the extreme of accuracy; the navigator only such a degree of precision as will enable him to determine his position within certain practical limits."[6]

The growth of astronomy was, however, not foremost on the minds of those pushing for an American nautical almanac. Rather, it was Maury's arguments of foreign independence and the relevance of Naval Observatory observations that Representative Frederick Stanton (Tennessee) used in the House Naval Affairs Committee, when he proposed an amendment to the Navy's appropriations bill on June 11, 1846 that would provide $5,000 for "computing and publishing . . . the American Nautical Almanac." Although John Quincy Adams, an unswerving friend of astronomy, supported funds for producing a more accurate almanac, other Representatives expressed fears that errors would put the American Navy in jeopardy, so the amendment was rejected. Secretary of the Navy Mason renewed his call for action on the almanac in 1847, and in 1848 submitted estimates of $6,000 "for calculating, printing and publishing the Nautical Almanac, including pay of superintendent of the same." Finally, in 1849 – in the closing days of Mason's tenure as Secretary of the Navy – the *Nautical Almanac* was approved. Maury's efforts had succeeded.[7]

The Naval appropriations Act of March 3, 1849 authorizing the preparation and publication of the *Nautical Almanac* was part of a paragraph relating to Maury's

---

[6] Anonymous, "Some Remarks Upon an American Nautical Almanac," AJSA, Series 2, **VII** (January, 1849), 123–125. Waff (ref.5 above, pp. 100–110) argues that the author is Benjamin Peirce. The argument for Gould as author is given in D. B. Hermann, "B. A. Gould and his *Astronomical Journal*," JHA, **2** (1971), 98–108. Gould had recently returned from three years of study in Europe, wanted the job as head of the Almanac Office, and even secured a recommendation from K. F. Gauss for the position. However, Gould was only 24 at the time, and Waff argues persuasively that the author was the more experienced Peirce.

[7] U. S. Congress, House, "Naval Appropriation Bill" debate, 29th Congress, first session, June 16, 1846, *Congressional Globe*, **XV**, 961–963 and 975; "Naval Appropriations" debate, 30th Congress, second session, February 1, 1849, *Congressional Globe*, **XVIII**, 428. Maury's renewed calls for action are in Mason, RSN, December 6, 1847, p. 12; December 4, 1848, pp. 614–615. For more details see Craig B. Waff (ref. 5 above), pp. 90–97.

Hydrographic Office. It provided only "That a competent officer of the Navy not below the grade of lieutenant, be charged with the duty of preparing the nautical almanac for publication;" the remaining clause referred to the other business of the Hydrographic Office.[8] As the wording made clear, however, the Nautical Almanac was to have its own Superintendent, and it was to be a Navy officer. This was no doubt a great disappointment to some civilians, notably B. A. Gould, who, on returning from Europe, had no job and actively lobbied for the position.

The actual choice of a particular Naval officer, not surprisingly, hinged again on politics. The day after Congressional approval of an American nautical almanac, the Whig administration of Zachary Taylor took over from the Democratic administration of President James Polk. There is no doubt that the change in Secretary of the Navy at just this time had its effect. Had Mason remained, the Office likely would have been put under the Naval Observatory immediately, for it was Mason who had both appointed Maury and nurtured the Observatory. But now, in a turn of events reminiscent of Maury's victory over Gilliss in gaining the Superintendency of the Observatory (then still a Depot) five years earlier, Maury was left out of the new plans. Restricted by the tenets of the appropriation bill to a Naval officer, Secretary of the Navy William Ballard Preston nominated Navy Lt Charles Henry Davis as the first Superintendent. This was perhaps at the behest of A. D. Bache, head of the Coast Survey, who thought very highly of Davis. In any case, it was neither Gould, who might have pushed astronomy, nor Maury, who would have pushed navigation, but Davis who was offered the job. Davis had an interest in both aspects of the Almanac, as is clear in a letter he wrote to Preston following an interview with the Navy Secretary: "The practical end of this work will be to supply the navigator with the elements required for determining his geographical position at sea by means of astronomical objects; its purpose in science is to predict for the astronomer the exact times and places of the principal heavenly bodies, used by him in his observations and computations. The first of these objects is already accomplished by the British Nautical Almanac, and though it may be a matter of proper national pride to be independent in this, as in all other commercial respects, yet our practical wants are now so perfectly supplied, that if this motive alone for publishing an American Almanac existed, it would hardly be considered sufficient to justify the necessary labor and expense." The Navy Secretary must have agreed with the spirit of Davis's letter; on July 11, shortly after the appropriation became available with the fiscal year beginning July 1, Preston officially placed Davis in charge.[9]

---

[8] Naval Appropriations Act, March 3, 1849, U. S. Statutes at Large, volume 9, pp. 374–375, Gustavus A. Weber, The Naval Observatory: Its History, Activities and Organization (Johns Hopkins Press: Baltimore, 1926), p. 27. The remainder of the clause reads, somewhat ungrammatically, "that the Secretary of the Navy may when, in his opinion, the interests of navigation would be promoted thereby, cause any nautical work that may, from time to time, be published by the hydrographical office, to be sold at cost, and the proceeds arising therefrom to be placed in the treasury of the United States."

[9] Davis to Preston, March 31, 1849, LC Manuscript Division, NHF Collections, Container 218, item 1, folios 1–3. Preston to Davis, July 11, 1849, LC Manuscript Division, NHF, Container 221, item 10, folio 4.

Figure 3.1. Charles Henry Davis, the first Superintendent of the American Nautical Almanac Office, in his later years. Davis was promoted to Admiral in February, 1863 after a victory at Memphis, Tennessee during The Civil War.

Though there is no documentary evidence that Lt Davis (1807–77) was "a leader and moving spirit in securing the appropriation" as Newcomb later put it, there is no doubt of his subsequent importance and his continued attention to astronomy as well as navigation. Davis (Figure 3.1) was born in Boston, educated at Boston Latin School, and graduated from Harvard in 1825, in the same class as Sears Cook Walker. He had left college in 1823, however, to enter the Navy, and, after some 17 years of sea duty, he was assigned to the U. S. Coast Survey in April, 1842. According to his son "our navigators and astronomers were still dependent on the British Nautical Almanac, a disadvantage which had long been apparent to Davis, and his Coast Survey work served to strengthen his conviction of the necessity of a national ephemeris." Convinced of the need for an American ephemeris and nautical almanac, Davis (according to his son) "threw the whole weight of his influence and energy into the accomplishment of this purpose. He was seconded by Bache and Henry, and by Maury."[10] Although this is rather an exaggeration with regard to his father's action, Davis certainly played a key role as the first Superintendent.

The Congressional statute said nothing about the establishment of a separate office, and not only was the Nautical Almanac Office formed separate from the Naval Observatory and Hydrographic Office, but also it was founded in an entirely different city. Though one might have thought that the new Office would immediately be

---

[10] C. H. Davis Jr, "Memoir of Charles Henry Davis," BMNAS, **4** (1902), 25–55; C. H. Davis Jr, *Life of Charles Henry Davis, Rear Admiral* (Houghton, Mifflin & Co.: Boston and New York, 1899), p. 86. Simon Newcomb, *Reminiscences of an Astronomer* (Houghton, Mifflin & Co.: Boston and New York, 1903), p. 63. Newcomb's estimate of Davis's role in the founding of the Office seems to have been based on these not totally objective sources.

associated with the Naval Observatory, or at least located in its proximity, there was considerable rationale for its location in Cambridge. Davis himself had lived in Cambridge (when not on sea duty) since 1840, engaged in the Coast Survey work. Harvard University was near, with Benjamin Peirce (Davis's brother-in-law) and other mathematical talent; and its library, enriched by the library of Bowditch, was important. The mathematical work of the Nautical Almanac Office differed significantly from the observational work of the Naval Observatory, requiring only the data from the latter and not a physical presence at the Observatory.[11] Furthermore, although Maury from the beginning had said that his observations would be useful for a nautical almanac, the two functions of observing and predicting could be separate without any loss. The relationship between the two Navy officers, Davis and Maury, may also have been important, with Maury an outsider in terms of his education and scientific contacts. Davis and Bache were part of the in-group of scientists known as "Lazzaroni;" Maury never was. While we may never know the exact interplay of forces, the incontrovertible fact is that, as of July, 1849, Davis set to work in Cambridge to produce the first volume of the *American Nautical Almanac*.

### 3.2    C. H. Davis: Organizing the almanac and the Almanac Office

*The controversy around the American prime meridian*

One of the first issues that had to be decided related to the suddenly contentious question of an American prime meridian. On July 31, 1849 Davis wrote the Secretary of the Navy a letter setting forth views they had already discussed privately. "The question is," Davis wrote, "whether, having a national observatory, and being about to publish an American Nautical Almanac, we shall still continue to count our longitudes from the meridian of Greenwich, or, whether it is preferable for convenience, for accuracy, or for other reasons, to establish a new prime meridian on this continent?" Davis pointed out that the ideal situation would be to have a single prime meridian for the world, a subject that had been broached four decades earlier even in the United States, but which was still prevented by obstacles that "render it distant and doubtful." In fact a variety of meridians were in use, with the French, Portuguese, and Dutch, for example, using prime meridians of their own countries. The United States had used the British meridian of Greenwich, sometimes for geographical positions, and always for nautical charts, setting chronometers, and astronomical calculations. Although such a familiar habit should not be changed lightly, Davis argued, "the scientific importance of assuming at present an American meridian is undoubted."[12]

---

[11] Newcomb, *Reminiscences*, p. 62, says that the Office was founded at Cambridge to "have the technical knowledge of experts," especially Benjamin Peirce; see also C. H. Davis Jr, *Life of Charles Henry Davis, Rear Admiral* (Houghton, Mifflin & Co.: Boston and New York, 1899), pp. 74–93. In 1842 Davis had married the youngest sister of Peirce's wife.

[12] Davis to SecNav William Ballard Preston, July 31, 1849, "American Prime Meridian," House Report No. 286 to accompany Joint Resolution No. 17, House of Representatives, 31st Congress, first session, May 2, 1850, pp. 2–7, hereafter cited as APM. On this controversy see Craig B. Waff, "Charles Henry Davis, the Foundation of the American Nautical Almanac, and the Establishment of an American Prime Meridian," *Vistas in Astronomy*, **28** (1985), 61–66.

The main problem, according to Davis, was that, as long as the United States depended upon a meridian separated from us by an ocean, accurate longitudes for the country remained illusory. In the United States the longitude of Boston was best determined: Bowditch had expended great labor on the project, numerous observations of solar eclipses, occultations, and Moon culminations had been collected for the purpose, and the transport of numerous chronometers between Boston and England had further pinned down the longitude with respect to that country. Nonetheless, Davis pointed out, the longitude of Boston was uncertain by three seconds of time. Should an American prime meridian be adopted, Davis argued, the telegraph could determine longitude differences within the country to a very high accuracy. Of paramount consideration, however, was convenience to the navigator. American vessels meeting British vessels on the high seas commonly compared their longitudes, a valuable custom that Davis argued must not be compromised. In addition, the common use of British charts and chronometers required an easy method of interchange between the Greenwich meridian and any American prime meridian. Because the meridian of Washington did not allow such an easy interchange (its longitude being some 5 h 8 m 16 s, or 77 degrees 4 minutes west of Greenwich) Davis proposed to establish the city of New Orleans as the American prime meridian. Its location at 6 h, or 90 degrees, from Greenwich would mean an easy conversion. Not only was it convenient, but also such a meridian "cuts the great valley of the west, and approaches the central line of our territory on this side of the Rocky mountains. It passes nearly through the centre of wealth and population of the great eastern slope of the continent, and enters the city of New Orleans, the mart and outlet of its products and trade. I propose to call it the meridian of New Orleans, in which city a spot is to be found having this suitable difference of meridian of six hours, or one-quarter of the circumference, from our present origin of Greenwich." If adjustments were needed later, the meridian mark in New Orleans would be moved, without changing the longitudes of the nation. Nor was there any scientific need, Davis argued, that Washington should be adopted as the American prime meridian because of its National Observatory. He would, however, in the astronomical ephemeris give the times of transit over Washington and New Orleans. Finally, Davis admitted to "a sentiment of American pride and gratification" that an American nautical almanac would be founded upon an American prime meridian.[13]

The Secretary of the Navy agreed that Davis should bring the matter before the meeting of the American Association for the Advancement of Science (AAAS) in Cambridge, Massachusetts. Here it was discussed and referred to a committee of 22 scientists, with Davis, Bache, and Maury forming a core subcommittee. Once out of the bag, Davis's proposal created a firestorm of opposition from those who would be most affected. J. Ingersoll Bowditch (the editor of Bowditch's *Navigator*) and chart

[13] APM, pp. 2–7.

seller George W. Blunt of New York orchestrated a "remonstrance" against Davis's proposal to change the meridian. One should recall here that Blunt was the American agent for the British almanac and British charts, and thus stood to lose money from the appearance of an American almanac and a changed meridian. However, the opposition was much broader than the commercial interests of one man. Organizations ranging from the Chamber of Commerce of the city of New York to the Baltimore Board of Trade, and even the Chamber of Commerce of New Orleans itself, argued that "the greatest confusion and perplexity would result to the navigator." More than 700 merchants, underwriters, and shipmasters from Nantucket, New York, Boston, Portsmouth, Salem, Portland, and Bath (334 from Boston alone) protested the change, finding no merit in the arguments and predicting chaos for navigation. The honor of the United States, they argued, was not compromised by use of the Greenwich meridian, any more than "our language, our arts, our names, and our very blood."[14]

The opinions of the AAAS committee were mixed, but in the end favored remaining with Greenwich. Some felt that an American prime meridian should pass through the National Observatory at Washington. Feeling the force of Davis's argument about the uncertainties of American longitudes, Maury argued for an American prime meridian, but he did not argue for Washington or accept Davis's suggestion of New Orleans. Uppermost in his mind was safe navigation, and the problem of changing from east to west longitudes dictated that the prime meridian be located further west in America than New Orleans, some 25 degrees from Washington and east of Santa Fe, so that ships in the Atlantic would always be at east longitude, and those in the Pacific at west longitude. The other concerns, he argued, were misplaced, and the remonstrants did not recognize the other great interests in the country aside from navigation, namely "perfecting the geography of the country, in the permanency of its landmarks, and in the general convenience of the people." The people, he believed, had already accepted longitudes reckoned from Washington, since that was the practice of most American atlases and schoolbooks. Not surprisingly, Davis was not swayed by the arguments of others, and in a final lengthy paper on the subject reiterated his views.[15]

The House Committee on Naval Affairs, with this voluminous documentation in hand, recommended to Congress a compromise solution. Although it perceived that the inconvenience to navigators was overrated, they sought to avoid all controversy by proposing the adoption of an American prime meridian for astronomy and geography, while retaining the Greenwich meridian for the navigational part of the

---

[14] Lt C. H. Davis, "Upon the Prime Meridian," *Proc. AAAS*, Second Meeting, held at Cambridge, August, 1849 (Boston, 1850), pp. 78–85. This article repeats much of Davis's argument in his July 31, letter to SecNav (ref. 12 above).

[15] Maury to Davis, September 26, 1849, and Maury to SecNav William Ballard Preston, November 30, 1849, in APM, pp. 18–20 and 25–32; Charles H. Davis, "Remarks upon the Establishment of an American Prime Meridian," APM, pp. 50–68.

Almanac. The prime meridian adopted was Washington. With the passing of this bill on September 28, 1850, about 15 months after the founding of the Almanac Office, a major hurdle had been removed with the goals of the Office still intact.[16]

As a direct result of this decision, the *American Ephemeris and Nautical Almanac* had a peculiar bipartite form, one part of more use to astronomers and geographers, and the other part to navigators. The ephemeris for the meridian of Greenwich gave the ephemeris of the Sun, Moon, and planets together with lunar distances. The ephemeris for the meridian of Washington gave the positions of the principal bright stars, the Sun, Moon, and larger planets, and other phenomena predicted and observed, including eclipses, occultations, and motion of Jupiter's satellites. This, of course, would be most useful for observers in the United States.[17]

### Organizing the Office

One can imagine the problems that Davis faced in organizing the Almanac Office. Not least was the problem of the scientific basis for the Almanac. Nevertheless, Davis made no little plans; he set about not only producing an Almanac, but also revising the theories of Sun, Moon, and planets on which it was based, including the theory of Neptune that "belongs, by right of precedence, to American science." It is thus important to understand that, right from the beginning, Davis considered the work of the Nautical Almanac Office broader than simply publishing pages of useful numbers. Most generally, Davis wished "to advance that which is, and has always been, the principal object of astronomy; and that is, in the language of Bessel, to supply precepts by which the movements of the heavenly bodies, as they appear to us from the earth, can be calculated." This, he considered, was the highest calling of astronomy, much more important than mere descriptive astronomy. Thus the Office had a broad role, one not only to improve the safety of navigation but also to contribute to astronomy, while compensating American mathematicians and astronomers for their often unsung labors, and proving a credit to the country that supported this highest form of intellectual endeavor. An Astronomical Ephemeris, Davis added, "was something more than a book of mere results of calculations based upon rules furnished elsewhere; it should itself help to investigate the theories it is obliged to employ."[18] The improvement of theories of solar, lunar, and planetary motion forms one of the central themes throughout the history of the Office.

Because this goal placed the Nautical Almanac Office squarely in a long tradition of classical celestial mechanics, it is important to realize where celestial mechanics stood at the founding of the American Office in relation to the production of tables of motion. Kepler's discovery in the early seventeenth century that each planet moved

---

[16] "Report," in APM, May 2, 1850, pp. 1–2, *U. S. Statutes at Large*, volume 9, p. 515.

[17] Newcomb, *Sidelights* . . . , pp. 203–205.

[18] Lt C. H. Davis, "On the Nautical Almanac," *Proc. AAAS*, fourth meeting held in New Haven, Connecticut, August, 1850 (Washington, 1851), pp. 56–60.

in an ellipse yielded much-improved tables of their motion, the first of which was published by Kepler himself as the *Rudolphine Tables* (1632). Kepler's predictions, however, were not perfect because they did not take into account the mutual perturbations of the planets, a concept that was conceivable only with Newton's theory of gravitation a half century later. Although Newton pointed out the likely cause of planetary perturbations, his geometric methods were inadequate to the task of predicting them. It was the continental mathematicians, especially Laplace in the third volume of his *Mécanique Céleste*, who conceived the method of finding algebraic expressions for the positions of the planets at any time, giving their celestial latitude, longitude, and distance from the Sun as functions of time. This method required that six elements of each orbit (such as period and orientation of the ellipse) be derived from observation. Even once these elements had been determined, no algebraic expression could give a rigorous solution. Instead, the expression was an infinite series of terms; by using more and more of the terms, one could approach mathematical exactness, but never reach it.

Even then, no general expression was applicable to all cases, so that one was needed for the inner planets, one for the Moon, one for Jupiter and Saturn, one for the minor planets, and so on. These expressions were in each case worked out by individual astronomers and mathematicians focusing on one case. Thus Charles Delauney at Paris Observatory, and Peter Hansen at Gotha, spent significant parts of their careers on lunar studies, the latter with the aim of representing with his theory all the Greenwich observations of the Moon since 1750 with no greater deviation than one arcsecond from observation. Bernhard August von Lindenau and Alexis Bouvard produced tables of the planets lasting through the first half of the nineteenth century, which were based on Laplace's formulae, and U. J. J. Leverrier at Paris Observatory undertook the next complete reconstruction of the planetary theory. The European ephemerides were based on tables now 15–40 years old, and the American office had the opportunity to undertake a full revision of the theories of the Sun, Moon and planets, and construct new tables from them.[19]

American interest in celestial mechanics was not completely lacking. Nathaniel Bowditch, for example, had translated Laplace's *Mécanique Céleste* during the last decade of his life from 1829–38, and that translation had found some readers. Nonetheless, even by 1849, hiring a staff to carry out the tasks necessary to the work Davis had set the Office was a challenge. "It has been necessary to train the computers for a work such as has never before been undertaken in this country," Davis wrote. Nevertheless, by 1852 Davis had recruited a variety of people, whose rank may be gathered from their pay (Figure 3.2).[20] The most highly paid employee of the new Office

---

[19] For a contemporary account of the state of celestial mechanics at the time of the founding of the American Nautical Almanac Office, see Robert Grant, *History of Physical Astronomy From the Earliest Ages to the Middle of the Nineteenth Century* (R. Baldwin: London, 1852).

[20] Davis to William Ballard Preston, RSN, October 2, 1849, pp. 443–444. See also Davis's reports in RSN, October 12, 1850, pp. 229–230; November 29, 1851, pp. 73–76; and December 4, 1852, pp. 345–348. Figures 3.2 and 3.5 are from the latter.

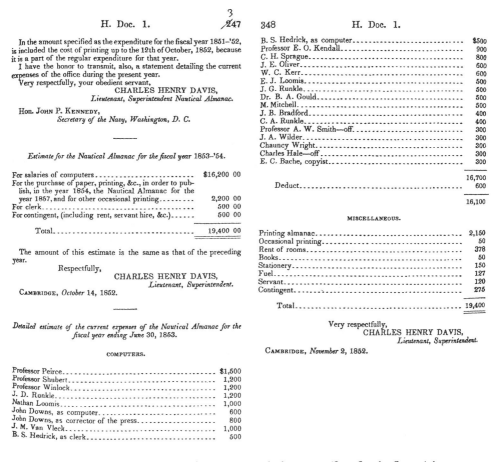

H. Doc. 1.                                            3
                                                   /247

In the amount specified as the expenditure for the fiscal year 1851–'52, is included the cost of printing up to the 12th of October, 1852, because it is a part of the regular expenditure for that year.

I have the honor to transmit, also, a statement detailing the current expenses of the office during the present year.

Very respectfully, your obedient servant,

CHARLES HENRY DAVIS,
*Lieutenant, Superintendent Nautical Almanac.*

Hon. JOHN P. KENNEDY,
*Secretary of the Navy, Washington, D. C.*

———

*Estimate for the Nautical Almanac for the fiscal year 1853–'54.*

| | |
|---|---|
| For salaries of computers | $16,200 00 |
| For the purchase of paper, printing, &c., in order to publish, in the year 1854, the Nautical Almanac for the year 1857, and for other occasional printing | 2,200 00 |
| For clerk | 500 00 |
| For contingent, (including rent, servant hire, &c.) | 500 00 |
| Total | 19,400 00 |

The amount of this estimate is the same as that of the preceding year.

Respectfully,

CHARLES HENRY DAVIS,
*Lieutenant, Superintendent.*

CAMBRIDGE, *October 14, 1852.*

———

*Detailed estimate of the current expenses of the Nautical Almanac for the fiscal year ending June 30, 1853.*

COMPUTERS.

| | |
|---|---|
| Professor Peirce | $1,500 |
| Professor Shubert | 1,200 |
| Professor Winlock | 1,200 |
| J. D. Runkle | 1,200 |
| Nathan Loomis | 1,000 |
| John Downs, as computer | 600 |
| John Downs, as corrector of the press | 800 |
| J. M. Van Vleck | 1,000 |
| B. S. Hedrick, as clerk | 500 |

348                                            H. Doc. 1.

| | |
|---|---|
| B. S. Hedrick, as computer | $500 |
| Professor E. O. Kendall | 900 |
| C. H. Sprague | 800 |
| J. E. Oliver | 600 |
| W. C. Kerr | 600 |
| E. J. Loomis | 500 |
| J. G. Runkle | 500 |
| Dr. B. A. Gould | 500 |
| M. Mitchell | 500 |
| J. B. Bradford | 400 |
| C. A. Runkle | 400 |
| Professor A. W. Smith—off | 300 |
| J. A. Wilder | 300 |
| Chauncy Wright | 300 |
| Charles Hale—off | 300 |
| E. C. Bache, copyist | 300 |
| | 16,700 |
| Deduct | 600 |
| | 16,100 |

MISCELLANEOUS.

| | |
|---|---|
| Printing almanac | 2,150 |
| Occasional printing | 50 |
| Rent of rooms | 378 |
| Books | 50 |
| Stationery | 150 |
| Fuel | 127 |
| Servant | 120 |
| Contingent | 275 |
| Total | 19,400 |

Very respectfully,

CHARLES HENRY DAVIS,
*Lieutenant, Superintendent.*

CAMBRIDGE, *November 2, 1852.*

Figure 3.2.  Budget estimates for the Nautical Almanac Office, for the financial year ending June 30, 1853. From the Report of the Secretary of the Navy, December 4, 1852.

(at $1,500 per year), was Harvard mathematician Benjamin Peirce (1809–80), the leading American mathematician since the death of Bowditch in 1838. Peirce was immediately signed on as a consulting astronomer, a position in which he served the Nautical Almanac Office until 1867, when he became head of the Coast Survey. Peirce was a Harvard man through and through, having secured degrees in 1829 and 1833, and taken positions as Professor of Mathematics and Natural Philosophy (1833–42) and Professor of Astronomy and Mathematics (1842–80). Peirce, who was considered one of the most influential American mathematicians of the century, had determined the orbit of Neptune and its perturbations on Uranus, and had come to the controversial result that Galle's discovery of Leverrier's predicted planet was only a "happy accident." Peirce not only helped plan the form of the Almanac, but also took charge of the revision of the theory of the planets. Davis believed that Peirce's involvement guaranteed "to the whole scientific world, that whatever is undertaken in theory will be creditably executed."[21]

Following Peirce in terms of pay and in mathematical skills were three men who received the salary of $1,200 each. John D. Runkle, the senior assistant in the Office, was associated with the Office from 1849 to 1884, but did much more. In addition to founding the *Mathematical Monthly*, he was one of the founders of Massachusetts Institute of Technology, Professor of Mathematics there from 1865, and served as its President from 1870–78. German-born Ernst Schubert, brought to the United States by Maury in 1849 at the recommendation of Schubert's mentor Johann Encke, introduced European techniques to the Office from its beginning. Joseph Winlock, of whom we will have more to say later, would become Davis's successor as Superintendent and later Director of Harvard College Observatory. Not far behind these three luminaries in salary were Nathan Loomis and J. M. Van Vleck. They were followed in salary seniority by a whole group of computers, including Gould himself, who, having not succeeded in his bid to head the office, nevertheless undertook computing work for it. Of equal rank in terms of salary was Maria Mitchell (Figure 3.3), of whom Davis spoke highly as "my distinguished and accomplished friend, Miss Maria Mitchell, of Nantucket, whose accuracy, fidelity, and learning, render her a most valuable assistant." Among the lowest-paid computers was Chauncy Wright, who remained associated with the Office for almost 20 years and devised means for speeding up Almanac calculations, so that more time could be devoted to philosophy. Wright's scientific philosophy, which was based on Darwinian thought, influenced William James and C. S. Peirce. Slightly later arrivals were Simon Newcomb (Figure 3.4), who in 1857 entered the happy ambience of the young Almanac Office that he described in his *Reminiscences*, and Truman H. Safford, whom Newcomb called "the most wonderful genius in the office." It is notable that many of the staff listed in Figure 3.4 were contractors not physically located in the Office at all; in this respect the organization and ambience of the Almanac Office was quite different from that of the Naval Observatory.[22]

The division of work of this staff may be seen in Figure 3.5. Already while waiting for a resolution of the problem of the meridian to which the almanac would be referred, and for the lunar and solar tables of Peter Hansen that would improve the predicted positions of the Sun and Moon, Davis had four computers begin a new set

---

[21] On Peirce see DSB, pp. 478–481. On the work of Peirce and Walker on Neptune, see Robert W. Smith and J. G. Hubbell, "Neptune in America: Negotiating a Discovery," JHA, **23** (November, 1992), 261–292. Davis, "On the Nautical Almanac," *Proc. AAAS*, **4** (1851), 56.

[22] Newcomb's *Reminiscences*, pp. 62–69, describes the Office during its early years in the chapter "The World of Sweetness and Light." Entries on many of these figures are also found in BDAS. See also Newcomb, "Aspects of American Astronomy," in *Sidelights on Astronomy*, pp. 290–291, for a description of the atmosphere of the office under Davis.

Figure 3.3. Maria Mitchell and her father William. Maria was employed at the Nautical Almanac Office from 1849 until 1868. She was influenced by her father, who Benjamin Peirce and William Bond had recommended to the Secretary of the Navy in 1845 for a civilian position at the Naval Observatory.

of tables of the planet Mercury based on the theory of Leverrier. By October, 1851, the lunar ephemeris was completed and ready for the press, the computation of lunar distances (divided between Schubert, Runkle, and Van Vleck) well advanced, and work under way on the reduction of the fixed stars, which, after the lunar ephemeris, constituted the bulk of the work. Maria Mitchell had nearly completed the ephemeris of Venus, Kendall that of Jupiter and its satellites, Mercury had been taken over by Winlock, and Smith of Connecticut and Coakely of Maryland were working "with great zeal and success."[23] In addition to helping with computations, Gould also produced a detailed table of geographical positions of the principal observatories of the world.

## 3.3 Opposition and success

The success of the American endeavor was by no means a foregone conclusion. However, because of the use of new theories and observations, there was good reason for hope. Of special interest were the tables of the Moon, for the method of lunar distances was still important, and this would serve as a crucial test of the skill of the new office. Already in his Annual Report for 1851 (and later in *Two Memoranda*), Davis boasted that the American results reduced to one third the average errors of the

---

[23] Davis, in AR, October 12, 1850, 229–230; ibid., 1851, 73–76.

Figure 3.4. Simon Newcomb in 1858 at age 23, when he was working at the Nautical Almanac Office. From *McClure's Magazine*, **35** (1910), 680.

Moon's place given in the British Astronomical Ephemeris. A crucial test was the solar eclipse of July 28, 1851, which the Navy Department authorized the Nautical Almanac Office to predict. Twenty days later Davis communicated his prediction to the AAAS annual meeting being held in New Haven, the same group as that which had taken up the prime-meridian issue two years earlier. In order to compare the predictions, Sears Walker observed the eclipse at Cambridge, Massachusetts, and Hubbard and James Ferguson observed in Washington. By both accounts the Americans won hands down: According to Davis, the British Almanac was 85 seconds in error at Cambridge and the American Almanac 20 seconds; at Washington the British Almanac was 78 seconds in error for the beginning of eclipse, 62 seconds for the end; whereas the American Almanac erred by only 13 and 1.5 seconds, respectively. Davis pointed out that the French and Berlin almanacs used the same tables as the British, and so were also in error by the same amount. In practical terms this meant a 15–20-mile error in determination of longitude at sea by lunar observations.[24]

Even as the first volume of the *American Ephemeris and Nautical Almanac* went to press in 1852 – and as the Navy Department published essential sections of Davis's translation of Karl Friedrich Gauss's classic *Theoria Motus Corporum Coelestium* – the skeptics (no doubt Blunt with his commercial interests among them) were still

---

[24] Letter of August 11, 1850 cited in AR (1851), p. 75; p. 2 of *Two Memoranda* (ref. 36 below). See also "Davis's Report" (ref. 26 below).

active.[25] In summer 1852 Davis was called to task by a member of the U. S. Senate "whose philanthropy is evidently more enlarged than his astronomy," as the *American Journal of Science and Arts* put it. By resolution of the Senate the Secretary of the Navy was instructed to inform the Senate where and at what observatory the observations and calculations for the Nautical Almanac were made, what progress had been made, what improvements would be made over the British almanac, what costs had been incurred, and whether the calculations for the Almanac would be extended indefinitely.[26] Davis's replies shed considerable light on the Office at the time. Distinguishing the calculations for predictions from the observations, Davis noted that the former were made principally at Cambridge, "the residence of the present superintendent, where the printing of the work can be conducted most expeditiously, most economically, and, what is still more important, most accurately; and where convenient reference can be had to the best scientific libraries of the country." At the same time he noted that the talent of the office was disbursed, in contrast to the Naval Observatory: Winlock was in Kentucky, Walker in Washington, Kendall in Philadelphia, Smith at Wesleyan University, and Mitchell at Nantucket, perhaps to show that salary monies were being distributed across a large number of states. He further remarked that cooperation with the "National Observatory" at Washington was being requested in order to make meridian observations of reference stars for the Greenwich observations of Mars; to test by observation the accuracy of the elements of the new minor planet Iris, to observe other newly discovered minor planets, and to furnish other data in advance of publication in the *Washington Observations*.

As to improvements over the British Almanac, Davis reserved his longest answer. He pointed to the ephemeris of the Moon and most of the planets; the error of one third of a minute of arc in the former, he noted, caused errors of ten miles of longitude at sea. He again referred to the solar eclipse of July 28, 1851, causing an error of 15–20 miles in longitude determinations. He noted also that one result of the American Nautical Almanac was that the price of the British Almanac had been reduced by one half.[27] The printing, Davis noted, was "far advanced," and the finished volume would appear in three or four months. After that, he estimated that the annual cost of production would be $19,400, including the computers "who must be gentlemen of liberal education and of special attainments in the science of astronomy."[28] In closing Davis appealed to the

[25] Karl Friedrich Gauss, *Theory of the Motion of the Heavenly Bodies Moving about the Sun in Conic Sections*, a translation of *Theoria Motus Corporum Coelestium*, by Charles Henry Davis (New York, 1857; reprinted by Dover: New York, 1963).

[26] Senate Documents, Ex. No. 78 (1852), reprinted in "Davis's Report on the Nautical Almanac," *AJSA*, second series, **14** (November, 1852).

[27] Davis (ref. 26), p. 330, noted that lunar observation was the only method employed in navigation when chronometers were lacking or untrustworthy, or required verification of their rates. The price reduction of the British Almanac was from 5 s to 2 s 6 d.

[28] Davis (ref. 26), p. 331, noted that this was somewhat higher than the 16–17 thousand dollars for the British Almanac, but argued that "intellectual labor commands a higher compensation in this country than in Great Britain."

scientific reputation of the country, "already established and widely extended by the coast survey and the national observatory." Furthermore, he took the opportunity to remind the Senate of the nature of the volume: to embrace all the information necessary to determine at any time the absolute and relative positions of the Sun, Moon, and planets, and some of the brightest stars; the phenomena for determination of longitude, including occultations, lunar distances, transits of the Moon and stars, and eclipses of Jupiter; also places of the minor planets, rules and tables for nautical astronomy, and tables of tides and geographical position. The geographical extent of the United States he argued, "makes it apparent that neither the authorities nor standards of Europe can satisfy our demands."[29] The work of the *Nautical Almanac*, Davis concluded, also serves the advancement of science and the diffusion of knowledge in the United States.

The criticisms must have been too little, too late. Even as Davis's report in answer to the Senate's queries was reprinted in the *American Journal of Science*, in early November, 1852 Davis reported that the first pages of the *Almanac* would issue from the press within the week. In January, 1853, the first volumes of a total print run of 1,000 copies were transmitted to Washington. The office had every reason to be proud of its work extending over three years in producing this volume. There, in the bipartite form dictated by the outcome of the prime-meridian controversy in 1850 (Figure 3.6), was everything needed to serve the needs of navigators, and astronomers and geographers. The navigational part, referred to Greenwich, consisted of the tables of lunar distances as well as the apparent places and rectangular equatorial coordinates of the Sun, and an ephemeris of Venus, Mars, Jupiter, and Saturn for every noon. The astronomical and geographical part, referred to the meridian of Washington, included a variety of astronomical phenomena. There was no doubt of the debt to European astronomers. The astronomical constants of precession, obliquity of the ecliptic, and equation of the equinoxes were those determined by Struve and Peters, and Struve's value was also used for the constant of aberration. The mean places of the fixed stars were derived from Airy's *Twelve Year Catalogue*. The ephemeris of the Sun was constructed from tables in the *Effemeride di Milano* (1833), and those of the Moon were based on the new lunar tables of Hansen, even though the latter would not be published in final form until 1857. However, there was equally no doubt of the American contribution. The planetary ephemerides were especially notable for their increase in precision, and most notably Winlock's tables derived from Leverrier's theory of Mercury, the ephemeris of Uranus with corrections for the gravitational perturbations of Jupiter, Saturn, and Neptune, as determined by Leverrier and Peirce, and Walker's ephemeris of Neptune for 1853.[30]

Not surprisingly, Gould, favorably reviewing the volume in the pages of the *Astronomical Journal*, acknowledged its navigational utility but also saw its place in the history of American astronomy and broader science. "It may not be an exaggerated

---

[29] Ibid., p. 335.
[30] Davis describes in detail the improvements of the *American Nautical Almanac* in his 1852 Report, pp. 322–330.

346  H. Doc. 1.

DIVISION OF WORK.

Professor Peirce—The general theory; planets generally; Mars par-
ticularly. Mr. J. B. Bradford, assistant.
Professor Winlock—Sun and Mercury, Astraea, Egina.
Mr. J. D. Runkle—Last ninety-two days of moon, Pallas. Mr. C.
A. Runkle, assistant.
Mr. Van Vleck—Second ninety-two days of moon, Hausen's theory of
Jupiter and Saturn. Mr. E. Loomis, assistant.
Mr. B. S. Hedrick—First ninety-one days of moon, Metis, Ceres.
Mr. W. C. Kerr, assistant.
Mr. C. Wright—Third ninety-one days of moon. Mr. J. G. Runkle,
assistant.
Mr. J. E. Oliver—Latitudes and longitudes; miscellaneous.
Mr. John Downs—Occultations, Saturn; proof-reading. Mr. J. A.
Wilder, assistant.
Miss M. Mitchell—Venus.
Professor E. Shubert—Iris and other asteroids.
Professor E. O. Kendall—Jupiter and Neptune.
Professor A. W. Smith—Flora.
Mr. C. Hale—Clio.
Dr. B. A. Gould—Vesta, Hygeia.
Mr. C. H. Sprague—Fixed stars.
Mr. Nathan Loomis—Star table.
Mrs. E. C. Bache—Copyist.
I transmit with this report a proof copy of the general preface to the
first number of the Nautical Almanac, for the approval of the depart-
ment.
In conclusion, I have the honor to inform the department that, not-
withstanding the slight delays referred to in the beginning of this report,
the general state and progress of the work under my charge is satis-
factory.
Very respectfully, your obedient servant,
CHARLES HENRY DAVIS,
*Lieutenant, Superintendent Nautical Almanac.*

Hon. JOHN P. KENNEDY,
*Secretary of the Navy, Washington, D. C.*

Figure 3.5. The division of work among Nautical Almanac Office computers in 1852.

estimate of its value," he wrote, "to believe that its influence upon the development of astronomy in the United States will be very important." The volume was, he said, "a pioneer of American science," worthy of taking its place among its predecessors published in Berlin, London, and Paris. However, the words the Americans longed to hear came from their British counterparts. The anonymous reviewer in the *Monthly Notices of the Royal Astronomical Society* called the "handsome volume" "fresh proof of the interest felt by our Transatlantic brethren in the advancement of astronomical science, and of their ability in carrying out views of practical utility."[31]

Subsequent volumes under Davis's tenure remained substantially unchanged in content. One of the changes was an increasing number of asteroids to keep track of. When the Office began in 1849 only ten such objects were known; by 1854 the number had increased to 31, including the recent discovery by Ferguson at the Naval Observatory. Already in May 1853 Davis offered the cooperation of his office with the British, French, and Germans in treating asteroids, but this did not come easily, and in 1854 Davis was hinting that, without foreign help, "I shall be obliged to ask for additional means to enable us to take charge of them alone." This is exactly what

---

[31] Review of *American Ephemeris and Nautical Almanac for the Year* 1855, in AJ, **3** (1853), 47–48;
Anonymous, MNRAS, **13** (1852–53), 168–171.

he did the following year, arguing that a knowledge of these orbits was not only "essential to the perfection of our work," but also that taking up their motion according to a uniform and comprehensive plan would be "gratifying to our pride, and will contribute to the progress of astronomy in America." The increase in appropriation was granted, and, beginning with the volume for 1860, ephemerides for an increasing number of asteroids were published.[32]

In 1854, after 31 years in the Navy and 23 in the grade of lieutenant, Davis was promoted to commander. This led to his being offered a command in November 1856, which he immediately accepted because, according to his son, "he could not contemplate a situation which obliged him to forego all hope of promotion, and resign the active life of the [navy] profession."[33] Although Davis would return to head the Office from 1859–61, his pioneering work was done. As the founding Director of the Office, he had placed his indelible stamp on the most creditable American mathematical feat to date, the *American Ephemeris and Nautical Almanac*.

Davis's successor as Superintendent in November 1856 was Joseph Winlock, who, except for a brief period in 1859–60, would head the office for a decade, including the Civil-War years. Winlock (Figure 3.7), a grandson of Revolutionary general Joseph Winlock and a graduate of Shelby College, Kentucky in 1845, was appointed Professor of Mathematics and Astronomy at his alma mater immediately upon graduation. Winlock met Peirce at the 1851 meeting of the AAAS in Cincinnati, and, at Peirce's recommendation, began work as a computer for the Almanac Office from his home base, before moving to Cambridge in 1852. In 1856 he became a Professor of Mathematics in the Navy and was assigned to the Naval Observatory, but in November took up duties as Superintendent of the Almanac Office. Winlock was a charter member of the National Academy of Sciences in 1863, and in 1866, was made Director of the Harvard College Observatory upon the death of George Bond. Here he would "exhibit a remarkable mechanical ingenuity and genius for invention," using and designing apparatus including the new photographic and spectroscopic instrumentation. Newcomb, who entered the Almanac Office a few months after Winlock took over, remembered him as "most companionable in the society of his friends," but "as silent as General Grant with the ordinary run of men."[34]

Compared with the battles and fundamental decisions of the Davis period, Winlock's tenure saw relatively smooth sailing, as the office settled down to the routine annual production of the *Almanac* volumes. In the interests of improving the theory of solar motion, Almanac staff were even occasionally sent out to observe solar

---

[32] AR, October 12, 1853, 395; AR, October 20, 1854, 416; AR, October 12, 1855, 77–78; AR, October 18, 1856, 480.  [33] C. H. Davis Jr., *Life of Davis* (ref. 10), p. 100.

[34] On Winlock see especially Joseph Lovering, "Memoir of Joseph Winlock, 1826–1875," BMNAS, 1 (Washington, 1875); *Nature*, **12** (1875), 191–192; *AJS*, **10** (1875), 159–160; Deborah Jean Warner, "Winlock," in DSB, 448–449. Newcomb, *Reminiscences*, p. 65. On his tenure at Harvard see Jones and Boyd, *The Harvard College Observatory* (Harvard University Press: Cambridge, Massachusetts, 1971), pp. 136–175.

# CONTENTS.

Figure 3.6. The bipartite division of *The American Ephemeris and Nautical Almanac* (Government Printing Office: Washington, 1852).

Figure 3.7. Joseph Winlock, courtesy of Harvard College Observatory.

eclipses, as in 1860 when Davis sent Newcomb and William Ferrel to Vancouver.[35] In an attempt to improve the speed of calculations for the Office, in 1858 Winlock experimented with a new "calculating engine" at Dudley Observatory, finding the machine ingenious, but not yet applicable to the production of the *Almanac*. Ironically, it was on Davis's temporary return as Superintendent that rough seas were encountered when the Office became embroiled in a controversy over control of the Dudley Observatory. Dudley's Director at the time was none other than B. A. Gould. Because Gould had previously worked for the Nautical Almanac Office, in the eyes of some in Congress the Office was guilty by association, no doubt made worse by the recent collaboration on the calculating engine. For such political reasons support for the Office was stricken from the Naval Appropriation Bill for 1859. Davis once again had to come to its rescue, arguing that "Hardly a single civilized nation considers its naval equipment complete without a Nautical Almanac. Six thousand copies of this year are spoken for; ten thousand will soon be the annual sale. The sale is constantly increasing, and the American is fast taking the place of the British Almanac in our own market."[36]

[35] Newcomb *Reminiscences*, pp. 88–95; AR, November 20, 1860, 225–229 has a detailed description of the eclipse expedition. At the request of Agassiz, Newcomb and Ferrel were accompanied by Samuel H. Scudder, from Harvard's Museum of Comparative Zoology. See chapter 5, ref. 54.

[36] Newcomb's, *Reminiscences*, p. 82, places the "Memorandum Concerning the Objects and Construction of a Nautical Almanac," 11–12, and "Memorandum on the American Ephemeris and Nautical Almanac, showing its special and peculiar merit and Utility," in context. Both were published as *Two Memoranda on the Objects and Construction of the American Ephemeris and Nautical Almanac* (Welch, Bigelow, and Co.: Cambridge, 1860). On the trials with the Dudley calculating engine see AR, September 11, 1858, 441, and on the Dudley Observatory controversy Mary Ann James, *Elites in Conflict: The Antebellum Clash over the Dudley Observatory* (Rutgers University Press: New Brunswick, 1987).

The Office once again survived, and Davis went on to bigger battles, both literally and figuratively, during the Civil War. While Winlock continued the Almanac work with his staff, in 1862 Davis, now a rear-admiral, was placed in charge of the newly formed Bureau of Navigation, under which the Nautical Almanac Office and the Naval Observatory were now placed. Events were converging that would finally place the Office in Washington, though not yet at the Naval Observatory, where Davis became Superintendent in 1865. With the end of the Civil War in the same year, the departure of Winlock to Harvard, and the move to Washington in early July, 1866, the Nautical Almanac Office entered a new era, which we shall pick up again in chapter 8 in connection with the work of Simon Newcomb.

# 4 Gilliss and the Civil-War years

On the memorable 15th of April, 1861, Commander Maury fled from his post at the Naval Observatory . . . A day or two later, orders were issued to Gilliss to assume the charge of the Institution, and poetic justice, though long deferred, was at last fulfilled. The sudden transformation which took place was like the touch of an enchanter's wand. Order sprang from chaos, system from confusion, and the hearts of the faithful few who had struggled on for years, hoping against hope, were filled with sudden joy.

B. A. Gould, 1867[1]

James Melville Gilliss took up his duties as Superintendent of the Naval Observatory on April 23, 1861, three days after Maury had headed South to join the rebel cause.[2] For Gilliss this must have been an emotional moment, though perhaps not as heart-wrenching as departure had been for Maury. It had been 25 years since Gilliss was first ordered as assistant to Wilkes at the Depot of Charts and Instruments on Capitol Hill, 19 years since he had successfully shepherded an authorization for a new Depot through the halls of Congress, and 17 years since Maury had been chosen as the first Superintendent, to Gilliss's great disappointment. At last he had the opportunity to put his personal imprint not only on the buildings of the institution, but also on its programs. He would lose no time in doing so. Within view of the half-finished Capitol dome (Figure 4.1), symbol of a nation with unfinished business, Gilliss had his own rebuilding to do.

---

[1] "Biographical Notice of James Melville Gilliss," BMNAS, 1 (1867), 54. See chapter 2, where I note Gould's obvious bias against Maury in the wake of the Civil War. As stated in chapter 1 (ref. 48), Gould is the best (if biased) source on Gilliss's life, whereas the best source on his genealogy and family is Frances Howard Ford Greenidge, *Ancestors of Raymond Oakley Ford and Frances Howard Ford* (privately published, 1994), volume 1, Naval Observatory Library. Another major source for this chapter, in addition to Gilliss's professional life documented in the Naval Observatory records of the National Archives, is Gilliss's correspondence with George Perkins Marsh during 1849–64, which is preserved at the University of Vermont. Marsh (1801–82) was a Congressman, diplomat, and philologist who served in Congress from 1843 to 1849. He later became minister to Turkey (1849–53) and Italy (1861–82), and became well known in intellectual circles especially for his *Man and Nature* (1864). His *Life* was written by his widow in 1888. See also David Lowenthal, *George Perkins Marsh, Versatile Vermonter* (New York, 1958). Transcriptions of the Gilliss–Marsh letters are deposited in USNOA.

[2] Gilliss's first letter as Superintendent was penned to Secretary of the Navy Gideon Welles, informing Welles that he had assumed charge of the Observatory on April 23, in accordance with the orders of the previous day. JMG to Welles, April 23, 1861, RG 78, Entry 1, volume 19. Maury's resignation letter to President Lincoln is dated April 20, 1861, RG 78, E 1. Gould in ref. 1 was therefore in error by a few days.

Figure 4.1. A balloon view of the City of Washington, summer, 1861. The partially completed Capitol Building is in the foreground. At the top right, on the near side of the Potomac River, are the partially completed Washington Monument, and the Naval Observatory, with a canal running between them. Wood engraving from *Harper's Weekly*, July 27, 1861.

## 4.1 Trials and triumphs

Gilliss, however, had not been idle in the years after his disappointment. In 1845 and 1846 he had prepared for the press his observations made at the old Depot, an effort that resulted in the publication of his star catalog virtually simultaneously with Maury's first volume of observations made at the new Observatory. During the next two years he was assigned to the Coast Survey under Bache, where he worked on reducing Bond's longitude observations (which had been made simultaneously with Gilliss's at the time of the Wilkes expedition) and tackled the increasing number of similar observations now being made under the Survey to determine longitude differences.[3]

---

[3] The Coast Survey annual reports for 1848 and 1849, under the title *Letter from the Secretary of the Treasury communicating the report of the Superintendent of the Coast Survey, showing the progress of that work*, detail Gilliss's duties during these years. At Walker's urging, after Gilliss had completed the reduction of his own observations, Bache applied through the Secretary of the Treasury to the Navy Department for the services of Gilliss. "The application was met in the spirit of encouragement to science which has characterized the action of the honorable Secretary," Bache wrote. Wilkes had turned over Bond's unreduced observations to the Survey in 1846, and in 1847 Gilliss reduced Bond's longitude observations of 1839, until Bond himself (apparently unexpectedly) furnished the reductions, which had probably been calculated by Peirce. Gilliss remained as "assistant U. S. Coast Survey" until November, 1848, when he began preparations for the Chilean expedition described below. See the Coast Survey reports for December 10, 1846, pp. 32 and 34; December 15, 1847, p. 50; December 12, 1848, p. 115; and December 27, 1849, p. 54.

SANTA LUCIA.

Figure 4.2. The observatory in Chile, which Gilliss founded and from which he made observations for determination of the solar parallax. From Gilliss, *The U. S. Naval Astronomical Expedition to the Southern Hemisphere during the Years* 1849–50–51–52 (Government Printing Office: Washington, 1855–56). This illustration is the frontispiece to volume 1 (1855). The observatory includes both buildings on the hilltop, one for the equatorial and the other for the transit circle.

In August, 1848, six years after Congress had funded the new Observatory, Gilliss once again showed his skill in shepherding appropriations through Congress. Having sought and won the support of the American Philosophical Society and the American Academy of Arts and Sciences, Gilliss succeeded in that year in extracting $5,000 for a Naval astronomical expedition to Chile.[4] The purpose was to determine the solar parallax by observations of Mars and Venus – an alternate method to the transits of Venus for determining the scale of the solar system, which had been proposed to Gilliss by Christian Gerling of Germany. From August 16, 1849 until its return on November 16, 1852 Gilliss headed this expedition in Chile (Figure 4.2), leaving on his departure the core for a Chilean National Observatory, thus having been instrumental, as Gould remarked, in the beginnings of two national observatories

[4] Under a clause of the act making appropriations for the Naval service, approved August 3, 1848. On January 26, 1849 $6,400 more was appropriated for the purchase of instruments, including equatorials of 6.5- and 4-inch aperture, a 4.5-inch Pistor and Martins transit circle, an astronomical clock, three chronometers, and magnetic and meteorological instruments. Reflecting the increasing expertise of American artisans, the mechanical parts of the 6.5-inch equatorial were made by William Young of Philadelphia, its lens from French material by Henry Fitz of New York, and its micrometer by William Würdemann of Washington.

thousands of miles apart. Much of the 1850s Gilliss spent preparing the elaborate multivolume set of results of this expedition, which included not only astronomical observations, but also observations of broader interest to science, geography, and politics.[5]

Because of his long shore duty without time at sea, Gilliss in 1855 suffered the same fate as Maury in being placed on the reserved list of the Navy with furlough pay.[6] He was only 44. However, his continuing keen interest in astronomy is made evident by his participation in observing the total solar eclipse in Peru in 1858,[7] and that in Washington Territory in 1860.[8] The Observatory was never far from his mind; indeed,

[5] Gilliss, *The U. S. Naval Astronomical Expedition to the Southern Hemisphere during the Years 1849–50–51–52* (A. O. P. Nicholson: Washington, 1855–56). These volumes were numbered 1, 2, 3, and 6. The first volume described Chile as a nation, and included a detailed narrative of the voyage. In volume 2 eminent scientists described the mineral, animal, and fossil collections returned by the expedition. Volume 3, almost 500 pages, is a detailed description and analysis of the astronomical observations to determine the solar parallax, including the origins and operations of the expedition. Volume 6 was devoted entirely to magnetic and meteorological observations. Volumes 4 and 5 were not published until much later: *A Catalogue of 1963 Stars Reduced to the Beginning of the Year 1850, together with a catalogue of 290 Double Stars* (Government Printing Office: Washington, 1870), and *A Catalogue of 16,748 Southern Stars* (Government Printing Office: Washington, 1895). For a discussion of this expedition see Wendell Huffman, "The United States Naval Astronomical Expedition (1849–52) for the Solar Parallax," JHA, **22** (1991), 208–220. This is part of Huffman's M. A. thesis, "James M. Gilliss and the American Determination of the Solar Parallax," (Norman, Oklahoma, 1987). On the founding of the Chilean National Observatory, see Philip C. Keenan, Sonia Pinto, and Hector Alvarez, *The Chilean National Astronomical Observatory (1852–1965)* (Universidad de Chile: Santiago, 1985).

[6] Gilliss's strong feelings about this action are detailed in his letters to Marsh beginning September 21, 1855. On that date he wrote Marsh of "the turmoil which has rendered me nervous beyond measure," arguing that shore duty was not his own choice: "I *never* applied for shore duty unless the application to Congress for the Chile expedition can be so considered. But the Department first ordered me as Assistant to the Depot: then kept me there to make observations for Wilkes against my will, and finally, when the Mexican War broke out and on one blessed Sunday morning I placed an application with the Secretary for active service, I was ordered to report to Prof. Bache [for the Coast Survey job]." In contemplating action to overturn the furlough ruling, Gilliss wrote a month later "Unquestionably Maury is a far stronger man with the mass [of people]; but I have the undivided support of a class who measurably repudiate him, the scientific men, and I have but to say the word and they will express their indignation most heartily." JMG to Marsh, November 21, 1855, USNOA. Because of these actions, Gilliss even contemplated leaving the Navy, and, in May of 1857, he was angling for the Directorship of an observatory proposed at the newly inaugurated St Louis University, with letters of recommendation from Humboldt, Peirce, and Argelander among others. JMG to Marsh, May 20 and 27, June 1 and 19, July 1, 13, and 24, 1857, USNOA. Gilliss held out hopes for this possibility for two years, and finally gave up hope in a letter to Marsh of August 18, 1859.

[7] Gilliss, "An Account of the Total Eclipse of the Sun on September 7, 1858, as Observed near Olmos, Peru," *Smithsonian Contributions to Knowledge*, **11** (1859), 1–18. This expedition required no special Congressional appropriation, but was undertaken with the cooperation of the Smithsonian, the Coast Survey, and the Naval Observatory, and with a small equatorial telescope supplied by Henry Fitz. Gilliss's detailed description is notable for its graphic description of the voyage, his severe illness, the eclipse itself, and the reaction of local residents. The sight of the corona, Gilliss wrote, "filled me with excitement and humble reverence." Gilliss was accompanied on this expedition by Mr Carrington and H. Raymond, a friend from New York who was a 17-year old student at Yale at the time. Nine years later Raymond married Gilliss's daughter Rebecca Melville Gilliss. Raymond's diary of the Peruvian trip, which was donated by Gilliss's great grandchildren Col Raymond O. Ford and Mrs Ralph M. C. Greenidge, is in the Naval Observatory library archives. On the origins of the expedition see also JMG to Marsh, June 10, 1858.

[8] "Solar Eclipse of 1860, July 18, observed near Steilacoom," AJ, **6** (1861), 155–157. The location of Gilliss's party was about 35 miles south of Seattle, in sight of Mt Ranier. Gilliss was no less struck by the sight of the corona during this eclipse. "Lifting my face from the box on which the time-keeper stood to the telescope, a most extraordinary scene was apparent! Over the moon's

in the late 1850s Gilliss and his friends contemplated a plan to take the Observatory from Maury. "I know that Peirce etc., that is the Trinity of Cambridge, Coast Survey and Smithsonian will sustain me either to get or to retain the observatory," Gilliss wrote his friend George P. Marsh in 1858. "But you say truly – it would be at the expense of incessant war against the ignorant bodies of higher rank who think themselves competent to conduct the establishment and yet for the life of them could not solve $2ab + 1cd - 2ab + 1cd$." A week later he again wrote Marsh: "Peirce and Gould say I shall go to the Observatory when M[aury] leaves it – if science has any weight in the land." After Gould had derived a new value for the solar parallax based on Gilliss's Chilean observations, the latter wrote "I suppose some of my astronomical friends will take occasion to say a word about it for the purpose of helping me to the observatory when Matthew [Maury] takes his dose of tar and ropes. Bache, Gould, Peirce, Henry and two or three others met at Philadelphia a week ago and [drafted?] a document to be used when the time comes. I have not seen it, but Henry tells me it could not well be stronger. He is wroth [?] at the suggestion that Wilkes and Goldsborough may demand the place and vowed he'd go right after the Secretary [of the Navy] . . . to tell him how extensive a humbug the Captain (Wilkes) is."[9]

Now, in April, 1861, Gilliss finally took on the job many, including Gilliss himself, felt he should have had all along. After 23 years as lieutenant, he was promoted to commander in July, and captain the following July. But he took on his new post with the country in desperate circumstances. On April 14 the Confederate flag replaced the American flag over Ft Sumter, South Carolina. The next day Lincoln called for 75,000 militiamen to be put into national service, and the Civil War had begun. "Our city is in a fearful state brought about by the policy of the administration in the Sumter and Pickens attempts contrary to the advice of Generals Scott and Fotten," Gilliss wrote on April 18, a few days before he assumed his post. "I fear we have a weak minded and timid man as our ruler who thinks more of his party and sharing the spoils among them than of the imminent perils of our country. What the result will be no living man can foretell," Gilliss wrote referring to President Lincoln. On the individual level the disaster of Civil War was no less pressing. "Tens of thousands, wholly unconnected with business, have ruin and poverty staring them in the face and the next twenty four hours will probably decide whether anything is left to us of my wife's patrimony and my twenty years savings."[10]

(footnote 8 cont.)
black disk colors of the spectrum flashed in intersecting circles of equal diameter with that body; and each apparently revolving toward the lunar center. These moving colors were not visible beyond the moon, but a halo of virgin white light encircled it, which was quite uniformly traceable more than a semidiameter beyond the black outline. This corona was composed of radial beams or streamers, having slightly darker or fainter interstices, rather than a disk of regularly diminishing or suffusing light. But the gorgeous appearance of the spectrum circles, with their incessantly changing bands of crimson, violet, yellow, and green, thoroughly startled me from the equanimity with which the preceding phenomena [of the eclipse] had been observed." Gilliss's son James accompanied him on this expedition.
[9] JMG to Marsh on January 25, 1958; JMG to Marsh, February 3, 1858; JMG to Marsh, February 22, 1858.    [10] JMG to Marsh, April 18, 1861.

Even under these foreboding circumstances – perhaps in part because of them – Gilliss would transform the Naval Observatory during the four years of his tenure – years that almost exactly matched the years of the Civil War. He did so in at least three ways: by effecting the rapid publication of its work, both past and present; by additions to the staff of young but promising astronomers who would usher in a golden era for the institution for the remainder of the century; and by setting the institution on a new course in astronomy by virtue of upgrades of instruments and programs. Finally, it is well to remember that the institution still functioned as a Hydrographic Office, and, although the wind and current charts no longer preoccupied the office, Gilliss was still responsible for providing the American Navy with the charts and nautical instruments now crucial for the Civil War effort. All of these tasks he performed in the midst of a nation with growing awareness of the importance of science, as evidenced by the founding of the National Academy of Sciences in 1863, an organization in which Gilliss too played his role.

Upon his arrival Gilliss found most of the instruments he had left 18 years earlier, though now altered in some details, placement, and accouterments. In the west wing stood the Ertel & Sons 5.3-inch meridian transit instrument, accompanied by a Parkinson and Frodsham clock, furnished with a mercury globule through which the pendulum passed in order to make the electrical connection to register the observations on a Morse fillet. Maury's much-ballyhooed Locke electro-chronograph stood unused, testimony to its apparent failure as an operational recording system. The east wing held both the mural circle, unused for several months, and the Ertel & Sons meridian circle, remodeled and furnished with new circles by William J. Young of Philadelphia, which also remained unused during 1861. A Frodsham clock similar to that in the west wing was also used with these instruments, with similar electrical hookups. In the south wing Gilliss found a third Frodsham clock and a prime vertical telescope that had not been used for several years, requiring considerable repairs before work with it could be resumed. The equatorial in the dome was much as it had been, but it was fitted with a chronographic apparatus with Bond's spring governor and cylinder, connected by battery to the clock in the south wing.[11]

On the nautical side Gilliss found the other room of the Observatory's east wing wholly devoted to the chronometer-rating function. Cases and shelves were filled with chronometers being rated or on trial, and present also was yet another Parkinson and Frodsham pendulum clock that was for the rating process "more satisfactory than that of any other one belonging to the establishment" except for a Kessels clock. A clock by Eggert and Son was also in this room, apparently the only American clock on the premises. In the central building were the rooms for the Superintendent, the Assistants, and the charts.

Only a few months after Gilliss had taken charge Hubbard wrote to his former

---

[11] *Astronomical and Meteorological Observations Made at the United States Naval Observatory, During the Year 1861*, (WO), Introduction, pp. vii–ix.

Observatory colleague J. H. C. Coffin in Rhode Island "You will be pleased to learn that matters here are rapidly tending toward a proper astronomical activity." The prime vertical and refraction circle were being rejuvenated, and "improvements are going on also in the chart and nautical instrument departments, and the introduction of order, energy and astronomical science, will doubtless soon bring us up to the position which such an institution as this ought to occupy."[12]

In his private correspondence Gilliss himself minced no words regarding his opinion of Maury and his administration of the institution. The Observatory, Gilliss wrote three months after he took over, "was in a sad state. Without order, system or responsibility of any kind. Some of its astronomical instruments literally buried under rubbish; another almost irreparably injured by neglect and two of the observing rooms used as store rooms for useless matter. $10,000 worth of instruments returned from ships and needing only slight cleaning or repair had been stowed away in order to purchase new ones and by the dispensation of public money inspire the recipients to subscribe for 'Maury testimonials.' I write in all sincerity and all sadness that a man in such position before the world should have left such evidences behind him."[13]

There is no doubt the new Superintendent had his work cut out for him, but where was he to start? Some of Gilliss's immediate actions on taking charge were dictated to a large extent by the Civil War that had vacated the position he now filled. Not only was Maury gone, but also most of the military employees had gone off to war on one side or the other, and one of the Professors of Mathematics (Thomas J. Robinson) had also joined the rebel cause, precipitating quick adjustments in the staff. Gilliss had his principal assistant appointed Professor, giving his son Jack the resulting position at $1,200. Gilliss also had a nephew of his wife appointed to another Navy position outside the Observatory, an event that would be insignificant except for Gilliss's revealing remark to Marsh "so you perceive that I have some influence with the powers that be. Indeed, there is the utmost kindness and cordiality between the Secretary, Harwood (now Chief Bureau Ordnance, etc. and my immediate chief) and all in the Department. [Secretary of the Navy] Mr. [Gideon] Welles intends to give me my rank too." These cordial internal circumstances undoubtedly account for much of Gilliss's success in carrying out his duties and restoring the Observatory. "Of course there was much for me to do," Gilliss wrote, "and there were left few hands to do it with for all the military men had gone to Norfolk and on other pressing service. But the Professors volunteered at once; I made Depots of instruments and charts at New York and Boston, to secure against obstruction to the navigation of the river and when mil-

---

[12] J. S. Hubbard to J. H. C. Coffin, May 20, 1861, RG 78, E 1, volume 19. The purpose of the letter was to inform Coffin that Yarnall had gone to Annapolis to dismount and pack for removal the astronomical instruments at the Naval Academy, and that the transit circle, equatorial telescope, clock, and telegraphic apparatus were being brought to the Naval Observatory. I have been unable to discover the reason for this removal, or the subsequent fate of the instruments.
[13] JMG to Marsh, July 19, 1861, USNOA.

itary duties demanding prompt and energetic action were completed, the Secretary and Harwood were brought here to examine the condition in which Maury had left matters. Authority was given me on the spot to restore the establishment to its original purposes and now I no longer blush when visitors come in. More than this, I restored harmony, on the spot, between Hubbard and Ferguson who had not spoken for years, and if I can only get help next year by the beginning of 1863 the Observatory will no longer rest under the stigma so long pressing upon it."[14]

These preliminaries out of the way, Gilliss's immediate goal was to address what many perceived as the chief failing of the institution under Maury. "One of my first efforts to remove the cloud will be by publishing the 10 years of Astronomical Observations and 19 years of Meteorological and Magnetic observations left as my legacy," Gilliss wrote. By the first summer of his tenure the new Superintendent had formed a commission to estimate the amount of incomplete astronomical work at the Observatory and to recommend how it could be completed. The solution may be characterized either as poetic justice or revenge, for with Maury gone and very little sympathy for him or his past work in the Congress, Gilliss approached Congress with the proposal to redress the astronomical backlog at the expense of the hydrographic work closest to Maury's heart. "Both Naval Committees agreed to allow me the unexpended balance to credit of 'Wind and Current' charts, for employing additional computers and it may have been put in as one of the amendments to the bill passed yesterday. I shall give Gould a contract for all the Astronomical work, so as to take it out of the observatory. The other work can be done here by young men of moderate capacity," Gilliss reported.[15] Even though Maury's charts were undoubtedly useful in the open ocean, with passions high and sights now set closer to home, further research and publication was halted, although distribution was continued. By July, 1861, Congress had authorized that the sums remaining in the treasury for "printing and publication of sailing directions, wind and current charts, astronomical observations and hydrographic surveys" be applied for the unfinished astronomical work left by Gilliss's predecessor.

In order to carry out the reductions, Gilliss proposed to Secretary of the Navy Welles that six young men at $50/month copy from the record books onto sheets all the astronomical observations; that Ferguson at the Observatory, Coffin at the Naval Academy, and Humphreys of the Topographical Engineers estimate the cost of the reductions and the time necessary to carry them out; and that B. A. Gould (Figure 4.3) be approached to carry out the work. With Gould's acceptance of this scheme in early October, 1861, Gilliss was well on his way to redressing the problem of unpublished

---

[14] Ibid.

[15] Ibid. Gilliss later estimated that $212,751 had been expended on Wind and Current Charts and Sailing Directions during the Maury era. In terms of man hours, he estimated 60 years for one lieutenant and 35.5 years for one midshipman added another $116,625, making the total cost to the government $329,376. JMG to C. H. Davis, RG 78, E 2, June 6, 1863.

Figure 4.3. B. A. Gould, seen here in 1864, helped analyze the backlog of observations. From the Mary Lea Shane archives of the Lick Observatory.

astronomical observations.[16] For months thereafter a steady stream of the copied sheets of observations passed between Washington and Gould in Cambridge. Gould, Maury's old enemy, was given two and a half years to complete the necessary computations and prepare the reduced observations for the printer, but the first volume of backlogged observations seems to have been published already in 1862.[17] Gilliss must have delighted in redressing one of Maury's failings literally at the expense of Maury's cherished wind and current charts. During his tenure the astronomical observations were consistently published within a year or two, even in the midst of war.

The publication of these backlogged observations began to redress the past, but Gilliss was looking to the future when he began a concerted effort to increase the staff beyond the initial shuffling necessitated by the onset of the war. The first of the appointments, occasioned by a vacancy among the Professors of Mathematics, was auspicious to say the least. In August, 1861, having been informed by Gould of a vacancy in the Corps of Professors of Mathematics at the Naval Observatory, young Simon Newcomb at the Nautical Almanac Office in Cambridge applied for the posi-

---

[16] JMG to Welles, July 25, 1861 and July 30, 1861; JMG to B. A. Gould, September 28, 1861. Gould accepted in a letter to JMG of October 3, 1861, and copies of the Commission's report and the correspondence with Gould were forwarded to Welles on October 7. The contract with Gould was dated October 9, 1861. RG 78, E 1, volume 19. Note that Gould had also reduced Gilliss's parallax observations from the Chilean expedition, JMG to Marsh, November 14, 1853.

[17] Although the title page carries the date 1867, both the Preface and the Introduction by Gilliss are dated November, 1862. Even with the possibility of delays due to the Civil War, it is likely that the actual publication date is 1862 rather than 1867. In a letter to Marsh of April 26, 1862 Gilliss speaks of the volume being "in press" and due out in a few months. In any case, the zone observations from 1846 to 1849 reduced by Gould were not published until 1872 and 1873, and the other meridian observations made from 1853 to 1860 reduced by Yarnall were not published until 1872, testimony to innate difficulties not peculiar to Maury or his tenure.

tion. Having given as references Gould, Davis, and Joseph Henry, and with his contacts at the Nautical Almanac Office, Newcomb apparently had to take no examination, and was surprised in the following month to receive his commission signed by Abraham Lincoln. On October 6, 1861 Newcomb began work at the Observatory, and over the next three decades would become America's foremost astronomer.[18]

The other appointments did not begin at such a high level, but some of the incumbents eventually rose in terms of scientific achievement to be nearly equal to Newcomb. Their positions were a direct result of the Civil War, which drew away military officers for duty and for which Gilliss succeeded in having Congress authorize four replacement "Aids." On April 13, 1862 Asaph Hall wrote from Cambridge that he had heard that four new assistants were to be hired at the Observatory. Hall was interested, but J. D. Runkle at the Almanac Office had told him that having a family might be a problem, and Hall now inquired of Gilliss what his chances might be. Gilliss responded that Congress had not yet approved the new positions, that he would accept no applications as yet, and that preference would indeed be given to those without families. Newcomb later recalled the details of Hall's appointment: ". . . the first aide was Asaph Hall; but before his appointment was made, an impediment, which for a time looked serious, had to be overcome. Gilliss desired that the aide should hold a good social and family position. The salary being only $1000, this required that he should not be married. Hall being married, with a growing family, his appointment was long objected to, and it was only through much persuasion on the part of Hubbard and myself that Gilliss was at length induced to withdraw his objections." Hall nevertheless applied in July, authority for the positions having been given in the meantime by Congress. Gilliss informed Hall that all applicants were "required to give evidence of theoretical intelligence of the construction and adjustment of instruments and such mathematical knowledge as will enable them promptly to apply formulae used in the reduction of every class of astronomical observations." Gilliss said that he would ask for Hall's appointment by the Secretary of the Navy provided that his examination "proves satisfactory." Apparently it did, and, on July 25, 1862, Hall received his appointment (to take effect August 1) as Aid.

On the same day as Hall received his appointment Gilliss dispatched letters of appointment to William Harkness at Fishkill Landing, New York and Rev Moses Springer of Minnesota, who had also applied for positions as Aids.[19] J. R. Eastman was the fourth Aid appointed. Within a year Hall and Harkness had advanced to Professors of Mathematics, followed by Eastman in early 1865.[20] In addition, Gilliss

[18] Newcomb, *Reminiscences of an Astronomer* (Houghton, Mifflin & Co.: Boston, 1903), pp. 97–99.
[19] JMG to Asaph Hall, April 17, 1862; JMG to Hall, July 17, 1862; JMG to Hall July 25, 1862. Newcomb, *Reminiscences*, p. 107.
[20] Hall was commissioned a Professor of Mathematics on May 2, 1863 upon the resignation of Hesse, Harkness was promoted in August, 1863 with the death of Hubbard, and Eastman in February 1865 with the death of Pendleton. In recommending to the Secretary of the Navy that Hall replace Hesse, Gilliss noted that Harkness was equally meritorious, but Hall was given preference only because he had a family dependent on him. JMG to Welles, April 27, 1863, RG 78.

succeeded in bringing more Professors of Mathematics to the Observatory; the four positions at the beginning of his tenure had grown to seven at its end.[21] Among them was Joseph E. Nourse, whose father had fought in the Revolutionary War and whose family had established itself as a prominent family in Washington, which was undoubtedly known by the Gilliss family. Nourse became a Presbyterian clergyman in 1849, was a professor of ethics and English studies and acting chaplain at the U. S. Naval Academy from 1850–64, and became a Professor of Mathematics in the U. S. N. in 1864. One of Gilliss's last acts was to hire Nourse in 1865 at the Observatory, where he remained until 1879. During this period Nourse would serve as the Observatory's librarian and historian; it was Nourse who composed the *Memoir of the Founding and Progress of the U. S. Naval Observatory* (1873) at the request of Superintendent Benjamin F. Sands. As we shall see in the following chapters, for the generation following Gilliss these Civil War appointees would dominate the work of the Naval Observatory.

Not only did Gilliss enjoy the complete support of his superiors and surround himself with a promising young staff, but also an administrative event beyond his control had a positive effect on the Observatory: in 1862 it would move from the Bureau of Ordnance and Hydrography to the newly created Bureau of Navigation. Although this was done at the recommendation of Gideon Welles in reports of July 4 and December 2, 1861, the motive force for this reorganization is revealed by Gilliss as political in origin. In early 1862 Gilliss noted that C. H. Davis, now playing an important role in the War, "expects at the end of the war to be Chief of the Bureau of Hydrography, or Longitude or Navigation if a title can be agreed upon and Congress persists in legislating for individuals as they propose in order to give Dahlgren Ordnance all to himself."[22] In fact the Bureau of Navigation was created as part of a larger reorganization of the Navy Department by an act of July 5, 1862; the Naval Observatory and Hydrographical Office and the Nautical Almanac Office were placed under it on August 31, and Davis became its permanent chief on November 15 of that year. Thus hydrography was separated from ordnance and placed under a more scientific bureau, headed by someone who knew a great deal about nautical science, which was surely a good move for the Observatory. At Davis's urging the Observatory was affected even more in 1866, when hydrography would be completely separated from the Observatory, and Davis himself became Superintendent.[23]

In the larger context of the Observatory's rejuvenation, the Civil War continued to dominate all aspects of individual and institutional life in Washington. Newcomb and Hall's son Angelo were among the many who described Civil-War Washington. "An endless train of army wagons ploughed [the] streets with their heavy wheels . . . After a rain, especially during winter and spring, some of the streets were much like

---

[21] Pendelton, Yarnall, Hubbard, and Hesse at the beginning; Yarnall, Beecher, Newcomb, Hall, Harkness, Nourse, and Eastman at the end.    [22] JMG to Marsh, April 26, 1862.

[23] Henry P. Beers, "The Bureau of Navigation, 1862–1942," *The American Archivist*, **6** (October, 1943), 212–252.

shallow canals. . . . By night swarms of rats, of a size proportional to their ample food supply, disputed the right of way with the pedestrian." In the summer of 1862, when Hall arrived, the city was in turmoil. "When I see the slack, shilly-shally way the Government has of doing everything, it appears impossible that it should ever succeed in beating the Rebels." Less than a month after Hall's arrival, the second Battle of Bull Run was fought a few miles away, and at the observatory the astronomers "could hear the roar of the cannon and the rattle of musketry." Hall and others went to hunt for wounded friends, finding his wife's step-brothers. After Antietam Hall took ill, was confined for six weeks indoors with jaundice, and did not fully recover for two years.[24]

Anxious as Gilliss was to reinvigorate astronomy by publishing past and present observations and adding new staff, it was the U. S. Navy's Civil-War needs that incessantly drove his activities. "As you may well suppose, the year past has been a busy one with me, but having the entire confidence of the Secretary and my immediate Chief (Harwood) . . . I am not hampered in any manner. To obtain supplies for 250 vessels against 30 previously at sea and execute the mere business writings inseparable therefrom taxed every ability I had. But it is a source of sincere gratification to say, no vessel was ever delayed an hour by me, and the money expended has been to our own people, who make me better instruments of every class, except spy glasses, and at considerably less cost, than was previously paid foreigners. Every appeal to Fitz and Alvan Clark to make spy glasses has so far been in vain. Yet I do not despair and when I go north next week I shall try them again, for I am determined so far as this establishment is concerned, to keep the money home."[25]

Gilliss's policy that as many naval instruments as possible should be of American manufacture was a prominent theme throughout his tenure, one that illuminates the growing capabilities of American artisans as well as the common theme among American scientists to build up American companies. He succeeded in this policy only with difficulty, as evidenced in the case of the spyglasses mentioned above. Writing both to Alvan Clark and to Henry Fitz in 1861, Gilliss reported that nearly all spyglasses and binoculars on American ships were imported, and none would define objects more than a mile distant. "It is discreditable to us as a manufacturing nation," he wrote. He needed 50 binoculars to start with, and "they must show a black rod one inch in diameter at the distance of a mile, have metal frames, covered with leather or shark's skin, and be jointed so as to fit eyes at different distances apart." When Clark declined a few months later Gilliss admonished him: "Occupying as you do the first rank among opticians, you do injustice to the genius and skill of our country." Gilliss wanted to be able to say not only that U. S. ships were provided with wholly U. S. instruments, but also that they were better than Europe's – something he felt at the

---

[24] See Newcomb, *Reminiscences*, chapter 12, "The Old and the New Washington," and Angelo Hall, *An Astronomer's Wife: The Biography of Angeline Hall* (Nunn and Co.: Baltimore, 1908), chapter 13, "Washington and the Civil War."

[25] JMG to Marsh, April 26, 1862. Gilliss claimed that Fitz "owes a good deal of his fame to the start I gave him," JMG to Marsh, March 25, 1858.

Figure 4.4. Alvan Clark (center) with his son Alvan Graham Clark (left) and George Bassett Clark (right).

end of 1861 he could say of all nautical instruments except for the optical ones.[26] Exasperated by his failed efforts with Clark and Fitz, the following year Gilliss wrote the British Astronomer Royal George B. Airy, through whom he ordered 30 telescopes from London.[27] By 1865, however, Gilliss had apparently persuaded Clark to manufacture binoculars for the Navy.[28] By this time he had come to know Clark quite well, both as a businessman and as a friend, the beginning of Alvan Clark and Sons' long relationship with the Observatory (Figure 4.4).[29] Whereas at the beginning of Gilliss's

[26] JMG to Clark and Fitz, August 19, 1861; JMG to Clark December 4, 1861, RG 78, E 1, volume 19.
[27] JMG to Airy, April 17, 1862, RG 78, E 1, volume 19. The telescopes were purchased from Elliott Brothers, Opticians at 30 Strand.
[28] On February 3, 1865 JMG wrote Clark thanking him for sending ten Navy telescopes. At this time Clark also sent a specimen of his first binoculars, which Gilliss found "very creditable," far better than Fitz's sent in 1863. However, Gilliss still pushed Clark to make better ones than the Europeans. RG 78, E 4, Miscellaneous Letters Sent.
[29] On the business side Gilliss in 1861 tried to purchase the 18.5-inch lens that eventually ended up at Dearborn Observatory in Chicago. Gilliss did not directly request the money from Congress, but asked Clark to have the subject brought to Congress through the Academy at Boston or Bowditch. The asking price was $11,187. JMG to Clark, December 12, 1861, RG 78, E 1, volume 19. This strategem did not succeed, but Gilliss did purchase a 3.9-inch lens to replace one stolen from the comet seeker, at a cost of $90. JMG to Clark, April 16 and 23, 1862, RG 78, E 1, volume 19. Clark also polished and improved the lenses of the transit instrument, mural circle, and equatorial telescope, for $25, $40, and $300, respectively. JMG to AC, October 2 and 14, November 10 and 13, 1862. RG 78, E 1, volume 19.

tenure he estimated that 80 percent of the appropriation for nautical instruments went to foreign workshops, his successor could report that all such instruments for the Navy were now of American manufacture, with the partial exception of spyglasses and binoculars, which were made at home but not yet in sufficient numbers, and not yet so well as abroad. No small contribution of Gilliss was therefore the stimulation of American commercial interests in the manufacture of nautical instruments, an impetus that had its beginnings during the Civil-War period.

Among other duties closely related to Civil-War operations, Gilliss had to supply charts to the Union Navy – not wind and current charts to decrease travel times, but coastal and river charts needed to outmaneuver the enemy. To name only two interesting examples, in which Gilliss dealt with past Naval Observatory figures now actively involved in the war, at the end of 1861 and early in 1862 Commander Gilliss forwarded to now-Admiral Louis M. Goldsborough, Commander of the North Atlantic Blockading Squadron, Bache's notes on the coast of the United States and charts of the coast from Point Judith to Cape Lookout, among others. Similarly, he sent to now-Admiral Wilkes, Commander of the James River Squadron, charts of that area, and later of the Potomac when Wilkes became Commander of the Potomac River Flotilla.[30] Gilliss obtained such charts from wherever he could, whether dealers such as Blunt in New York or the Coast Survey Office itself, where he worked closely with Superintendent Bache. Of the 20,000 sheets of charts distributed during the years 1861–62, 18,500 came from the Coast Survey, causing Gilliss to say of his former mentor and supervisor "the prompt and unceasing efforts of Prof. Bache to supply the demand for information resulting from operations under his directions, and so indispensable to vessels of the navy, cannot be too highly appreciated."[31] By 1864 the Observatory was distributing more than 33,000 charts to naval vessels.[32]

The rating and supply of chronometers, normally the most crucial aspect of the nautical duties, took a back seat to the need for charts and optical aids to spot the enemy. Chronometers still had to be supplied to ships, even though the determination of longitude was not a crucial problem for navigation during the Civil War, but the supply was not plentiful. "Owing to the unusual demand for chronometers during the last autumn and winter and the small number of really reliable instruments in the country, the Department reduced the number to be furnished ships," Gilliss wrote to Wilkes, who had complained about his ration of chronometers. "With respect to rating," Gilliss continued, "I would advise that the instrument be left on board the respective vessels and the masters come to the observatory with their comparing

---

[30] JMG to LMG, December 18, 1861 and January 6, 1862; JMG to Wilkes, July 16, 1862 and Septrember 2, 1862, RG 78, E 1, volume 19.

[31] *Astronomical and Meteorological Observations . . . for 1861*, p. x.

[32] In 1864, aside from charts already on hand, Gilliss obtained 16,500 from the Survey and 4,200 from the British Hydrographic Office. The latter included not only original surveys of the British Admiralty, but also reproductions of continental surveys, which the British apparently published and distributed quickly. JMG to CHD, "Report of 20 September, 1864 on Operations of the Observatory during the last year," RG 78, E 1, volume 19.

watches as frequently as they desire. It is scarcely probable that a rate given at the Observatory will remain precisely the same after removal to the ship."[33] In order to remedy the situation Gilliss wrote his superior that the Navy needed to find out more about the facilities at William Bond & Sons in Boston, James Munroe in New Bedford, Thomas Negus & Co., and Bliss and Creighton in New York, and ascertain to what extent these companies were willing to use their facilities for the Navy. After visiting these facilities and others, Gilliss reported that, with Navy patronage, both the Negus firm and the Bond firm could successfully compete with the best foreign establishments, but "they are the only firms upon which the Navy can safely rely for the necessary supply of chronometers." During most of the Civil War the ordinary one year trial of chronometers at the Observatory was waived. Makers reported to the Observatory the rates of their best instruments over 4–6-month periods, and none was accepted whose rate varied by more than one second during the whole interval.[34]

In addition to using correct time for rating ship's chronometers, the Observatory continued the practice of giving time to the city of Washington via a time ball. In 1863, the supply of balls having been exhausted, Gilliss ordered three others to be constructed. As in the past, they were constructed at the Washington Navy Yard and consisted of hoops of hickory three feet in diameter, covered with canvas painted black, and fitted with a loop at each extremity of a diameter, the latter undoubtedly for attachment to the flagstaff atop the Observatory dome.[35]

In the midst of wartime activities, the Observatory had its lighter moments. It was about 1863 that Hall had an unexpected visit that has become part of Naval Observatory folklore. Mrs Hall "took her little boy to one of Lincoln's receptions, and one night Lincoln and Secretary Stanton made a visit to the Naval Observatory, where Mr. Hall showed them some objects through his telescope. . . . Now the great War President, who signed his commission in the United States Navy, talked with him face to face." That was not all: "One night soon afterward, when alone in the observing tower, he heard a knock at the trap door. He leisurely completed his observation, then went to lift the door, when up through the floor the tall President raised his head. Lincoln had come unattended through the dark streets to inquire why the moon had appeared inverted in the telescope. Surveyors' instruments, which he had once used, show objects in their true position." Hall explained to the war-weary President why the optics of an astronomical telescope caused inverted images.[36]

On at least one occasion the Civil War came too close for comfort. In July, 1864,

---

[33] JMG to Wilkes, September 9, 1862, RG 78, E 1, volume 19.
[34] JMG to C. H. Davis, October 7 and 23, 1863; November 14, 1863. RG 78, E 4. In New York Gilliss visited Thomas Negus & Co., John Bliss & Co., D. Eggert & Son, and P. L. de Mory Gray; in New Bedford James Munroe, and in Boston Bond & Son.　　[35] JMG to CHD, August 10, 1863.
[36] Angelo Hall (ref. 24), p. 91. A similar account is repeated in G. W. Hill's *Biographical Memoir of Asaph Hall, 1829–1907* (Judd and Detweiler: Washington, 1908), p. 256. In C. Percy Powell, *Lincoln Day by Day: A Chronology 1809–1865* (Washington, 1960), volume 3, p. 203 we find under the entry for August 22 "In evening President, John Hay, and Mrs. Long visit observatory, 23d and E Sts. NW." See also "Lincoln's Visit to Observatory," *Lincoln Herald*, xli (May, 1939), 23, which has Lincoln returning about 2 o'clock the same night. This brief article also states that "Lincoln's son, Robert, was also interested in astronomy and during his residence in Washington he

Confederate General Robert E. Lee sought to divert the Union's strength by sending a corps of men to attack Washington itself. This corps, led by Confederate General Jubal Early, approached Washington on July 12, when firing was heard north of the city. Newcomb, Hall, Eastman, and Rogers were ordered to report to Admiral Goldsborough at the Washington Navy Yard, while Harkness was sent to Ft Slocum to act as surgeon. "Every officer at the Observatory was anxious to confront the rebel enemy," Gilliss wrote, but he forbade those less physically able to go.[37] At the Navy Yard a brigade was formed by mid-afternoon, and they marched to Ft Lincoln to aid in the defense of the city. Newcomb still recalled his depression during this march, in view of the uncertainty and "extreme gravity" of the situation. "But this depression wore off the next day, and I do not think I ever had a sounder night's sleep in my life than when I lay down on the grass, with only a blanket between myself and the sky, with the expectation of being awakened by the rattle of musketry at daybreak." Newcomb and his fellow astronomers were undoubtedly not disappointed when the rebel troops did not attack the following day. They remained that day and another night, but when the regular army corps arrived, Early and his troops retreated. In the end the attack on Washington did not take place; the real action was elsewhere and the astronomers of the Naval Observatory saw no battle action.[38]

In the midst of these war activities Gilliss and his newly assembled staff, young and anxious, carried out their astronomy. The astronomical observations for 1862 included Newcomb's discussion of the longitude of Washington, Harkness's drawings of Comet II, 1862 (Figure 4.5) and of Mars (Figure 4.6) at one of its periodic 15-year close approaches. Hall and Ferguson busied themselves with an investigation of the solar parallax from observations of the red planet; on the next close approach in 1877 Hall would discover the two moons of Mars.

Aside from the routine observations of the Sun, Moon, and planets now being systematically made, reduced, and published, and meteorological observations that continued unabated, yet another of Gilliss's achievements that would far outlast his Superintendency was the design and purchase of a new 8.5-inch transit circle, an instrument on which Ferguson reported Gilliss "placed much of his affection" and took "much pains in preparing for it a proper site and position, and arranging the details of its future use."[39] Gilliss first made a recommendation for its purchase in his Annual

---

frequently visited the Naval Observatory to make observations." Newcomb's *Reminiscences*, p. 342 describes the public receptions given by Lincoln.

[37] JMG to Davis, Chief Bureau of Navigation, July 12, 1864, RG 78, E 1, volume 19. For the context see James McPherson, *Battle Cry of Freedom: The Civil War Era* (Pergamon Press: New York and Oxford, 1988).

[38] This episode is described in Newcomb, *Reminiscences*, pp. 339–342, and Angelo Hall (ref. 24), p. 91. When Hall did not return home, Mrs Hall took her small son Asaph to the observatory, where she found a note from her husband: "Dear Angie: I am going out to Fort Lincoln. Don't know how long I shall stay. Am to be under Admiral Goldsborough. We all go. Keep cool and take good care of little A."

[39] [James Ferguson], obituary of Gilliss, AN, **64**, no. 1525, 199–202. A copy of the obituary, giving Ferguson as the author, is in RG 78, E 4, Letters to Navy Officers and Secretary of the Navy, February 23, 1865.

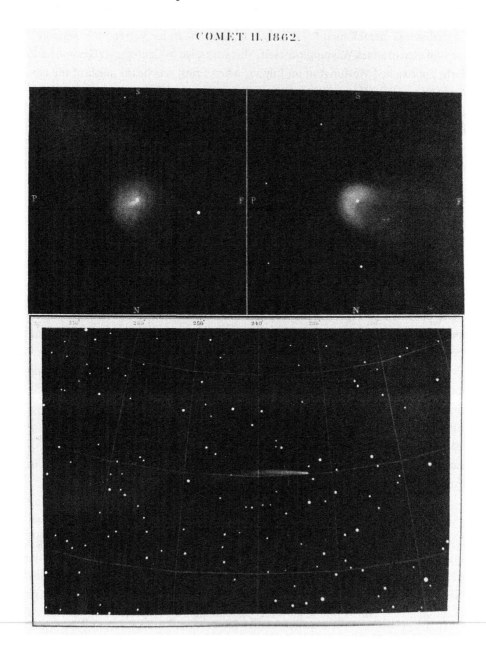

Figure 4.5. Comet II, 1862, drawing by Harkness. From *Washington Observations for 1862* (1863), p. 511.

Report of September 5, 1863, following a statement by Airy that the observation of asteroids was becoming too onerous at Greenwich. The Observatory had been observing asteroids with the transit instrument and mural circle, but only 12 of 81 could be seen with the small lens apertures of those instruments. Gilliss decided that he needed a new instrument to track these asteroids, and, although asteroids were only of peripheral

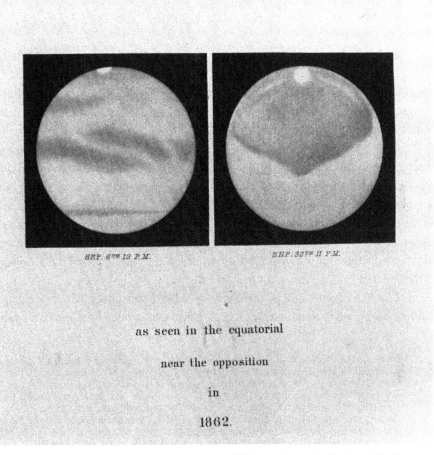

VIEWS

of the

# PLANET MARS,

SEP. 6ᵀᴴ 13 P.M.            SEP. 30ᵀᴴ II P.M.

as seen in the equatorial

near the opposition

in

1862.

Figure 4.6. Mars at the close approach of 1862, as viewed with the 9.6-inch refractor and drawn by Harkness. From *Washington Observations for 1862* (1863), p. 513.

interest to the Navy, Admiral Davis, now Chief of the Bureau of Navigation, immediately approved his request. After deciding on the dimensions for the new instrument, Gilliss traveled in November, 1863 to the Ann Arbor and Dudley Observatories to view their transit circles before deciding on final specifications. "I wanted an object glass which would show all the asteroids and less than 8″ (French) would not do it," Gilliss recalled in 1864. "The focal length will be 12 [feet] necessitating the breadth of building and height of ceiling preparing for it. It ought to be the very first instrument of its class in the world; for though Airy and LeVerrier have nearly similarly sized ones I have avoided their defects and profited by the practical knowledge neither of them ever had opportunity to acquire." Much as Gilliss would have liked to buy one of American make, the order for such an important and complex instrument went to Pistor and Martins in Berlin. The arrival of this instrument, Newcomb later reported, was in his eyes "the greatest event in the history of the Observatory," though subsequent usage showed the difficulties associated with using such "monster" transit-circle telescopes.[40]

## 4.2 End of an era

Events came quickly in 1865, for the Observatory and the nation. The end of the Civil War was in sight, as the Southern forces were worn down, and headed toward that inevitable climax that took place on April 9, 1865 when Lee surrendered to Grant at the Appomattox Court House.

Closer to home, on January 24 Gilliss lent a helping hand during a disastrous event at the Smithsonian. Joseph Henry and his assistant were working in Henry's office in the main tower of the Smithsonian in Washington when a piece of ceiling tile fell on Henry's desk and they realized that the Smithsonian was going up in flames. As the Franklin No. 1 Company and other fire-fighters converged on the institution, Gilliss arrived by horse and buggy from the Observatory. He helped Henry and others carry material from the museum. The fire destroyed the upper floors of the main part of the building.[41]

Closer to home still, on February 8 Gilliss's son James returned from a Confederate prison to be reunited with his family.[42] Gilliss must have looked forward

---

[40] Newcomb, *Reminiscences*, pp. 108–109. The 1863 Annual Report is found in RG 78, E 2, volume 1, p. 185, and the 1864 Report on p. 385. On the trips to Dudley and Ann Arbor, see JMG to J. C. Watson, November 1, 1863, Watson Collection, Bentley Historical Library, University of Michigan, Box 1. For the quotation on Airy and Leverrier and the "monster" label applied by the Germans and Americans alike to the instrument see JMG to Marsh, November 2, 1864, USNO Library archives. It should be noted that the design of the 8.5-inch instrument was very similar to that of the telescope constructed by Pistor and Martins for the Navy's Chilean expedition under Gilliss; see volume 4 of the Chilean expedition, p. 7. On subsequent difficulties with this telescope, see chapter 5.

[41] Patricia Jahns, *Matthew Fontaine Maury & Joseph Henry: Scientists of the Civil War* (Hastings House: New York, 1961), pp. 249–252.

[42] The younger James Gilliss had been taken prisoner on October 19, 1864. He had commanded batteries C and J of the 5th regiment in Hancock's corps, crossing the Rapidan with six guns and 180 men, but at the time of his capture was between commands awaiting orders. "When the battle commenced on the morning of 19th October having no actual duties to do," the elder Gilliss wrote, "he started to help a Massachusetts battery of their corps which appeared to be in trouble. Both he and his companion Lieut. Brewerton (also of 5th Regt.) were almost instantly captured and we have seen the announcement of their arrival at Richmond. He sent a card to me by a Surgeon who was captured and released stating that they were unharmed. Stanton gave me a special exchange on the instant and we hope James will be back by an early flag of truce." JMG to Marsh, November 2, 1864, USNO Library archives. Gilliss's son Jack was an engineer in charge of all Union defenses from Louisville to the Virginia line and between the Ohio and the Cumberland. Gilliss reported in the same letter that "The season has been especially pestilential and fevers have prevailed from Soldiers' Home to the banks of the Potomac."

to becoming reacquainted with his son after his absence at war, but, on the morning of Thursday, February 9, shortly after rising, Gilliss collapsed and died, "from a stroke of apoplexy," at the age of 53. "On the evening preceding his death," Naval Observatory astronomer James Ferguson wrote, "he seemed even in better than his usual health, – elated and happy at the return of his eldest son, a Captain of Artillery, who had been since October a prisoner in the hands of the rebels, and had returned only the evening before. He retired to rest at the usual hour among thoughts at once hopeful and happy, and breathed his last soon after awaking in the morning."[43] The certificate of death also noted that "Captain Gilliss had been stationed at the Naval Observatory for some years, a locality noted for its insalubrity. During the last summer and fall he was frequently attacked with intermittent and, on one occasion, with remittent fever, which left him in a weak condition; this, combined with excessive mental labor incident to his position, no doubt caused his death in the line of duty." In death, as in life, Gilliss affected the future of the observatory he had founded; 12 years later his death certificate was used as part of the argument for removing the Observatory to a new location.[44] Gilliss's last letter, dated February 7, was penned to Clark regarding the number of optical instruments the Navy wished to order.[45]

Though Gilliss still had much to accomplish, Gould in his eulogy refused to call his colleague's death untimely. He had during his career founded not only the Naval Observatory, but also what would become the Chilean National Observatory a hemisphere away. After years of disappointment he had finally seen his services recognized. Furthermore, he had lived long enough to see the likely survival of the Republic he had served for a lifetime, and the likely flourishing of the institution he had founded.

Gilliss's death marks the end of the founding era for the Naval Observatory. By one of those many coincidences of history, the end of an era came almost simultaneously for the Harvard College Observatory and, most importantly, for the United States. Eight days after Gilliss's death, George P. Bond died at Harvard, having written one of his final letters to Asaph Hall saying "I had planned to accomplish something considerable, and this is the end." Thus 25 years of Harvard's Observatory under the Bonds came to a close.[46] Eight weeks after Gilliss's death, on April 9, General Robert E. Lee surrendered at Appomattox, bringing the Civil War to a halt. Then, in one of the greatest tragedies that the country would experience, President Abraham Lincoln was

---

[43] James Ferguson, obituary of Gilliss (ref. 39). The cause of death is given by Gould in the same volume of AN, 94.

[44] The death certificate is reprinted in *Reports on the Removal of the United States Naval Observatory* (United States Naval Observatory: Washington, 1877), p. 4.

[45] JMG to Clark, RG 78, E 4, Miscellaneous Letters Sent.

[46] Edward S. Holden, *Memorials of William Cranch Bond: Director of the Harvard College Observatory, 1840–1859, and of his Son, George Phillips Bond, Director . . . 1859–1865* (C. A. Murdock & Co: San Francisco, 1897), p. 214; Bessie Zaban Jones and Lyle Gifford Boyd, *The Harvard College Observatory: The First Four Directorships, 1839–1919* (Belknap Press of Harvard University Press: Cambridge, Massachusetts, 1971), p. 136.

shot, a few blocks from the Observatory where he had once inquired of Asaph Hall about the inverted image of the moon.

Of the other main characters in the founding era, only C. H. Davis would yet play a role at the Naval Observatory, capping his career as its Superintendent following Gilliss's death. Maury, in the new America wrought after the Civil War, would depart the Earth in 1873; in another coincidence of history, Wilkes, Goldsborough, and Davis died in Washington in 1877 within two weeks of each other. The torch had been passed to a new generation, one that would bring a golden era to the Naval Observatory.

Part II

# The golden era, 1866–93

# 5 Scientific life and work

The Naval Observatory continues its career of usefulness in the science of astronomy, which is the basis of navigation. Its annual volumes of observations . . . show the faithfulness of the work under its present organization.

Daniel Ammen, 1872[1]
Chief, Bureau of Navigation

It is highly important that such a spread of standard time should take place; and any method by which it can be done will receive the intelligent support of the Observatory. We now furnish the standard time to your company daily; and here our actual performance ceases: it rests with your company to disseminate it.

C. H. Davis to Western Union, 1890[2]

In the decades following the Civil War, the Naval Observatory entered a golden era during which it stood at the pinnacle of the world's astronomical observatories by virtue of high scientific accomplishment. This status is in one sense ironic, for the Observatory continued to be plagued by problems ranging from its bad location at "Foggy Bottom" on the Potomac River to persistent conflict over naval versus civilian control. Other factors, however, favored such a golden era, not least the blossoming talents of astronomers whom Gilliss had hired during his brief tenure. It was during this time that Simon Newcomb, Asaph Hall, William Harkness, and J. R. Eastman became among the most respected astronomers of the century, a point that the Navy did not fail to emphasize to its detractors who wanted a civilian in charge. The separation of the Hydrographic Office from the Observatory in 1866 left the institution free to concentrate its energies on astronomy, even if still largely in the service of navigation. The purchase of new instruments, including a 26-inch refractor that for several years was the largest telescope in the world, gave astronomers the material means for forefront research. Finally, during this era the positional astronomy in which the Observatory specialized could still lay claim to being the central and most exciting part of astronomy, a claim that by the end of the nineteenth century was strongly challenged by the rise of the "new astronomy" known as astrophysics.

In Part II of our study, we therefore concentrate on the scientific work at the Naval Observatory, placed in the context of its physical and administrative setting as

[1] Daniel Ammen to Secretary of the Navy George M. Robeson, AR (1872), 89.
[2] C. H. Davis to N. Green, December 5, 1876, USNOA, AF.

well as in the larger context of developments in astronomy. So important was the work of Hall, Harkness, and Newcomb that we devote separate chapters to them, centered on the primary accomplishments of each. The work of other less-prominent staff we describe in this chapter on general scientific life and work at the Naval Observatory and its administrative context during the last third of the nineteenth century, the post-Civil-War era in which the United States was rebuilding its physical, social, and intellectual fabric. While the moons of Mars and the transits of Venus are colorful and reputation-enhancing episodes, the work of positional astronomy and timekeeping detailed in this chapter were the Observatory's bread-and-butter programs, indeed, the reason for which the Navy funded astronomy in the first place.

## 5.1    Administrative concerns

As the Observatory built its scientific reputation, it could not help but be affected by administrative concerns, most notably the relationship between the Superintendent and the civilian staff, its new freedom from hydrographic duties while still holding to a rather strict Navy mission, and the deteriorating physical conditions at its Foggy Bottom site. All of its scientific and mission-related accomplishments were overlaid, and in some cases strongly impacted, by these concerns.

Following the death of Captain Gilliss the Navy was immediately faced with the problem of choosing his successor. The most natural choice by virtue of experience was Rear Admiral C. H. Davis, first Superintendent of the Nautical Almanac Office and now Chief of the Bureau of Navigation. This Bureau, which Davis had nurtured as its head since late 1862, was founded "to bring under one head all the scientific departments of the navy relating to hydrography, astronomy, navigation and surveying," including the Nautical Almanac, the Naval Observatory, and the Naval Academy, the last of which had been equipped with an astronomical observatory from which scientific investigation "was confidently expected."[3] The old Bureau of Ordnance and Hydrography not only lost two of its agencies by the creation of the Bureau of Navigation, but also lost the hydrographic function when Davis finally succeeded in 1866 in having legislation passed to establish a separate Hydrographic Office. While it was seemingly a step downward for Davis to head the Observatory, which had been only one of three agencies under his control, it is some measure of the importance that both Davis and the Navy attached to the Observatory that he in fact became its new Superintendent on April 27, 1865. Although it was a subordinate position, this move "carried him back to the field of scientific usefulness, where his taste and inclination

---

[3] C. H. Davis II, *Life of Charles Henry Davis, Rear Admiral, 1807–1877* (Houghton, Mifflin & Co.: Boston and New York, 1899), p. 283. Davis took charge of the Bureau of Navigation on November 15, 1862. According to Davis's son, the Bureau never reached its potential after Davis went to the Observatory, never having a scientific chief, and concentrating instead on detailing officers; pp. 284–285. On the history of the Bureau of Navigation see Henry P. Beers, "The Bureau of Navigation, 1862–1942," *The American Archivist*, 6 (1943), 212–252.

Figure 5.1. President Lincoln signs the charter of the National Academy of Sciences, March 3, 1863, surrounded by Senator Henry Wilson, who introduced the bill establishing the Academy, and the founding members. Left to right: Benjamin Peirce, Alexander Dallas Bache (first President of the Academy), Joseph Henry, Louis Agassiz, President Lincoln, Senator Wilson, Admiral C. H. Davis, and B. A. Gould. The apocryphal group portrait was painted by Albert Herter in 1924 for the dedication of the Academy Building, and now hangs in the Academy's Board Room. Courtesy of the National Academy of Sciences.

really lay."[4] Certainly he was qualified, both scientifically and for his connections, by virtue of his work for the Coast Survey and the Nautical Almanac, among other duties. During the Civil War Davis had served with Joseph Henry and A. D. Bache on the Permanent Commission that led to the founding of the National Academy of Sciences (Figure 5.1); as one of the best-connected men of science in the country, his appointment to head the Naval Observatory was above reproach. Although during his tenure Davis was, by Congressional resolution, drawn into side issues having to do with explorations of the Arctic and Isthmus Canal, he continued Gilliss's emphasis on reviving astronomy at the Observatory. It was under Davis that the 8.5-inch transit instrument, which had been ordered from Pistor and Martins by Gilliss, began its observations.[5]

---

[4] Ibid., p. 315. By Act of Congress dated June 21, 1866, the Hydrographic Office separated from the Naval Observatory (Statutes at Large, volume 14, p. 69). The separation actually took place on August 1 of that year, when Commander T. S. Fillebrown was detached from the Naval Observatory, where he had charge of charts, and was appointed first Hydrographer. On this date the charts, books, and instruments (except for chronometers) were removed to quarters known as the "Old Octagon House," at New York and 18th St, NW. On the subsequent history of the Hydrographic Office see Gustavus Weber, The Hydrographic Office: Its History, Activities and Organization (Johns Hopkins Press: Baltimore, 1926); Lt W. S. Hughes, Founding and Development of the U. S. Hydrographic Office (Government Printing Office: Washington, 1887), and Thomas G. Manning, U. S. Coast Survey vs. Naval Hydrographic Office: A 19th-Century Rivalry in Science and Politics (Government Printing Office: University of Alabama Press: Tuscaloosa, 1988).

[5] Rarely mentioned in the context of the Naval Observatory are Davis's two publications, Report on Interoceanic Canals and Railroads between the Atlantic and Pacific Oceans (Government Printing Office: Washington, 1867), and Davis, ed., Narrative of the North Polar Expedition, U. S. Ship Polaris, Captain Charles Francis Hall, Commanding (Government Printing Office: Washington, 1876).

When, in the spring of 1867, Davis was ordered to sea, the problem of a new Superintendent was again thrown open, but with a new dimension. No one else with Davis's knowledge of astronomy could be found in the naval service, and the time was therefore ripe for the civilians to act. Newcomb, already pushing his career at the tender age of 31, favored himself and Nautical Almanac Office Superintendent J. H. C. Coffin as candidates. Smithsonian Secretary Joseph Henry urged Navy Secretary Gideon Welles to appoint William Chauvenet, B. A. Gould, or Naval Observatory astronomer James Ferguson. Others, however, argued that a purely administrative head should be appointed, a position the Navy could easily supply. By appointing Commodore (later Rear Admiral) Benjamin F. Sands, Welles chose the latter strategy.[6]

As a lieutenant Sands (Figure 5.2) had been attached to the Depot of Charts and Instruments prior to its removal to the Foggy Bottom site, and it was Sands who, in Maury's absence, transferred the chronometers, instruments, and charts to the new buildings in 1844. Although he did not have the scientific credentials of Davis, Sands's appointment was for him a kind of homecoming, even if he was not immediately welcomed with open arms. "I found serious obstacles meeting me at the outset in my new labors," Sands recalled. The Observatory "had begun to expand its work into spheres of usefulness not strictly pertinent to the object intended in its establishment . . . so that astronomy became not simply and purely its sole province. The frequent changes of officers appeared also to militate against its interests in preventing that uniformity in its work that was so very essential to its success; consequently, astronomers and other scientists began to murmur at the misapplication of its functions. This may be called the beginning of the opposition to a naval administration of its affairs – an opposition which met me at the threshold of my work."[7]

Sands found that one source of animosity was lack of credit given by Navy management to astronomers for their own work, and this he determined to change. The policy of Superintendent Sands toward the Professors, Newcomb later wrote, "was liberal in the last degree. Each was to receive due credit for what he did, and was in every way stimulated to do his best at any piece of scientific work he might undertake with the approval of the superintendent. Whether he wanted to observe an eclipse, determine the longitude of a town or interior station, or undertake some abstruse

[6] On the issue of civilian control in the wake of Davis's departure see Arthur Norberg, "Simon Newcomb's Early Astronomical Career," *Isis*, **69** (1978), 220–221. It is notable that Commodore Thornton A. Jenkins, Head of the Bureau of Navigation and a friend of Newcomb, leaned toward the idea of a civilian superintendent. Davis's son later noted (ref. 3, p. 288) that, by placing Admiral Mouchet in charge of the Paris Observatory, the French experimented with the American system, thus placing fewer demands on the scientists.

[7] Benjamin F. Sands, *From Reefer to Rear Admiral: Reminiscences and Journal Jottings on Nearly Half a Century of Naval Life (1827–74)* (Frederick A. Stokes Company: New York, 1889), pp. 281–283. Sands was at the Depot from 1844 to 1847. He remarks that the only staff member he knew at the time of beginning his duties at the Observatory was Ferguson, with whom he had worked at the Coast Survey. He praised the astronomer's "prompt and cordial cooperation," but Ferguson died a few months after Sands had assumed charge. A famous photograph of Sands and Newcomb with the 26-inch telescope (Figure 6.3 in this book) appears after p. 292 of Sands's autobiography.

Figure 5.2. Admiral B. F. Sands (1811–83), Superintendent, 1867–74. He rose to the rank of Commodore (one star as shown here) in 1866, and became a Rear Admiral in 1871.

investigation, every facility for doing it and every encouragement to go on with it was granted him. Under this policy the observatory soon reached the zenith of its fame and popularity."[8] Sands's evaluation was much the same: "Thus was all intramural difficulty entirely done away with, and our work went on smoothly and successfully," he recalled three decades later. Sands enjoyed recounting the remark by Benjamin Peirce, Superintendent of the Coast Survey, that the joint Naval Observatory–Coast Survey eclipse parties sent out under Sands received universal acclaim in Europe. Still, some external complaints continued; to Sands's mind the chief complaint was instigated by an astronomer whom he did not name, but who clearly was Cleveland Abbe. Abbe had briefly been an aid at the Observatory, and then became Director of the Cincinnati Observatory, where he "was stirring up antagonism towards the naval administration of our Observatory by sending out circulars to the directors of the several astronomical observatories" trying to get up a petition to Congress to change the Naval Administration. This attempt failed, and Abbe not only admitted its failure, but also sent Sands the letters from astronomers opposing the idea of civilian control.[9]

It was under Sands's administration (1867–74) that the Great Refractor was purchased, the eclipse parties of 1869 and 1870 carried out, and preparation begun for the Transit-of-Venus expedition for 1874. Sands counted as one of his chief accomplishments ensuring that the observatory concentrate on astronomy, rather than being sidetracked into other areas such as meteorology. When Leverrier proposed that the Naval Observatory undertake national meteorological observations, for example, Sands recommended that the Army's Signal Office be given the task instead. Although the Observatory still carried out its own meteorological observations, it did not take on the job of coordinating those from around the country. Sands retired by law from active duty at age 62 in 1874, characterizing his time there as "seven long and happy years." These years were followed by equally happy years under Davis's second administration (1874–77), during which the first of the famous transit-of-Venus expeditions were actually dispatched.

A similar policy was carried out under the administration of Rear Admiral John Rodgers (1877–82), who was appointed Superintendent upon Davis's death, and whom Newcomb judged second only to Farragut in his fame in the Navy.[10] At the time of his death Davis was senior Rear Admiral on the active list, and Rodgers was next to him in seniority. When Rodgers arrived at the Observatory he found a staff of some 20 men (Figure 5.3), including four line officers, seven astronomers who were Professors of Mathematics, a computer, an instrument maker, and a clerk, as well as watchman and laborers. Asaph Hall never forgot his first encounter with Rodgers: "My first contact with him was rude. He came near taking off my official head, because of a

---

[8] Newcomb, Reminiscences, pp. 111–113.     [9] Sands (ref. 7), pp. 287–289.
[10] On Rodgers, one of the founding members of the National Academy of Sciences, see Robert E. Johnson, Rear Admiral John Rodgers, 1812–1882 (United States Naval Institute: Annapolis, 1967), and Asaph Hall, Biographical Memoir of John Rodgers (Judd & Detweiler: Washington, 1906).

Figure 5.3. Professional staff of the Naval Observatory, 1879. Clockwise from the top: Asaph Hall, William Harkness, J. R. Eastman, Edgar Frisby, A. N. Skinner, H. M. Paul, Lt-Cdr George Pigman, Thomas Ide, Henry S. Pritchett, Thomas Harrison, Edward S. Holden, and J. E. Nourse. At the center is Admiral John Rodgers (1812–82), Superintendent, 1877–82. He had been a Rear Admiral since 1869. Pritchett, who arrived at the Observatory in 1878, would become President of MIT. Harrison had been at the Observatory since Maury's time, and would become one of the longest-tenured civil servants in U. S. government history. The collage must date from 1879, after Yarnall's death in February, and before Pritchett's departure in 1880.

complaint presented to him that I had not done my duty. He was a man of action, but he said, 'We will wait a day or two,' and in this time the accusation proved to be false. This led to a more intimate acquaintance. I found the Admiral one of the grandest men I have ever known. His character was open, frank and noble. He went straightforward in the path of duty perfectly fearless. He was ready to hear opposing argument with patience – in fact, rather liked it – and made up his mind fairly." However, Rodgers was apparently not quite as accommodating to the astronomers as Sands had been. According to Newcomb, Rodgers largely agreed with (though "did not fully commit himself to") a policy composed during his administration, namely that "The superintendent of the observatory should be a line officer of the navy, of high rank, who should attend to the business affairs of the institution, thus leaving the professors leisure for their proper work."[11] Nevertheless, the benefits of a naval head were demonstrated by the fact that Rodgers was responsible for setting in motion the plans for a new Observatory site, a task for which someone of his Navy stature was needed. (Even for Rodgers it proved a difficult task. Barely begun by his death in 1882, it would not be fully realized until 1893, when the new Observatory was occupied.) He was obviously, like Sands, a strong supporter of astronomy; under his tenure the most elaborate eclipse expedition in the Observatory's history was organized in 1878, and the second round of the transit-of-Venus expeditions was dispatched in 1882.

After Rodgers's death there followed a period of some four years that Davis's son (himself a Superintendent with the rank of Captain at the turn of the century) termed "the decadent period." Rodgers's senior assistant, Cdr William T. Sampson, filled the job temporarily for two months, but even though Newcomb called Sampson "one of the most proficient men in practical physics that the navy has ever produced," the Navy policy had to favor rank rather than scientific knowledge, particularly when the two could not be combined in one person as they had been with Davis. Four Superintendents were subsequently appointed within four years, three of them at the rank of Admiral, and one at the rank of Commodore.[12] Vice Admiral Stephen C. Rowan stayed less than a year, and Rear Admirals Robert W. Shufeldt and S. R. Franklin, as well as Commodore George E. Belknap, did no better. Franklin, whose

[11] Hall, *Biographical Memoir*, p. 92; Newcomb, *Reminiscences*, p. 121. There are two types of officers in the Navy: line officers and staff officers. Line officers fall into two categories, restricted line and unrestricted line. Unrestricted line officers are officers who are eligible to command U. S. Naval vessels. Restricted line officers are generally specialty officers, eligible to command bases like the Naval Observatory, but not Naval vessels. Ship drivers, submariners, and seals are all unrestricted line officers. Oceanographers, intelligence officers, etc. are restricted line officers. Officers in supply corps, civil engineers, chaplains, and officers in the medical corps are examples of staff officers, as are the Professors of Mathematics themselves. Hamersly, *A Naval Encyclopedia* (Philadephia, 1881), pp. 443 and 770.

[12] Newcomb notes that Cdr William T. Sampson was appointed assistant to the Superintendent under Rowan, and believed that this was done to see whether any improvement could be made over the liberal policy of work. Newcomb relates the controversy over Sampson on pp. 121–122 of his *Reminiscences*. Davis, USNOA, AF, file labeled "C. H. Davis II, 1901," 6, believed that "Sampson was preeminently qualified to be Superintendent, but was opposed [by civilian astronomers] on account of his very fitness."

chief qualifications were extreme age, high rank, and great reputation, ordered the work of the Observatory to be planned by a board including the Superintendent, the senior line officer, and the senior Professor. Even this seemingly innocuous arrangement did not go well: according to Davis's son looking back several decades later, following Rowan's departure, "constant changes were brought about by a relaxed discipline, a condition of intrigue, discord and jealousy on the part of the astronomical staff and a want of harmony and cooperation in the astronomical work, coupled with the demand for an officer of high rank who should act as a mere figure-head. Officer after officer took up the task, and was either forced out, like Sampson, or laid down the office in disgust."[13] During this time, Davis's son believed, the civilian scientific element could easily have captured the institution if it had united on a leader. No such unanimity existed, and beginning in 1886, a series of officers at the ranks of Commodore and Captain took over, restoring order and continuity. For the next 15 years only three officers (R. L. Phythian, F. V. McNair, and C. H. Davis, Jr) served as Superintendent, wrestling not only with the rambunctious astronomers, but also with other administrative problems. Foremost among those problems was the search for a new site, and extended calls for a civilian Director. Because these problems did not come to a head until the new site was actually occupied in 1893, we shall detail them in Part III.

In the meantime, the Observatory had to make do with its old site. Figure 5.4 shows the two chief changes to the Observatory about 1878, a new building to house the 8.5-inch transit circle, and the dome for the 26-inch refractor. The 8.5-inch transit circle was originally mounted in 1865 in the west wing of the Observatory building, formerly occupied by the 5.3-inch transit instrument. Because the brick construction of that building was not ideal for observing, in 1869 the new building was constructed adjacent to the west wing.[14] The 26-inch telescope was mounted in 1873. Figure 5.5 shows the plan of the entire grounds, including the outlying buildings such as the horse and cow stables.

Newcomb has given some indication of the working environment. The program for the new transit circle called for observations 24 hours a day, a duty that had to be spread among four observers working from 9 o'clock in the morning until midnight or even dawn. Moreover, "No houses were then provided for astronomers, and the observatory itself was situated in one of the most unhealthy parts of the city. On two sides it was bounded by the Potomac, then pregnant with malaria, and on the

---

[13] See document in USNOA, AF, file labeled "C. H. Davis II, 1901," 5–6. In Davis's opinion (p. 8), the Professors of Mathematics, with one exception (probably Newcomb), were loyal to the Navy. He had especially good things to say about Asaph Hall, who "kept wholly clear of intrigue, quarreled with nobody, criticized nobody, and supported always earnestly and faithfully the existing administration."

[14] Figure 5.4 first appeared as the frontispiece to *Instruments and Publications of the U. S. Naval Observatory, 1845–76*, which was published as Appendix I to *WO for 1874* (1877). It was also published in *WO for 1875*, which was published in 1878. On the new transit-circle building, see Eastman, *Second Washington Catalogue* (ref. 20 below), p. v.

Figure 5.4. A view of the Observatory, 1877. The dome housing the 9.6-inch refractor is atop the central portion of the building, and the time ball and mast can be seen above it. The dome for the 26-inch refractor is visible in the background. The Superintendent's residence is on the extreme left; the pavilion for the 8.5-inch transit circle is on the extreme right. The central portion and east and west wings were built in 1843; the Superintendent's quarters in 1847; the transit-circle pavilion in 1866, and the 26-inch portion of the south wing in 1873. Frontispiece from *Instruments and Publications of the U. S. Naval Observatory* (Government Printing Office: Washington, 1845–76).

other two, for nearly half a mile, was found little but frame buildings filled with quartermaster's stores, with here and there a few negro huts. Most of the observers lived a mile or more from the observatory; during most of the time I was two miles away. It was not considered safe to take even an hour's sleep at the observatory. The result was that, if it happened to clear off after a cloudy evening, I frequently arose from my bed at any hour of the night or morning and walked two miles to the observatory to make some observation included in the programme." This routine was the cause of some consternation to Newcomb, who had hesitated to apply at the Naval Observatory to start with because of an aversion to night work. It was a far cry from previous observing practice that Newcomb carried out with his colleagues Mordecai Yarnall and J. S. Hubbard, which he characterized as "free and easy," whereby the observer would select what he thought best to observe from the *Nautical Almanac* or a catalog of stars, make observations as long as he chose, and then "vote it cloudy and go out for a plate of oysters at a neighboring restaurant."[15]

With an idea of the Observatory's physical and administrative setting, as well

[15] Newcomb, *Reminiscences*, pp. 101–102 and 110–111.

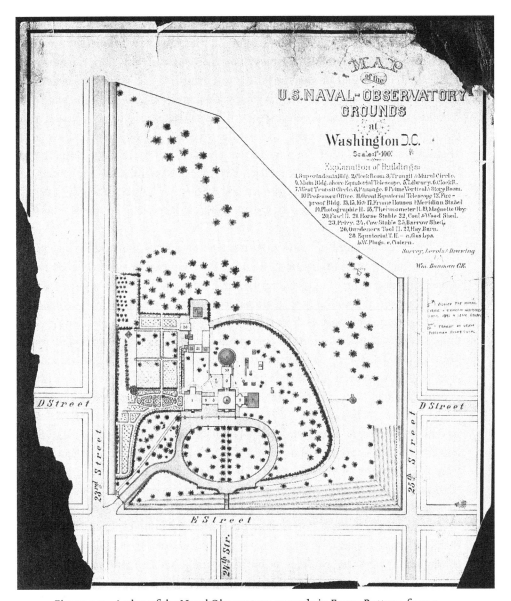

Figure 5.5. A plan of the Naval Observatory grounds in Foggy Bottom, from a hand-colored single sheet in the Naval Observatory Library. A similar version was published in Daniel Ammen *et al.*, *Report on the Commission on Site for the Naval Observatory* (Government Printing Office: Washington, 1879).

as its *dramatis personae* in the last third of the nineteenth century, we now turn to the science itself.

## 5.2 Charting the heavens

During the last third of the nineteenth century, the Naval Observatory threw itself into a variety of activities, ranging from matters of time to solar-eclipse and transit-of-Venus expeditions, and observations with equatorial telescopes. The most time-consuming activity of the Observatory during this period, however, was the observation of star positions with meridian instruments, followed by analysis and publication. We recall from chapter 4 that, during the Civil War, Gilliss had begun the process of publishing the backlog of observations made during the early tenure of Matthew Maury. For a half-dozen years after Gilliss's death this effort continued, with the publication during the early 1870s of the meridian-transit and meridian-circle zones observed during 1846–49, as well as the results from 1853 to 1860.[16] In addition, Yarnall and Eastman published catalogs of an enormous number of stars compiled from many years of observations at the Naval Observatory. Finally, the observations of selected objects continued, and were published annually in the *Washington Observations*.

In order to understand the importance of this activity, we need to place positional astronomy in an international context. Driven in part by practical needs, star catalogs of varying precision were being produced at observatories large and small around the world, with the great national observatories of Greenwich and Pulkovo setting the standard. Most of these catalogs were simply "observed catalogs" based on observations made with one instrument during a single year or period of years. Thus, catalogs of star positions from meridian instruments were published almost every year in the *Washington Observations* during the last third of the nineteenth century. Greenwich under G. B. Airy from 1836 to 1881 opted for catalogs embracing several years (generally six to ten), whereas Pulkovo Observatory under the Struves published its catalogs at intervals of 20 years, the epochs of the catalogs having been 1845, 1865, and 1885. In this sense the catalogs of Yarnall (1873 and 1878) and Eastman (1898) were similar to the Greenwich and Pulkovo catalogs in embracing many years of observations, during which complications such as the proper motions of the stars had to be taken into account.[17] These observed catalogs are labelled Level 1 in Table 5.1.

There was, however, much more to positional astronomy during this period than catalogs of observations made at a single observatory. In the quest for greater accuracy and systematic coverage of the sky, during the last half of the nineteenth century astronomers began to conceive more grandiose plans, some requiring international cooperation. These efforts went into two distinct but related directions: first, to produce larger and more uniform star catalogs of relatively good accuracy, and secondly to produce small star catalogs of the utmost possible accuracy, to which the larger catalogs could be tied. The major thrust for this work came from Germany, where in 1867 F. W. Argelander proposed to the Astronomische Gesellschaft a plan for determining the positions of all the stars in the northern hemisphere to the ninth

---

[16] WO, volume XVI, Appendix IV (1872); WO, volume XVII, Appendix I (1873); WO, volume XVII, Appendix II (1872).

[17] For an overview of nineteenth-century catalogs see Newcomb, *Compendium of Spherical Astronomy* (The Macmillan Company: New York, 1906), chapter 13.

Table 5.1. *A hierarchy of star catalogs begun or completed in the nineteenth century*[a]

| Level 4: | Photographic catalogs | | |
| | Astrographic Catalogue (published in zones by a variety of observatories beginning in 1890s) | | |
| Level 3: | International reference star catalogs | | |
| | AGK1 (published in zones by a variety of observatories beginning in 1890s) | | |
| Level 2: | Compiled fundamental catalogs | | |
| | BAC 1845 8,377 stars | PGC L. Boss 1910 6,188 stars | |
| | | N1 Newcomb 1872 32 stars | N2 Newcomb 1898 1,257 stars |
| | | FC Auwers 1879 N 1883 S 539 N stars 83 S stars | NFK Peters/Auwers 1907 925 stars |
| Level 1: | Absolute observed catalogs Including Washington, Greenwich, Russian, Cape (South Africa), French, and German observations | | |

*Notes:*
[a] Dates are dates of publication, not epoch of observation. For catalog names see list of abbreviations.

magnitude, except for the 10 degrees around the pole. This project was eagerly undertaken almost immediately by dividing the sky into zones observed by 12 observatories in six countries.[18] The resulting catalog of positions for 144,000 stars, known as the AGK1 (Astronomische Gesellschaft Katalog 1), was not completed until 1910, and its extension to 23 degrees south of the equator was finished only in 1924.

The second goal of astronomers at this time, namely to produce for a select group of stars the most precise positions possible by critically examining and combining from many catalogs those stars most often observed, also emanated largely from Germany. It was with the idea of eliminating systematic errors in the larger catalog that Auwers at the Berlin Observatory in 1879 produced a catalog known as the AG, or FC (Fundamental Catalog) system, in order to refer the numerous stars of the AGK1 to "fundamental stars." These twin activities – the production of large star catalogs, and the production of smaller very precise "fundamental" catalogs to which the large

[18] For a list of participating observatories and a discussion of the catalog, see H. Eichhorn, *Astronomy of Star Positions* (Academic Press: New York, 1974), pp. 228 ff.

catalogs were referenced – began a task that has continued through the twentieth century. The FC catalog launched a German series of fundamental catalogs that would be followed by the NFK (1907) and others, seen at Level 2 in Tables 5.1 and 10.3. The AGK1 also began a series at Level 3 of reference catalogs (Tables 5.1 and 10.3). At Level 4 in these tables catalogs of even larger numbers of star positions would make use of photography, a technique not seen as useful for star positions until the end of the nineteenth century, when the enormous Astrographic Catalog was begun. In the 1860s these goals were only dimly perceived, but during the 1870s the Americans would contribute not only to observed catalogs, but also to fundamental catalogs via the work of Newcomb at the Naval Observatory and Lewis Boss at the Dudley Observatory.

It is in this context that events at the Naval Observatory must be judged. During the last half of the nineteenth century, star cataloging was still dominated by observations made with the transit circle, and the Naval Observatory is a case study of the trials and tribulations of astrometry during a period when astrometry still reigned supreme. Looking back at his years at the Observatory, Simon Newcomb wrote that "In October, 1865, occurred what was, in my eyes, the greatest event in the history of the observatory. The new transit circle arrived from Berlin in its boxes. Now for the first time in its history, the observatory would have a meridian instrument worthy of it, and would, it was hoped, be able to do the finest work in at least one branch of astronomy." Superintendent Davis agreed, stating that the use of this instrument at the Observatory "constitutes a new era in its progress, and restores it to the rank of a first class institution."[19] The importance of this instrument (Figure 5.6) to the Naval Observatory during this era may be seen from Harkness's statement, looking back from 1898, that observations with it "absorbed the labors of about two-thirds of the Observatory staff for more than thirty years." When the combined results, known as the Second Washington Catalog, were published by J. R. Eastman in 1899, they showed 72,914 observations made with the 8.5-inch transit circle alone from 1866–91. If Yarnall's catalog "forms an epoch in the history of the Observatory," as he himself put it, then Eastman's catalog represented the next generation of meridian instruments and methods.[20]

[19] Newcomb, *Reminiscences*, p. 108. See also pp. 108–127. The work of mounting the instrument began October 23, and it was in position by November 8. During November and December its errors and scales were determined, and observing began in 1866; AR (1866), 135. The instrument had been ordered by Gilliss in 1863, and is described in detail in *WO for 1865* (Washington, 1867), Appendix I, pp. 1–47.

[20] John R. Eastman, "The Second Washington Catalogue of Stars, together with the Annual results Upon Which It is Based. The Whole Derived from Observations made at the U. S. Naval Observatory, with the 8.5 inch transit Circle, during the Years 1866 to 1891, and reduced to the Epoch 1875.0," Appendix I to *WO for 1892* (Washington, 1899). Harkness's statement is in the Preface. The first Washington Catalogue was Yarnall's *Catalogue of Stars* (1873), cited in ref. 8 of Table 2.3 . For Yarnall's statement see the Introduction to the second edition of his *Catalogue* (Washington, 1878). Instruments used for star positions in Yarnall's catalog include the 5.3-inch transit instrument, mural circle, 4.5-inch meridian circle and prime-vertical transit. The 4.5-inch Ertel & Son meridian circle does not appear to have been used after 1862, when Yarnall and Newcomb were in charge of it; it was later sold to Yale according to Yarnall. Yarnall noted that the Ertel Circle was well known to be inferior to the other instruments, and was only given half the weight in his catalog.

Figure 5.6. The 8.5-inch Pistor and Martins Transit Circle, mounted 1866. The focal length is 12 feet. Plate V from *Instruments and Publications*.

The excitement of this event represented progress in astronomical instrumentation echoed at observatories around the world. The transit circle, which in the case of the Naval Observatory superseded the 5.3-inch meridian transit and the 4-inch mural circle dating from 1844, as well as the even older 4.5-inch Ertel meridian circle, provided a state-of-the-art single instrument that could observe in both coordinates, and to a much fainter magnitude. At Greenwich the famous Airy transit circle (defining what became the world's prime meridian) had superseded the mural circle and transit instrument in 1850. At Pulkovo, however, Struve still preferred using the transit instrument and vertical circle to measure each coordinate separately.

Behind this progress in instrumentation, the quest for accuracy was clearly an important goal. When the senior Davis laid out the plan for the instrument in 1866, he was quick to emphasize that it would be used to determine more accurately the positions of the stars contained in the *American Ephemeris*. The Naval Observatory staff, he stated, would also use the new transit circle to undertake a systematic series of observations of all the bodies in the solar system. Both institutional pride and nationalism are evident in Davis's statement that "No other American observatory has the means of making and publishing such a series; and the very few European observatories at which they are made, are too far north for the advantageous performance of the work." Part of this program would be observations of asteroids, especially the fainter ones, and those that by virtue of their southerly positions were more difficult to reach from the famous European observatories. Another item of immediate practicality was observation of the Coast Survey list of some 350 stars needed for their work in the determination of latitude. Yet another use is implied by Yarnall's statement that Ferguson had made constant demands on him for "comparison stars" (reference stars) for objects observed with the equatorial; though Ferguson died in 1867, one can be sure that demands for reference stars continued with other equatorial observers.[21]

These stated goals were somewhat piecemeal compared with the grand prize of a large number of stars with accurate positions, tied to a system of stars whose positions were even more accurately known. Newcomb grasped the latter early on when he mapped out an important objective for himself and the Observatory, foreshadowing a lifetime of concern for "standards" in astronomy, and preceding the German program of fundamental reference catalogs. Newcomb wanted to carry out a "fundamental" program (meaning one independent of any previous observations) of observing equatorial stars used as reference stars for the motions of the Moon and planets. The goal was to determine errors in right ascension suspected in previous catalogs, thus putting them on a homogenous system. This work, which Newcomb saw as doing for right ascensions what Auwers had done for declinations, was carried out through 1869, and, according to Newcomb, succeeded in its goals. It employed 12 catalogs of stars, ranging from Auwers's reductions of Bradley's 1755 observations, to Bessel, Struve, Greenwich, and the *Washington Observations* of 1862–67, including the first years of the 8.5-inch results.[22] Newcomb later extended this so-called N1 catalog of 32 stars to the catalog of 1,098 "standard stars" and finally to the N2 Fundamental Catalog of 1,257 stars (in 1898), based on 43 independent observed catalogs, forming

---

[21] AR, (1866), 135–136; Yarnall, Introduction to the first edition of his *Catalogue* (ref. 20 above).
[22] Newcomb, "On the Right Ascensions of the Equatorial Fundamental Stars, and the Corrections necessary to reduce the Right Ascensions of Different Catalogues to a Mean Homogeneous System," *WO for 1870* (1872), volume XVI, Appendix III.

Figure 5.7. J. R. Eastman, Professor of Mathematics, U. S. N, and cataloger of stars.

an American series of fundamental catalogs.[23] One of the impressive features of the Newcomb catalog was the derivation of the zero point for right ascension (the equinox), using an original method involving observations of the Sun from 1750 to 1869. As an indication that astrometric work had not reached a level of total cooperation, even within one country, Lewis Boss published the first installment of his own fundamental system in 1877, which eventually led to the *Preliminary General Catalogue* of 1910 (see Table 5.1).

With the completion of his research program using the observations of the 8.5-inch, Newcomb decided to turn to the theory of the Moon's motion, which he judged "much more important work than making observations." Although Hall, Harkness, and others would inherit the instrument for more routine work, the astronomer in charge of the 8.5-inch transit circle for most of its career was John Robie Eastman (1836–1913), a member of a prominent Massachusetts family that had come to Salem in 1638, and a graduate of Dartmouth in 1862. Eastman (Figure 5.7) was one of those "aides" hired by Gilliss and quickly promoted to Professor of Mathematics in 1865, a post he would retain at the Observatory until his retirement in 1898. He had broad interests, as indicated by his positions as Vice President of the AAAS in 1887 and 1892, president of the Philosophical Society of Washington in 1889, president of the Cosmos Club in 1892, and first president of the Washington Academy of Sciences in 1898. His interests in astronomy were also broad, as is indicated by his participation

---

[23] H. R. Morgan, "Astronomy of Position," PA, **51** (December, 1943), 527. The positions from the *Fundamental Catalogue* were used in the national ephemerides until 1925, and formed the basis of the revision of the observations of the Sun, Moon, and planets used by Newcomb, G. W. Hill, and E. W. Brown in their new tables.

in many solar-eclipse expeditions, the transit-of-Venus expedition to Florida in 1882, and publications on subjects ranging from stellar motions to meteoric astronomy. He was for much of this period also in charge of publishing the meteorological observations at the Observatory. However, his *magnum opus* was the *Second Washington Catalogue of Stars*, of which about one quarter of the observations were made by himself using the new transit circle.[24]

Curiously, the program undertaken by Eastman and his colleagues with the 8.5 inch did not include participation in the international AGK1 project, although W. A. Rogers at Harvard and Lewis Boss at the Dudley Observatory did participate. Only with the decision in 1887 to extend the AGK1 to the southern hemisphere did the Naval Observatory become involved, observing a zone from $-4$ to $-8$ degrees, the results of which were published in 1908 by A. N. Skinner. Appropriately, Eastman's enormous catalog of stars appeared in the last volume of *Washington Observations*, culminating the series at the Foggy Bottom site begun under Matthew Maury 50 years before. His feelings on this publication of a lifetime's work were ambiguous; in a frank address as retiring Vice President of the AAAS in 1892, Eastman argued that fundamental astronomy, to which his catalog of star positions should have been a major contribution, was being neglected.[25] Moreover, despite all the high hopes, in the end, the Pistor and Martins transit circle was a disappointment. Newcomb commented many years later that the errors he discovered in the star positions in prior catalogs were due more to the method of observation than to the excellence of the instrument. In fact, the instrument "proved to have serious defects which were exaggerated by the unstable character of the clayey soil of the hill on which the observatory was situated. Other defects also existed, which seemed to preclude the likelihood that the future work of the instrument would be of a high class."[26] So great were these defects that the 8.5 inch would be greatly modified, including the installation of a new 9-inch lens. It was transported to the new Observatory site in 1893, where it would be used for another 50 years along with a new-generation transit circle.

In summary, a tremendous effort was put into the determination of star positions at the Naval Observatory during this period. The perceived goals were practical, if piecemeal, yet Benjamin Boss wrote in 1937 that, toward the end of the nineteenth century, "it became increasingly and discouragingly apparent that the outlook for positional astronomy was sorry indeed unless new developments were forthcoming."[27] Astrophysics was undergoing "brilliant and breathtaking development," while

---

[24] On Eastman see the obituary by R. H. Tucker in *PASP*, **26** (1914), 41–42; DAB, 303–304; Marc Rothenberg, *The Educational and Intellectual Background of American Astronomers, 1825–1875*, Ph. D. dissertation (Bryn Mawr, 1974), pp. 169–170. Eastman was in charge of the meridian work of the Observatory from 1874 until 1891. Edgar Frisby had ten years of service on the instrument, A. N. Skinner 19, and H. M. Paul ten, among others. A list of observers and their lengths of service is given in Table 29 of the *Catalogue*.

[25] J. R. Eastman, "The Neglected Field of Fundamental Astronomy," *Proceedings of the AAAS*, **XLI**, (1892).    [26] Newcomb, *Reminiscences*, pp. 114–115, see also pp. 109–111.

[27] Benjamin Boss, *General Catalogue of 33342 Stars for the Epoch 1950* (Carnegie Institution of Washington: Washington, 1937), Introduction, p. 2.

the old positional astronomy languished. Under the stress of that situation, Boss believed, his father Lewis Boss and others had found a new goal for astrometry. While astrophysics would investigate the formation and evolution of the stars, classical astronomy would study the structure of the universe, as revealed by the motions of the stars, which in turn depended on obtaining precise knowledge of their positions. It was this program that gave rise to the formation of Boss's *General Catalogue* of more than 33,000 stars, which was eventually completed and published by his son. That catalog would make use of hundreds of compiled catalogs (although not individual yearly ones). Among those, Eastman would be happy to know, were not only his own "Washington 75" catalog, but also the third edition of Yarnall's Washington 60, and Gilliss's star observations at Santiago, Chile during 1849–52, the last of which were not published until 1895. Determining the structure of the universe was a laudable goal, one in which the Naval Observatory played its part, even if its immediate goals under Navy patronage were the more practical goals of navigation and timekeeping.

## 5.3    Time balls and telegraphs: Time for the nation

One of the applications of precise star positions is the determination of time. Although large star catalogs were not needed for this enterprise, a small number of fundamental stars, known as "clock stars," had been observed so well that any differences in their right ascensions were assumed to be due to errors in the clock rather than in the star position. During the last third of the nineteenth century, the Naval Observatory not only maintained an essential interest in time determination, but also expanded its influence in the area both by the development of subsidiary clocks linked to the Observatory's master timekeepers, and by the telegraphic dissemination of time via the nation's telegraph companies. The classic need for time to rate chronometers, determine longitude, and measure the right-ascension coordinate of star positions remained and required constant improvements in accuracy. However, it was the telegraphic dissemination of this commodity that occasioned considerable rancor, even as it greatly expanded the reach and prestige of the Observatory. Moreover, the Observatory became deeply involved in issues of more general temporal interest, including the need for an international prime meridian. The latter issues culminated in the 1880s and 1890s with decisions that still affect us today.

The telegraph, which, as we have seen in chapter 2, revolutionized the process of longitude determination, was capable of having a similarly revolutionary effect on time service. As Naval Observatory Assistant Superintendent Cdr Allan D. Brown put it looking back late in the century, electrical distribution of time encompassed "the automatic electrical synchronization of one or more clocks by a clock at a central station; with this synchronization may also be associated the registering of the signal upon a chronograph, the ticking of a telegraphic instrument, the firing of a gun, the ringing of a bell, the discharge of a flashing signal, or the dropping of a time ball; in a complete system any or all of these means will be used according to the circumstances of the case under consideration." The central station, he emphasized, should be at an observatory, and the product thus provided was called a "time service."[28]

Through the tenure of Gilliss in early 1865, the Naval Observatory time service for Washington had been limited to dropping a ball at noon from the flagstaff on the Observatory dome. Toward the end of Gilliss's tenure, however, the potential for more extensive local dissemination of time became available. In particular, Harkness recalled that, soon after the fire-alarm telegraph had been introduced in the City of Washington in 1864, he had proposed to Gilliss, and then to the mayor, that the Observatory be connected by wire with the fire-alarm-telegraph office, and the fire bells struck daily at 7 am, noon, and 6 pm by a signal from the Observatory. This plan was brought into effect in late summer of 1865. Because the State Department building had a direct wire to the fire-alarm office, it also received the time signal. "The Western Union Telegraph Company also had wires in the State Department building, and their operators almost immediately fell into the habit of transferring the time signals from the fire alarm wire to their own lines, and thus sending them to the main office of the Company in Washington. That, I believe, was the origin of the Western Union Company's connection with our time service, and it has continued ever since," Harkness recalled in 1890, when the issue of time as a commodity became contentious.[29]

Once it became available to a telegraph company, time could be telegraphed to any office its wires reached. By 1867 time was telegraphed "at noon, by the different lines of wires, to the northward, eastward, and westward, and as far southward as Texas." Because there were not yet any time zones in the United States, every city was on its own local time, and had to adjust accordingly to the Washington noon-time signal (Figure 5.8). By 1869 the Observatory distributed signals that "serve to regulate the clocks of nearly all the railroads in the southern States." Beginning in 1871, the need for accurate timing for meteorological observations by the U. S. Signal Service greatly extended the daily distribution of time, "from the Atlantic to the Pacific and from the great lakes to the Gulf of Mexico." Yet, in the early 1870s, the focus was still on the railroads, rather than cities and citizens: "The immediate object of these signals is to furnish accurate and uniform time to the railroads: and, throughout the whole of the vast territory in question, there is scarcely a train whose movements are not regulated by the Observatory clocks."[30]

---

[28] Allan D. Brown, "The Electrical Distribution of Time," *Journal of the Franklin Institute*, **125** (1888), 462–474, and **126** (1888), 14–24; the quote is on p. 463. Based on a lecture delivered before the Franklin Institute on January 13, 1888, which was also reprinted in *Scientific American Supplement*, **26** (August 18, 1888), 10,521–10,524.

[29] "Statement respecting early history of the time service of the Observatory," Harkness to Superintendent, USNO, May 15, 1890. RG 78, E 6, Letters Sent. Admiral Davis gave the order for purchase of the telegraph instruments at the Observatory on July 15, 1865, the bill was paid on August 22, and the service began "very promptly" thereafter. It should be noted that Harkness did not arrive in Washington until 1862, and might not have been aware of possible earlier telegraphic time distribution by the Naval Observatory.

[30] AR (1867), 133; AR (1869), 56; AR (1871), 120; AR (1873), 94. In a letter from B. F. Sands to the Superintendent of the Western Union Telegraph Office in Washington, dated March 15, 1871, Sands suggested that Western Union send signals to specific scientific applicants, but this did

Figure 5.8. Before time zones, each city had its own local time, which had to be taken into account for the noon signal coming from Washington. Detail, *New Map of the State of New York* (H. H. Lloyd and Company: Washington, 1869), Library of Congress.

not lead to any general distribution of time to cities. Bartky, *Selling the True Time* (Stanford University Press: Stanford, 2000), points out that the statement about railroads is exaggerated, since time was also disseminated from other observatories. See also Carlene Stephens, "The Most Reliable Time: William Bond, the New England Railroads, and Time Awareness in 19th-Century America," *Technology and Culture*, **30** (1989), 1–24, and Stephens, "The Impact of the Telegraph on Public Time in the United States, 1844–1893," *IEEE Technology and Society Magazine* (March, 1989), 4–10.

The idea of telegraphically distributing Naval Observatory time to cities (as opposed to railroads) was ostensibly precipitated by requests from citizens who had no access to a local observatory. At least that is the reason Admiral Davis gave in a letter to a Vice President of Western Union in December, 1876, where he cited a Leavenworth, Kansas jeweller who had requested a signal from the Naval Observatory in order to regulate the town's time. However, as Ian Bartky has shown in *Selling the True Time*, no evidence exists for such a groundswell among citizens; Davis likely wanted to expand the Observatory's peacetime activities by selling Washington time throughout the United States. Thus, after citing the jeweler's inquiry in his letter to Western Union, Davis wrote that "it is highly important that such a spread of standard time should take place; and any method by which it can be done will receive the intelligent support of the Observatory. We now furnish the standard time to your company daily; and here our actual performance ceases: it rests with your company to disseminate it." It is important to note that by "standard time" Davis had in mind not the modern definition of zone time, but the more basic definition of a single standard of uniform time (Washington local time, based on the Naval Observatory's meridian), which could be distributed by telegraph and corrected at each locale by taking into account the known differences in longitude. "There is a general desire in the country to receive the standard time, and to pay for it, if necessary," Davis wrote, and he suggested that Western Union "form a scale of prices at which daily, weekly and monthly transmissions of our noon signal could be made to various points."[31] In return Davis offered to help in whatever way he could, including providing a statement of the differences of time between Washington and specific points, and would "endeavor to spread the system, the adoption of which seems so desirable."

Having initiated contact with Western Union, Superintendent Davis ordered astronomer E. S. Holden to travel to Western Union's New York office for negotiations. The negotiations were successful, for, by April 2, 1877, Western Union had printed a circular "To Shipmasters, Underwriters, Chronometer Makers and others," announcing the arrangement entered into with the Observatory to disseminate time to cities from Maine to Louisiana and from the Atlantic to the Pacific. The announcement included a list of cities with populations of 20,000 or more to which the company considered it might practicably deliver a time signal. Although the effect was the dissemination of Naval Observatory time to various parts of the country, the only duty of the Observatory, the Superintendent emphasized, was to furnish the correct time every 24 hours.[32]

---

[31] Davis to N. Green, December 5, 1876, USNOA; Ian Bartky, *Selling the True Time*, pp. 104–107 discusses Davis's letter in some detail. Davis also quoted the procedures and prices for the service offered by the Post Office Department in England, which transmitted the Greenwich Observatory signal to London and surrounding areas. Davis intended his proposal to result in a plan, which would then be submitted to the Navy Department for approval.

[32] Holden to Lt Cdr C. H. Davis, March 1, 1877; N. Green to Superintendent, USNO, February 27, 1877; Davis to Chief of the Bureau of Navigation (Ammen), April 10, 1877; AR (1877), 117. The younger Davis was then in charge, since his father had died two weeks before.

Figure 5.9. William F. Gardner, instrument maker and inventor of the Naval Observatory System of Observatory Time and the Gardner System of Observatory Time.

Once the time signal had been disseminated to a particular city, the problem was that of how to disseminate it within the city. This had been done within the city of Washington by time ball since 1845 and by fire-alarm telegraph since 1865. These methods were supplemented by the use of government clocks under constant control via electrical pulses from a master clock at the Naval Observatory, beginning with the Navy Department in 1868, the Army Signal Office in 1871, and the Treasury Department in 1873. By the late 1870s reliable synchronizers had been developed, so that clocks were being regulated via periodic signals that reset their displays. By 1884, 20 public buildings, including the Executive Mansion and the Senate wing of the Capitol, had such clocks installed, and were linked by a time wire to the Observatory. By 1885 the number of periodically synchronized clocks in various public offices had increased to 84, by 1886 to 200 and by 1888 to 347. The two systems – controlled clocks and regulated clocks – could be used in any city, regulated in terms of the "correct" time, which was still defined as local time. Such local time might be obtained from a local observatory, some of which in fact made money from selling time. It might also be obtained via telegraph from Washington, Washington time being appropriately corrected by taking into account the differences in longitude, which were accurately known for any number of cities.[33]

The technical genius behind the Naval Observatory's system of synchronizer clocks in public buildings and controlled by its signals was William F. Gardner (Figure 5.9). Gardner had been employed by the Observatory since 1864, and was promoted

---

[33] Bartky, "Naval Observatory Time Dissemination before the Wireless," in *SOJ*, and Bartky, *Selling the True Time*, chapters 3 and 4.

to instrument maker in 1866. Working under Harkness, Gardner kept the astronomical instruments in fine working order, and also arranged for the installation of a telegraph to the Navy Department. He ensured that the stream of pulses wired to a controlled clock there, would have the clock "beat in unison with the standard time-keeper at the observatory." Over the following years Gardner devised a reliable clock synchronizer and components, which were used both by the Naval Observatory and in private clock systems, the former called the "Naval Observatory System of Observatory Time," the latter the "Gardner System of Observatory Time." The system allowed the periodic resetting of the second and minute hands of a clock.[34] The Naval Observatory System of Observatory Time, which had continuously been improved by Gardner, was exhibited in 1889 at the Exposition Universelle in Paris, winning a "Grand Prix" (highest honors) gold medal for the Observatory; Gardner himself, who was in charge of the exhibit, also was awarded a gold medal and made an "officier d'académie."

Crucial to the dissemination of accurate time was the chronograph. As the public demand for more accurate time from an authoritative source grew, the Naval Observatory instituted in-house clock comparisons via the chronograph (Figure 5.10). Stellar observations determined the (master) sidereal clock's precise time, and it was then compared with the mean clock via chronograph. The mean-time clock was in turn compared with the transmitting clock, which was adjusted so that it gave a time signal exactly at noon (Figure 5.11).

Electrical apparatus and the telegraph also gradually came to be used for dropping time balls. The time ball at the Naval Observatory was dropped manually until sometime during the 1850s, when "an electrical circuit was completed through a magnet on the roof, to which the hallyards of the ball were attached at one minute before noon, the ball having been hoisted nine minutes previous and the observer at the clock pressed a key and released the ball at the second of twelve." In 1879 a transmitting clock was set up in the clock room, and the circuit passed through it, so that the ball was released automatically. As of July 11, 1885 a non-automatic time ball was dropped from the State, War and Navy Department building instead of the Naval Observatory, because the growth of Washington had rendered the Observatory time ball of "little account."[35] Meanwhile, in New York City, following up on discussions of Holden with Western Union, a time ball was dropped by telegraphic signal to the telegraph company's office beginning September 17, 1877 (Figure 5.12).[36] The Navy's

[34] Brown (note 28), 22. On Gardner (who died December 11, 1898) see Bartky, *Selling the True Time*, pp. 173–177; PA, **8** (1900), 109; AR (1867), 133.

[35] *New York Herald*, January 21, 1882, USNOA, SF, "Time, 1866–93" folder; AR (October 5, 1885), 7.

[36] The New York City time ball was to be mounted on the flagstaff of the East (Broadway) tower of the Western Union building, some 230 feet above the street, with the iron flagstaff adding another 50 feet. It was to be made of copper-wire netting painted black, and 3.5 feet in diameter. The ball was hoisted at 11:55 New York Time, and dropped by electric signal at noon. The company agreed to the expense of about $500 only if the ball were dropped at local noon. This was 12 h 0 m 0.00 s, or 11 h 47 m 49.53 s am in Washington or 4 h 56 m 1.65 s at Greenwich. The ball could be seen "by every vessel in New York harbor and at the Brooklyn wharf from Quarantine to the Harlem river and to Spuyter Suyorl [?] creek, as well as by a large proportion of

Chronographic Record of Standard Clock and Transmitter.

Chronographic Record of the Observation of a Star.

Chronographic Comparison of a Mean Time Clock with a Siderial Clock.

Figure 5.10. Three uses of the chronograph for time. The cylinder of the chronograph is made to revolve once per minute by a clockwork mechanism controlled by a governor. For determining sidereal time the second beats of the sidereal clock are registered automatically with each oscillation of the pendulum. The times of passage of a star over the micrometer wires are recorded when the astronomer presses a key in his hand. The resulting recording is seen in the middle three lines above. For transmitting mean time, both clocks are recording their beats. The chronograph compares the mean time clock with the sidereal clock (three lower lines) to determine any error in mean time. The mean standard clock is also compared with the transmitting clock to determine any error (top four lines). From Taylor (1890, ref. 48).

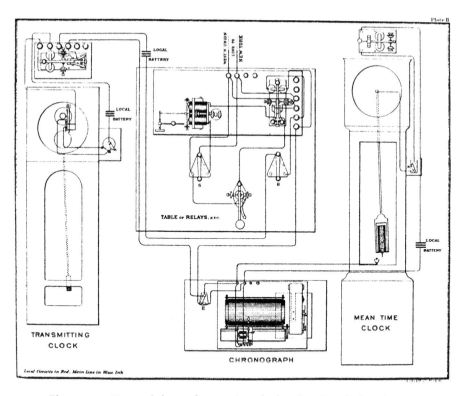

Figure 5.11. Transmitting and mean time clocks, showing clock and chronograph connections, and the connection to the Western Union line at top. From *Notes on Navigation and the Determination of Meridian Distances for the Use of Naval Cadets at the U. S. Naval Academy* (Government Printing Office: Washington, 1882), plate II. A simplified version is found in Taylor (1890, ref. 48). Most of the circuits were local, with "main line" circuits within the block marked "Table of Relays."

<hr/>

(footnote 36 cont.)

the citizens of New York." For a vessel's chronometer, which kept Greenwich time, any difference from 4 h 56 m 1.65 s was the chronometer's error, and daily observation of the time ball would give its rate. Various versions of the New York time ball are described in *Scientific American*, **39** (1878), 335 and 337, and E. S. Holden, "On the Distribution of Standard Time in the United States," *The Popular Science Monthly*, **11** (June, 1877), 174–182. As Bartky (*Selling the True Time*, pp. 108 and 250, note 21) has pointed out, there were two failed time balls before the operational one on September 17. For a step-by-step statement of the method for dropping the New York time ball from the Naval Observatory as of 1882 (the year before Standard Railway Time was inaugurated), see John Rodgers, "Method for Dropping the New York Time Ball used at the Naval Observatory," in *Notes on Navigation and the Determination of Meridian Distances for the use of Naval Cadets at the U. S. Naval Academy* (Government Printing Office: Washington, 1882), pp. 92–93 and Plate II.

NATIONAL OBSERVATORY, AT WASHINGTON.

Figure 5.12. Dropping the New York City time ball by telegraphic signal from the chronometer room of the "National Observatory," from S. P. Langley, "The Electric Time Service," *New Monthly Magazine*, **56** (April, 1878), 665–671. The telegraph wires are clearly seen attached to the transmitting clock.

Hydrographic Office and the Army's Signal Service also became involved in time-ball drops; an 1881 Signal Service report provides some of the most detailed accounts of time-ball operations at the time.[37] By 1888 the Observatory signal via Western Union was regulating (without charge) time balls at Woods Hole, Massachusetts; Newport, Rhode Island; New York City; Philadelphia; Baltimore; Hampton Roads; Savannah; and New Orleans, in addition to the ball dropped on the Navy Department Building in Washington just mentioned. These, the Assistant Superintendent of the Naval

---

[37] Winslow Upton, compiler, *Information Relative to the Construction and Maintenance of Time Balls* (Government Printing Office: Washington, 1881), Professional Papers of the Signal Service, No. 5. Although it was prepared under the direction of General W. B. Hazen, Cleveland Abbe was more immediately responsible for the report, and in turn delegated the compiling of the information to Winslow Upton. The monograph consists of reports concerning time balls around the country.

Observatory remarked in that year, were "but the beginning of what it is hoped will be a comprehensive system embracing the whole coast, so that in every commercial city there shall be a like installation."[38] In the same year Heinrich Hertz was experimenting with radio waves, and within 15 years the "wireless" transmission of time would be a reality, eventually to completely supersede the telegraphic determination of time that was so dominant at the end of the nineteenth century. Time balls would continue to be used for a while, however. In 1905, 19 time balls were being dropped by signals in the principal ports of the Atlantic, Pacific, and Gulf of Mexico coasts, and several Great Lakes ports, the signals coming via Western Union, the Postal Telegraph Company, and American Telephone and Telegraph Co.[39]

Meanwhile, the telegraphic dissemination of time had raised the issue of what "standard time" should mean. As long as one stayed in a single city, there was no problem; standard time could mean simply the same time throughout the city or its immediately surrounding region. However, the issue was especially crucial to the increasingly far-flung empires of the railways, and already in 1877 Holden wrote that "In the opinion of many experienced and prominent railway-officials in the United States, it is quite feasible and very desirable for all railways to be operated by one common time, and the first step toward this is plainly the certainty that the time signals which are now regularly sent from the Naval Observatory shall reach each railway station once daily, at least."[40]

Other reasons existed aside from the needs of the railways, and, according to Bartky, it was not railway needs, but "scientific pursuits requiring simultaneous observations from scattered points" that were the issue leading to the adoption of standard time as it exists today. In particular, he points to geophysical observations such as of aurorae, and Cleveland Abbe's role in leading the process of adoption beginning in 1875.[41] Even if scientific interests were the fundamental cause of the adoption of uniform time, the implementation of it for railways was absolutely essential to its widespread adoption. When North American railroads adopted the new system at noon on November 18, 1883 the Naval Observatory changed its time signals to the new system of "standard time" according to which the meridians of 75, 90, 105, and 120 degrees became the time meridians of Eastern, Central, Mountain, and Pacific

[38] Brown (note 28), 10,524. On November 23, 1893 the time ball (installed by Gardner and operated by the Chicago branch of the Hydrographic Office) was first dropped in Chicago. Its operation is described in detail in "The Time Ball Service," *The American Jeweler*, **15** (February, 1895), 48–53.

[39] Edward Evertt Hayden, "The Present Status of the Use of Standard Time," PUSNO, volume 4, Appendix IV (Washington, 1905), G 9.

[40] Holden (ref. 36), 181. Holden noted that the United States was far behind the U. K. in this respect, where 500 railway stations received a daily signal. The lead of the U. K. he ascribed to its small geographical extent, and to the fact that the telegraphs were owned by the government.

[41] Ian R. Bartky, "The Adoption of Standard Time," *Technology and Culture*, 30 (January, 1989), 25–56. Abbe was made chair of the American Metrological Society's Committee on Standard Time, and Bartky cites the Committee's "Report on Standard Time" as "the key document in the process leading to uniform time for the United States." (p. 37). See also Bartky, *Selling the True Time*, chapter 11.

Standard Railway Time.[42] Thereafter, the New York time ball was dropped at noon Eastern Time, and other cities were regulated by the hourly differences.

Because the new "Standard Railway Time," as it was called, was indexed to the Greenwich meridian, the 1883 event was an important step in the adoption of Greenwich as the prime meridian of the world.[43] Nevertheless, as Ian Bartky has pointed out in *Selling the True Time*, Standard Railway time was not America's legal time in the sense of being enshrined in law, nor did it yet make Greenwich the official prime meridian. The latter was the sincere intention at the International Meridian Conference held in Washington from October 1 to November 1, 1884, which was attended by 41 delegates from 25 countries. The purpose of the meeting, authorized by act of Congress dated August 3, 1882, was for "fixing upon a meridian proper to be employed as a common zero of longitude and standard of time-reckoning through-out the globe."[44] Although S. R. Franklin, the Superintendent of the Naval Observatory, attended (representing Colombia!), and although the U. S. delegates included Navy officers Rear Admiral C. R. P. Rodgers (Superintendent of the Naval Academy) and Cdr W. T. Sampson (as well as Lewis M. Rutherfurd, W. F. Allen of the Railway Time Conventions, and Cleveland Abbe of the U. S. Signal Office), the Naval Observatory involvement was minimal (Figure 5.13).[45] The pressure for such a meeting had been brought by the American Metrological Society and others, and the scientific issues had been discussed thoroughly at the International Geodetic Conference in Rome the previous October. The delegates in Washington overwhelm-ingly approved Greenwich as the prime meridian, produced a substantial proceedings of their meeting, and left Washington satisfied in a job well done. However, as Bartky has shown, "world-wide adoption of a prime meridian and a time coupled to it pro-ceeded at a snail's pace, a country-by-country process unaffected by the Washington deliberations and recommendations. In this sense, the International Meridian Conference of 1884 failed."[46] Even in the United States, Greenwich-indexed timekeep-ing was not mandated by law until World War I, when Daylight Saving Time was

---

[42]  Hayden (ref. 39), p. 5; AR (1884), 7. AR (1884) notes that the District of Columbia did not adopt standard time until March 1, 1884, and that, up to that time, the Observatory ball was dropped, and the fire bells struck, at noon of Washington mean time. After that all Washington signals were based on 75th-meridian time. Federal legislation for Standard Time was not adopted until 1918.

[43]  In speaking of a prime meridian, it is important to recognize that, throughout history, there have been many levels of "prime meridian," including international, national, and local. Moreover, one always needs to ask the question of whether it is a prime meridian for navigation, time, longitude, or geographical positions; they have not always been the same. See Derek Howse, *Greenwich Time and the Discovery of the Longitude* (Oxford University Press: Oxford, 1989).

[44]  *International Conference held at Washington for the Purpose of Fixing a Prime Meridian and a Universal Day. October, 1884. Protocols of the Proceedings* (Gibson Bros: Washington, 1884). The act of Congress is reprinted in Annex I, p. 209. The meeting is discussed in Derek Howse, *Greenwich Time and the Discovery of the Longitude* (Oxford University Press: Oxford, 1989), pp. 138–151. See also the Centennial Proceedings, Longitude Zero, 1884–1984, ed. Stuart Malin, A. E. Roy, and P. Beer, in *Vistas in Astronomy*, **28** (1985).

[45]  Newcomb and Hall were admitted as invited observers (p. 15), and Newcomb spoke briefly at one point (ref. 44), pp. 59–62.    [46] Bartky, *Selling the True Time*, chapter 12, p. 152.

Figure 5.13. Delegates to the International Meridian Conference, held in Washington in 1884. Circled left to right are Cleveland Abbe, L. M. Rutherfurd (directly in front of Abbe), Commodore S. R. Franklin (representing Colombia), Janssen from France, J. C. Adams from Great Britain, the President of the Conference, Admiral C. R. P. Rodgers, and Commander W. T. Sampson. Courtesy of the National Archives, where full identifications are available.

instituted. Still, the Washington meeting was a symbolic landmark in adopting Greenwich as the prime meridian for the world.

During the quarter century from 1865 to 1890 citizens across America became much more aware of time because its accurate dissemination had been revolutionized, thanks largely to the Naval Observatory and Western Union. In keeping with the nation's westward expansion, in 1884 the Naval Observatory established a station at the Navy Yard at Mare Island, California, to maintain for the Pacific coast the same system of time signals as for the East and South regions, and for the care and rating of chronometers (Figure 5.14). Already in 1886 the transmitting clock at the Mare Island station was dropping time balls at Mare Island and San Francisco; Western Union itself paid for the cable across the Mare Island straits. Writing in 1899 Ensign Everett Hayden wrote that "The principal object in view was then, and is now, the storage and supplying to naval vessels of accurately rated chronometers and the furnishing to shipmasters in any of our ports facilities for comparing and rating their own chronometers, either by means of time-balls, dropped electrically at noon, or by

Figure 5.14. The Observatory's Mare Island Station at the Mare Island Navy Yard in California. It was established in 1884 as a west coast time service. T. J. J. See stands in the doorway.

visits to the local telegraph offices, where signals are received."[47] As the Navy lieutenant in charge of Time Service wrote in an article entitled "U. S. Government System of Observatory Time," by 1890 not only were the clocks in government offices set to accurate time, but also numerous time balls were being dropped around the country and other time signals given to cities and the railways via telegraph. Furthermore, "the whole operation is done automatically, except the closing of the circuits, which is performed by the officer in charge of the service."[48]

By 1890 too, apparently precipitated by a business decision of Western Union to expand its time service, American observatories began to complain bitterly about the relationship of the Naval Observatory with Western Union. While defending this practice against an inquiry of 23 observatories, many of which were selling time,

---

[47] Edward Everett Hayden, "Clock-rates and barometric pressure as illustrated by the mean-time clock and three chronometers at Mare Island Observatory: with a brief account of the Observatory," PASP, **11** (June 1, 1899), 100–104. The Observatory was located on the crest of a hill near the northern extremity of the island. As of 1899 it consisted of two brick buildings, one with a transit-of-Venus 5-inch Alvan Clark telescope, the other with a clock, chronometer, and chronograph; the two rooms were separated by a wooden structure housing a 2.5-inch Stackpole transit instrument, also from the transit-of-Venus expeditions. Three Howard clocks completed the instrumentation. See "Mare Island Naval Observatory" folder, USNOA.
[48] Lt Hiero Taylor, "U. S. Government System of Observatory Time," *Jewelers' Circular and Horological Review*, **21** (February, 1890), 82, 85–86, and 88. On the establishment and early years of the Naval Observatory station on Mare Island see AR (1885), 7 and AR (1886), 8. The time ball was dropped daily at noon of the 120th meridian.

Figure 5.15. Commodore R. L. Phythian, who served three times as Superintendent of the Observatory during a period when crucial decisions were made about expansion of the Navy's time service.

Superintendent R. L. Phythian (Figure 5.15) noted that, although Western Union might charge for a time signal to corporations or individuals, the company "does not enjoy, nor does it claim, the sole right to the Observatory signal." Any company, he emphasized, could obtain the same signal by making the necessary connections. While the Observatory recognized that the "local time service patronage of observatories" would be injured by the aggressiveness of Western Union in seeking this business, it emphasized that the determination of time by any observatory was relatively inexpensive compared with the means of distribution financed by Western Union – an extensive system of wires, operators, and other apparatus. Moreover, he pointed out that Western Union could itself determine the time and put observatories out of the time business. In arguing that the Naval Observatory signal should not be cut off from Western Union, Phythian was quite clear: If it were cut off "the result would be the breaking up, for a time, of the services now established. The struggle would then begin for the reestablishment of what is known beyond question to be a lucrative business. Can anyone doubt who would be the victor in such a contest? Not only would the Observatories fail, but they would bring upon themselves the active enmity, embittered by the controversy, of the Western Union Company, and their patronage would be totally destroyed." Phythian called the matter of "grave importance," since the convenience and business of people in hundreds of cities, as well as the railroads, would be disrupted. Phythian suggested that a commission, composed of the Smithsonian Secretary, one or two observatory heads, and a representative of the Navy Department, be composed to meet with Western Union.[49]

---

[49] Phythian to Dewey, Chief of Bureau of Equipment, May 15, 1890, USNOA. See Bartky, *Selling the True Time*, chapter 15, for details of this controversy. For an official statement of Observatory policy on time dissemination, see AR (1893), 4.

In the end the Naval Observatory–Western Union alliance survived, but not without considerable fallout. One example is the case of Harvard Observatory. On March 31, 1892, Harvard discontinued its time service, some 20 years after it had begun regular transmission of signals for compensation. In terminating this service, Observatory Director E. C. Pickering complained bitterly that Harvard could not match the publicly funded Naval Observatory for determining time, or Western Union's low prices in disseminating it. The loss was more than financial for Pickering, who pointed out that "One of the greatest advantages of the time-service to the Observatory has been that it kept before the public the practical value of astronomical work. Many thousands of persons who take no interest in work of a purely scientific character recognize the great financial value to the public of an accurate system of time. The Observatory desires to confer this benefit on the public, and it would be ready to do so even at a financial loss." The Naval Observatory–Western Union arrangement, he complained, precluded this, even though signals sent long distances ran the risk of frequent interruption. Thus embittered, there was no doubt which side Pickering would be on in the question of civilian control of the Naval Observatory.[50]

In the midst of the telegraphic dissemination of time, it must not be forgotten that chronometer rating still remained a major function of the Observatory. In 1866, 220 chronometers were on trial, which were wound and compared daily against the standard clock. In 1890 there were some 115 chronometers in use, ready for issue, or under repair at the Observatory; 156 more were issued to ships, 37 to shore stations, and 40 to the Navy Yard on Mare Island.[51] All of this constituted a considerable program of important, if routine, work at the Observatory, especially since the chronometers were subject to elaborate procedures for testing at various temperatures. The importance of these procedures is evidenced in the construction of an additional chronometer room in 1883, on the north side of the Observatory connecting with the original chronometer room (Figure 5.16).[52] Rigorous tests (Figure 5.17) showed that the chronometers supplied by Negus of New York City and William Bond & Son of Boston were most accurate, and during the last decades of the century Negus supplied the chronometer of choice for new Navy purchases. Nor did the Observatory's responsibility end with the rating process. After rating they were transported by hand from the Observatory to their destination while still running, but taken from their gimbals,

---

[50] E. C. Pickering, "Time Service," December 26, 1891, National Archives, RG 37, Naval Hydrographic Office, E 57, Correspondence and Reports Files, 193.11; file also in USNOA "Time, 1866–1893" folder. This article gives Pickering's history of Harvard's time service, and gives the origination date as prior to 1856 (see, however, Bartky, *Selling the True Time*, pp. 197–198 and Appendix, p. 211).

[51] SecNav Report dated December 3, 1866, with AR dated October 15, 1866, 137; AR ending June 30, 1890, 95.

[52] Lt E. K. Moore, U. S. N., "Method of Testing Chronometers at the U. S. Naval Observatory," USNIP, **10**, no. 2 (1884), 171–186, including eight plates. See also Everett E. Hayden (Ensign), "Clock-Rates and Barometric Pressure as Illustrated by the Mean-Time Clock and Three Chronometers at Mare Island Observatory, with a Brief Account of the Observatory," PASP, **68** (June, 1899), 101–114; and E. E. Hayden (Lt Cdr) "Chronometer Rates," USNIP, **29**, no. 3 (c. 1903?), 729–742.

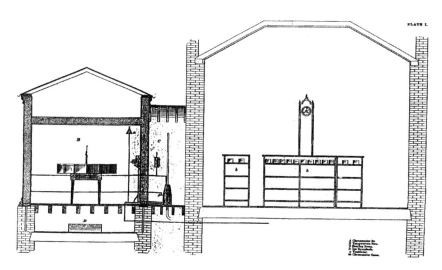

Figure 5.16. The chronometer house for testing chronometers. The chronometer room is on the right, the temperature room on the left. Chronometers were subjected to a wide range of temperatures to determine their rates. From Moore (1884, ref. 52).

wrapped in paper, and placed level in a basket of cotton. The supervision of the Observatory ceased only when chronometers had been delivered to the navigation officer of the Navy yard or the vessel, but began again when they were returned to the Observatory for rating.[53]

By the end of the nineteenth century time for the Navy was as important as ever, and time for the public had expanded from the Washington time ball to the systematic dissemination of telegraphic time around the country, based on the Observatory's noon time signal. The full measure of its economic importance is to be found not only in the few observatories selling time, or in Western Union's dissemination of it, but in the commercial interests that increasingly depended on accurate time. That dependence would only grow, and in unforeseen ways, in the next century.

## 5.4 Solar-eclipse expeditions

We have seen that solar eclipses were the object of sporadic attention in the U. S. Navy prior to the Civil War, notably with Gilliss's expeditions to Peru in 1858 and Washington Territory in 1860. Though it was not yet attached to the Naval Observatory, the Nautical Almanac Office had used the solar eclipse of July 28, 1851 as a crucial test of its new lunar tables prior to its publication of the first volume of the *American Ephemeris*. In 1860 the Office sent an expedition to Vancouver, including Simon Newcomb and William Ferrel, again primarily to test the theories of the motion

---

[53] Moore (ref. 52), 186. A detailed description of the care of chronometers is given in *Notes on Navigation* (ref. 36 above), 89–108.

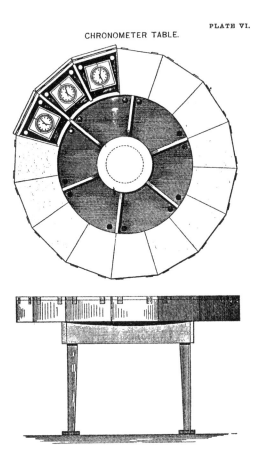

PLATE VI.

CHRONOMETER TABLE.

Figure 5.17. A chronometer table for testing. A view from above of item B in Figure 5.16, showing placement of chronometers. From Moore (1884, ref. 52).

of the Sun and Moon. The Office issued charts and elements of these solar eclipses, as well as that of 1854.[54]

The last third of the nineteenth century and the first third of the twentieth century, however, became the golden age of eclipse expeditions, as the U. S. Navy sponsored far-flung journeys to observe these relatively rare occurrences, and reported them in detail in its publications (Table 5.2). Why they should have done so can be justified in part in terms of navigation (the theory of the Moon for lunar distances), and in part in terms of science. The benefits for science were not always made explicit to Navy officials; in his reports to the Navy Department, Superintendent Sands spoke of unspecified "important results" as the object of the 1869 eclipse observations. For the eclipse the following year Sands wrote that "Failing to obtain an appropriation for the purpose of engaging the services of observers outside of our own

[54] Newcomb, *Reminiscences*, pp. 88–93. Newcomb also cites a book on the expedition by Samuel H. Scudder, an assistant at the Museum of Comparative Zoology who accompanied them at the request of Professor Agassiz. This book is *The Winnipeg Country, or Roughing it with an Eclipse Party* (Cupples, Upham & Co.: Boston, 1886). Scudder is not identified on the title page, where the author is given only as "A. Rochester Fellow."

Table 5.2. *Total solar-eclipse expeditions of the U. S. Naval Observatory, 1869–1929*

| Year (date) | Location | Participants | Descriptions[a] |
|---|---|---|---|
| 1869 (August 7) | Des Moines, Iowa | Newcomb, Harkness, and Eastman | *Reminiscences*, p. 113<br>AR (1869) (coronal line)<br>WO for 1867 (1869) |
| | Behring Strait | Hall and Rogers | AR (1869)<br>WO for 1867 (1869) |
| 1870 (December 22) | Gibraltar<br>Syracuse,<br>Sicily | Newcomb<br>Eastman,<br>Hall, and Harkness | *Reminiscences*, p. 113<br>AR (1870), 46–47<br>WO for 1869 (1871) |
| 1878 (July 29) | Colorado<br>Wyoming, Texas | Most USNO staff | AR (1878), 156–160<br>WO for 1876 (1880) |
| 1880 (January 11) | California | Frisby | WO for 1876 (1880) |
| 1900 (May 28) | Georgia<br>North Carolina | Updegraff *et al.*<br>Skinner *et al.* | PUSNO, volume 4 (1905) |
| 1901 (May 17) | Sumatra | Skinner, Littell,<br>Eichelberger *et al.* | Ditto |
| 1905 (August 30) | Spain | Eichelberger,<br>Littell *et al.* | PUSNO, volume 10 (1924) |
| 1918 (June 8) | Baker, Oregon | Hammond | Ditto |
| 1923 (September 10) | California and Mexico | Navy Aviators | Ditto<br>From U. S. Navy aircraft |
| 1925 (January 24) | Long Island, New York | Watts, Littell *et al.* | PUSNO, volume 13 (1930)<br>From Navy dirigible<br>U. S. S. *Los Angeles* |
| 1926 (January 14) | Sumatra | Littell, Peters *et al.* | Ditto |
| 1929 (May 9) | Phillipines | Sollenberger *et al.* | Ditto |

Notes:
[a] AR, *Annual Report of the USNO*; WO, *Washington Observations*; and PUSNO, *Publications of the USNO*.

institution, the [Navy] Department, ever ready to contribute all in its power to the advancement of science, ordered, at the request of the Superintendent of the Observatory, four of its professors for that duty."[55] However, there is no doubt that total solar eclipses held out the promise of advancing science in at least three ways: by testing theories of motion of the Sun and Moon, by creating optimal conditions to search for a putative new planet near the Sun, and for physical studies of the still mysterious solar "protuberances" and especially the solar corona.

Together with the transit-of-Venus expeditions discussed in chapter 7, these expeditions may be viewed as part of the age of exploration, the celestial analog of the American push westward. They are of historical importance today not only for their astronomical data, but also because they shed light on the status of techniques in photography and spectroscopy, and for their human interest. Moreover, the eclipse of

[55] AR (1870), 46.

1878 is of particular interest because it marked a high point of cooperation between the Naval Observatory and American astronomers, made possible by an $8,000 Congressional appropriation.[56]

The official reports of the eclipses of 1869, 1870, and 1878 reveal the goals of the observations. For the 1869 event, Newcomb, Harkness, and Eastman were sent to Des Moines, Iowa, while Hall was sent to the Bering Strait. Although Hall was troubled by clouds, the Des Moines party had favorable weather. For Newcomb the "main object" of observation was to search for anything near the Sun that normally could not be seen without an eclipse. "More especially was it determined to search in the neighborhood of the sun for an immense group of very minute intra-mercurial planets, the existence of which had been rendered so probable by the researches of LeVerrier on the motion of Mercury." Failing to detect any new planets, Newcomb then concentrated on studying the corona. Another important task for Newcomb was to compare the *times* of totality observed with the predictions from tables, again checking the theory of the motion of the Sun and Moon; Newcomb found discrepancies on the order of several arcseconds, or several tenths of a second of time. Finally, Newcomb had issued a call for observations by "intelligent citizens" of the duration of totality within 1–10 miles of the limits of totality. The several dozen reports for which Newcomb gave data testify to the avid interest of the public along the path of totality.[57]

Harkness and Eastman set up at a site in the city separate from Newcomb (Figure 5.18). Harkness had decided to use photographic techniques to obtain the times of contact, and "attached very little importance to the making of optical observations of the times of contact." His prime object was spectroscopic observations to learn more about the physical constitution of the corona, which was visible only during an eclipse. The nature of the corona was not at all understood in those days. There were those who still believed it was an effect due only to sunlight passing through the Earth's atmosphere, or through a supposed lunar atmosphere. Using a spectroscope built by Desaga of Heidelberg for chemical purposes but greatly altered for his astronomical observations (Figure 5.19), Harkness obtained a coronal spectrum that was continuous except for a single bright green line, later known as the coronal line K 1474. He concluded from this single spectroscopic observation, as well as visual and photographic observations (Figure 5.20), that the corona was "a highly rarefied self-luminous atmosphere surrounding the sun, and, perhaps, principally composed of the incandescent vapor of iron." The true origin of the line, the first of many to be discovered, remained a mystery for seven decades until B. Edlén and others

---

[56] AR (1869), 57; AR (1870), 46–47; AR (1878), 156–160. For background on solar-eclipse history, see especially S. A. Mitchell, *Eclipses of the Sun* (Columbia University Press: New York, 1923). On the 1878 eclipse, see especially Mitchell's chapter 9.

[57] "Reports on Observations of the Total Eclipse of the Sun, August 7, 1869," Appendix II of *WO for 1867* (1870). Newcomb's report is on pp. 7–22 of the 218-page report. See also Newcomb, *Reminiscences*, p. 113, on the 1869 and 1870 eclipses.

J.Bien, lith.                                                    Dr.E.Curtis, U.S.A.,photo.

U.S. Eclipse Observatory; Des Moines, Iowa.

Figure 5.18. A general view of the observing building (measuring 32 feet by 16 feet) for the 1869 solar-eclipse expedition in Des Moines, Iowa. Harkness and Eastman observed from here, Newcomb from a mile south. From *Washington Observations for 1867* (Government Printing Office: Washington, 1870), Appendix II, plate 1.

found that these lines resulted from iron atoms at high ionization levels at coronal temperatures exceeding 2 million degrees Kelvin.[58]

The 1869 eclipse was also an early instance of the use of photography in astronomy, in some ways a preparation for the transit-of-Venus photographs five years later. In charge of the photographic telescope (Figure 5.21) was Assistant Surgeon Dr Edward Curtis, who was skilled in photography and had been sent on the expedition

[58] Ibid., pp. 25–96; especially pp. 59, 65–66, and, for spectroscopic observations, 27–30 and 60–67. Harkness actually measured the line at 531.6, which was in units of millionths of a millimeter (AR (1871), 119). The K 1474 designation was based on Kirchoff's map of the solar spectrum then in general use; Ångström's map had just been published in 1868 and would soon come into general use, followed by Rowland's, where the line was designated 5316.9. C. A. Young is considered the co-discoverer of this coronal line, which he observed while on the Naval Observatory expedition in Burlington, Iowa. He gave credit to Harkness for the coronal K line in his volume *The Sun* (Appleton: New York; 1881), pp. 224–225, where he notes that all doubts about the reality of the line were erased in 1870 and 1871 by additional spectroscopic observations during totality. Young further discusses this line in the chapter on the corona in the 1895 edition of his book, pp. 257–263. For the context see Karl Hufbauer, *Exploring the Sun: Solar Science Since Galileo* (Johns Hopkins University Press: Baltimore, 1991), pp. 62 and 112. In addition to assisting Harkness, Eastman was responsible for meteorological and photometric observations during the eclipse, and issued his own detailed report.

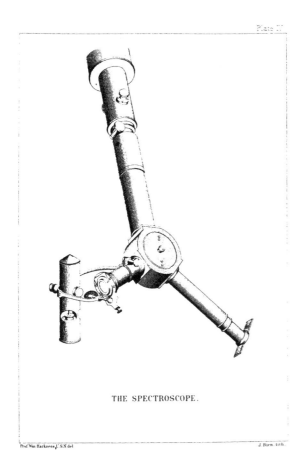

Plate

THE SPECTROSCOPE.

Prof. Wm Harkness U.S.N del                    J. Bien. lith.

Figure 5.19. The spectroscope used by Harkness to discover the coronal K line. It was a single-prism spectroscope made by Desaga of Heidelberg, which had originally been intended for chemical purposes, but was altered for astronomy. Harkness attached it to his personal 3-inch telescope. From *Washington Observations for 1867* (Government Printing Office: Washington, 1870), Appendix II.

especially for that purpose. Curtis described the telescope and the photographic process in great detail in his report in the same volume of *Washington Observations*, giving a colorful picture of the state of the art at the time. He used the 7.75-inch Alvan Clark refractor, which had been loaned by the Naval Academy for the expedition, and altered by instrument-maker William F. Gardner for photographic purposes. Making use of the experience of De la Rue with the 1860 eclipse in Spain and Vogel's photographs of the 1868 eclipse in Arabia, Curtis obtained 113 photographs of the partial phase, and two "exquisite" negatives of totality, so excellent in quality that Curtis believed that they marked "an era in eclipse photography." Curtis's conclusions about the corona agreed with those of Harkness: "difficult and perplexing though it may be to conceive of an atmosphere that will exist above a layer of extremely rarefied hydrogen gas, yet the evidence afforded by these photographs that the corona is such an atmosphere seems incontestable."[59]

In 1870 the shadow of a solar eclipse raced across Europe. Once again

---

[59] The Curtis report is on pp. 123–156 of the *WO* for 1867; the quote is on p. 145. J. Homer Lane, known for his later contributions to solar physics, was also a member of the Des Moines party.

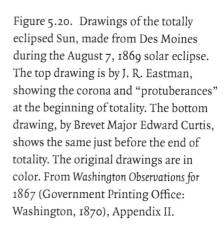

Figure 5.20. Drawings of the totally eclipsed Sun, made from Des Moines during the August 7, 1869 solar eclipse. The top drawing is by J. R. Eastman, showing the corona and "protuberances" at the beginning of totality. The bottom drawing, by Brevet Major Edward Curtis, shows the same just before the end of totality. The original drawings are in color. From *Washington Observations for 1867* (Government Printing Office: Washington, 1870), Appendix II.

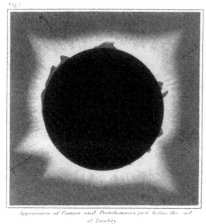

Newcomb, Harkness, Hall, and Eastman were dispatched to observe it, Newcomb (who was already on assignment in Europe) to Gibraltar and the latter three to Syracuse, Sicily. Again Harkness made his spectroscopic observations, confirming the bright coronal line, and possibly two fainter lines, and reiterating his conclusions about the physical nature of the corona. Hall concentrated on visual observations of the corona, and Eastman once again specialized in meteorological observations, as well as using a polariscope, and offered his observations of the corona.[60] Since no special appropriation had been secured for the eclipse, no photographic expert was sent, and no photographic work was undertaken by any of the Naval Observatory employees.

The 1869 and 1870 expeditions were mere preludes to the 1878 eclipse. S. A. Mitchell, in his book *Eclipses of the Sun*, estimates that some 12 stations and 100 astronomers participated, including much of the Naval Observatory staff, which this time

[60] *WO for 1869* (1872), Appendix I, "Reports on Observations of the Total Solar Eclipse of December 2, 1870."

Figure 5.21. The 7.75-inch telescope with camera attached for observing the solar eclipse of August 7, 1869 in Des Moines, Iowa. It was located in the building seen in Figure 5.18. The telescope was built in 1857 for the U. S. Naval Academy. The chronometer is seen strapped to a tripod stand close to the camera box, where the observer is timing an exposure. From *Washington Observations for 1867* (Government Printing Office: Washington, 1870), Appendix II, p. 126.

had secured an $8,000 appropriation from Congress (undoubtedly thanks to the political clout of the new Superintendent, John Rodgers). Three Naval Observatory parties went to Colorado: Hall to La Junta, Eastman to West Las Animas, and Holden to Central City (Figure 5.22). Harkness went to Creston, Wyoming, and Newcomb, in his first year as Superintendent of the Nautical Almanac Office, also traveled under Naval Observatory auspices to Separation, Wyoming. In addition, the Congressional appropriation allowed for numerous other parties: Allegheny Observatory Director S. P. Langley and others occupied the summit of Pike's Peak; G. W. Hill observed the eclipse from Denver; Cincinnati Observatory Director Ormond Stone and W. Upton of Harvard College Observatory were a few miles east of Denver; and J. C. Watson was at Separation, Wyoming. The observations of all of these parties were published in an enormous volume of *Washington Observations* exceeding 400 pages and 30 plates.[61]

One notable difference since the 1870 solar eclipse was the advance in photographic knowledge due to the intervening transit-of-Venus in 1874. For the eight American expeditions (see chapter 7) eight photoheliographs had been constructed specially. Although the photoheliographs themselves were not suited to studying the

[61] "Reports on the Total Solar Eclipses, July 29, 1878 and January 11, 1880," Appendix III to *WO for 1876*, part II (1880), 421 pages and 30 plates. The vast bulk of the report deals with the 1878 eclipse.

U. S. Naval Observatory Eclipse Party,

Figure 5.22. The 1878 solar-eclipse party, Central City, Colorado. On the left (west wall) front, is Professor Edward Holden with legs crossed, and (rear) James E. Keeler behind the instrument; at the front chimney, Lt S. W. Very is behind the instrument and Professor Charles S. Hastings to its side; toward the rear, Professor Edgar W. Bass is behind the tripod and Mr C. H. Rockwell at the table near Bass.

corona (one was set up experimentally at Newcomb's station), the experience gained by their use was invaluable. Accordingly, Harkness ordered two equatorial cameras of six-inch aperture and 36-inch focal length from Dallmeyer of London, the camera tubes and finders from Stackpole of New York, and both wet- and dry-plate holders from the American Optical Company of New York. These two cameras were used by the parties of Harkness and Hall. In the end only dry plates were used, giving satisfactory results. Alvan G. Clark and Aaron Skinner (assistant astronomer at the Naval Observatory) in the Harkness party obtained six photographs of the corona "thought to be at least as extensive and rich in detail as any ever taken." Joseph A. Rogers, a former employee of the Observatory who had manufactured the dry plates, obtained five photographs of the corona at the Hall-party site. In Central City a local photographer also obtained a good image of the corona in connection with the Holden party.[62]

[62] *Ibid.*, on the Harkness party photographs, pp. 51–56; the Colorado party, pp. 198–200; the Hall party, pp. 282–294; AR (1888), 156–160. Mrs Angeline Hall accompanied her husband on the

Only the photoheliograph taken under the personal supervision of Newcomb "inexplicably" failed to produce any images at all!

Among the most spectacular claims to emerge from the 1878 eclipse was the discovery of at least one intra-Mercurial planet claimed by J. C. Watson and Lewis Swift.[63] Newcomb, Hall, and Holden also searched, but in vain, casting doubt on the claimed discovery. Hall, fresh from his discovery of the moons of Mars less than a year before, gives a good feel for the circumstances and technique: "The day at La Junta was all that could be desired for observing a total eclipse. My own special work during totality was searching for an intra-mercurial planet, the supposed Vulcan, indicated by the researches of Leverrier on the orbit of Mercury. Before the eclipse I studied the configuration of the stars as they are laid down on the chart published by the Observatory, and during totality a copy of this chart was placed a few feet in front of me, so that I could refer to it instantly. As soon as totality began I turned my shade to the free opening and commenced sweeping above the sun and near the ecliptic. My sweeps extended from the brighter part of the corona to a distance of about ten degrees from the sun. The magnifying power was so great that the sweeping could not be done very rapidly. In this part of the sky I saw nothing but the stars laid down on the chart."[64] For the discoverer of the moons of Mars, the detection of another planet would have been spectacular, but it was not to be. Nevertheless, so great were the reputations of Swift and Watson that only after a quarter century of further observing did astronomers give up on the supposed intra-Mercurial planets. By that time Einstein's theory of relativity had explained the motion of the perihelion of Mercury, which had given rise to the search for intra-Mercurial planets.

Aside from a minor effort to observe the solar eclipse of 1880 in California, the memorable event of 1878 was the last Naval Observatory eclipse effort until 1900. The expeditions are an example of naval interest in, and Congressional funding for, science quite aside from its practical applications for navigation. Moreover, in the decade 1869–78 Congress appropriated not only a total of $13,000 for solar eclipse expeditions, but also $50,000 to build the world's largest telescope, and $170,000 for the 1874 transit-of-Venus expeditions discussed in the following chapters, very large amounts for the time. None of these were strictly necessary for navigation, and the same liberality would not always mark the Naval Observatory's history. The theme of pure versus applied science, which runs throughout the history of American science, is thus seen in microcosm at the Naval Observatory.

---

trip, and her brief description of the eclipse is given on pp. 257–258 of the report, along with her drawing in Plate 15. The well-known astronomy artist E. L. Trouvelot and his son observed near the Harkness party, and also provided a detailed report.

[63] Watson's report is *Ibid.*, pp. 117–124, especially 119 ff; and Swift's pp. 226–232, especially 228ff. The proposed planet was also known as Vulcan. On the search for intra-mercurial planets see Richard Baum and William Sheehan, *In Search of Planet Vulcan* (Plenum: New York and London, 1997).

[64] George W. Hill, "Biographical Memoir of Asaph Hall, 1829–1907," BMNAS, **6** (Washington, 1908), 264.

# 6 Asaph Hall, the Great Refractor, and the moons of Mars

> I am so far from disbelieving the existence of the four circumjovial planets, that I long for a telescope, to anticipate you, if possible, in discovering *two* round Mars, as the proportion seems to require, *six or eight* round Saturn, and perhaps *one* each round Mercury and Venus.
>
> Johannes Kepler, 1610[1]

> They have likewise discovered two lesser stars, or satellites, which revolve about Mars.
>
> Jonathan Swift, *Gulliver's Travels*, 1726[2]

Asaph Hall's discovery of the moons of Mars is among the better-known episodes in the history of astronomy. Nevertheless, it has seldom been discussed in the context of Naval Observatory history, in terms of the history of the telescope used to make the discovery, or even in the context of Hall's much broader career. In this chapter we examine the origins of the 26-inch refractor, the largest in the world at the time of its construction; we analyze the events in Hall's career leading to the discovery of the two tiny moons orbiting Mars; and we follow the aftermath of the discovery in the context of Hall's career and the research program of the 26-inch telescope.

## 6.1    The Great Refractor

When, in September, 1867, James Ferguson died after observing for 20 years on the 9.6-inch equatorial, it was the end of an era. Ferguson was one of the old school of astronomers; Newcomb, who was placed in charge of the 9.6-inch for a few months, and Asaph Hall, who assumed charge after Newcomb, represented a new generation. Probably both made a case to Superintendent Benjamin F. Sands that the 9.6 inch was no longer worthy of a national institution. Pulkovo Observatory, after all, had boasted a 15-inch Merz and Mahler refractor since 1840, as had Harvard since 1843; both Greenwich and Paris possessed 13-inch-class refractors, and numerous telescopes in the 11–12-inch range were being used by universities and even private owners. Moreover, the Dearborn Observatory in Chicago had embarked on a program with the world's largest refractor, an 18.5-inch achromat figured by Alvan Clark & Sons.

---

[1] Kepler to Wachenfels, cited in Asaph Hall, *Observations and Orbits of the Satellites of Mars, with Data for Ephemerides in 1879* (Government Printing Office: Washington, 1878), p. 44.

[2] *Gulliver's Travels* (London, 1726), part 3, chapter 3.

It was therefore not difficult to make a convincing case for a larger telescope, and, in the same month as Ferguson's death, Newcomb wrote Alvan Clark inquiring about the possibilities of a 25-inch telescope. Clark replied that such an instrument would require three to five years to build, and cost $40,000 in gold, of which about $27,000 would be for the lens, including the glass from Chance Company in England. In summarizing the 9.6-inch observations the following year Sands reported to his superior, the Chief of the Bureau of Navigation, that "The comparatively small size of the instrument prevents it from entering into competition with many telescopes of other observatories in the observation of faint objects . . . At the present time the deficiency of the observatory which would be noticed with most surprise is the absence of a telescope at all comparable with many owned by colleges, observatories and private individuals throughout the country. This will seem the more remarkable since the most successful living constructor of telescopes is an American – Alvan Clark, of Cambridge, Massachusetts. Mr. Clark has constructed not only nearly all the best instruments lately erected in America, but some of the finest in Europe. It is understood that he has been for some time desirous of receiving an order for the largest refractor in the world; but the great cost – $40,000 in gold – has hitherto deterred individuals from giving the order."[3]

It is notable that Sands did not argue for such an instrument on practical grounds for Navy purposes, but rather on the grounds of the reputation of the institution as a national observatory: ". . . [I]t seems eminently desirable and proper that the government of the United States should employ at its observatory such an instrument of the highest power. Considering that any smaller instrument than that proposed would soon be superseded, that several institutions of learning in the country are endeavoring to procure one of this high character, and considering that not more than one such can probably ever be undertaken by Mr. Clark, it will be seen that delay endangers our being able ever to command it." Sands did, however, realize that $40,000 was a great deal of money – more than the entire appropriation for the new Observatory and its instruments 25 years before. He therefore emphasized that "as the construction of the instrument will occupy some four years, it is not necessary that more than one-fourth of the cost should be appropriated in any one year."

Commodore Thornton Jenkins, the chief of the Bureau of Navigation, did not forward this request to Secretary of the Navy Gideon Welles, however, and the following year Sands more specifically requested of the new chief of the Bureau (Commodore James Alden) that Congress be approached for funding. This request

---

[3] Clark to Newcomb, October 1, 1867, Simon Newcomb Papers (hereafter SNP), Library of Congress, General Correspondence, Clark folder; Sands, AR (1868), 82. Unless otherwise specified, the *Annual Reports* were written in October of each year. Newcomb states in his *Reminiscences of an Astronomer* (Houghton, Mifflin & Co.: Boston and New York, 1903), p. 130, that he had submitted the thrust of these paragraphs to Sands. Although I have in most cases gone back to original sources in this section, I have also profited much from Robert W. Rhynsburger's article, "A Historic Refractor's 100th Anniversary," *Sky and Telescope*, **46** (October, 1973), 208–214.

went nowhere because of the policy of Secretary Welles to recommend no increases in budget so soon after the Civil War. It was an accident of history, therefore, that the subject came to the attention of Congress, in a way typical of Washington politics. As Newcomb recalled, "very different might have been a chapter of astronomical history, but for the accident of Mr. Cyrus Field, of Atlantic cable fame, having a small dinner party at the Arlington Hotel, Washington, in the winter of 1870. Among the guests were Senators Hamlin and Casserly, Mr. J. E. Hilgard of the Coast Survey, and a young son of Mr. Field, who had spent the day in seeing the sights of Washington. Being called upon for a recital of his experiences, the youth described his visit to the Observatory, and expressed his surprise at finding no large telescope. The only instrument they could show him was much smaller and more antiquated than that of Mr. Rutherfurd in New York." When Hilgard confirmed the young boy's statement, one of the Senators said "This ought not to be," and "Why is it so?" Hilgard pointed out that Congress was reluctant to appropriate money for the purpose, and the Senator (Hamlin) promised that he would see any such proposal through the Senate. This he did, in fact using Sands's *Annual Report* as the basis for doing so.[4]

Losing no time, Hilgard and Newcomb immediately approached Superintendent Sands on the subject, Sands "earnestly promoted it," and Hilgard even gathered a petition from leading men of science. Three or four days after the dinner party Senator Hamlin had the petition in hand; the subject was soon considered by the Committees on Naval Affairs and Appropriations, and the request was adopted without opposition as an amendment to the naval appropriations bill. In the House, however, the Appropriations Committee recommended against the amendment. In retrospect this seems surprising, since the chairman of the committee was Mr Cadwallader C. Washburn, who a few years later funded the Washburn Observatory of the University of Wisconsin. However, in the conference committee the telescope prevailed and $50,000 was appropriated (about $675,000 in 2001 dollars).[5]

Thus section 18 of the Naval Appropriations bill, approved July 15, 1870, contained the provision "That the superintendent of the Naval Observatory be, and he is hereby, authorized to contract for the construction of a refracting telescope of the largest size, of American manufacture, at a cost not exceeding fifty thousand dollars."[6] The Clarks were to provide not only the lens, but also all other parts of the telescope, including a German mounting, divided circles reading to seconds of time

---

[4] Sands, AR (1869), 53; Newcomb, *Reminiscences*, pp. 128 ff.

[5] Newcomb, *Reminiscences*, pp. 128 ff. In conference Senator Drake and Representative Niblack were instrumental in obtaining approval for the telescope.

[6] Naval Appropriations Act, July 15, 1870, *Statutes at Large*, volume 16, p. 334. Only $10,000 of the $50,000 was appropriated in this Act. See also E. S. Holden, "The XXVI-Inch Equatorial", in *Instruments and Publications of the United States Naval Observatory, 1845–76*, Appendix 1 to WO for 1874 (Washington, 1877), pp. 26–45. This appendix was also published separately. See also the same volume of observations, pp. lxvii–lxxi and 26. The earliest published contemporary account of the 26-inch instrument is in the *WO for 1873* (Washington, 1875), pp. ciii–cvi, which also includes Newcomb's first observations with the instrument (pp. 177–178) and his extended discussion of "The Uranian and Neptunian Systems."

in right ascension and tenths of a minute of arc in declination, a clock drive that would run continuously for at least three hours, two filar micrometers, and a spectroscope. Accordingly, on 13 August, 1870 a contract was consummated with Alvan Clark & Sons for $46,000, including $1,000 for the spectroscope.[7] The contract gave the Clarks four years to complete the job from the time of first payment, which was to be made when the glass was in hand ready for grinding.

Newcomb, charged with negotiating the contract with the Clarks, had to overcome a final hurdle. Even as the Naval Observatory appropriation was still pending in Congress, Mr L. J. McCormick also contracted with the Clark firm for the world's largest telescope. According to Newcomb, this might have proved a hindrance had the Clarks not decided they could make the two instruments almost simultaneously. When the U. S. Government refused to pay in gold, the Clark's accepted paper currency only because McCormick was paying on a gold basis.[8]

There was also the matter of determining the exact size of the new lens. As Newcomb recalled, the Clarks would not have been willing to go much beyond their 18.5-inch Dearborn Observatory lens for fear of failure had it not been for the 25-inch refractor constructed by Thomas Cooke & Sons for R. S. Newall in England. The Clarks were willing to go slightly beyond this telescope to 26-inches, but, despite Newcomb's pleas, would not go beyond the 26-inch size McCormick had ordered. McCormick generously allowed the Naval Observatory to have the first lens, a move not without benefit to McCormick. Not only would the first lens be a kind of guinea pig, but also, in the end, the Naval Observatory received a 26-inch lens, whereas McCormick's lens was 26.25 inches of clear aperture, a distinction that Newcomb did not like to make, but that McCormick and University of Virginia astronomers certainly did.[9]

After careful consideration the Clarks contracted with Chance Bros & Co. of Birmingham, England for the glass, which had to be perfectly homogeneous disks, one of crown and the other of flint, in order to obtain an "achromatic" focus. By late January, 1871, when Alvan G. Clark visited the Chance company in Birmingham after observing an eclipse, the flint disk was "ready but of poor quality" and the crown disk

---

[7] The contract is reprinted in full in Holden's description (ref. 6 above), pp. 26–27. The Naval Observatory reserved the right to dispense with the spectroscope and discount $1,000, but this did not happen. The spectroscope is seen in the photograph facing p. 33, and is described briefly on p. 38.

[8] Newcomb, *Reminiscences*, pp. 132–133. Clark reported to Newcomb on September 12, 1870 that "Mr. McCormick has given us instructions to make him a telescope of 26 inches clear aperture." SNP, General Correspondence, Clark folder. He also reported in this letter that the Naval Observatory's 8.5-inch Pistor and Martins lens had successfully been refigured.

[9] Clark to Newcomb, June 17, 1870, SNP, General Correspondence, Clark folder. On the Newall, McCormick, and other telescopes at this time see Henry C. King, *The History of the Telescope* (Dover: New York, 1979; first published 1955), especially chapter 12. On the history of the McCormick telescope see S. A. Mitchell, *Leander McCormick Observatory of the University of Virginia* (Charlottesville, 1947). For the Naval Observatory and McCormick telescopes in the context of Clark's other work see Deborah J. Warner and Robert B. Ariail, *Alvan Clark & Sons: Artists in Optics* (second edition, Wilman-Bell: Richmond, Virginia, 1995).

was still in the annealing oven. By summer the senior Clark was already fretting about the glass, writing Sands that "We hear nothing further from Chance Bros. and Co., though it is about time. We have offered them every inducement in our power to make an effort, paid them one half, and offered them the whole, that we might not fail at this essential point." On July 28, Chance reported that, although the flint disk was ready, the crown disk was not, and they feared that it would come out of the annealing process broken. There was no way to predict when success would come. Finally, some time after mid-October, Alvan G. Clark again traveled to Birmingham, this time to examine and approve the glass disks. In December the two disks were received in Cambridgeport, and by the following June the Clarks reported that "we are getting on nicely with it except we have to put it in the tube every good evening and keep it out till about midnight so the glass and atmosphere shall be even temperature." In October 1872 the Superintendent reported to his superiors that the lens "may be regarded as finished, the artist not having been able, for some weeks past, to detect certainly any imperfection of figure." Work proceeded at the Clark establishment on the mounting, and when, in early summer, 1873, it was finished, Newcomb journeyed to Cambridgeport to test the lens, which was mounted in a temporary tube in the yard of the Clark establishment. "I have had few duties which interested me more than this," Newcomb recalled. "I was filled with the consciousness that I was looking at the stars through the most powerful telescope that had ever been pointed at the heavens, and wondered what mysteries might be unfolded." Happening upon a faint cluster of stars perhaps never before seen, Newcomb "could not help the vain longing which one must sometimes feel under such circumstances, to know what beings might live on planets belonging to what, from an earthly point of view, seemed to be a little colony on the border of creation itself."[10]

The telescope was clearly far ahead of schedule, and attention turned to completion of the building to house it, which had been commenced already in mid-1872 with a $10,000 appropriation. There was no question of placing the giant telescope in the old 9.6-inch dome atop the observatory, so the south wing of the building was extended to accommodate a new dome, some 41 feet in diameter, covered on the outside with galvanized iron and with canvas on the inside (Figure 6.1).[11] By September 11 the Clarks wrote that they expected to send the telescope within four weeks. On Saturday morning October 25, 1873 the massive parts of the telescope arrived in Washington, having been transported by water from Boston to Norfolk and up the Potomac River. The trip was not uneventful, and, while he was briefly stranded in Norfolk (Figure 6.2), Clark wrote to ask whether the boat might land near the Observatory, commenting that there had already been some damage to the equipment

[10] Alvan Clark to Newcomb, May 31, 1871, SNP, Clark file; Alvan Clark to Sands, July 31, 1871, USNOA, SF, Clark Letters file; Chance Bros to Clark, July 28, 1871, SNP, Clark file; Alvan Clark to Newcomb, August 17, 1872, SNP, General Correspondence, Clark folder; Holden (ref. 6 above); AR (1872), 92; Newcomb, *Reminiscences*, pp. 135–136.

[11] The details of the dome mechanism are given in Holden (ref. 6 above), pp. 27–29.

Figure 6.1. The south wing of the Naval Observatory, with the 26-inch Great Equatorial dome, photographed *circa* 1888. At the same time as the dome was built in 1873, the two-windowed section immediately to the right of the dome was added to the original south wing and used as Professors' offices. The next section with the door and awninged window was a storeroom, and the remaining portion, with the double shutters, housed the 6.6-inch Ertel transit circle known as the "Refraction Circle" (left shutter) and the 5-inch Prime Vertical (right shutter).

in transit, and that "some of the pieces are very heavy to load and we have had much trouble with them." All went well with the unloading, however, and on November 12 celestial objects were first viewed through the telescope. Three days later the Observatory was receiving the mandatory distinguished visitors to view the telescope; Newcomb, Holden, and Sands entertained guests including Postmaster General Creswell, General Sherman, Governor Shepherd, Professors Hilgard, Baird, and Nourse, Commodore Rodgers, Admiral Bailey, and General Garfield. A steady flow of visitors continued. On February 16, 1874 an entry in the 26-inch logbook by Newcomb or Holden reads "Pres. Grant, Mrs. Grant and Miss Nellie Grant, the representatives of the American people, did the Nebula of Orion, the Cluster in Perseus and the planet Uranus the honor to augustly gaze at them."[12]

<hr />

[12] *Washington Evening Star* for November 17, 1873, cited in Rhynsburger, 4 (ref. 3 above); 26-inch logbook, 8:30 pm, February 16, 1874, USNO Library.

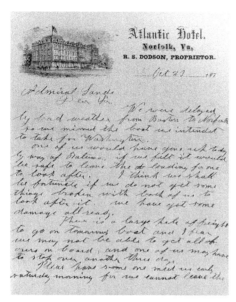

Figure 6.2. A letter from Alvan Clark to Admiral Sands dated October 23, 1873 giving the status of shipment of the 26-inch telescope. The letter reads, in part, "We were delayed by bad weather from Boston to Norfolk so we missed the boat we intended to take for Washington. One of us would have gone today by way of Baltimore if we felt it would be safe to leave the loading for one to look after. I think we shall be fortunate if we do not get some things broken with both of us to look after it. We have got some damage already."

The telescope itself (Figure 6.3) was indeed a wonder to behold. The steel tube was 32 feet long, and 31 inches in diameter at its widest. At the objective end, the two carefully shaped components of the lens weighed in at 110 pounds. At the eye end (see Figure 6.4, in which the clock drive and spectroscope are also visible) was a brass plate, into which the sliding tubes for the eyepieces and one of two filar micrometers could be inserted. Along with two finder scopes, four microscope tubes ran up the sides of the great refractor to read the right ascension and declination circles. The cast-iron harp-shaped mounting for the telescope rested on a two-ton block of sandstone, which in turn rested on a pier of stone and brick extending 18 feet below the surface. For coarse pointing the massive telescope was maneuvered by hand through a series of gears, aided by a rope. On the east side of the dome stood the sidereal clock and chronograph, which were essential for the precise measurements of position. Much could go wrong, and, once the instrument was in place, the senior Clark remained the anxious craftsman: on returning home after installing the telescope he wrote "I shall remember my visit to Washington with great interest and satisfaction especially should you be able to publish some discovery or achievement creditable to the abilities of the opticians."[13]

A creditable discovery was, of course, exactly what Newcomb hoped for. On November 15, the same day that the distinguished visitors arrived, Newcomb was officially placed in charge of the telescope. Although Hall had been in charge of the 9.6-inch refractor, Newcomb with his slight seniority had done more to bring the telescope into being, and Hall was shut out of even an assistant's role. Instead

[13] Clark to Newcomb, November 27, 1873, SNP, General Correspondence, Clark folder. The telescope is described in Holden (ref. 6 above).

Figure 6.3. The 26-Inch Great Equatorial Telescope, soon after its mounting in 1873. Simon Newcomb is at the eyepiece and Admiral Benjamin F. Sands, Superintendent, is seated in the background. Plate IV from *Instruments and Publications*.

Newcomb chose as his assistant on the Great Refractor E. S. Holden (Figure 6.5), whom we have already seen in the previous chapter in connection with time balls and solar-eclipse expeditions, and who had arrived at the Naval Observatory only in March, 1873. Although he had little formal training in astronomy, Holden had astronomical connections; he was a cousin of Harvard astronomer George P. Bond, lived

Plate V.

Heliotype.                                                      James R. Osgood & Co., Boston.

DETAILS OF CLOCK-WORK, ETC., OF XXVI-INCH EQUATORIAL.

Figure 6.4. Detail of the eye-end of the Great Equatorial Telescope, showing the clockwork, filar micrometer (attached to the instrument), and eyepieces. The spectroscope rests on the table. Plate V from *Instruments and Publications*.

Figure 6.5. E. S. Holden, Newcomb's protégé, later Director of Lick Observatory. He is seen here in 1873, shortly after he had joined the staff of the Naval Observatory. Courtesy of the Mary Lea Shane Archives of the Lick Observatory.

with astronomer William Chauvenet during his student days at Washington University in St Louis, and married Chauvenet's daughter Mary in 1871. He graduated from West Point in 1870. Holden seems to have been hired specifically to assist Newcomb on the 26-inch refractor, although his duties gradually expanded. Unlike Hall, Holden was a kind of apprentice whom Newcomb could take under his wing; under these circumstances Newcomb could make his observations without competition from the more formidable Asaph Hall.[14]

Together, Newcomb and Holden would monopolize the Great Refractor during its first 18 months. When observations began on November 20, one of Newcomb's first goals was to use the largest telescope in the world to spot a new companion to the bright star Procyon, similar to Alvan G. Clark's detection of the predicted companion of Sirius in 1862 when testing the new 18.5-inch objective. Although several faint companions had already been observed, a closer companion was suspected due to perturbations observed in the motion of Procyon, and Otto Struve had reported the existence of such a companion the previous March. However,

[14] On Holden see W. W. Campbell, "Biographical Memoir of Edward Singleton Holden, 1846–1914," BMNAS, **8**, 347–372, including a bibliography. For shorter biographies see PASP, **26**, 77–87, and C. A. Whitney, in DSB, 470–472. Holden became one of Newcomb's favorites, and it was with Newcomb's help that Holden went on to become Director of Washburn Observatory in 1881, President of the University of California in 1885, and first Director of Lick Observatory on its opening a few years later. He was also one of the founders of the Astronomical Society of the Pacific. On Holden's controversial tenure at Lick see Donald Osterbrock, John Gustafson, and W. J. Shiloh Unruh, *Eye on the Sky: Lick Observatory's First Century* (University of California Press: Berkeley, 1988).

neither Newcomb nor Holden, nor a host of others later – including Hall, Eastman, Alvan G. Clark, and C. H. F. Peters – could see a close companion with the 26-inch refractor (though some of them did report a new more distant companion and confirm several others). Fifteen years later, neither the keen-eyed Burnham nor E. E. Barnard spotted Struve's close companion with the 36-inch refractor at the new Lick Observatory (perhaps because it was too close to Procyon at the time), but in 1896 it was finally discovered by J. M. Schaeberle, another Lick astronomer.[15]

Although the discovery of a new companion to Procyon would have been a modest triumph with which to inaugurate the new telescope, Newcomb, of course, had a more substantial program in mind. As we shall see in chapter 8, his goal – "long before mapped out as the greatest one in which I should engage" – was to produce new tables of the motions of the planets, a goal that required an exact knowledge of their masses. Because this knowledge could best be determined (via Kepler's third law) by observations of planetary satellites, and because the masses of the outermost planets were the most uncertain of all (by a factor of a third to a half of their masses), Newcomb turned his attention to Uranus and Neptune. He was especially interested in discovering a new moon of Neptune, which at that time had only one known satellite very close to the planet (compared with the four satellites then known to be circling Uranus), and which was therefore very difficult to measure for purposes of determining the mass of Neptune. In this he was unsuccessful (a second satellite was not discovered until 1949); however, in 1875 Newcomb published his observations of the Uranian and Neptunian systems, which stretched over the first 18 months of work with the 26-inch telescope. The first great work of the instrument was thus not any unexpected discovery, but a set of routine observations. Nevertheless, the data significantly increased our knowledge of the masses of the outer planets.[16]

In addition to the focus on Uranus and Neptune, observations were occasionally made of the satellites and surface features of Saturn and Jupiter, as well as double stars, and the Omega and Orion nebulae. Also, in an era during which it was believed that "a well-trained eye alone is capable of seizing the delicate details of structure and configuration of the heavenly bodies," the artist E. L. Trouvelot used the 26-inch to make several pastel drawings of Saturn, Orion, and the Horseshoe Nebula, part of a much larger number made with many other telescopes.[17] If the public could not fully appre-

---

[15] Newcomb, *Reminiscences*, pp. 137–140; AR (1876), 95. Hall and Holden summarized the 26-inch observations of Procyon beginning with Newcomb on November 29, 1873 and ending in January, 1876, in "The Companions of Procyon," AN, **87** (1875–76), columns 241–246. Admiral Davis reported in his *Annual Report* that Holden, Newcomb, Watson, Peters, and Todd had seen some of the fainter companions.

[16] Newcomb, *Reminiscences*, p. 141. "The Uranian and Neptunian Systems, investigated with the 26-inch Equatorial of the United States Naval Observatory," Appendix I, WO for 1873 (1875). These observations were also reported in MNRAS, **35**, 49, and in AN, no. 2061, **86**, column 321.

[17] Trouvelot, who had been invited by the Superintendent for the purpose, remained at the Naval Observatory from September 21 to October 1, 1875, less time than required for "perfectly satisfactory representations" of the objects, according to Holden. Over the next few years, Trouvelot made some 7,000 drawings at various observatories, from which he selected 15 of the

Figure 6.6. The drawings of Trouvelot, made using the 26-inch Great Equatorial, were displayed at the U. S. Centennial Exhibition in Philadelphia in 1876. The Orion Nebula is seen at the top left, Saturn in the center, and the Horseshoe Nebula at the top right. Among other items displayed are a variety of chronometers (bottom left), observatory publications, and photographs from various naval expeditions.

ciate minute measures of satellites and double stars, it could appreciate the visually stunning art of Trouvelot, which was prominently displayed in the Naval Observatory section of the U. S. Centennial Exhibition in Philadelphia in 1876 (Figure 6.6).

As with any new telescope, shakedown problems were inevitable. Although the overall operation of the telescope was described as "eminently satisfactory," the driving clock, a system of weights originally wound up by a water wheel in the cellar powered by water drawn from the Potomac, "proved much too powerful for the

best. In 1881 Charles Scribners published these as lithographs, and in the following year Trouvelot presented a copy to Asaph Hall. In the 1930s the Hall family donated these to the Naval Observatory, where they remain along with the three originals. Holden reported on the drawings at the October 9, 1875 meeting of the Washington Philosophical Society, abstract in *Bulletin of the Philosophical Society of Washington*, volume II, pp. 51–52, and in AR (1875), 75–77. On Trouvelot (who is notorious for having accidentally brought the gypsy moth to America) see Jan K. Herman and Brenda G. Corbin, "Trouvelot: From Moths to Mars," *Sky and Telescope* (December, 1986), 566–568, where the quotation is to be found. Four of the Trouvelot drawings are reproduced in WO for 1874 (Washington 1877), Appendix I. Other engravings are reproduced in volume 8 of the *Annals of the Harvard College Observatory*. Some of the other early observations of the 26-inch instrument, with references, are cited in Holden (ref. 6), pp. 43–45.

delicate regulating apparatus." Hall and the Observatory instrument-maker, William F. Gardner, altered it in July, 1876. The effect of rapid temperature changes on the lens occasionally caused delays in observing. By 1876 "it was thought that the performance of the objective was not quite so good as when it was first mounted," and in spring of that year the Clark firm repolished both surfaces of the flint lens, resulting in a "decided improvement." The graduated circles had painted black-on-white lines that gave errors of 15–20 minutes of arc; Gardner finally replaced them in June, 1879. The original wooden shutter on the dome soon became warped, and Hall's assistant, George Anderson, rigged a sail-cloth shutter with a rope-and-pulley system. The 42-foot-diameter dome was so difficult to move that lists of north and south objects were compiled so as to lessen the need for turning the dome. This problem was resolved in 1884 when a small gas engine was installed to turn the dome.[18]

By late spring of 1875 Newcomb was tiring of observing, which was never his favorite occupation. He and Holden had accomplished a good deal with their observations of the satellites of Uranus and Neptune, but had not made the hoped-for spectacular discovery. When Newcomb turned the 26-inch telescope over to his colleague Asaph Hall, it was still the largest telescope in the world, and would be for a few more years, until the Imperial Observatory of Vienna mounted a 27-inch lens by Grubb (1880), McCormick dedicated its 26.25 inch (1883), and Pulkovo mounted its 30-inch Clark telescope.[19] There was still time for spectacular discovery along with routine work, and Asaph Hall knew it.

## 6.2 Asaph Hall and the moons of Mars

On June 16, 1875 Asaph Hall was put in charge of the Great Refractor when Newcomb went on to pursue more theoretical interests.[20] Newcomb's protégé, E. S.

---

[18] For technical specifications and measurements made when the lens was removed from its cell for repolishing, see Holden's "Investigation of the Objective and Micrometers of the Twenty-Six Inch Equatorial," Appendix I, WO for 1877 (1881). The work on the lens was done in the dome between April 20 and May 4, 1876. See WO for 1876, pp. xci–xcii. Hall wrote (see Appendix VI, WO for 1877, p. 13) that the reason for the slight refiguring of the flint lens was because "after having been in use two years the form of the lenses seemed to have undergone a slight change." On the other problems, see Hall, WO for 1877, pp. 6 ff, WO for 1888, pp. E5 ff., and AR (1874), 67.

[19] In August, 1879 Otto and Hermann Struve visited the Naval Observatory, and, after examining the 26-inch refractor, placed an order for the 30-inch instrument from Alvan Clark. AR (1879), 123. Newcomb also played a role in negotiating with the Clarks, who offered to grind the lens for $15,000, if furnished the glass. Clark to Newcomb, April 17, 1879, SNP, Clark file. With the advent of these larger telescopes, Superintendent Rodgers suggested that the 26-inch instrument might be moved for five or ten years to a branch observatory in the central part of the country, in order to see "what is to be gained by observing at a station which is above a large part of our atmosphere." AR (1877), 113. Nothing came of this suggestion, but Asaph Hall noted after his observations of the solar eclipse of 1878 from Colorado that "I cannot but think that those elevated plains offer advantages for astronomical observations that have not hitherto been made use of," and he further suggested that "an experienced astronomer with a good telescope" should be put there for a few years. Hill, BMNAS (ref. 22), 264–265.

[20] Many years later Newcomb wrote to Hall that, by spring of 1875, he had finished all the work with the 26-inch instrument in which he was seriously interested, and that, although he might have retained charge of the instrument, he let Hall have it to make amends for past secondary assignments to instruments, assignments in which Newcomb had apparently played a role.

Figure 6.7. Chloe Angeline Stickney Hall, wife of Asaph Hall. From her obituary in the *Journal of the National Science Club* (March, 1899).

Holden, remained as astronomical assistant on the telescope. We recall that Hall, one of Gilliss's recruits, had arrived at the Observatory in 1862, about a year after Newcomb and in the position of Aid, but was elevated to the same level as Newcomb (Professor of Mathematics) the following year. As the eldest son of a clockmaker from Goshen, Connecticut,[21] young Asaph learned carpentry before entering Central College in McGrawville, New York. This school was not renowned for its teaching, and, when Hall left in 1855 after little more than a year, perhaps the most valuable asset he took with him was Chloe Angeline Stickney (Figure 6.7), who, as a student two years ahead of Hall, tutored him in mathematics. In 1856 they were married in Wisconsin, where they had gone to look for work near Miss Stickney's relatives. Not finding any, they moved on to Ann Arbor, Michigan. Here, having apparently decided to become an astronomer, Hall studied under Franz Brünnow for several months. Lacking resources to continue, he then headed east with his wife, and (having made some money applying his carpentry skills) in 1857 entered the Lawrence Scientific School at Harvard. Here he attended Benjamin Peirce's lectures briefly before deciding that they were too theoretical for his interests. He worked for

---

[21] The future astronomer's father, sometimes designated Asaph Hall II, died in 1842. His grandfather, who served in the Revolutionary War, was Asaph Hall I. The astronomer was therefore Asaph Hall III, and his son (also to become an astronomer) was Asaph Hall IV, although neither used the nomenclature. As we shall see in chapter 10, Asaph Hall IV was appointed an Assistant Astronomer at the USNO on July 27, 1889. He would become a Professor of Mathematics in February, 1908.

William C. Bond making astronomical observations, coming to the Naval Observatory in 1862.[22]

We have already seen in chapter 4 some of the events in Hall's life at the Observatory during the Civil War, including the visit of Lincoln, his two-day stand in defense of the city, and some of his early scientific work on minor planets and comets. One aspect of Hall's life that needs to be emphasized here is the influence of Mrs Hall, who was always proud of her husband's abilities, even while cultivating her own. When a permanent position opened at the Observatory in 1863, she wrote Superintendent Gilliss, "I received a letter from Mr. Hall this morning saying that Prof. Hesse has resigned his place at the Observatory. I wish Mr. Hall might have the vacant place. If the question is one of ability, I should be more than willing that he with all other competitors should have a thorough and impartial examination. I know I should be proud of the result. If on the other hand the question is who has the greatest number of influential friends to push him forward whether qualified or unqualified, I fear, alas! that he will fail. He stands alone on his merits, but his success is only a question of time. I, more than any one, know of all his long, patient and faithful study. A few years, and he, like Johnson, will be beyond the help of some Lord Chesterfield. Mr. Hall writes me that he shall do nothing but wait. I could not bear to have his name at least proposed." Two weeks later Hall wrote his wife "Yesterday afternoon Capt. Gillis [sic] told me to tell you that the best answer he could make to your letter is that hereafter you might address me as Prof. A. Hall. . . . You wrote to Capt. Gillis, did you? What did you write?"[23] This episode, which marked the beginning of Hall's permanent 30-year career in the U. S. government, illustrates the importance of Mrs Hall's influence on her husband's career. Even as she pioneered in women's rights, and raised four sons at their home in Georgetown, she was a source of constant encouragement to her husband, even for his most famous discovery.[24]

---

[22] The best biographical source for Hall is George W. Hill, "Biographical Memoir of Asaph Hall, 1829–1907," BMNAS, **6** (Washington, 1908), 243–309, which also contains a nearly complete list of his publications compiled by Naval Observatory Librarian William D. Horigan. See also Owen Gingerich, "Asaph Hall," DSB, 48–50, and the biography by Hall's son Percival Hall, *Asaph Hall: Astronomer, A Biographical Sketch of Asaph Hall*, 3d (1945), privately printed and distributed by its author. The latter reprints some interesting correspondence. Among the numerous obituaries see especially Henry S. Pritchett, PA (February, 1908), 66–70; DAB, **8**, 117–118; MNRAS, **68** (February, 1908), 243. For a view of Asaph and his wife, see the biography of Mrs Hall by their son Angelo Hall, *An Astronomer's Wife: The Biography of Angeline Hall* (Nunn and Company: Baltimore, 1908), as well as Percival Hall (above), and *Journal of the National Science Club* (March, 1899), 1–6. The Asaph Hall papers, comprising six boxes of material and approximately 1,000 items, are located at the Library of Congress, and are hereafter cited as AHP.

[23] Angeline S. Hall to Gilliss, April 17, 1863, and Asaph Hall to Angeline Hall, in Angelo Hall (ref. 22 above), p. 89. The original letter from Angeline to Gilliss is in AHP, Family Correspondence, 1857–69. Hall's letter of appointment as Professor of Mathematics is dated May 2, 1863, and points out that he takes rank after Simon Newcomb, AHP, General Correspondence, U. S. Navy Department file.

[24] See Angelo Hall (ref. 22 above). The Halls lived on I Street between 20th and 21st Street NW until moving to No. 18 Gay Street in Georgetown in November, 1867 (pp. 89 and 96). According to Angelo Hall (p. 99) the Eastmans were Hall's next-door neighbors for 13 years. Among Naval Observatory folklore is the story that the Halls gave their children – Asaph Jr, Samuel, Angelo, and Percival – names whose initials would spell 'Asap." Angelo Hall commented in 1908: "It has

In light of his subsequent association with Mars, it is notable that, during the year 1862, Mars was in one of its periodic 15-year close approaches to the Earth – the closest it would be to Earth until Hall's great discovery of 1877. Only a few weeks after Hall's arrival in August, 1862, he and James Ferguson began a series of observations of Mars for the purpose of determining the solar parallax – the same kind of observations as those Gilliss had made in 1847 from Chile, the ultimate purpose of which was to determine the scale of the solar system. Although Gilliss had already determined the observing program and sought the cooperation of observatories around the world, it was Hall who in 1863 deduced a value for the solar parallax from the observations of Mars he made with the equatorial at Washington, and others made at Santiago and Upsala. From his earliest career at the Naval Observatory, therefore, Hall had a special interest in the planet Mars.[25]

Mars was, however, only one of many interests. In his early years at the Observatory, Hall co-authored numerous articles with James Ferguson, many of which reported their observations of comets and asteroids with the 9.6-inch refractor. During 1864–70 he cataloged the positions of 151 stars in the Praesepe cluster. Nor was his work confined to the equatorial; in 1866–67 he assisted Newcomb on the new transit circle. In 1869 Hall observed the solar eclipse from the east coast of Siberia; he returned to Siberia in 1874 as chief of the transit-of-Venus party at Vladivostok (see chapter 7). Hall (Figure 6.8) had thus undertaken a variety of astronomical projects by 1875, including occasional duties observing on the Great Refractor during its planetary-satellite program. However, from 1868 to 1875 his main activities were related to duties as astronomer in charge of the 9.6-inch equatorial.

Once Hall had been placed in charge of the giant telescope, reports of his observations appeared regularly in the pages of the *Astronomische Nachrichten*, beginning with his study of the satellites of Saturn from June to December, 1875; continuing with a summary of 26-inch attempts to find a close companion to Procyon; followed by observations (with Holden) of the satellites of Uranus and Neptune made from September 1875 to April 1876. During the last six months of 1876 Hall again observed the satellites of Saturn, and it was while making these observations that he

been humorously maintained that it was my parents' design to spell out the name "Asaph" with the initials of his children. I am inclined to discredit the idea, though the pleasantry was current in my boyhood, and the fifth letter, – which might, of course be said to stand for Hall, – was supplied by Henry S. Pritchett, who as a young man (and an aid at the Naval Observatory) became a member of the family, as much attached to Mrs. Hall as an own son;" Hall, pp. 97–98. Pritchett later became a Director of the U. S. Coast and Geodetic Survey and president of MIT.

[25] Asaph Hall, "Solar Parallax, Deduced from Observations of Mars with Equatorial Instruments," *WO for 1863* (1865), Appendix A, pp. lx–lxiv. Also published in AN, **68** (1866–67), columns 235–236. An author is not listed, but it is listed as item 40 in Horigan's "Published Writings of Asaph Hall," in Hill's biography of Hall (ref. 22 above). For the context see, in the same volume of *Washington Observations*, "Solar Parallax," pp. xlv–lix, and for a complementary effort "Solar Parallax, Deduced from Observations with Meridian Instruments," pp. lxv–lxxii. The Naval Observatory sent a circular in March, 1862 to all observatories requesting cooperative observations. Observations were made August 27–November 9. Hall deduced a solar parallax of 8.8415 seconds of arc, with a probable error "not likely to exceed one-tenth of a second." See also Hill (ref. 22 above), p. 256.

Figure 6.8. Asaph Hall about 1875.

detected a new brilliant white spot on Saturn. He used it to determine a new rotation period, differing by 15 minutes from the previously accepted value. Spring of 1877 found him observing the companion of Sirius, and the satellites of Uranus and Neptune, and continuing a series of double-star observations.[26]

In these observations, aside from Holden, Hall was assisted at the Great Refractor by George Anderson (Figure 6.9), whom he often cited as important to his work, even though Anderson was a laborer rather than an astronomer. Anderson was born in Scotland June 21, 1839, came to the United States about 1860, and served as a soldier in the Union Army throughout the Civil War. Anderson's outfit spent much of its time "chasing and running away from Mosby," as Hall put it, referring to John S. Mosby, one of the notorious Southern guerrillas whose squads attacked Union outposts and wagon trains. On the strength of his having served in the Civil War, Anderson came to the Naval Observatory as a laborer in 1870, and in 1873 was assigned as assistant on the Great Equatorial, where he remained for 27 years. However, Anderson was more than a laborer; he discovered some close double stars,

[26] Asaph Hall, "Observations of the Satellites of Saturn," AN, **87** (1875–76), 177–190; Asaph Hall and E. S. Holden, "Companions of Procyon," AN, **87** (1875–76), columns 241–246, also in *American Academy Proceedings*, **11** (1875–76), 185–190; "Observations of the Satellites of Neptune and Uranus," AN, **88** (1876), columns 131–138; "Elements of Hyperion, with Ephemeris for 1877," AN, **90** (1877), columns 7–12; "Observations of the Satellites of Saturn (June–December, 1876)," AN, **90** (1877), columns 129–138; "On the Rotation of Saturn," AN, **90** (1877), columns 145–150, also in MNRAS, **38** (1877–78), 209–210, reviewed in *Nature*, **16** (1877), 363–364; and "Observations made with the 26-inch Refractor," AN, **90** (1877), 161–166.

Figure 6.9. Asaph Hall (left) and his assistant, George Anderson, in front of the building for the 26-inch equatorial telescope at the Naval Observatory's new site, sometime after its completion in 1893.

which were measured by Hall, as well as faint companions to Vega, A Leonis and other bright stars. "Mr. Anderson was exceedingly efficient in managing the dome and the driving clock," Hall's friend, the great mathematician G. W. Hill, recalled. "Professor Hall threw aside all ceremony in his intercourse with Mr. Anderson. Two brothers could not have been more intimate. Professor Hall would call Mr. Anderson to the eye-piece of the telescope. 'Well, George, what do you see?' George would describe. 'Well, how is it situated?' George would again describe in his homely way. 'Well, I think we may enter on the observing book that we both saw it'."[27]

For Hall, Holden was a potential rival; Anderson a trusted, if uneducated, ally. It was no accident that, when Hall began observations for what would be his most famous discovery, it was Anderson rather than Holden who was in the dome with him. In fact, Hall later wrote "In the case of the Mars satellites there was a practical diffi-culty of which I could not speak in an official Report. It was to get rid of my assistant.

---

[27] Hill (ref. 22 above), p. 274. Hill was one of Hall's most intimate friends, and stayed with the Halls for several years when he came to Washington in the fall of 1881. For one of Anderson's discoveries see "On the Recently Announced Companion of alpha Lyrae," AJ, **18** (1898), 30, where Anderson signs himself "Skilled Laborer." After his trip to European observatories, Albert Winterhalter commented that "The canvas shutter for the great dome, as arranged by Mr. George Anderson, is the simplest and most practical for its purpose." Albert G. Winterhalter, *The International Photographic Congress and A Visit to Certain European Observatories and Other Institutions* (Government Printing Office: Washington, 1889), p. 322. Anderson died October 21, 1900, and was buried in Arlington National Cemetery. For a brief obituary see T. J. J. See, AN, **154**, 299.

It was natural that I should wish to be alone; and by the greatest good luck Dr. Henry Draper invited him to Dobb's Ferry at the very nick of time. He could not have gone much farther than Baltimore when I had the first satellite nearly in hand."[28]

What led Asaph Hall in the summer of 1877 to search for the moons of Mars, and to find them where others had failed, is an intriguing scientific detective story. There had been speculation about Martian moons since the time of Galileo and Kepler, and in his *Gulliver's Travels* (1726) Jonathan Swift had spoken of the scientists of Laputa discovering "two lesser stars, or satellites, which revolve about Mars."[29] Observation, however, had failed to find them. While one might have expected that the general background of speculation, and the approaching favorable opposition of Mars, coupled with the new 26-inch refractor and the need for an improved mass of Mars, would be all the incentive needed, the decision to search was also an immediate result of Hall's work on Saturn, which he had commenced two years earlier on the 26-inch refractor. Consciously retracing his steps, Hall wrote within a few months of that discovery that Saturn was the principal factor:

> In December, 1876, while observing the satellites of Saturn I noticed a white spot on the Ball of the planet, and the observations of this spot gave me the means of determining the time of the rotation of Saturn, or the length of Saturn's day, with considerable accuracy. This was a simple matter, but the resulting time of rotation was nearly a quarter of an hour different from what is generally given in our text books of astronomy: and this discordance, since the error was multiplied by the number of rotations and the ephemeris soon became utterly wrong, set before me in a clearer light than ever before the careless manner in which books are made, showed the necessity of consulting original papers, and made me ready to doubt the assertion one reads so often in the books, "Mars has no moon."[30]

The crucial seed of doubt planted, the opposition of Mars fast approaching, and considering his plan of work for the coming year, Hall began in the Spring of 1877 seriously to ponder the possibility that Mars might have a moon. A letter from C. H. F. Peters, Director of the Litchfield Observatory of Hamilton College, warned that, in the

---

[28] Hall to E. C. Pickering, February 14, 1888, in Gingerich (ref. 29).

[29] On earlier speculation about Martian moons, see Owen Gingerich, "The Satellites of Mars: Prediction and Discovery," *JHA*, **1** (August, 1970), 109–115; also S. H. Gould, "Gulliver and the Moons of Mars," *Journal of the History of Ideas*, **vi** (1945), 91–101. Hall had considered how the mass of Mars could be determined from asteroid observations in "On the Determination of the Mass of Mars," *AN*, **86** (1875), columns 333–336.

[30] Hall, "Discovery of the Satellite of Mars," manuscript in U. S. Naval Observatory Archives, read before Washington Philosophical Society February 16, 1878, p. 1. The manuscript was first discussed in Steven J. Dick, "Discovering the Moons of Mars," *Sky and Telescope* (September, 1988), 242–243. Hall published parts of this manuscript in *Observations and Orbits of the Satellites of Mars, with Data for Ephemerides in 1879* (Washington, 1878), and in *MNRAS*, **38** (1877–78), 205–209. These published reports are the account-of-record, but the manuscript gives a fuller and more informal account. Modern accounts of the discovery of the moons of Mars are given by Gingerich (ref. 29 above); by Gingerich and Dan Pascu in a centennial volume *The Satellites of Mars*, in *Vistas in Astronomy*, **22** (1978), and in the *Sky and Telescope* article cited above, from which part of the following account is taken.

following year, a powerful new 27-inch refractor would be mounted in Vienna, and this warning, Hall recalled, quickened his attention. So slight did the chances seem at the outset, Hall later wrote, "that I might have abandoned the search had it not been for the encouragement of my wife."[31] Guided by a summary of Martian observations given by Frederik Kaiser in the *Annals of the Leiden Observatory* for 1872, Hall was surprised that almost no one had searched for Martian moons since William Herschel in 1783, with the possible exception of Sir John Herschel in 1830, and the definite exception of Heinrich d'Arrest, Director of the Observatory at Copenhagen, who had made a search during the opposition of 1862. The failure of the Herschels and d'Arrest was discouraging, but "remembering the power and excellence of our glass there seemed to be a little hope left," though this hope was further diminished by the southern declination of Mars at this time, a circumstance that favored the giant new 4-foot reflecting telescope at Melbourne.[32]

With this faint hope, Hall began his search in early August. Hall clearly realized from d'Arrest's theoretical considerations that it would be useless to search beyond 70 arcminutes from the planet. Hall does not say where he started, only that his attention was first directed to faint objects "at some distance from the planet," all of which turned out to be fixed stars. On August 10 Hall began to examine the region within the glare of light of the planet. "This was done by keeping the planet just outside the field of view, and turning the eyepiece so as to pass completely around the planet." On this night the image of the planet was "blazing," the sky "too thick," and Hall saw nothing. Hall tried again early the next night and found nothing, "but trying again some hours later I found a faint object on the following side and a little north of the planet. I had barely time to secure an observation of its position when fog from the River stopped the work. This was at half past two o'clock on the night of the 11th." In his logbook Hall wrote "Seeing good for Mars. The edge of the White spot [the Martian South Pole] has two notches near the center of its outline. (A faint star near Mars)." Although Hall later wrote in ink "this proves to be satellite 1," with only one set of measures he had no way of knowing at the time whether the object was moving with Mars or was just a faint background star.[33]

One can imagine Hall's state of mind as cloudy weather intervened for the next several days. On the 15th Hall slept at the Observatory, and when the sky cleared at 11 o'clock he returned to the telescope. After observing Saturn's ring, and the Saturnian moons Titan and Iapetus, he returned to Mars. But the seeing was very bad, and (as Hall later discovered) the moon was so close to the planet as to be invisible, so again

---

[31] Hall, *Observations and Orbits* (ref. 30), p. 5. As it turned out, while the mechanical parts of the 27-inch Howard Grubb telescope for Vienna were completed by 1878, it was not operational until 1880.

[32] Hall, *Observations and Orbits* (ref. 30), p. 5. The report of the Melbourne Observatory for this period shows that its "Great Reflector" was used in a search for satellites of Mars, but the observers did not see them, MNRAS, **38** (1877–78), 188–189.

[33] Hall, *Observations and Orbits* (ref. 30), p. 6; Hall manuscript (ref. 30 above), p. 4; 26-inch log book, August 10–11, 1877, Naval Observatory Library.

Hall went home disappointed. Here he received continuing encouragement. Son Angelo recalled that "Mrs. Hall was full of enthusiasm. Each night she sent her husband to the observatory supplied with a nourishing lunch, and each night she awaited developments with eager interest. I can well remember the excitement at home. There was a great secret in the house, and all the members of the family were drawn more closely together by mutual confidence."[34]

Finally, on Thursday the 16th the weather cooperated and Hall was able to try again. Mars rose about 8:25 in the evening, preceded 20 minutes before by Saturn. At 11:42 pm he noted a 12th-magnitude "star near Mars" and measured its distance from Mars as 7.804 revolutions of the micrometer screw. By 1:30 am, after more micrometer measures, he had concluded that it was moving with the planet, which was itself moving at the rate of 15 arcseconds per hour and therefore would quickly have left behind any fixed stars.

Although he still referred to it in the logbook as the "Mars Star," Hall could barely contain his excitement. "Until this time I had said nothing to anyone at the Observatory of my search for a satellite of Mars, but on leaving the Observatory after these observations of the 16th, at about three o'clock in the morning, I told my assistant, George Anderson, to whom I had shown the object, that I thought I had discovered a satellite of Mars. I told him also to keep quiet as I did not wish anything said until the matter was beyond doubt."[35]

Anderson did faithfully keep quiet, but not Hall: "the thing was too good to keep and I let it out myself. On August 17th between one and two o'clock, while I was reducing my observations, Professor Newcomb came into my room to eat his lunch and I showed him my measures of the faint object near Mars which proved that it was moving with the planet." Hall himself spilled the beans, and to none other than his arch-rival, Simon Newcomb! By examining Hall's measures, Newcomb estimated its period of revolution around Mars, and pointed out that, if it were a satellite, it would not be visible most of the following night, but would reappear toward morning. Consulting an ephemeris, Newcomb had also found that the minor planet Europa (not to be confused with the Jovian moon) was at that time only two or three degrees from Mars. If the ephemeris were erroneous, Newcomb reasoned, Hall might be observing Europa. This, however, was easily disproven, since an asteroid would not follow Mars, whereas the moons would.[36] As we shall see, this brief interaction became the subject of considerable contention between Hall and Newcomb, with Hall being somewhat cautious and Newcomb claiming that Hall did not realize what he had.

[34] Angelo Hall (ref. 22 above), p. 112.
[35] Hall manuscript (ref. 30 above), pp. 4–5; 26-inch log book, August 15–16, 1877.
[36] Hall manuscript (ref. 30 above), p. 5; Newcomb, Appendix to *Popular Astronomy* (1878), and *Reminiscences*, p. 142. A modern calculation shows that the asteroid 52 Europa (discovered in 1858 by Goldschmidt at Paris) was within a half degree in declination, and 1.5 degrees in right ascension, of Mars at the time of Hall's search in August, 1877. Phobos was a tenth-magnitude object when it was discovered, compared with magnitude 12 for Deimos at the time of its discovery.

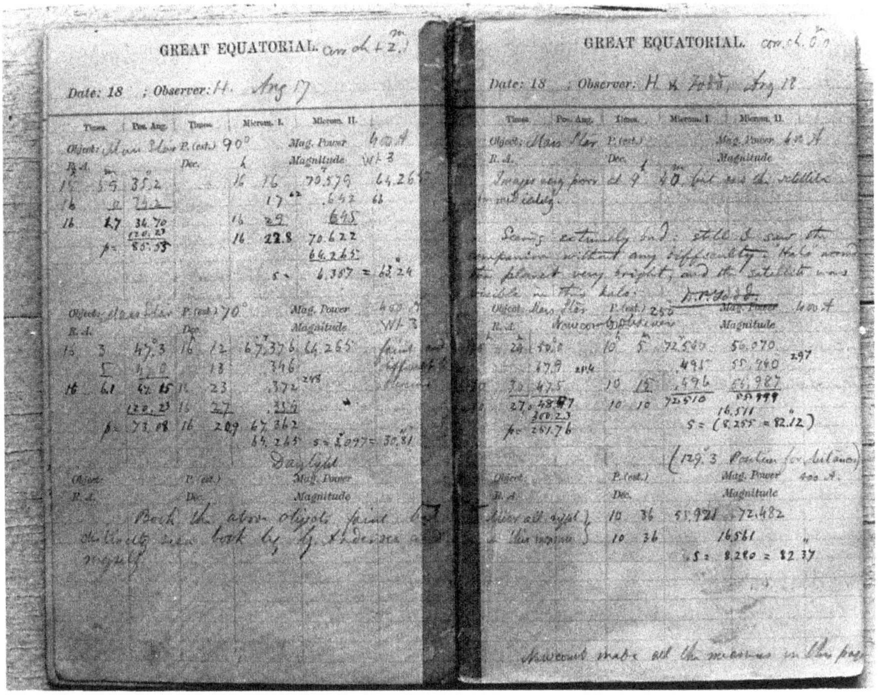

Figure 6.10. Hall's logbook of the Martian observations with discovery of two Martian satellites, August 17, 1877. At the bottom left, the log reads "Both the above objects faint but distinctly seen both by G. Anderson and myself." On the right, the next night David P. Todd and Newcomb have both seen the newly discovered moon.

The night of Friday the 17th was therefore filled with uncertainty and surprise. Early in the night Hall measured a "Mars Star" that was "a fixed star and not the object observed last night;" as Newcomb had predicted, the Martian satellite was not visible. At four o'clock in the morning, however, not only did the satellite emerge from behind the planet as predicted, but also, while watching for the outer moon, Hall discovered the inner one. Within a few minutes Hall had measured the separations and position angles of both, still referred to as "Mars Star," and, with daylight quickly approaching, noted in his logbook (Figure 6.10) "Both the above objects faint but distinctly seen by G. Anderson and myself."[37]

On Saturday the 18th notice of the discoveries was telegraphed to Alvan Clark & Sons at Cambridgeport, who that night verified Hall's discovery, as did Pickering and his assistants with the Harvard 15-inch refractor. That same night Newcomb, astronomer David P. Todd, and William Harkness joined Hall and Anderson in the dome. Already at 9:40 Hall saw the satellite immediately. Todd wrote "seeing extremely bad: still I saw the companion without any difficulty. 'Halo' around the

[37] 26-Inch logbook, August 17, 1877, USNO Library.

planet very bright, and the satellite was visible in this halo." First Newcomb and then Harkness measured the satellite positions, with Hall writing down the micrometer readings for separation and position angle. With Harkness's observations, "Mars Star" becomes for the first time "Mars satellite." The measures of this and the preceding evening had "put beyond doubt the character of these objects." Not one to waste a good night, after midnight Hall turned to observations of Saturn and its moons, returning to Mars at 2:30 to check on his newly discovered satellite.[38]

August 19, normally a quiet Sunday, was the day to tell the world. His observations finished, about four o'clock in the morning, Hall wrote Superintendent Rodgers, "From my observations made tonight I think that I can state with certainty that the planet Mars has at least two satellites," even though he could not yet state the period of the second satellite. The discovery was announced to the scientific world and the general public by Admiral Rodgers, via a communication to the Smithsonian Institution, which in turn released the following dispatch: "Two satellites of Mars discovered by Hall, at Washington. First, elongation west, August 18, eleven hours, Washington time. Distance eighty seconds. Period, thirty hours. Distance of second, fifty seconds." Still, Hall remained puzzled by the inner moon; it would appear on different sides of the planet on the same night, leading Hall to believe that there might be two or three inner moons. On the nights of the 20th and 21st he watched the inner moon for long intervals, "and saw that there was in fact but one inner moon which made its revolution around the primary in less than one third the time of the primary's rotation – a case unique in our solar system." On August 20 Hall wrote in his notebook "The observations of Mars Sat. 2 of tonight show that this satellite has a period of about 6 h 20 m. I had a good opportunity to compare the brightness of the two satellites. The inner one is the brighter."[39]

It was not until the 21st that Admiral Rodgers sent the Observatory's official announcement to the Secretary of the Navy, forwarding Hall's observations from August 11 to 20, including those of Newcomb and Harkness on the night of the 18th, and Newcomb's preliminary elements of the orbit.[40] The news took both the public and the astronomical world by storm. The *Washington Evening Star* for August 20, under the heading of "Political News from the Executive Mansion and Departments" of the government, reported "Glorious news from the skies – Brilliant Discovery with the Great Telescope – Two satellites of Mars found by Prof. Hall." The British journal *Nature* briefly announced the "extraordinary discovery" in its August 23rd issue (having heard from the Smithsonian via Leverrier in Paris), and its September 6th

[38] 26-Inch logbook, August 18, 1877, USNO Library; Hall, *Observations and Orbits* (ref. 30 above), p. 6.

[39] Hall, *Observations and Orbits*, p. 6; Hall to Rodgers, August 19, 1877, USNOA. Next to the date, Hall wrote "1600 hours," which, since the astronomical day was then reckoned from noon, refers to 4 am.

[40] John Rodgers, "Letter to the Hon. R. W. Thompson, Secretary of the Navy, announcing the discovery of Satellites of Mars," AN, **90** (1877), 173–176. The first public announcement via the Smithsonian was printed in AN, **90** (1877), 189–190.

issue published Newcomb's calculations of the first ten days' measurements. *Nature* characterized the discovery as "in the highest degree an honour to American science," the instrument, of American construction, being "a masterpiece of mechanical skill." Similarly, *Scientific American* called the discovery "a triumph both for Professor Hall and for Mr. Alvan Clark," and looked forward to the even more colossal telescope under consideration for Lick Observatory. Leverrier, in the last few weeks of his life, called the find "one of the most important discoveries of modern astronomy." Visitors great and small streamed to the telescope even more than before, including the Vice President of the United States, who obligingly came on November 3 on a night, Hall duly reported, "good for naught but visitors."[41]

Remarkably, aside from workers at the Great Reflector in Melbourne, no one else had seriously attempted to search for Martian satellites, and there were no rival claims to the discovery. There were, however, claims of more satellites, including one from Holden, whom, we recall, had been away from the Observatory for the entire episode. On August 28, Holden wrote from Dobbs Ferry, New York to Superintendent Rodgers that "Dr. Henry Draper and myself have (probably) detected a third satellite to Mars, on the 26th and 27th of August."[42] This and a claim from one "Alexander," however, were not confirmed. In his logbook for August 28, Hall wrote "Observed two stars preceding Mars for Alexander but both prove to be fixed stars. Looked carefully for the Draper–Holden moon, but could not find it." To Hall would go the credit for discovering the two – and only two – satellites of Mars.

Hall continued his observations of the new satellites with the 26-inch refractor until October 31; these observations alone gave enough data for a good determination of their orbits. By this time observations had been made at many other observatories, including those of E. C. Pickering and his assistants with the 15-inch refractor at Harvard and of H. S. Pritchett with the 12-inch refractor at the Morrison Observatory in Missouri. Hall and his Naval Observatory colleagues J. R. Eastman and H. M. Paul even saw the outer satellite with the Observatory's 9.6-inch refractor. Six months after the discovery, several observatories in Europe had seen the outer satellite, but, so far as Hall knew, not the inner one. By this time photometric observations at Harvard had already shown that the diameter of the outer moon was about six miles, and that of the inner moon seven miles. Hall calculated the period of the outer satellite at 30 h 18 min at a distance from the surface of Mars of 12,500 miles, and the period of the inner satellite as 7 h 39.5 min at a distance of 3,760 miles. When they were discovered they were only 85 and 34 arcseconds from Mars, with the inner and outer satellites at magnitudes 10 and 12, respectively. Hall delighted in describing the peculiar appearance that the two moons would present to a hypothetical Martian. The inner moon

[41] *Nature*, **16** (August 23, 1877), 341; (September 6, 1877), 397–398; (September 13, 1877), 427–428; *Scientific American* (September 8, 1877), 1.

[42] Gingerich (ref. 29), 113. The Melbourne reflector was in many ways a failure, one which G. W. Ritchey lamented set back the era of reflecting telescopes a generation.

would rise in the west and set in the east, and pass through all its phases in 11 hours. Such a moon could be of very practical use, Hall the practical astronomer notes, but Martian astronomers might be hindered by a dense atmosphere and large refraction effects.[43]

Hall chose to name the inner satellite Phobos (Flight) and the outer satellite Deimos (Fear), on the suggestion of a Mr Madan of Eton, England. The names were taken from the 15th book of Homer's *Iliad*, "He spake, and summoned Fear and Flight to yoke/ His steed, and put his glorious armor on." In the context of the poem, Ares (Mars in Roman mythology) is preparing to descend to the Earth to avenge the death of his son, and Fear and Flight (the horses that drew the chariot of Mars) mean Terror that frightens and Speed that overtakes. Hall noted that, as personified by Homer, Phobos and Deimos are the attendants, or sons, of Mars.[44]

The question of why the moons of Mars had not been discovered long before was one that naturally interested Hall. He calculated that, for 42 days, from September 4 to October 16, during the favorable opposition of 1862, the moons were brighter than they were in 1877, when they had been viewed with a small 9.6-inch refractor like the one at the Naval Observatory. Moreover, the planet was 15 degrees further north in 1862 than it had been in 1877, and so higher in the sky in the Northern Hemisphere. During this time, Hall pointed out, the moons were within easy reach of the Harvard 15-inch refractor. "The reason therefore why they were not found before is that astronomers did not search for them in the right place and in the right manner. Probably every one who searched looked too far away from the planet, and the fact appears to be that in recent times very few searched at all." A good telescope was certainly helpful in the discovery of the moons of Mars, but even more essential were curiosity, persistence and the need to doubt conventional wisdom.

Such a remarkable discovery was bound to raise controversy, and for this we need look no further than Hall's colleague, Simon Newcomb. There was considerable hard feeling between Hall and Newcomb regarding Newcomb's role in the whole affair, stemming especially from the August 17 lunch encounter and Newcomb's subsequent actions. As Newcomb recalled, "One morning Professor Hall confidentially

---

[43] Clearly the discovery caught the imagination of the normally staid Hall. Noting that the angle subtended from Earth by the outer satellite of Mars was about 0.031 seconds of arc, Hall calculated that, at the distance of our Moon, this angle would correspond to 187 feet on the Moon's surface. "Hence, if we assume that the diameter of six miles is nearly correct it appears that the proposition of a German astronomer to establish on the plains of Siberia a system of fire signals for communicating with the inhabitants of the Moon is by no means a chimerical project." In the year that Schiaparelli would begin the great Mars controversy with his observations of *canali*, it is interesting to see Hall sizing up the prospects of this scheme, attributed to K. F. Gauss, for communication with lunar inhabitants!

[44] Hall, *Observations and Orbits* (ref. 30), p. 6. While one might think (as most astronomers do) that Phobos is "fear" and Deimos "flight", Jurgen Blunck's *Mars and its Satellites: A Detailed Commentary on the Nomenclature* (Second edition, Exposition Press: Smithtown, New York, 1982), p. 151 has the definitions the other way around. In fact, classicist sources such as *Crowell's Handbook of Classical Mythology* agree that Phobos means "panic,", "flight," or "rout," while Deimos means "fear" or "terror." Thus there is some confusion between astronomers and classicists on this matter. Richard Tresch Fienberg to Steven J. Dick, private communication, September 23, 1988.

showed me his first observations of an object near Mars, and asked me what I thought of them. I remarked, 'Why that looks very much like a satellite.' Yet he seemed very incredulous on the subject; so incredulous that I feared he might make no further attempt to see the object. I afterward learned, however, that this was entirely a misapprehension on my part. He had been making a careful search for some time, and had no intention of abandoning it until the matter was cleared up one way or another." Nevertheless, it was Newcomb who computed the first preliminary orbits, but not because of any knowledge Hall lacked: "The labor of searching and observing was so exhausting that Professor Hall let me compute the preliminary orbit of the satellites from his early observations."

Newcomb's role became so ambiguous that he felt it necessary to explain himself, first on September 10 by letter to the Superintendent and then in print a few weeks later (by which time he was Superintendent of the Nautical Almanac Office). Writing in *Nature* for September 27, he recalled that "When on the morning of August 17 Prof. Hall showed me his observations, the communication was purely confidential and friendly, and was not made either in the line of duty or because he failed to recognise the signification of his observations, or because any special skill he did not possess would aid in interpreting them. I suggested that, from the few measures he had made, it was possible to estimate the time of revolution of the satellite, if the object really were one; and thus ventured the prediction that it would be hidden during most of the following night, but would reappear toward morning near the position in which it was seen the night before. The fulfillment of this prediction facilitated the establishment of the true character of the object, but, without it, an equally certain hold on the satellite would very soon have be obtained by Prof. Hall alone. The credit of sole discovery is therefore due to him."[45]

As late as 1901 Hall and Newcomb were still battling over the incident. Responding to a letter from Hall about his role, Newcomb wrote to Hall that "Your letter was the greatest surprise to me I can now recall, and came near costing me a sleepless night . . . Of all the acts of my life I do not think there is one of which I have been more repentant or more ashamed than that of having given a newspaper an account of your discovery in which I used language implying that I had first recognized the satellites of Mars as such. The fact is I was completely misled by your seeming indifference and incredulity on the subject, and, in addition, had completely lost my head, as I am apt to do on a sudden and unexpected occasion. But this is no sufficient excuse, and I have always felt that you must have felt it to be a very mean proceeding under the circumstances." His only comfort, Newcomb admitted, was his letter to *Nature*. In 1904 Hall wrote to his son that Newcomb "acted like a crazy man, said it was the greatest discovery of the age, and rushed around full of excitement . . . he was greedy and unscrupulous." Many years later Percival Hall reprinted some of the

---

[45] Newcomb, *Reminiscences*, pp. 141–143; *Nature*, **16** (September 27, 1877), 456. The Newcomb letter to Superintendent Rodgers on September 10, is reprinted in Percival Hall (ref. 22), p. 39.

letters in his biography of his father, just to make certain that the record was straight on credit for the discovery.[46] The controversy was apparently widely known, at least in the scientific world. Hall tells of being summoned by Joseph Henry on his death bed and requested to resolve the dispute between the Halls and the Newcombs. The two did finally come to some understanding late in life; in 1901 Hall wrote to Newcomb that the statement in his *Reminiscences* denying any credit "puts you in a different light from what I have seen you for the last 24 years . . . After the excitement was over I could see that you and Holden must naturally feel a great deal of disappointment and I should have felt the same had we changed places."[47]

For Hall himself, the discovery was a turning point in his career. From the French he received the Lalande Prize (1877), from the British the Gold Medal of the Royal Astronomical Society (1879), and from the Academy of Sciences at Paris the Arago Medal (1893). In 1896 became a chevalier of the French Legion of Honor. In awarding the Gold Medal of the Royal Astronomical Society, Lord Lindsay praised Hall as one "combining the skill of the observer with the labours of the mathematician." Offering Hall as an example of "what perseverance and determination may effect in overcoming even the most adverse circumstances," he recounted Hall's major accomplishments, the singular discovery of the moons of Mars, and the subsequent refinement to the mass of Mars, which was revised considerably and finally given a more certain value.[48] The long hours of study and observation, beginning in Goshen, continuing at the small college in New York where he met his wife, and at Michigan, Harvard, and finally the Naval Observatory, had paid off.

## 6.3  Aftermath: Hall and the 26-inch refractor after the discovery

In discovering the moons of Mars, Asaph Hall had done what Newcomb had hoped to do by observing Procyon – make a triumphant discovery that vindicated the largest telescope in the world and as a major byproduct bring personal fame. Hall's

---

[46] Newcomb to Hall, August 21, 1901, in Percival Hall, *Asaph Hall*, pp. 40–41; the original letter is in AHP, General Correspondence, with a followup letter from August 27, in which Newcomb tries to make further amends; Gingerich, pp. 113–114, citing Newcomb papers, *loc. cit.* The younger Hall also reprints correspondence opposing Todd's claim that he was the first to spot the inner moon. Hall was at odds with Newcomb over a number of issues, including his popularization of astronomy. Hall always declined invitations to write a popular astronomy book, such as Newcomb did. "There is enough of such trash afloat without my adding to it," G. W. Hill quoted him as saying, in BMNAS (ref. 22), 248.

[47] Hall to Newcomb, August 23, 1901, SNP, Hall folder.

[48] "Address Delivered by the President, Lord Lindsay, on Presenting the Gold Medal of the Society to Professor Asaph Hall, U. S. Navy," MNRAS, **39** (1879), 306–318. Lindsay notes that 1/m (expressed as a fraction of the Sun's mass) had varied from 1/2,812,526 to 1/2,994,790 even in Leverrier, and before that was considered to be 1/3,200,900 in Hansen, whereas Hall now gave 1/3,093,500, with an error of only 3,295. It is now (2000) known to be 1/3,098,710, so Hall's value has stood up very well. Hall's medals were displayed at the Naval Observatory during the centennial of the Martian moons' discovery. The exhibit and photographs of some of the medals are described in Brenda G. Corbin, "Asaph Hall and the Moons of Mars: An Exhibition at the U. S. Naval Observatory," *Vistas in Astronomy*, **22** (1978), 211–217. The Hall family donated these medals to the Observatory in ceremonies in 1997–1998. They are now on permanent display in the Naval Observatory Library.

discovery had not quite proven the necessity for a large telescope, since it turned out that much smaller telescopes could have made the discovery, but it was a triumph nonetheless, one that helped place the Naval Observatory at the center of the astronomical world during the last third of the nineteenth century. The discovery had other byproducts; the increase in public visibility helped generate support for moving the Observatory from its Foggy Bottom site. The discovery of the moons of Mars found the 26-inch refractor barely four years into its life, and Asaph Hall only at the midpoint of his career. A great deal remained to be done on both accounts.

Although the principal work of the telescope remained the observation of satellites for the determination of planetary masses, when the instrument was not needed for this primary duty Hall and Holden undertook a variety of other observations. Before his departure for Washburn Observatory in 1881, Holden put together his extensive observations of the central parts of the Orion Nebula, a monograph that almost four decades later W. W. Campbell still called "the most extensive paper on the subject that has ever been published." However, it was Hall who would make and publish the bulk of the observations. Foremost among them was a vigorous program of observing double stars. Hall had first become involved with this work in 1863, when Gilliss asked him to observe them for comparison with observations the latter had made in Santiago, Chile. Only in 1875 did Hall again take up these observations, this time with the 26-inch refractor, after having written to Struve and been spurred on by certain discrepancies in Struve's work. Using the Alvan Clark & Sons filar micrometer furnished with the telescope, Hall began a series of observations that extended 16 years until his retirement in 1891. During that time thousands of observations were made of the separations and position angles of hundreds of known double stars. Had Hall not discovered the moons of Mars, he would have been known primarily as a double star astronomer.[49]

Among the more sporadic programs of the 26-inch refractor was the attempt to determine stellar parallaxes for several stars, a direct measure of their distances away from Earth. The hazards of measuring the extremely small angles involved are evident in the fact that observations of Vega with the 9.6-inch refractor, begun by J. S. Hubbard in 1862 and continued by Newcomb, Harkness, and Hall until 1867, had yielded a negative parallax, indicating the occurrence of some systematic error. Anxious to try again with the new telescope and filar micrometer, between May 1880 and December 1881 Hall made observations to determine the parallax of alpha Lyrae (Vega) and 61 Cygni. The results, which were published in 1882, were shown to be

[49] Asaph Hall, "Observations of Double Stars," Appendix VI, *WO for 1877* (Washington, 1881), and "Observations of Double Stars, Part Second, 1880–1891," Appendix I, *WO for 1888* (1891). The first publication contains mainly Hall's double-star observations from 1875–80, but also (pp. 129–136) the 1863 observations made with filar micrometer on the 9.6-inch instrument, shortly after the objective had been refigured by Alvan Clark. Holden's observations of Orion were published as Appendix I in *WO for 1878* (1882); on Campbell's comment see Campbell (ref. 14), 349.

erroneous when Peters pointed out that the correction for temperature had been applied with the wrong sign. Suitably chastised, between 1883 and 1886 Hall continued his observations of several more stars, and in the latter year published these results, as re-reductions of Vega and 61 Cygni. Hall's results in general were smaller than the same parallaxes found by other observers, especially in the case of 61 Cygni, for which he found only about half the values found by others; and, in one case, he still calculated a negative parallax! Disheartened, Hall could only note that "I regret this discordance, and can only say that I have given as much care as possible to the work, and that the above results appear to be the best that can be derived from the observations." Neither he, nor anyone else, undertook any more observations for parallax with the 26-inch refractor. This is in stark contrast to its twin, the McCormick telescope at the University of Virginia, which went on to become one of the standard instruments for parallax determination.[50]

Despite these ventures into the realm of the stars, Hall's attention was continuously on the solar system and planetary masses. The fruits of Hall's observations of planetary satellites over the decade 1875–84 appeared in a series of monographs published in 1885–86 in the Appendices to the *Washington Observations*. In the search for Saturn's mass, G. W. Hill – no mean critic – called Hall's memoir on the orbit of Iapetus "among the most admirable pieces of astronomical literature . . . after the lapse of a quarter of a century, it is scarcely superseded." Hall also treated the six inner satellites of Saturn in the same way, discovered a 20-degree retrograde motion in the peri-Saturnium of Hyperion due to Titan, and summarized his observations of Saturn and its ring from the period 1875–89. Hall had also observed Oberon and Titania – the outer satellites of Uranus. Shortly before the 26-inch refractor was mounted, Lassell had reported a suspected new satellite of Neptune. Newcomb and Holden had searched for it, as did Hall, the latter during 1881–84. Never again would Hall discover a planetary satellite, but his continued observations of the known satellites did much to refine the masses of the outer planets, including that of Neptune, the most distant known planet at time.[51]

During all this time, Hall was involved in a variety of broader activities, even as he carried out his scientific work. In 1875, even before his discovery of the moons of Mars, he was elected a member of the National Academy of Sciences, subsequently serving as its Home Secretary for 12 years and its Vice President for six. He was President of the American Association for the Advancement of Science, and received the degree of LL. D. from Yale and Harvard. He served on the board that recommended

[50] Asaph Hall, "The Parallax of alpha Lyrae and 61 Cygni," Appendix I, *WO for 1879* (1882), and "Observations for Stellar Parallax," Appendix II, *WO for 1883* (1886). The parallax program with the 26-inch instrument at the University of Virginia was initiated by S. A. Mitchell the year after his arrival in 1913. It was a product of Mitchell's own background and interest, and new developments in photographic techniques. See Mitchell (ref. 9), pp. 12–16.
[51] Hall, "Orbit of the Satellite of Neptune," Appendix II, *WO for 1881* (1885), p. 6; Hill, BMNAS (ref. 22), 266.

Figure 6.11. Hall in the building for the 26-inch equatorial telescope at the new Naval Observatory site, August, 1899, with a Mars globe.

the new site for the Naval Observatory, and, during his last six months of Observatory employment, served as "assistant to the Superintendent in conducting the astronomical work of the Naval Observatory," thus taking a hand in shaping the circumstances of the next generation of Naval Observatory astronomers.[52] Almost all of these non-research activities brought increased recognition not only to Hall, but also to his institution.

By law Hall was retired on October 15, 1891, having reached the age of 62. The Superintendent, however, tendered him the use of the 26-inch telescope at the favorable opposition of Mars in the summer of 1892.[53] It was perhaps Hall's last glimpse of the satellites he had discovered, at least with the Great Refractor. His observations complete, Hall (Figure 6.11) remained in Washington until 1894, frequently visiting the Observatory's library. In that year he moved back to his native Goshen, Connecticut, and purchased a Revolutionary-War-era house, which he renovated using his long-ago-learned carpentry skills. He was invited by Harvard to teach celestial mechanics, which he accepted for five years (1896–1901). He then retired to his rustic home, and died on November 22, 1907, just after the next favorable opposition of Mars.

The landmarks of Hall's life followed the 15-year cycle of close oppositions of

---

[52] The appointment as assistant to the Superintendent was made in McNair to Hall, April 4, 1891, AHP, General Correspondence. Hall's long reply set forth his idea of what work the Naval Observatory should continue at it new site, and also renewed his call, first made under Admiral Rodgers, for a Board of Visitors to examine the work of the Observatory annually. Hall to McNair, April 6,1891, AHP, General Correspondence.
[53] Hall, "Observations of Mars," AJ, **12** (1892–93), 185–188.

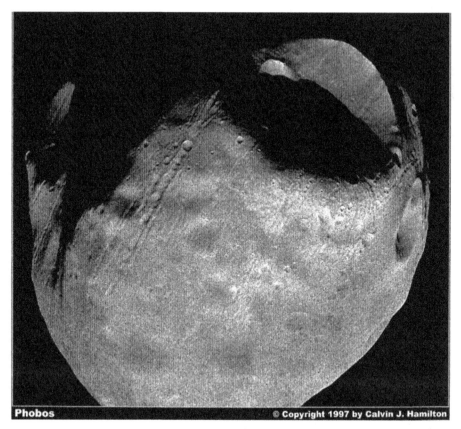

Phobos
© Copyright 1997 by Calvin J. Hamilton

Figure 6.12. A high-resolution mosaic of Phobos created from three Viking orbiter images. The striking feature in this image is the giant crater Stickney, named after Asaph Hall's wife. Grooved fractures caused from the impact that created Stickney extend away from the crater. Phobos is 17 miles × 12 miles in size, Stickney is six miles in diameter, and crater Hall (not seen here) is less than four miles across. Copyright Calvin J. Hamilton.

Mars. In 1862 he came to the Naval Observatory; in 1877 he made his famous discovery, exactly half-way through his career at the Observatory; in 1892 he made his last observations and endured the death of his beloved wife Angeline;[54] and in 1907 he himself died, in the midst of Mars furor precipitated by Lowell and in which Hall took no part. Although he had undertaken many observations in the course of his career, it was the Martian discovery that caught the public imagination, earned the scientific esteem of his peers, and made of Hall a reluctant celebrity. The fame of his Martian discovery far outlived his death, however. When the Mariner 9 spacecraft approached

[54] Angeline died of a stroke at age 62, according to her son Angelo Hall (ref. 22), p. 106. Her death notice (AHP, Miscellany, Printed Matter folder), gives her date of death as July 3, 1892. She was born November 2, 1830, and died in North Andover, Massachusetts, "passing away as peacefully as a little child falling asleep."

Mars in 1971, the detailed nature of the satellites was revealed, including their elongated "potato" shape and numerous craters. In another bow to history, the largest crater on Phobos was named Stickney, in honor of Hall's wife, who had insisted that her husband continue his search for the Martian moons (Figure 6.12). The crater "Hall" is also located near the south pole of Phobos. Hall and his discovery have even been novelized in recent years.[55]

Hall's work as a whole was made all the more notable by his humble background, tenacity and cautious temperament. "What a world of gratitude we owe him for his heroic, herculean perseverance!" eulogized the eminent mathematician and celestial mechanician G. W. Hill. "I do not suppose that he ever suffered from hunger, but he had only to look out of the window to see the wolf at the door." Hall, he recalled, "had a wholesome dread of 'subjectivities.' He knew that Sir William Herschel had announced four satellites of Uranus that turned out to be 'subjectivities,' and at Pulkova they had long observed a 'subjectivity' as a companion of Procyon. He determined, if possible, to escape such mistakes." Hall occasionally made mistakes, but they were far outweighed by his achievements, which place him near Newcomb as one of the nineteenth century's most prominent astronomers.

With his long program of observing planetary satellites and double stars, Hall set the 26-inch refractor's agenda for the rest of its career. While the McCormick telescope at the University of Virginia would concentrate on parallaxes, and other observatories would specialize in spectroscopy with their instruments, the Naval Observatory's 26-inch refractor concentrated on bodies orbiting one another, whether in the realm of the solar system or in the realm of the stars. Shunning spectroscopy, its mission remained within the classical bounds of determining the positions and motions of heavenly bodies, basic data on which knowledge both of the solar system and of stellar evolution depend.

---

[55] Thomas Mallon, *Two Moons* (Pantheon Books: New York, 2000).

# 7 William Harkness and the transits of Venus of 1874 and 1882

We are now on the eve of the second transit of a pair, after which there will be no other till the twenty-first century of our era has dawned upon the earth, and the June flowers are blooming in 2004. When the last transit season occurred the intellectual world was awakening from the slumber of ages, and that wondrous scientific activity which has led to our present advanced knowledge was just beginning. What will be the state of science when the next transit season arrives God only knows. Not even our children's children will live to take part in the astronomy of that day. As for ourselves, we have to do with the present . . .

William Harkness, 1882[1]

Twice during the last third of the nineteenth century the planet Venus passed in front of the Sun as seen from the Earth, a very rare event causing considerable excitement in the astronomical world. Unlike Hall's discovery of the moons of Mars in 1877, the passage of Venus across the face of the Sun in 1874 and 1882 had long been predicted, and had eagerly been awaited since the last pair in the mid-eighteenth century, when astronomy (especially in America) was indeed in a different age. The reason for the excitement was not simply the unusual spectacle of the silhouetted ball of Venus moving across the Sun; the timing and position of Venus as it moved across the Sun were linked to one of the great unsolved problems in the history of astronomy – the accurate determination of the distance of the Sun from the Earth, and thus the scale of the solar system. Accurate time and position were well-known specialties of the Naval Observatory, and when, in 1874 and 1882, many nations of the world dispatched expeditions to observe the transits of Venus, the Naval Observatory naturally became deeply involved in leading the observations in the name of American science.

In this chapter we describe the high hopes and preparations for these events, the far-flung expeditions to make the elusive observations, and the debate over the worth of the result, a subject on which Simon Newcomb and his colleague at the Naval Observatory William Harkness differed. Aside from their scientific results, the expeditions are of historical interest for the international disagreements over techniques and instruments, as an early example of international cooperation in astronomy, and for their place in two broader historical trends: the determination of the fundamental astronomical constants, and the great scientific voyages of the nineteenth century.

---

[1] Harkness, "Address by William Harkness," *Proceedings of the AAAS 31st Meeting . . . August, 1882* (Salem, 1883), p. 77.

We shall see that, although it was Newcomb who initiated interest in the transit of Venus in the United States, it was Harkness who carried through with the difficult analysis of the observations and produced the final result – long after Newcomb had given up on the method. So great was Newcomb's authority that even today it is widely believed that the expeditions yielded no result at all, a view that does no justice to a significant part of Harkness's life work.

## 7.1 Motives and preparations: The U. S. Commission on the Transit of Venus

The determination of the distance from the Earth to the Sun was one of the great classical problems of astronomy. From Aristarchus, Hipparchus, and Ptolemy through the seventeenth century, attempts to find this value were based on geometric methods that were very inaccurate. In the third century BC Aristarchus determined a solar distance of about 19 times that between the Earth and the Moon, equivalent to a parallax of about 180 arcseconds (hereafter written 180″); by the seventeenth century Kepler's method implied a solar parallax of about 60″. Since the actual distance is about 390 times the Earth–Moon distance (the modern value for the solar parallax being 8.794148″), Aristarchus's value for the distance was almost 20 times too small, and Kepler's seven times too small. Unlike that of the Moon (whose parallax of 3,422″ had been determined accurately by Nicolas Lacaille and others in the mid-eighteenth century), the solar parallax could not be measured directly because of the undefined roiling nature of the solar limb, not to mention that the Sun's much greater distance made the parallactic shift correspondingly smaller and therefore more difficult to measure.[2]

Nevertheless, the solar parallax could be indirectly measured, at least crudely, via the parallaxes of Mars and Venus. Not only were Mars and Venus sometimes closer to the Earth than the Sun (therefore producing larger and more easily measurable parallaxes), but also their relatively small disks were much easier objects of parallax measurement than that of the bright Sun. Since the proportions of the distances between the Sun and the planets were known from Kepler's third law (which gave the proportions between the known period and the unknown absolute distances), the parallax for the Sun could be determined from the parallax of a planet. From the parallax of Mars observed simultaneously in 1672 from South America and the Paris Observatory, the French astronomer Giovanni Domenico Cassini claimed to have determined a solar parallax of 9.5″, corresponding to a Sun–Earth distance of some 86 million miles. From the same occurrence, British Astronomer Royal John Flamsteed deduced 10″ (81,700,000 miles) and Jean Picard 20″ (41 million miles). Historians now question whether any solar parallax was actually measured in the

---

[2] For a history of cosmic dimensions from antiquity to the seventeenth century, see Albert van Helden, *Measuring the Universe: Cosmic Dimensions from Aristarchus to Halley* (University of Chicago Press: Chicago and London, 1985). On modern determinations, which are made by radar with independent confirmation from spectroscopy, see Stephen H. Knowles, "Determination of the Astronomical Unit from Hydrogen-Line Radial Velocity Measurements," dissertation (Yale, 1968). Chapter 2 gives a history of astronomical-unit determinations. The 180″ estimate for Hipparchus comes from dividing 3,422″ by 19. Ptolemy also arrived at about 180″ using lunar eclipses.

seventeenth century, arguing that the instrumental errors probably masked the parallax itself: "The conviction that the parallax of Mars was actually measured in 1672 has been misplaced . . . an upper limit was placed on the Sun's parallax (meaning a lower limit in its distance) . . . Cassini put this upper limit at about 10"; most other astronomers at the forefront of their profession at that time put the limit at something like 15"." The uncertainty was therefore very great, until the eighteenth century transits of Venus succeeded in measuring the parallax to an accuracy of several tenths of a second of arc, still an uncertainty of several million miles.[3]

We can therefore see exactly what was at stake in the nineteenth-century transits of Venus, namely, reducing an uncertainty in the Earth–Sun distance (and thereby the scale of the entire solar system) by several million miles (Table 7.1). Every difference of one hundredth of an arcsecond translated into approximately 100,000 miles. The transits of Venus afforded a method for determining the Earth–Sun distance by determining the relative parallaxes of Venus and the Sun, a measurement that could best be undertaken when the two were viewed together during a transit (Figure 7.1). The transit could be viewed in either of two ways: by precisely noting the times of contact between the limbs of Venus and the Sun, or by observing the position of Venus on the Sun's disk. The contact method had in the past proved largely a failure, because of the "black-drop" phenomenon (Figure 7.2), in which the limbs of Venus and the Sun appeared to cling together for several seconds, making accurate measurement of contact times impossible. Observing the position of Venus on the Sun's disk, however, offered more promise, with the additional benefit that the entire transit need not be observed. The principle here was simple. In Newcomb's words "if the planet is actually between us and the sun, so as to be seen projected on the sun's face, the apparent distance of the planet from the centre or from the limb of the sun may be found with considerable accuracy. Moreover, this distance will be different as seen from different parts of the earth's surface at the same moment, owing to the effect of parallax; that is, different observers will see Venus projected on different parts of the sun's face. But the change thus observed will be only that due to the differences of the parallaxes of the two bodies; while both change their directions, that nearest the observer changes the more, and thus seems to move past the other." In other words, the Sun was used as a reference against which to measure the parallax of Venus, from which the solar parallax could then be deduced. In theory the same technique could be used during the much more frequent transits of Mercury, which occurred about 13 times per century, but, because Mercury was closer to the Sun and further from the Earth, the parallax was too small to be measured.[4]

---

[3] Van Helden (ref. 2), p. 161. On the parallax determinations of 1672 see van Helden's chapter 12, and the conclusion in chapter 14.

[4] Newcomb, *Popular Astronomy*, (Harper and Brothers: New York, 1884), pp. 171–172. The Naval Observatory and others nevertheless sponsored many transit-of-Mercury observations, not for the determination of the astronomical unit, but for comparing the theory of Mercury's orbit with observations. See WO for 1876 (1879), PUSNO, **6** (1911), Appendix 2, and PUSNO, **15** (1942) part 2.

Table 7.1. *Solar-parallax determinations (selected values)*

| Source | Date | Method | Parallax in arcseconds (error) | Mean Sun–Earth distance (miles) |
|---|---|---|---|---|
| Aristarchus | Third century BC | Geometric | 180 | |
| Kepler | Seventeenth century | Eclipses/parallax of Mars | 60 | |
| Cassini *et al.* | 1672 | Meridian parallax of Mars | 9.5 | 87,000,000 |
| Flamsteed | 1672 | Diurnal parallax of Mars | 10 | 81,700,000 |
| Picard | 1672 | Parallax of Mars | 20 | 41,000,000 |
| Lalande | 1771 | Transit of Venus (1769) | 8.55 to 8.63 | 95,600,000 to 94,700,000 |
| Pingré | 1772 | Transit of Venus (1769) | 8.80 | 92,885,000 |
| Encke | 1824 | Transits of Venus | 8.5776 | 95,250,000 |
| Gilliss & Gould | 1856 | Meridian parallax of Mars | 8.495 | |
| Hall | 1865 | Meridian parallax of Mars | 8.842 | |
| Newcomb | 1867 | Meridian parallax of Mars | 8.855 | |
| Todd | 1881 | Transit of Venus (American photos, 1874) | 8.883 (0.034) | |
| Proctor | 1882 | Transit of Venus | 8.80 | 92,885,000 |
| Obrecht | 1885 | Transit of Venus (French photos, 1874) | 8.81 (0.06) | |
| Harkness | 1889 | Transit of Venus (American photo, 1882) | 8.842 (0.0118) | 92,455,000 (123,400) |
| Newcomb | 1891 | Eighteenth-century transits | 8.79 (0.051) | |
| Harkness | 1894 | System of constants | 8.809 (0.00567) | 92,797,000 (59,700) |
| Newcomb | 1895 | Venus on solar disk (1874 and 1882 transits) Venus solar contacts (four eighteenth-and nineteenth century transits) | 8.857 (0.016) 8.794 (0.023) | |
| Newcomb | 1895 | System of constants | 8.800 (.0038) | |
| Spencer Jones | 1941 | Eros campaign | 8.790 (0.001) | |
| Modern | 1976 | Radar | 8.794148 (0.000007) | 92,955,859 |

Notes:
Errors (except for 1976) are "probable errors," 74% of the "mean error" or "standard error" used today.

Transits of Venus occur only four times every 243 years, at very predictable intervals of 105.5 years, 8 years, 121.5 years, and 8 years. A transit of Venus was first predicted by Kepler for 1631, and was first observed in 1639 by Jeremiah Horrox in England. The eighteenth-century transits of Venus, occurring in 1761 and 1769, were the subject of much more serious observations. The 1761 expeditions resulted in values ranging from 8.3″ to 10.6″ for the solar parallax, but the experience of that

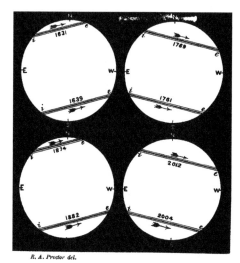

Figure 7.1. Four pairs of transit-of-Venus events, showing points of ingress and egress, from the frontispiece of Richard A. Proctor, *Transits of Venus: A Popular Account of Past and Coming Transits* (second edition, London, 1875).

Figure 7.2. The black-drop phenomenon, seen here when Venus makes "second contact" with the Sun as seen from Earth. The three images are shown in sequence from left to right. The phenomenon rendered contact times imprecise by seconds or even tens of seconds. Sketched by F. Allerding at Sydney, Australia on December 9, 1874 using a 3.5-inch refractor, and printed in H. C. Russell, *Observations of the Transit of Venus Made in New South Wales* (Sydney, 1892).

transit allowed the 1769 observers (including David Rittenhouse in America) to reduce the range from 8.43″ to 8.80″, equivalent to about 93–97 million miles. When, in 1824, Encke analyzed the results of both eighteenth-century transits, he settled on a value of 8.5776″, corresponding to a mean distance from the Earth to the Sun of 95.25 million miles. Thirty years later, however, the great German celestial mechanician Peter Andreas Hansen argued, on the basis of the motion of the Moon, that the Sun must be considerably closer, a claim supported when measurement of the parallax of Mars in 1862 gave values between 91 and 92.5 million miles. On the eve of the nineteenth-century transits of Venus, therefore, the distance to the Sun was still a value of considerable uncertainty. Interest in its accurate determination went far beyond the Sun, for it was the "standard measure for the universe . . . the great fundamental datum of astronomy – the unit of space, any error in the estimation of which is multiplied and repeated in a thousand different ways, both in the planetary and sidereal systems." It was, British Astronomer Royal George B. Airy said at mid-century, "the noblest problem in astronomy."[5]

It is therefore not surprising that already in 1857 Airy had formulated a general plan for observing the 1874 transit of Venus, and by 1870 Britain was constructing the necessary instruments. Similar plans were under way in other parts of the scientific world, and, when the great event came in 1874, Russia launched no less than 26 expeditions, Britain 12, the United States eight, France and Germany six each, Italy three, and Holland one. "Every country which had a reputation to keep or to gain for scientific zeal was forward to co-operate in the great cosmopolitan enterprise of the transit," wrote nineteenth-century-astronomy historian Agnes Clerke. The complete history of these expeditions is beyond the scope of this chapter; we concentrate here on the American expeditions and results, placed in their international context.[6]

In the United States, where Gilliss, Hall, and Newcomb had independently attempted to determine the solar parallax by various methods,[7] it was Newcomb who began the discussion of the transits of Venus with a paper in 1870, and who introduced

---

[5] On the history of determinations of the Sun's distance from Earth in the nineteenth century see Agnes Clerke, *History of Astronomy during the Nineteenth Century* (fourth edition, A. and C. Black: London, 1902), chapter 6, from which this quote is taken. For a recent treatment of the transits of Venus see Eli Maor, *June 8, 2004: Venus in Transit* (Princeton University Press: Princeton, 2000). For a similar treatment during the last transits of Venus see R. A. Proctor, *Transits of Venus: A Popular Account of Past and Coming Transits* (fourth edition, Longmans: London, 1882). For the eighteenth-century expeditions see Harry Woolf, *The Transits of Venus: A Study of Eighteenth-Century Science* (Princeton University Press: Princeton, 1959); on the observation of the 1769 transit of Venus in America, see Brooke Hindle, *The Pursuit of Science in Revolutionary America* (W. W. Norton and Co.: New York, 1974), pp. 146–165. Van Helden (ref. 2), p. 161, concludes that the eighteenth-century transit-of-Venus expeditions gave "the very first positive determination of a parallax other than the Moon's." Newcomb reviewed the eighteenth-century transit-of-Venus results in his "Discussion of Observations of the transit of Venus in 1761 and 1769," *Astronomical Papers prepared for the use of the American Ephemeris and Nautical Almanac*, volume 2 (Government Printing Office: Washington, 1890), pp. 259–405. Newcomb further discusses the subject in his *Popular Astronomy* (New York, 1884), pp. 178–184. See also van Helden, "Measuring Solar Parallax: The Venus Transits of 1761 and 1769 and their Nineteenth-Century Sequels," in R. Taton and C. Wilson, eds., *Planetary Astronomy from the Renaissance to the Rise of Astrophysics, Part B, General History of Astronomy*, volume 2 (Cambridge University Press: Cambridge, 1995).

[6] G. B. Airy, "On the Means which will be Available for Correcting the Measure of the Sun's Distance, in the Next Twenty-five Years," *MNRAS*, **17** (1857), 208; Clerke (ref. 5 above), p. 234. For details of the Russian method and instruments see Struve to Newcomb, March 16, 1872, RG 78, E 18, Box 40, item 24. Struve notes that he had sent copies of the Russian observing protocols to the Naval Observatory the previous summer. As of March, Struve wrote that the principal instruments had already been ordered, and were expected within a year. Archives related to the preparations, observations, and analysis of American transit-of-Venus parties for 1874 and 1882 are found in 21 boxes, labeled Box 27–Box 47, RG 78, E 18, "Records of Astronomical Observations made chiefly in and Near Washington, with Subsequent Computations and Compilations, January 1845–June 1907." Not all of the transit-of-Venus observations, of course, were made "in and near Washington."

[7] Gilliss's measurements, reduced by Gould, are to be found in "The Solar Parallax, Deduced from Observations of the U. S. Naval Astronomical Expedition, under Lt. J. M. Gilliss," in *The U. S. Naval Astronomical Expedition to the Southern Hemisphere*, volume 3 (Government Printing Office: Washington, 1856), pp. lix–cclxxxviii. Hall's measurements are in his "Solar Parallax, deduced from Observations of Mars with Equatorial Instruments," *WO for 1863*, Appendix A. Newcomb's treatment is in "Investigation of the Distance of the Sun and of the Elements which Depend upon it," *WO for 1865*, Appendix II.

a resolution before the National Academy of Sciences on April 16 of that year.[8] This is somewhat ironic, since Newcomb by his own account was not an enthusiastic advocate, and his hopes of measuring the Sun's distance by that method "not at all sanguine." In any case the resolution provided for a committee to report on necessary steps for American astronomers to observe the upcoming transit. Superintendent of the U. S. Coast Survey Benjamin Peirce, U. S. Naval Observatory Superintendent Commodore B. F. Sands, and the latter's boss, Rear Admiral C. H. Davis (of *Nautical Almanac* fame), were appointed to the Committee, but no report was made until Congress, via a clause in the Naval appropriations bill approved March 3, 1871, approved $2,000 for "preparing instruments." It also provided "That this and all other appropriations made for the observations of the Transits of Venus shall be expended, subject to the approval of the Secretary of the Navy, under the direction of a commission to be composed of the super-intendent and two of the professors of mathematics of the Navy attached to the Naval Observatory, the President of the National Academy of Sciences, and the superintendent of the coast survey, for which service they shall not receive any compensation." In creat-ing a government commission, the United States followed the lead of Germany (1869) and France (1870), while in Britain the responsibilities were shared by the Royal Observatory at Greenwich and the Royal Astronomical Society. The U. S. Transit of Venus Commission as it was originally constituted included Sands, Peirce, Joseph Henry (President of the National Academy at the time), Newcomb, and William Harkness. Only the latter two would survive the full term of the Commission, which held its last meeting and effectively disbanded in 1891. Of those two, only Harkness believed to the end that the method produced a valuable result.[9]

Among the first and most crucial decisions to be made by the American Commission, and by other countries, were the method of observing and the instru-ments required. The measurement of the relative position of the center of the planet and the center of the Sun was the method of choice, but there was nothing to lose in attempt-ing to measure the exact moment when the planet came into contact with the limb of the Sun, despite the famous black-drop problem. Either method might be attempted visually or photographically, and, with photography still in its infancy, especially as applied to astronomy, that decision was by no means a foregone conclusion.[10] However, at the urging of Newcomb, from early on the U. S. Commission was drawn to the photo-

---

[8] Newcomb, "On the Mode of Observing the coming Transits of Venus," *American Journal of Science*, **50**, second series (July–November, 1870), 74–83; Newcomb, *Reminiscences*, pp. 160–161. See P. M. Janiczek, "Transits of Venus and the American Expedition of 1874," *Sky and Telescope*, **48** (December, 1974), 366–371, and Janiczek, "Remarks on the Transit of Venus Expedition of 1874," in SOJ, 53–73.

[9] Naval Appropriations Act, March 3, 1871, *Statutes at Large*, volume 16, p. 529, under "Naval Observatory." Harkness records that he was appointed to the Commission on November 13, 1871; see his "Biographical Memorandum," reprinted in the obituary by A. N. Skinner, *Science*, **17** (April 17, 1903), 601–604. In February, 1874 Admiral Davis replaced Admiral Sands, and in May, 1877 Admiral Rodgers replaced Admiral Davis soon after the latter's death. In 1874 C. P. Patterson replaced Peirce, and in 1879 William B. Rogers replaced Henry, who had died in May, 1878.

[10] John Lankford, "Photography and the 19th-Century Transits of Venus," *Technology and Culture*, **28**

graphic method, although it hedged its bet by also making visual observations with small refractors. Already in January, 1872 Sands wrote to Lewis M. Rutherfurd, a pioneer in applying photography to astronomy, soliciting his advice on what photographic method should be used for the transit of Venus. Rutherfurd replied by describing the method currently in use at his own observatory for photographing the Sun, using a 13-inch equatorial refractor. Rutherfurd cautioned that "the sun has no sharply defined outline, even to the eye, but, in its best state, is an irregular, seething ever-restless object, utterly unfit to be the starting point for measures of precision," but he still believed that the ability to measure a photographic plate accurately would counteract the disadvantages. Although Rutherfurd offered to superintend the construction of the instruments required, his proposal was not adopted, and, apart from further correspondence related to instrumentation, this was the extent of his involvement.[11]

The photographic method adopted, as proposed by Newcomb, was far different. Unlike most of the European participants, who also opted for the photographic method but devoted their attention to securing the best photographs, the American method proposed by Newcomb concentrated also on the problem of measuring the photograph. Because the measurements on the photograph were made in inches and fractions, Newcomb reasoned, and because the quantity to be determined was in minutes and seconds of arc, a precise knowledge of the scale factor was necessary in order to convert from linear to angular measurement. This conversion, according to Newcomb, was "the greatest difficulty which the photographic method offered." To meet it, Newcomb proposed an instrument conceived by Joseph Winlock and already in operation at Harvard College Observatory. This was a fixed horizontal telescope of nearly 40-foot focal length, through which sunlight was directed by a heliostat (Figure 7.3), a slowly turning mirror that kept the Sun's image stationary with respect to the telescope. The lens and heliostat mirror – a piece of finely polished but unsilvered glass that reflected about 1/20th of the sunlight into the lens – were mounted on a 4-foot-high iron pier embedded in concrete. The lens, designed to give the best *photographic* image (rather than the best *visual* image as was usually the case), formed that image 4 inches in diameter about 38.5 feet away on the photographic plate. The plate

---

(1987), 648–657. On the rise of photography in astronomy, see J. Lankford, "The Impact of Photography on Astronomy," in Owen Gingerich, ed., *Astrophysics and Twentieth-Century Astronomy to 1950, General History of Astronomy,* volume 4A (Cambridge University Press: Cambridge, 1984), pp. 16–39.

[11] Sands to Rutherfurd, January 19, 1872; Rutherfurd to Sands, January 23, 1872; Rutherfurd to Sands, February 11, 1872, in U. S. Commission on the Transit of Venus, *Papers Relating to the Transit of Venus in 1874,* Part I (Government Printing Office: Washington, 1872), pp. 8–13. The original Rutherfurd letter is found in NA, RG 78, E 18, Box 40, items 2, 11,15, 19, and 31. On Rutherfurd see Deborah Jean Warner, "Lewis M. Rutherfurd: Pioneer Astronomical Photographer and Spectroscopist," *Technology and Culture,* **12** (April, 1971), 190–216. Part II (Government Printing Office: Washington, 1872) of the transit-of-Venus papers consists of charts and tables predicting details of the transits of Venus, by G. W. Hill, working out of his home for the Nautical Almanac Office. Part III, by Newcomb, was his "Investigation of Corrections to Hansen's Tables of the Moon, with Tables for their Application" (Government Printing Office: Washington, 1876), to be used in analyzing the occultation observations for determining the longitudes of the various stations.

Figure 7.3. The fixed horizontal telescope known as a photoheliograph, in which a weight-driven heliostat directs the Sun's rays through a lens, which focuses the image onto a photographic plate 38.5 feet away. The long focal length was necessary because the scale factor needed to be known very accurately in order to obtain maximum accuracy for the parallax. The method was used by American and French observers. From Simon Newcomb, *Popular Astronomy* (Haughton and Mifflin: New York, 1878), p. 186.

itself was held vertically on another iron stand next to a grid that was overlaid on each photograph for purposes of measurement. A special device consisting of 5-foot lengths of pipe was used to measure the distance from the lens to the plate to within a hundredth of an inch – the crucial measurement on which the scale factor depended. To complete the setup, a transit instrument was to be used to align the system north–south, as well as for other purposes. This photographic method was to be used for photographs both of contact and of Venus as it moved across the face of Sun, with the latter believed to hold the most promise.[12]

Before any further action could be undertaken, funding was necessary, and, in March, 1872, Sands asked the Secretary of the Navy to request from Congress

[12] Newcomb, *Popular Astronomy*, pp. 185–190; Newcomb, "On the Application of Photography to the Observation of the Transits of Venus," in *Papers*, Part I (ref. 11), pp. 14–25. The pipes were screwed together and one end came within six inches of the photographic plate, and the other end to within six inches of the lens, with the intervening distances measured by a micrometer designed by Harkness. S. Newcomb, ed., *Observations of the Transit of Venus, December 8–9, 1874 made and reduced under the direction of the Commission created by Congress*, Part 1 (Government Printing Office: Washington, 1880), p. 30.

$150,000, in annual installments of $50,000, to observe "one of the rarest and most interesting phenomena in astronomy, a transit-of-Venus across the disk of the sun." In doing so, less than two years after $50,000 had been appropriated for the Observatory's Great Refractor, Sands again argued not on any practical basis, but for reasons of national patriotism. The astronomers of Great Britain, Russia, and Germany had received liberal grants for the purpose, he noted, and the transit "will afford our countrymen a peculiarly favorable opportunity to exercise their inventive ingenuity in the introduction of improved modes of observation. Their successful introduction of two of the most important appliances in practical astronomy – astronomical photography and the electro-chronograph, both of which are now widely adopted in Europe, – their successful competition with Europeans in producing some of the finest classes of astronomical instruments, and the rapid advances which practical astronomy has made in this country within the last few years, all warrant the belief that they can take a leading position in making the observations in question."[13] As a result, $50,000 was appropriated in the Sundry Civil Bill of June 10, 1872 for the purchase and preparation of instruments, and another $100,000 in the Navy's appropriations request in 1873 for the actual expeditions. (The following year another $25,000 was appropriated to complete work and return parties home, for a grand total of $177,000 for the 1874 event alone, not including salaries and the use of Navy facilities and ships.) This was a munificent sum indeed, one from which the country might well expect a decent scientific return.[14]

With the first $50,000 in hand, the Commission held its first meeting on July 22, 1872, and went to work in earnest on the matters of instruments, choice of stations, and organization of the expeditions and their personnel. The Naval Observatory, with the majority of the members of the Transit of Venus Commission including its President (the Superintendent) and Secretary (Newcomb), was authorized to take charge of the details of the expeditions.[15] Harkness (Figure 7.4) drew up the specifications for most of the instruments, eight sets of which had to be manufactured, since the Commission had decided that the appropriation was enough to equip eight American parties. Although instruments were constructed by a variety of makers (Table 7.2), for the most crucial of these instruments the Commission turned to Alvan Clark & Sons, who were then constructing the Observatory's Great Refractor. The firm made not only the 5-inch refractors for the visual observations (Figure 7.5), but also

---

[13] Sands to SecNav George M. Robeson, March 5, 1872, in *Papers*, Part I (ref. II above), pp. 7–8. For correspondence relating to drafting the "Memorial to Congress" and other issues, see RG 78, E 18, Box 40.

[14] "History of the Operations," in Newcomb, ed., *Observations* (ref. 12 above), pp. 9–10. The laws are to be found in Sundry Civil Acts of June 10, 1872, *Statutes at Large*, volume 17, p. 367; March 3, 1873, *Statutes at Large*, volume 17, p. 514; and June 23, 1874, *Statutes at Large*, volume 18, part 3, p. 210. As extraordinary expenses, all transit-of-Venus appropriations were made under Sundry Civil Acts rather than the usual Naval Appropriation Act.

[15] The National Archives holds 152 items of "letters received" related to preparations for the 1874 transit-of-Venus expeditions, many addressed to Newcomb or Sands. RG 78, E 18, Box 40, covering the period January 23, 1872–June 28, 1873. Box 41 holds correspondence dated May 1875–February 1878, as well as letters turned over to the USNO by Newcomb.

Figure 7.4. William Harkness, who led the American efforts for the 1882 transit of Venus, and almost single-handedly achieved the final American result.

Table 7.2. *Transit-of-Venus instruments for the American Expeditions*[a]

| Instrument | Maker | Cost |
|---|---|---|
| **Photographic** | | |
| 40-Foot photoheliograph | Alvan Clark & Sons (Cambridge, Massachusetts) | $525 each |
| **Visual** | | |
| 5-Inch refractor | Alvan Clark & Sons | $1,200 each |
| **Timing** | | |
| Portable transit | Stackpole Bros (New York) | $1,360 each |
| 2.5-Inch lens and rectangular prism for portable transit | Alvan Clark & Sons | $80 for eight sets |
| Chronograph | Alvan Clark & Sons | $500 each |
| Observing key for chronograph circuit | Edward Kahler (Washington) | |
| Pendulum clock | E. Howard & Company (Boston, Massachusetts) | $275 each |
| **Alignment** | | |
| Theodolite | Stackpole Bros | $200 each |
| Engineers level | Stackpole Bros | $175 each |
| **Magnetic** | | |
| Dip circle | Edward Kahler | |

Notes:
[a] Eight sets of equipment were constructed, one for each of the eight stations outfitted.

Figure 7.5. A 5-inch Alvan Clark refractor, one of eight manufactured for the American transit-of-Venus expeditions. The portable telescope was adjustable to any latitude from the north pole to the south pole. Several remain on display at the Observatory today. Pictured here is astronomer Alfred Mikesell with the instrument in 1965.

the 5-inch 40-foot photoheliograph lenses and the heliostat mirrors crucial for the photographic method, as well as the chronographs for precise registration of time.[16] The polishing of the seven-inch mirrors, Newcomb recalled, was the most difficult part of the whole apparatus. The accuracy needed to be such that "if a straight edge laid upon the glass should touch at the edges, but be the hundred-thousandth of an inch above it at the centre, the reflector would be useless."

By contrast to the American system, the English and Germans decided on a more conventional photoheliograph, designed by the wealthy English amateur astronomer Warren de la Rue and used for solar observations since 1858. Unlike the horizontal instrument of the Americans, it was equatorially mounted and produced an image only 1.3 meters (about 4 feet) away. In August, 1873 Newcomb attended a meeting of the German Commission in Hanover (chaired by Hansen, in the last year of his life) and pointed out the impossibility of obtaining a satisfactory result with this type of equipment. Unfortunately, by then it was too late; although the Germans did significantly modify two of their four photoheliographs to yield a better color correction, their equipment was basically the same as the British. Among the Europeans, only the French adopted a system similar to the Americans, also a long-focus horizontal refractor that they had used for solar eclipses.[17]

While the visual 5-inch refractors and the 40-foot photoheliographs were the most important equipment, a broken-tube transit instrument and sidereal clock were also crucial, as well as sidereal and mean-time chronometers, and a chronograph for registering time. The transit instrument (Figure 7.6), designed by Harkness and built by Stackpole, was used not only for determining latitude, longitude, and time but also

[16] As early as February, 1872, the Clark firm gave Sands a rough estimate of $1,000 for each 5-inch refractor, and $350 for each photoheliograph. Clark to Sands, February 14, 1872, RG 78, E 18, Box 40, item 12. The actual contract was executed May 9, 1873 for instruments to be delivered by March 1, 1874. The eight 5-inch achromatic telescopes were to be 70 inches in focal length, "mounted on portable universal equatorial stands, furnished with driving clocks, divided circles, finders, eye-pieces, and micrometers. The declination circles to read to single minutes of arc, the R.A. circles, to 5 seconds of time." The cost was $1,200 each. The chronographs, which were made from the design of Harkness, cost $500 each. For another $80, the firm also supplied eight 2.5-inch lenses with 30-inch focal length and eight rectangular prisms for the portable transit instrument, without mounting. The same contract specified eight 40-foot photoheliographs, but, because the Clarks had never before constructed this instrument, a price was to be agreed upon only after the first instrument had been constructed. RG 78, E 18, Box 40, item 134. Item 138 (dated May 24, 1873) is a contract with Edward Kahler for eight dip circles and eight observing keys for making or breaking a circuit. He also provided the eyepieces for the portable transit instruments at $30 each, and the portable declinometers. The contract with Stackpole Bros of New York, dated June 18, 1873, provided for eight small theodolites at $200 each; eight portable transit instruments, with optics furnished by Clark, price to be determined after the first had been made; and eight engineer's levels at $175 each. Ibid., item 145. The history of each of these instruments to the end of the nineteenth century is given in a manuscript record, A. N. Skinner, "Schedule of U. S. Naval Observatory Instruments in Store and Not in Use" (1898), USNOA, SF. The subsequent history of the 5-inch refractors is compiled in Ted Rafferty, "Five Inch Equatorial Refractors made by Alvan Clark & Sons," USNOA, SF.
[17] Lankford (ref. 10 above); AR (1872), 94; (1873), 94; (1874), 68–69. Newcomb, Popular Astronomy, p. 188; Newcomb, Reminiscences, p. 167. Faye had independently invented the horizontal telescope in France. The American photographic instrument is described in more detail, with figures, in Simon Newcomb, ed., Observations (ref. 12 above), chapter III, pp. 25 ff.

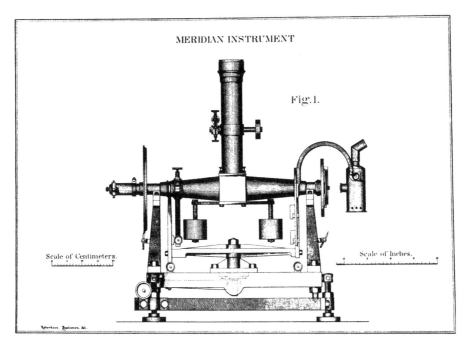

Figure 7.6. The Stackpole "broken-tube" transit instrument, designed by Harkness. The viewing eyepiece and setting circle are on the left, and an oil lamp for illumination on the right.

to insure that the central vertical line of the photographic plate holder would be very near the meridian. The clock used with the transit instrument (Figure 7.7), built by the Howard Clock Company of Boston, was designed for the rugged fieldwork and therefore not particularly elegant. One of the best photographs of the layout of all this instrumentation was taken at the U. S. Centennial Exhibition in Philadelphia, where a full-scale display of the transit-of-Venus equipment was mounted, a showcase for the progress of astronomy in the United States (Figure 7.8).[18]

Once the methods had been decided it was necessary to choose the stations. Although the entire transit would last about 4 hours, a very long time compared with the few minutes of totality for a solar eclipse, optimal weather was a prime consideration. To observe the parallax effect, both Northern- and Southern-Hemisphere stations were required, and, after studying weather records, it was decided to have three northern and five southern stations. In order to choose the stations, Newcomb began heavy correspondence with U. S. consulates and astronomers around the world. At the suggestion of Struve, Vladivostok was chosen as one northern station, with Nagasaki

---

[18] "History," in *Observations* (ref. 12), pp. 14–15. The contract for the Howard clocks, at $275 each, is dated May 1, 1873. E 18, item 115. They were to be "provided with gravity escapements" made in accordance with drawings furnished by the Commission, with "Mercurial compensation pendulum each weighing not less than forty pounds." Note that all instruments were contracted for May and June 1873.

Figure 7.7. A recent photograph of the E. Howard & Company transit-of-Venus clock, still running at the Naval Observatory. It was of sturdy but simple construction for field work, one of eight made for the Observatory. The clocks have gravity escapements with mercurial compensation pendulums holding 45 pounds of mercury. They were used as sidereal clocks and were employed for a variety of purposes long after the transits of Venus.

Figure 7.8. The transit-of-Venus exhibit at the U. S. Centennial Exhibition in Philadelphia, 1876. The transit instrument is located in the building with the wall and roof slits on the far right. On the same north–south axis is the photoheliograph, which focuses the Sun's image on a glass photographic plate 40 feet away in the photographic house on the far left. The equatorial telescope, for visual observations, is housed in the central building, with its slit open.

and Peking the other two.[19] The southern stations chosen were Crozet Islands, Kerguelen Island, Hobart Town (Tasmania), Bluff Harbor (New Zealand), and the Chatham islands off New Zealand. In the end the Crozet Islands site was abandoned when the ship could not land due to severe weather, and two American stations would be located in Tasmania (Figure 7.9).

The personnel of each station (Table 7.3) consisted of one chief of party, one astronomer, one chief photographer, and two assistant photographers, with a few parties having an additional astronomer and, in one case, an instrument maker. The members of the parties were chosen with the greatest care, especially the chiefs and the all-important chief photographer. Two of the chiefs-of-party were from the Naval Observatory, two from the Coast Survey, one from the Army, one from the Navy, and two from outside the government. The chief photographers were all professionals in photography, but the assistants were for the most part "young gentlemen of education, recent graduates of different colleges, who had been practiced in chemical and photographic manipulation."[20]

For practicing visual observations, beginning in May 1873 an artificial Sun and Venus apparatus was mounted on a building near the War Department, about two thirds of a mile from the Observatory. Using the 9.6-inch refractor (sometimes stopped down to 5 inches), a 5-inch telescope, and a 4-inch comet seeker from the dome atop the Observatory, Newcomb, Harkness, and Hall repeatedly observed the small black dot representing Venus impinging on the artificial Sun, which was a white circular disk. In the Spring of 1874 many of the participants gathered on the grounds of the Observatory to practice (Figure 7.10), with the goal of improving the accuracy with which contacts could be observed. At the same time, the photographic apparatus was set up and the photographic process rehearsed, with Henry Draper offering his services to the Commission for several weeks.[21]

By the end of May, as the planet Venus sped inexorably toward its rendezvous with the Sun, the American expeditions began their journeys around the world. At

---

[19] Regarding the Vladivostok site, Struve wrote (probably to Newcomb) on February 4, 1873 that "we astronomers would be very glad if the American commission would occupy a station in the neighborhood of ours on the Russian Coast of the Pacific Ocean. A very valuable control on the difficult methods of obtaining heliographic determinations would thus be gained. Our government is also well disposed in this regard, and the Grand Duke Constantine, with whom I spoke on the subject, has authorised me to propose that the coast region [by] Wladivostok should be chosen as principal station on the part of the Americans." Vladivostok was recommended for its favorable weather, the short period during which the harbor was closed in winter, and accurately known geographic position. Struve also wrote "I am rejoiced to learn from your letter that you have obtained so satisfactory results from the application of the long telescope. This method is also recommended by the comparative cheapness of the necessary apparatus. Nevertheless I do not think I have sufficient reason to give up the English method (Warren De la Rue's)." Translated excerpt from Struve's letter, RG 78, E 18, Box 40. Other items related to the choice of stations are filed in Box 40.

[20] "History," in Observations (ref. 12), p. 16. Much of the correspondence in E 18, Box 40 relates to choice of personnel.

[21] Newcomb, Reminiscences, pp. 170–171; AR (1874), 69; "History", in Observations (ref. 12), pp. 120–134, where the artificial apparatus is described in more detail. Records of contact observations are given in Observations (1880), pp. 145–157.

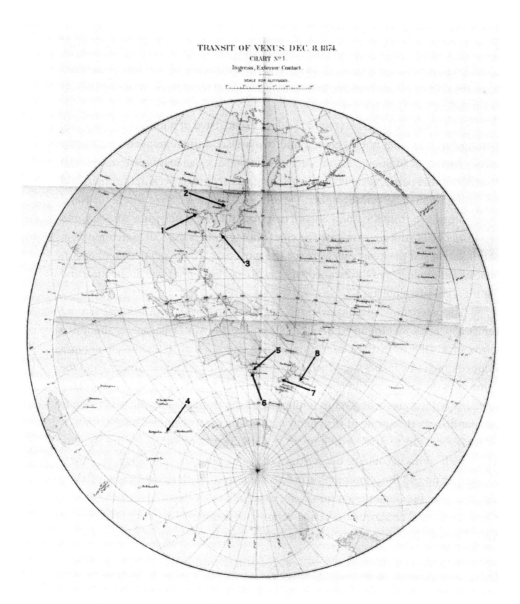

Figure 7.9. Locations of the three northern and five southern 1874 American transit-of-Venus stations are indicated on this map of times of first contact. (1) Peking, (2) Vladivostok, (3) Nagasaki, (4) Kerguelen Island, (5) Hobart Town, Tasmania, (6) Campbell Town, Tasmania, (7) Queenstown, (8) Chatham Island. The map was prepared by G. W. Hill under direction of J. H. C. Coffin as a supplement to the *American Ephemeris and Nautical Almanac*, but appeared in *Atlas to Part II of the Papers Relating to the Transit of Venus* (Government Printing Office: Washington, 1873).

Table 7.3. *American Transit-of-Venus parties, 1874*

**Vladivostok**

| | |
|---|---|
| Professor Asaph Hall, U. S. N. | Chief of Party |
| Mr O. B. Wheeler | Assistant Astronomer |
| D. R. Clark, T. S. Tappan, George J. Rockwell, and F. M. Lacey | Photographic Assistants |

**Nagasaki**

| | |
|---|---|
| Professor George Davidson, USCS | Chief of Party |
| Mr O. H. Tittman, USCS | Assistant Astronomer |
| Mr W. S. Edwards, USCS | Second Assistant Astronomer |
| S. R. Seibert, H. E. Lodge, and Frank H. Williams | Photographic Assistants |

**Peking**

| | |
|---|---|
| Professor J. C. Watson (University of Michigan) | Chief of Party |
| Professor C. A. Young | Assistant Astronomer |
| W. V. Ranger, E. Watson, and B. J. Conrad | Photographic Assistants |

**Molloy Point, Kerguelen Island**

| | |
|---|---|
| Cdr G. P. Ryan, U. S. N. | Chief of Party |
| Lt-Cdr C. J. Train, U. S. N. | Assistant Astronomer |
| D. R. Holmes, G. W. Dryer, and Irvin Stanley | Photographic Assistants |

**Hobart Town, Tasmania**

| | |
|---|---|
| Professor William Harkness, U. S. N. | Chief of Party |
| Leonard Waldo | Assistant Astronomer |
| John Moran, W. H. Churchill, and W. B. Devereux | Photographic Assistants |

**Campbell Town**

| | |
|---|---|
| Captain C. W. Raymond, Corps of Engineers | Chief of Party |
| Lt S. E. Tillman, Corps of Engineers | Assistant Astronomer |
| W. R. Pywell, J. G. Campbell, and Theodore Richey | Photographic Assistants |

**Queenstown, New Zealand**

| | |
|---|---|
| Dr C. H. F. Peters (Hamilton College) | Chief of Party |
| Lt E. W. Bass, Corps of Engineers | Assistant Astronomer |
| C. L. Phillippi, Israel Russell, | Photographic Assistants |
| E. B. Pierson, and L. H. Aymé | |

**Whangaroa, Chatham Island**

| | |
|---|---|
| Mr Edwin Smith, USCS | Chief of Party |
| Albert H. Scott, USCS | Assistant Astronomer |
| Louis Seebohm, Otto Buehler, and W. H. Rau | Photographic Assistants |
| Sumner Tainter | Instrument Maker |

stake was not only the scientific reputations of the nation, but also those of Newcomb and Harkness, as well as the head of each party, and indeed each member responsible for a specific duty. No less on trial was the usefulness of the nascent technique of photography to astronomy.

## 7.2 The 1874 expeditions and their results: Newcomb's frustration

On May 30, 1874, three days after the practice sessions ended at the Naval Observatory, the instruments for the Southern-Hemisphere transit-of-Venus stations were placed aboard the USS *Gettysburg* and shipped to New York. There, on June 7, the

Figure 7.10. Spring, 1874 practice for the transit of Venus on the grounds of the
Naval Observatory. Standing to the left nearest the photoheliograph is Admiral
C. H. Davis, founder of the American Nautical Almanac Office and President of
the Transit of Venus Commission at this time. Standing in front of him are Henry
Draper and C. H. F. Peters (with hat); seated is Simon Newcomb. Asaph Hall is
the tall man in front of the ladder, with hat; on the far right (with stovepipe hat)
is the Observatory's instrument maker, William F. Gardner. From a set of stereo
views of Washington published by J. F. Jarvis, Washington.

parties boarded the USS *Swatara*, lying off the Battery of New York Harbor, for the
journey to the far-flung southern stations. Sailing eastward via Bahia and Cape Town,
the *Swatara* anchored in Three Island Harbor of Kerguelen Island on September 7, and
on October 1 discharged the Harkness and Raymond parties at Hobart Town, Tasmania,
the first to remain in Hobart Town, the latter to go on to Campbell Town. The next port
was New Zealand's Bluff Harbor on October 16, from which C. H. F. Peters led his party
to Queenstown over government railway free of charge. Thirteen days later the final
southern party, that of Edwin Smith of the Coast Survey, was dropped off on the deso-
late Chatham Islands.[22] On November 13 the ship anchored back in Hobart Town, to
await the transit observations and then retrace its track to return the parties home.

[22] For the chronology of the *Swatara*, see "History," in *Observations* (ref. 12), pp. 17–20.

Meanwhile, two months after the departure of the southern parties, the three northern parties departed from San Francisco to Nagasaki by the Pacific Mail steamships. Coast Survey George Davidson's party remained in Nagasaki, while Asaph Hall's party sailed to Vladivostok on the USS *Kearsarge*, arriving September 9; Professor James Watson's party arrived on the same day in China for the trip to Peking.

Newcomb wrote detailed instructions for all parties to follow, and these, as well as subsequent reports, allow us to view the activities of each party upon arrival. The first order of business was the location of the site, which needed a good foundation for the instruments, a nearly level line 60 feet from north to south for the photographic telescope and transit instrument, shelter from the prevailing winds, and, of course, a clear view of the Sun during the entire duration of the transit. Piers were erected for the transit instrument, the photographic objective and heliostat, and the plate holder. All the instruments were carefully mounted, and observations were made for determining instrumental errors at least twice a week. Occultations of stars by the Moon were observed for longitude determinations, using both a chronometer and clock, and the chronograph. For determination of local time, "azimuth" and "clock" stars were observed every night, weather permitting. The parties also continued practicing the photographic operations from start to finish, using the prescribed formula for developing the glass plates, and following every aspect of the developing process. On December 9, the day of the transit, each of the equatorials was pointed toward the Sun, the clockwork set running and the highest-power eyepiece placed in the telescope tube. For the photographic operations, break-circuit chronometers were brought into the dark room, and all was prepared for the fateful event.[23]

We cannot here detail all the events at each station, but an idea of the activities at each station may be gleaned from the example of the Hobart Town, Tasmania station, where William Harkness, the Transit of Venus Commissioner in whose hands the analysis of all the observations would eventually be placed, headed the party. Having arrived on Thursday, October 1, the following day the U. S. consul introduced Harkness to the premier of Tasmania, who offered the help of his government for any need. Saturday the scientific equipment was lodged in space provided by the government, and the party took up residence at a hotel. After a careful reconnaissance, on Monday (October 4) a site was chosen half a mile from the hotel, which the Tasmanian government dutifully enclosed, put under police guard, and granted Harkness exclusive possession of for the duration of the expedition (Figure 7.11). On October 7 a meridian line was determined by observations of the Sun, and on October 9 the work of erecting the piers begun. The transit house was ready by October 14, the photographic house by October 16, and the equatorial house by October 17, exactly two weeks after the equipment had been taken off the ship; within a few days the instruments were mounted in them. On October 23

---

[23] Commission on the Transit of Venus, *Instructions for Observing the Transit of Venus, December 8–9, 1874* (Government Printing Office: Washington, 1874). 28 pages.

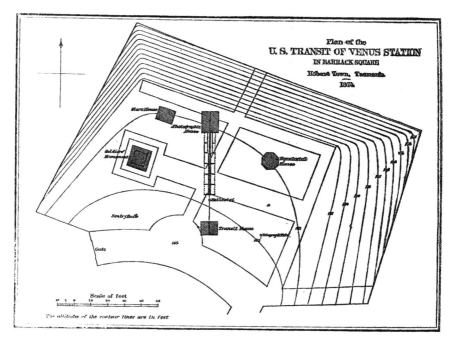

Figure 7.11. The layout of the Hobart Town transit-of-Venus station in Tasmania.

the first experimental photograph was obtained, by October 26 the focus and all other adjustments had approximately been made, and by November 12 all was considered to be in readiness. Harkness traveled to Melbourne, Australia to make arrangements for telegraphic determination of longitude differences with the Melbourne Observatory, the government of Tasmania having linked one of its wires to the observers' transit house. The weather was not auspicious; from Harkness's return to Tasmania on November 23 to the day of the transit there was not a single clear night, and the days were not much better. The instruments could not be used for anything but "dummy practice."[24]

On the morning of December 9, Harkness wrote, "the heavens were black and lowering." The first contact was predicted to occur at 11:40 am, but there was no trace of the Sun. Only at 2:37 – three hours into the transit – did the sky clear, remaining so until 3:14. During this time 41 photographs were secured. As third contact approached, Harkness rushed to the equatorial house to make micrometer measures, but just as he was readying himself to make the measures, a cloud hid the Sun. It cleared just in time for him to see that Venus was about to make third contact, when

[24] Harkness's account is to be found in *Observations* (ref. 12), Part II, page proof only, USNO Library, pp. 269–372.

Table 7.4. *American Transit-of-Venus photographs, 1874*

| | Number of plates | Contacts | | | |
|---|---|---|---|---|---|
| | | 1 | 2 | 3 | 4 |
| **Northern stations** | | | | | |
| Vladivostok, Siberia | 13 | x | x | x | |
| Peking, China | 90 | x | x | x | x |
| Nagasaki, Japan | 60 | x | x | x | |
| Total northern | 163 | | | | |
| **Southern stations** | | | | | |
| Kerguelen Island | 26 | x | | | |
| Hobart Town, Tasmania | 39 | | | | |
| Campbell Town, Tasmania | 55 | | | x | |
| Queenstown, New Zealand | 59 | x | x | | |
| Chatham Island | 8 | | | | |
| Total southern | 187 | | | | |
| Grand total northern and southern | 350 | | | | |

another cloud intervened. When that cleared, third contact had been made, and Harkness missed the opportunity to make his measurement. "Respecting the amount of work accomplished here between October 1 and December 15," Harkness wrote Admiral Davis, "allow me to say that it has fallen far short of my expectations. I regret this deeply, but it is due solely to the unusually cloudy weather which prevailed during that period. Day after day and night after night I watched the sky, and no opportunity of observing which occurred before midnight was ever allowed to pass unused."[25]

At each of the other American stations, these events were repeated with their own variations. For all their trouble Harkness and his party secured some 39 photographs, two having been spoiled by clouds. However, it was for just this eventuality that so many stations had been established. As can be seen from Table 7.4, two of the southern stations fared worse than this, and two better, so Harkness's expedition experienced average success for the southern stations. Of the northern stations, Asaph Hall's fared worst, securing only 13 photographs, while Peking and Nagasaki also did better than any of the southern stations.[26]

The great event over, Harkness traveled via the *Swatara* to the German transit-of-Venus station on the Auckland Islands and on to the American station on Chatham Island. He returned a final time to Tasmania to make telegraphic measurements for longitude differences with Melbourne Observatory, before sailing again to Melbourne, where he left the *Swatara*. On March 13 he sailed from Sydney on the steamer *Mikado*, and returned to the United States via the Hawaiian Islands and San

---

[25] Ibid., 274.
[26] AR (1875), 80; for the details of the photographs see *Observations*, Part I (ref. 12), pp. 78–84.

Francisco, thus having made a round-the-world trip. When he returned to the Naval Observatory on June 22, more than a year after his departure, Harkness carried only half the precious transit photographic plates with him, his colleague Leonard Waldo having returned by separate route with the other half and a copy of all the observations made.

Now began the saga of analyzing the observations and producing a result. In October, 1875 Newcomb still expressed optimism based on what he knew at that point about the observations. The optical observations of contacts made by the observers of all nations would, he believed, "by their combination give a value of the solar parallax of which the probable error will lie between 0.02″ and 0.03″." The American photographs alone, he further felt, "will give a result at least as accurate as this, and probably more so." While he held out no hope for the European photographs, he believed the heliometer measures made by the Germans, the Russians, and Lord Lindsay, taken together with other optical measurements of internal contact and photographs in which Venus was partly on the Sun, held promise of giving an independent result of equal weight, which, when taken with the others, might push the probable error to 0.01″. However, Newcomb cautioned, "it is not to be disguised that there is a possibility of unforeseen perturbing causes being brought to light by a comparison of all the observations which will upset all our *a priori* estimates of probable error."[27]

Newcomb's cautionary statement proved prophetic. Reporting on the results of the 1874 expeditions eight years later, Harkness recalled that, after the parties had returned, attention was first turned to the visual-contact observations as the easiest to analyze, but "it was soon found that they were little better than those of the eighteenth century." Around the world the result was the same: The problem was that "the black drop, and the atmospheres of Venus and the Earth, had again produced a series of complicated phenomena, extending over many seconds of time, from among which it was extremely difficult to pick out the true contact. It was uncertain whether or not different observers had really recorded the same phase, and in every case that question had to be decided before the observations could be used. Thus it came about that within certain rather wide limits the resulting parallax was unavoidably dependent upon the judgment of the computer, and to that extent was mere guesswork."[28]

The photographic observations were thus all the more important, but here again disappointment was widespread. Harkness recalled that "it soon began to be whispered about that those taken by European astronomers were a failure." The offi-

---

[27] AR (1875), 80–81. On Lindsay see Lord Lindsay and David Gill, "On Lord Lindsay's Preparations for Observations of the Transit of Venus, 1874," MNRAS (November, 1872).

[28] Harkness, "Address" (ref. 1), pp. 85–86; Newcomb, AR (1875), 80–81. The physical cause of the black-drop phenomenon has been the subject of considerable controversy. Bradley Schaefer reviews the controversy and provides a solution in "The Transit of Venus and the Notorious Black Drop Effect," JHA, **32** (2001), 325–336. He concludes that the phenomenon is not caused by diffraction or refraction of light around Venus, nor by an optical illusion, but by smearing due primarily to the Earth's atmosphere, diffraction in the telescope, and other secondary smearing effects.

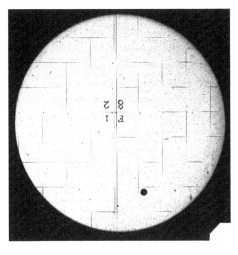

Figure 7.12. An image from a photographic plate of Venus crossing the face of the Sun. The horizontal and vertical lines are about half an inch apart. Wet bromo-iodide plates were used in 1874, but, by 1882, dry collodion emulsion plates were available. This photograph is one of only 11 plates surviving from the American 1882 expeditions; none of the plates from the American 1874 transit expeditions has survived. This photograph has been oriented with north at the top and east on the left, and the planet is following the path from left to right shown for 1882 in Figure 7.1.

cial British report declared that "after laborious measures and calculations it was thought best to abstain from publishing the results of the photographic measures as comparable with those deduced from telescopic view." The problem was the Sun itself: "however well the sun's limb on the photograph appeared to the naked eye to be defined, yet on applying to it a microscope it became indistinct and untraceable, and when the sharp wire of the micrometer was placed on it, it entirely disappeared."[29]

All hope focused on the American expeditions, which had returned with about 220 measurable plates in 1874 taken with the long-focus photoheliographs (Figure 7.12). In June, 1875 the Commission charged Harkness with measuring these plates, and he devised a machine especially for this purpose (Figure 7.13). At first it seemed that the American photographs were plagued by the same problem as the European ones. But this proved to be a problem with the microscope, and, this having been repaired, Harkness reported that 221 photographs yielded "excellent results" for the period between second and third contact when the planet was on the face of the Sun. However, those taken between first and second, and again between third and fourth contacts "proved of no value" because of the infamous black-drop problem. In other words, while even long-focus photographic contact observations were no better than visual ones, there was reason for hope in obtaining results from the photographs of Venus fully upon the face of the Sun.[30]

By the end of 1877 Harkness reported that the measurement of the

---

[29] Harkness, "Address", p. 86.
[30] Harkness, "Address, "pp. 86–87; AR (1876), 97; AR (1877), 114. The French also reported their photographic results in 1885, publishing a value of 8.81 ± 0.06. See A. Obrecht, "Discussion des résultats obtenus avec les épreuves daguerriennes de la Commission française du passage de Venus de 1874," CR, **100** (1885), 227–230; see also the same volume pp. 341–343 and (for the final result) p. 1,121.

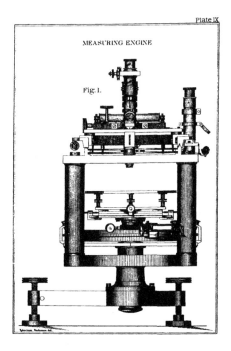

Figure 7.13. The Rutherfurd-type "micrometrical Engine" for measuring transit-of-Venus photographs, one of the earliest measuring machines for the budding field of astronomical photography. For linear measurements the reading microscope was provided with two movements, at right angles to each other, each of which had a range of 152 millimeters. For angular measurements a circle, 305 mm in diameter, was read by means of two verniers to within 10 seconds of arc. A later version of this machine is seen in Figure 10.28.

photographs was complete. However, this was only one aspect of the analysis. For the even more laborious task of determining the longitudes of the various stations from observations made at each station, the "lightning calculator," T. H. Safford, was employed. In the end, however, no result of the 1874 American transit-of-Venus expeditions was ever officially published, although in 1881 D. P. Todd (then an assistant in the Nautical Almanac Office) did publish a brief three-page article in which he determined a provisional value of 8.883 ± 0.034″. This he did by forming equations of condition based on data published in Part I of the *Observations*, the "General Discussion of Results" published in 1880. None of the other projected parts was published.[31]

Newcomb, who as Secretary of the Transit of Venus Commission was charged with analyzing the observations and publishing the result, felt compelled in his *Reminiscences* to explain his failure to publish official results. In 1876, he pointed out, confusion about funding deprived the Commission of the $3,000 required to reduce the observations, causing Newcomb to discharge his computers. Then, upon the death of Admiral Davis in February 1877, legal questions as to the powers of the Commission delayed action and required Newcomb to discharge his computers a second time. In 1879 concurrent resolutions were passed in the House and Senate providing for the printing of the results, but, because of a dispute over control of the

---

[31] D. P. Todd, "The Solar Parallax as derived from the American Photographs of the Transit of Venus, 1874," *American Journal of Science*, **21** (1881), 491–493. Todd pointed out that, in order to reach his value, provisional values had been adopted for longitudes of the stations, as well as other small corrections.

funds shortly before the 1882 transit, Newcomb had to let his computers go a third time. Completely frustrated, he turned the work over to Professor Harkness.[32]

By the eve of the 1882 transit the American results remained uncertain, and it was a heated question whether parties should even be dispatched for the 1882 transit. Despite his desire to see the 1874 results published, Newcomb had apparently made up his mind that the method itself was flawed; he believed that such parties were not worthwhile "in view of the certain failure to get a valuable result for the distance of the sun by this method," a view supported by E. C. Pickering at Harvard but few other astronomers. In Newcomb's view, given many years later in his *Reminiscences*, after the return of the 1874 expeditions, "it did not take long for the astronomers to find that the result was disappointing, so far, at least, as the determination of the sun's distance was concerned. It became quite clear that this important element could be better measured by determining the velocity of light and the time which it took to reach us from the sun than it could by any transit of Venus." In fact Newcomb undertook such velocity of light experiments in Washington during 1880–82 (see chapter 8), and, in view of this turn of events, he may not have been altogether reluctant to relinquish control of the 1874 transit results, even if he maintained some influence as a member of the Transit of Venus Commission.[33]

One thing was clear: If there were going to be any result published from the 1874 expeditions, or any 1882 expeditions launched from America, both tasks would be on the shoulders of William Harkness.

## 7.3  The 1882 expeditions and their results: The work of William Harkness

While Newcomb had given up on the transits of Venus as an optimum method for determining the solar parallax, others had not, among them William Harkness. Although Newcomb had initiated American interest, it was Harkness who had designed much of the equipment and personally led one of the parties, and it was Harkness who, after Newcomb's resignation as Secretary in 1882, would not only be the driving force behind the 1882 expeditions but also (in sharp contrast to Newcomb) produce a final result. Bringing the transit-of-Venus observations to fruition became a major goal of Harkness and a landmark in his career.

We recall from chapter 4 that Harkness, who had earned his A. B. (1858) and

---

[32] Newcomb, *Reminiscences*, pp. 177–180. The *Statutes at Large* show that $3,000 had been appropriated for reducing observations in January, 1875; $5,000 in March, 1877; $1,000 for illustrations in 1878, and $950 in February, 1881 "to finish the computations of the second part of the transit of Venus observations, and to complete them for publication."

[33] The results were reported in Newcomb, "Measures of the Velocity of Light made under direction of the Secretary of the Navy, 1880–82," *APAE*, **2**, 107–230. Volume 1 of this series, begun by Newcomb, included Michelson's experimental determination of the velocity of light. That Newcomb wished to maintain some role is apparent in the fact that the Urgent Deficiency Act of March 6, 1882 provided "that the Superintendent of the Nautical Almanac be, and he is hereby, created an additional member of the said commission." W. D. Horigan, "Memorandum for the Superintendent concerning the origin and operations of the U. S. Transit of Venus Commission," May 27, 1922.

A. M. (1861) from the University of Rochester and his M. D. in 1862, was one of three "aids" recruited by Gilliss in 1862, along with Asaph Hall and Moses Springer. Harkness was promoted to the status of Professor of Mathematics in 1863, almost two years after Newcomb had attained that rank; Newcomb was therefore technically his senior. In addition to the traditional positional astronomy practiced at the Naval Observatory, Harkness developed an interest in photography and spectroscopy, discovering the coronal K line during the 1869 solar eclipse. Harkness was also fundamentally interested in instrumentation, and this combination of interests undoubtedly weighed heavily in his appointment to the Transit of Venus Commission in 1871. From that time he was heavily involved in the preparations for the 1874 event, in observing the transit from Tasmania, and then in analyzing the observations. In pursuance of the latter, in 1877 he published a lengthy paper on the theory of the horizontal photoheliograph, and developed the instruments and methods for measurement of the photographs. We have seen that, by the end of 1877 (during which he was briefly sidetracked by events surrounding Hall's discovery of the moons of Mars), his measurement of the photographs was complete, but not the final result for solar parallax. The photographic observations of the transit of Mercury of 1878, reduced under his supervision in 1880 and 1881, gave information on the usefulness of the technique and the orbit of Mercury, but did not yield a solar parallax.[34]

It was in this transit-of-Mercury work that Harkness was engaged as the 1882 transit-of-Venus approached. Although the English and Germans had decided not to use photography on the basis of their 1874 experience, and although Newcomb and Pickering were urging the U. S. Transit of Venus Commission to pursue the same course, Harkness actively argued for its continuance. In 1881 he wrote a paper comparing the accuracies of different methods for determining solar parallax with the object "to show that the photographic method employed by the United States Transit of Venus Parties in 1874 is among the most accurate known, and should not be neglected in observing the transit of 1882." Here Harkness noted that the reductions of the U. S. photographs were not quite completed, and therefore the degree of accuracy they would yield unknown, but the transit of Mercury observed in 1878 with the same instruments gave reason for hope. In any case, he argued, "the astronomers of the twentieth century will not hold us guiltless if we neglect in any respect the transit of 1882." His analysis showed that the parallax was somewhere between 8.75″ and

---

[34] The most detailed statement on the work of Harkness is found in his own "Biographical Memorandum," published after his death as part of A. N. Skinner's obituary of Harkness in *Science*, **17** (1903), 601–604. On Harkness see also Nathan Reingold's entry on Harkness in DSB, 119, and the obituaries by Frank Bigelow, *Popular Astronomy*, **11** (1903), 281–284; PASP, **15** (1903), 172–177, and R. S. Dugan, DAB, 266. Because Harkness was never elected a member of the National Academy (perhaps because of differences with Newcomb), there is no biographical memoir in the Academy series. Harkness's published works are listed in volumes 7, 10, and 15 of the Royal Society Catalogue. His largely mathematical photoheliograph paper is "Theory of the Horizontal Photoheliograph, including its Application to the determination of the Solar Parallax by means of Transits of Venus," MNRAS, **43** (1877), 129–155.

8.90″, "and is probably about 8.85″." Putting the situation in the best light, Harkness noted that, while at the beginning of the eighteenth century the uncertainty in the parallax was some 2″, that uncertainty in light of the latest observations and analysis (despite the lack of American results) was now narrowed to 0.15″, about half of the uncertainty after the eighteenth-century transits of Venus. Left unsaid was that in 1874 astronomers had hoped to determine the solar parallax to within 0.01″, so this estimate of the uncertainly was cause for disappointment.[35]

Harkness's hopeful conclusions about the usefulness of photography had taken so long to reach, however, that his views were not known when an international convention of astronomers was held in Paris to consider how the 1882 transit should be observed. Most of the Europeans (except for the French) declared photography a failure, and returned to the also-doubtful method of visual contacts, while at least some Americans, led by Harkness, continued to regard photography "as the most hopeful means of observation." Harkness's paper on the relative accuracies of different methods was translated into French, with the result that France also decided to continue with photography. But mostly "The Europeans condemn photography, and trust only to contacts and heliometers; the Americans observe contacts because it costs nothing to do so, but look to photography for the most valuable results."[36]

As a result of the agitations of Harkness and others, in August 1882 the U. S. Congress appropriated $10,000 for improving the instruments and $75,000 for sending the parties. The instruments themselves would be basically the same, but there was one important technical improvement; whereas the 1874 expeditions had used wet bromo-iodide photographic plates, the 1882 photoheliographs would imprint the images on dry emulsion plates.[37] This time the transit would be visible from the United States, and the four northern parties were all situated within that country, while the four southern stations were all located at different sites than in 1874, including South Africa, Patagonia, Chile, and the North Island of New Zealand. This time, Harkness observed from Washington, D. C. rather than Tasmania; Hall from San Antonio, Texas rather than Vladivostok; Davidson from New Mexico rather than Nagasaki; and Eastman traveled to Florida (see Table 7.5). Newcomb, who as Secretary of the Commission in 1874 had not observed the earlier transit, this time traveled all the way to the Cape of Good Hope, despite his grave doubts about the usefulness of the observations. In addition two parties independent of the Transit of Venus Commission observed at Princeton (under C. A. Young) and Lick Observatory

[35] Harkness, "On the Relative Accuracy of Different Methods of Determining the Solar Parallax," *American Journal of Science*, **22** (November, 1881), 375–394.

[36] Harkness, "Address," pp. 87–88; Skinner, "William Harkness," 604; Newcomb, *Popular Astronomy*, p. 191.

[37] Newcomb, *Reminiscences*, pp. 173–174; AR (1885), 12; Sundry Civil Act, August 7, 1882, *Statutes at Large*, volume 22, p. 323, under "Navy Department." Correspondence related to the 1882 American transit-of-Venus observations is found in RG 78, E 18, Box 40, folder marked "Transit-of-Venus, 1882, September to December, 1882." Unlike the 1874 correspondence, this consists largely of letters from the public asking for information about observing the transit.

Table 7.5. *American Transit-of-Venus parties, 1882*

| | |
|---|---|
| **U. S. Naval Observatory, Washington** | |
| Professor William Harkness, U. S. N. | Chief Astronomer |
| Joseph A. Rogers, A. H. Buchanan, and Lt-Cdr C. H. Davis, U. S. N. | Assistants |
| **Cedar Keys, Florida** | |
| Professor J. R. Eastman, U. S. N. | Chief Astronomer |
| Lt. John A. Norris, U. S. N. | Assistant Astronomer |
| George Prince and George F. Maxwell | Photographers |
| **San Antonio, Texas** | |
| Professor Asaph Hall, U. S. N. | Chief Astronomer |
| R. S. Woodward | Assistant Astronomer |
| D. R. Holmes and George H. Hurlbut | Photographers |
| **Cerro Roblero, New Mexico** | |
| Professor George Davidson, USC&GS | Chief Astronomer |
| J. S. Lawson, USC&GS, and J. F. Pratt, USC&GS | Assistant Astronomers |
| D. C. Chapman, USC&GS, and T. S. Tappan | Photographers |
| **Wellington, South Africa** | |
| Professor Simon Newcomb, U. S. N. | Chief Astronomer |
| Lt Thomas L. Casey Jr and Ensign J. H. L. Holcombe | Assistant Astronomers |
| Julius Ulke | Photographer |
| **Santa Cruz, Patagonia** | |
| Lt Samuel W. Very, U. S. N. | Chief Astronomer |
| O. B. Wheeler | Assistant Astronomer |
| William Bell and Irvin Stanley | Photographers |
| **Santiago, Chile** | |
| Professor Lewis Boss | Chief Astronomer |
| Miles Rock | Assistant Astronomer |
| Theodore C. Marceau and Charles S. Cudlip | Photographers |
| **Auckland, New Zealand** | |
| Mr Edwin Smith, USC&GS | Chief Astronomer |
| Professor Henry S. Pritchett | Assistant Astronomer |
| Augustus Story and Gustav Theilkuhl | Photographers |

*Source: From the Observatory's Annual Report for 1882, 265–266.*

(under David P. Todd). Finally, instructions for optical observations of contacts were extensively distributed throughout the country, and Western Union supplied the Observatory's time signals to anyone in the country intending to observe the transit. As a result of this effort, which "resulted in bringing nearly every telescope in the country into action," the Transit of Venus Commission received 93 reports of observations from individuals.[38]

Once again, each party has its own interesting history, although detailed published descriptions are more difficult to find than for the pioneering 1874 observa-

[38] Many of these reports are found in RG 78, E 18, Box 39, bound mss, including those from E. E. Barnard and others. At the age of 14, George Ellery Hale, later renowned for building the world's largest telescopes, made his first astronomical observations during the 1882 transit of Venus. Henry Norris Russell observed the transit at the age of five.

Table 7.6. *American Transit-of-Venus photographs, 1882*

| | Number of plates | | Contacts | | | |
|---|---|---|---|---|---|---|
| | Exposed | Measurable | 1 | 2 | 3 | 4 |
| **Northern stations** | | | | | | |
| Washington | 53 | 49 | x | x | x | x |
| Cedar Keys, Florida | 176 | 165 | | x | x | x |
| San Antonio, Texas | 204 | 121 | | | x | x |
| Cerro Roblero, New Mexico | 216 | 216 | x | x | x | x |
| Princeton, New Jersey[a] | 190 | 127 | | | | |
| Lick Observatory, California[a] | 23 | 115 | | | | |
| Total Northern Hemisphere | 962 | 793 | | | | |
| **Southern stations** | | | | | | |
| Wellington, South Africa | 236 | 200 | x | x | | |
| Santa Cruz, Patagonia | 224 | 204 | x | x | x | x |
| Santiago, Chile | 204 | 152 | x | x | x | x |
| Auckland, New Zealand | 74 | 31 | | x | x | |
| Total Southern Hemisphere | 738 | 587 | | | | |
| Total both hemispheres | 1,700 | 1,380 | | | | |

Notes:

[a] Though they were not officially part of the transit-of-Venus expeditions, these stations used the same procedures, and their numbers were officially reported in the Observatory's *Annual Report for 1882*, 267, from which this table is taken.

tions.[39] The end result was that the parties had much more luck in terms of weather than they had had in 1874 and correspondingly better data. Whereas in 1874 the number of measurable photographs had been 221, this time some 1,700 photographs were taken (including those by the two independent northern teams), of which 1,380 were measurable, with 793 from the Northern Hemisphere, and 587 from the Southern Hemisphere. All four contacts were observed by Harkness's party in Washington, D. C. and by Davidson's in New Mexico, as well as by two southern teams in Patagonia and Chile (Table 7.6).[40] If ever a transit of Venus could be made to successfully yield the solar parallax, it was now.

[39] On the New Mexico party see Vincent Ponko Jr, "19th Century Science in New Mexico: The 1882 Transit of Venus Observations at Cerro Roblero," *Journal of the West*, **33** (October, 1994), 44–51. On the Florida party see Vincent Ponko Jr, "Cedar Key, Florida, and the Transit of Venus: The 1882 Site Observations," *Gulf Coast Historical Review*, **x** (1995), 47–65. On the San Antonio party, Steven J. Dick "The American Transit of Venus Expedition of 1882, including San Antonio," *BAAS*, **xxvii** (1995), 1331. On Newcomb's experiences see his *Reminiscences*. On his departure Newcomb left standing the two iron pillars at the site of his observations, and entertained the "sentimental wish" that, with the help of these markers, the next transit of 2004 would be observed from the same site. However, by the 1930s the piers had disappeared, and, in 1937, an iron post was erected to mark the site. (See "Astronomical Relics at Touws River, C. P.", *South African Journal of Science*, **33** (March, 1937), 127–129.) Even this iron post no longer exists, but Willie Koorts of the South African Astronomical Observatory has tracked down the site (Koorts to Steven J. Dick, 21 July, 1997, private communication).

[40] AR (1882), 117–119; AR (1883), 264–268. As it turned out, the latter report, and that of 1884, constitute the most detailed reports published on the results of the American 1882 expeditions.

Harkness fully intended to publish the results of the 1882 observations. Under his direction, by October, 1883 the photographs from Washington, Florida, and Texas had been measured, the New Mexico measurements were in progress, and "considerable progress" had been made in reductions based on these measurements. By August of the following year the measurement of all the plates (932 northern and 639 southern) had been completed. This, however, was only the first step. The measurements for time (longitude) and latitude at each of the stations remained to be reduced, as well as the determination of the scale of the photographs, and the many steps of calculations preceding a final result for the solar parallax. There were also worries about the comparability of the wet plates versus the dry plates, but tests revealed no systematic differences. All these developments were detailed in the Observatory's *Annual Reports*, and, by September of 1887, Harkness reported that the reductions were "fast drawing to a close, and it is believed that if no unforeseen obstacle arises the value of the solar parallax deducible from them will be obtained about the end of the present calendar year."

Finally, six years after the second transit and 14 years after the first, Harkness reported on October 11, 1888 that the value for the solar parallax deduced from 1,475 photographs was $8.847 \pm 0.012''$. This corresponded to a distance of the Earth from the Sun of 92,385,000 miles, with a probable error of 125,000 miles. The following year this was refined to $8.842 \pm 0.0118''$, or 92,455,000 miles, with a probable error of 123,400 miles (Table 7.1). Harkness remembered well the exact date of his achievement: Speaking of himself in his *Biographical Memorandum*, Harkness wrote, "The work of reducing all the observations obtained by the various parties was assigned to him [Harkness], and with the aid of a small corps of assistants he completed it in a little more than six years, the final result for the value of the solar parallax from the photographs being obtained on February 13, 1889." Although he had taken a few shortcuts with regard to the longitude of the observing stations, Harkness wrote to a correspondent in 1891 that "The value of the solar parallax from the 1882 transit of Venus . . . is not likely to be much affected by any possible changes in the longitudes of the stations."[41]

Although Harkness finally had his value from the 1882 American expeditions, he did not stop there. The solar parallax, he pointed out, was not an independent constant, and treating it as such resulted in "a mass of discordant values, all of which are more or less affected by constant errors, and none of which commands anything like universal assent." In fact, he insisted, the solar parallax was inextricably entwined with the lunar parallax, the constants of precession and nutation, the parallactic

---

[41] Harkness to John Tebbutt, February 27, 1891, in Horigan, "Memorandum," p. 28; AR (1883), 267–268; AR (1884), 9–12; AR (1885), 12–13; AR (1886), 19; AR (1887), 15–17. The final result is given in AR (1888), 17–18, and AR (1889), 424–425. Among the personnel working on the reductions was Asaph Hall Jr. The Legislative Act of July 7, 1884 provided $5,000 for reductions.

inequality of the Moon, the masses of the Earth and Moon, and the velocity of light among others. He set about treating these constants as a system, and his result, the crowning achievement of a lifetime of work, was published in 1891 as his monograph *The Solar Parallax and Its Related Constants*.[42] In his Presidential Address "On the Magnitude of the Solar System," presented before the American Association for the Advancement of Science in 1894, Harkness concluded that the best estimate for the solar parallax was $8.809 \pm 0.0059''$, yielding a Sun–Earth distance of 92,797,000 miles, with a probable error of 59,700 miles.[43]

The publication of the official reports of the American Transit of Venus Commission, however, was another matter. Recalling the impressive tomes published by other nations, Harkness argued that the full publication of the American results was essential: "The meager statement of the results attained has been made in the report of the Superintendent of the Observatory for the year 1889, but until the observations are published in the detailed form adopted by other great nations the benefits accruing from the money already expended cannot be fully realized." His hope in this respect, however, was not realized. Aside from the *Papers Relating to the Transit of Venus*, and the *Instructions for Observing* it in 1874 and 1882, only Part I of a projected four parts of *Observations* was published. It consisted of 250 copies of a "General Discussion" of 157 quarto pages, printed and distributed in 1880. Of this general discussion the first 117 pages analyze the photographic work without conclusion, while the last 40 pages discuss the optical observations. Part II, two volumes of reports of the eight 1874 parties and consisting of some description and much data, but no results, reached the page-proof stage, and today exists as only a single copy of 564 pages in the Naval Observatory Library. Parts III and IV, which were supposed to be the results, were never published.

The reasons for this failure to publish again were bureaucratic. With the passing of the 1882 expeditions, the Transit of Venus Commission decided that it would be best to combine the results of both transits into one report. In 1886 Harkness wrote that "The work is now at a stage where it will soon be desirable to insert the observations of the transit of Venus in December 1882 . . . by combining the two much economy will be effected, both in the office work and in printing." Harkness asked that the 1879 House and Senate resolutions regarding printing the results be so amended, and estimated that the final report would cover between 1,200 and 1,500 pages. The Commission adopted Harkness's recommendation, and approved a draft of a new Resolution. By 1891, however, despite a recognition of the need "that the United States keep pace with other governments in publishing the results of its

---

[42] "The Solar Parallax and its Related Constants," *WO for 1885* (1891), Appendix III. Harkness also presented the result to the Philosophical Society of Washington on October 13, 1888.

[43] Harkness, "On the Magnitude of the Solar System," *Astronomy and Astro-Physics*, **13** (October, 1894), 605–626.

Table 7.7.  *Results of solar-parallax determinations, with Newcomb's weights*

|  | Solar parallax (arcseconds) | Weight |
|---|---|---|
| From the mass of the Earth resulting from the secular variations of the orbits of the four inner planets | 8.759 ± 0.010 | 9 |
| From Gill's observations of Mars at Ascension Island | 8.780 ± 0.020 | 2 |
| From Pulkovo determinations of the constant of aberration | 8.793 ± 0.0046 | 40 |
| From observations of contacts during the transits of Venus | 8.794 ± 0.018 | 3 |
| From the parallactic inequality of the Moon | 8.794 ± 0.007 | 18 |
| From determinations of the constant of aberration made elsewhere than at Pulkovo | 8.806 ± 0.0056 | 28 |
| From heliometer observations on the minor planets | 8.807 ± 0.007 | 20 |
| From the lunar equation in the motion of the Earth | 8.825 ± 0.030 | 1 |
| From measurements of the distance of Venus from the Sun's center during transits | 8.857 ± 0.023 | 2 |

Source: From Simon Newcomb, *The Elements of the Four Inner Planets and the Fundamental Constants of Astronomy* (Washington, 1895), p. 157. Solar-parallax values are arranged in order of magnitude.

observations of these important transits," no further action had been taken.[44] The failure to publish an official American report, however, must be distinguished from the American result, which was not only obtained but also published and discussed in the context of the other astronomical constants by Harkness in the *Washington Observations* for 1885.

How important, then, were the transit-of-Venus observations? In answering this question we need to recall that the transits of Venus were only one method for determining the solar parallax. As we shall see in the next chapter, the man whose system of astronomical constants, including the solar parallax, would be used for most of the twentieth century was Simon Newcomb. Newcomb, as we have seen, gave the Venus transits very low weight. Just how low may be seen from Table 7.7, in which the photographic observations of 1874 and 1882 are given a weight of two, compared with a weight of three for observations of visual contact during the eighteenth- and nineteenth-century transits, and 40 for the solar parallax derived from Pulkovo Observatory determinations of the constant of aberration. Harkness must have been disappointed by Newcomb's low ranking of his work, a ranking in which Newcomb admitted he had some leeway: "The weights assigned are convenient small integers,

[44] "Memorandum concerning the Printing of the Report upon the Observations of the Transits of Venus of 1874 and 1882," typescript, U. S. Naval Observatory, December 22, 1891. In 1896 Superintendent R. L. Phythian, NAS President Wolcott Gibbs, Newcomb, and Harkness wrote a memorandum agreeing that the transit-of-Venus results should be published as appendices or supplements to the *Washington Observations*, and thus at no expense to the Commission. This too did not happen. RG 78, E 18, Box 37, envelope marked "Authority for publishing Transit of Venus Observations in the Observatory Volumes, April, 1896."

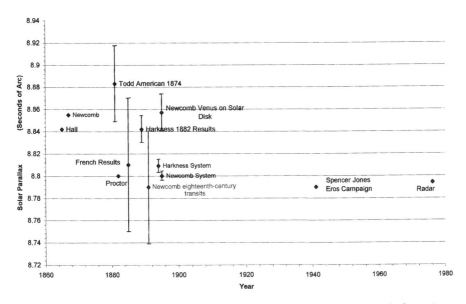

Figure 7.14. Selected solar-parallax determinations, 1860–1976. Aside from the results of Hall and Newcomb in 1865 and 1867 from observations of the meridian parallax of Mars, all nineteenth-century observations shown were derived from the transits of Venus. Two twentieth-century results are shown for comparison; their error bars are so small that they cannot be seen at this scale.

generally such as to make the weight unity correspond to the mean error ±0.30 [arcseconds], allowance being made, however, for doubt as to what value should be assigned to the mean error and for the different liabilities to systematic error." In the end Newcomb, not Harkness, had the final say, for it was his system of constants that was approved for international use; the value for the solar parallax adopted at the 1896 Conférence Internationale des Etoiles Fondamentales, held in Paris, was Newcomb's 8.80 arcseconds, not Harkness's 8.809. Considering the probable errors, Newcomb's system and Harkness's system actually overlapped in their values for the solar parallax (Figure 7.14), and Newcomb came closest to overlapping the modern value of 8.794 146.[45]

By act of Congress dated July 26, 1886, the instruments and records of the Transit of Venus Commission were turned over to the Secretary of the Navy. In that year the instruments, valued at some $30,000, were made the property of the Naval Observatory, and put into the hands of the Observatory's instrument maker, William

---

[45] Newcomb, *The Elements of the Four Inner Planets and the Fundamental Constants of Astronomy* (Government Printing Office: Washington, 1895), especially p. 157. On the controversial events leading to the introduction of Newcomb's constants, see Arthur L. Norberg, "Simon Newcomb's Role in the Astronomical Revolution of the Early Nineteen Hundreds," in SOJ, 74–88. On the importance of astronomical constants and Newcomb's role in defining them, see chapters 8 and 12.

Figure 7.15. The memorial plaque at the Queenstown, New Zealand site. "Otago" is the province in which Queenstown is located.

Gardner. On April 25, 1891 the Commission held its last meeting, at which Harkness reported on the status of the reductions. Captain McNair, Superintendent of the Observatory, was elected President of the Commission. However, the move to the new Naval Observatory was imminent, and Harkness was named to the newly created position of Astronomical Director in 1892. The Commission never met again, and, along with the effective demise of the Commission, an interesting episode in astronomy passed into history. Today some of the instruments and records remain at the Naval Observatory, and memorial plaques still mark several of the sites of the observations, of which one is shown in Figure 7.15.[46]

The nineteenth-century transits of Venus thus took their place in the long history of attempts to determine what is today known as the astronomical unit, one of the fundamental constants of astronomy. Without vastly superior methods, Harkness cautioned in his 1894 paper, the value of the solar parallax was not likely to be improved from where it stood after the nineteenth-century transits of Venus. Harkness did live to see the close passage of the minor planet Eros to Earth in

[46] Among the artifacts remaining at the Naval Observatory in 2001 are five of the eight 5-inch Clark refractors, three Howard clocks, two Stackpole portable transits, and one of the photoheliographs. The photoheliographs were used until the early 1970s for taking photographs of the Sun to measure sunspots. Although most of the records are in the National Archives, the Naval Observatory Library still holds metal boxes marked with the name of each 1874 observing site, which contain some of the record books.

1900–1901, a method that gave a considerable improvement then, and again in 1930–31. Nonetheless, he would surely marvel that radar methods have now determined the solar parallax to six significant figures, and that the mean distance to the Sun is known to within a few kilometers. Newcomb's reservations notwithstanding, Harkness's work on the transits of Venus represents the best work that could have been done on the subject during his time.

# 8 Simon Newcomb and his work

> There are tens of thousands of men who could be successful in all the ordinary walks of life, hundreds who could wield empires, thousands who could gain wealth, for one who could take up this astronomical problem [of an astronomical ephemeris] with any hope of success. The men who have done it are therefore in intellect the select few of the human race – an aristocracy ranking above all others in the scale of being. The astronomical ephemeris is the last practical outcome of their productive genius.
>
> Simon Newcomb, 1903[1]

One man dominates the history both of the Naval Observatory and of the Nautical Almanac Office – and indeed American astronomy – during the last quarter of the nineteenth century: Simon Newcomb. He has been called "the most honored American scientist of his time," and is widely regarded as having wielded an unparalleled influence both on professional astronomy and on popular astronomy.[2] A nineteenth-century combination of Albert Einstein and Carl Sagan, Newcomb's wide-ranging interests and energy were legendary in his own time. Newcomb, it seemed, had a hand in almost everything going on in the astronomy of his day. He stood, however, astride an age of great change in his field; even as the excitement of astrophysics overwhelmed the older positional astronomy, Newcomb's scientific work, if not his broader interests, remained firmly entrenched on the side of classical astronomy.

In previous chapters we have noted Newcomb's early work at the Almanac Office in Cambridge, Massachusetts, as well as his early efforts at the Naval Observatory during the 1860s and 1870s with the transit circle and the 26-inch refractor, and his leading role in the 1874 American transit-of-Venus expeditions. Newcomb's career, however, may be understood only in terms of the central driving force of his last 30 years: placing planetary and satellite motions on a completely uniform system, and thereby raising solar-system studies and the theory of gravitation to a new level. This effort could be carried out under government funding because it meant reforming the entire theoretical and computational basis of the *American Ephemeris*, which he carried out as Superintendent of the Nautical Almanac Office from 1877 to 1897. Thus Newcomb's seemingly disparate work on the transits of Mercury

---

[1] Simon Newcomb, *Reminiscences of an Astronomer* (Houghton & Mifflin: Boston and New York, 1903), p. 63.    [2] B. Marsden, "Newcomb," DSB, 33.

and Venus, the velocity of light, the constant of nutation, lunar motion, and many other subjects may be understood only as part of this grand scheme, which encompassed reform of the system of astronomical constants, determinations of the elements of planetary orbits, and the production of tables of motion of the Moon and planets based on the new data. "To endeavour to build up the theory of our whole planetary world on an absolutely homogeneous basis of constants was an almost superhuman task," the Director of the Paris Observatory, Maurice Loewy, remarked in 1899. "One would have been inclined to predict the failure or, at least, only partial success of such a scheme," the mathematician G. W. Hill wrote on Newcomb's death, "but Professor Newcomb, by his skilful management, came very near to complete success during his lifetime; only tables of the Moon were lacking to the rounding of the plan."[3] Through sheer perseverance – and a good deal of help from dedicated colleagues like Hill – Newcomb largely succeeded in his life's goal.

Despite the universal acclamation for Newcomb's work, and probably because of the technical nature of much of it, little has been written about the details of his accomplishments. The major exception is the largely unpublished work of historian Arthur Norberg. In a broader sense, Newcomb's varied interests and his role as a spokesman for science have been emphasized recently in a book-length study.[4] In this chapter we shall see in outline the progression of his life's work in astronomy, while also glimpsing something of Newcomb the man. Because this is an institutional history, our primary focus is on the work Newcomb carried out under the auspices of the Naval Observatory (1861–77) and the Nautical Almanac Office (1857–61, 1877–97), rather than his much broader work in areas outside of astronomy. Even with Newcomb's retirement in 1897, he did not completely sever ties to the Navy, and continued a productive career. Indeed, in the first years of the twentieth century, he remained the undisputed doyen of American astronomers.

## 8.1    Nautical Almanac computer and Naval Observatory astronomer

The story of how Newcomb found his way from "the world of cold and darkness" into "the world of sweetness and light" is told in his autobiography, *Reminiscences*. Newcomb was born (like Mark Twain) during the 1835 apparition of Halley's comet, and died in 1909, one year shy of Twain's feat of also departing the Earth with the return of that comet. He was born in Nova Scotia, and underwent two years of unsettling adventures beginning at age 16 as an apprentice to what turned out

---

[3] Loewy, "Simon Newcomb," *Nature*, **60** (May 4, 1899), 1–3; George W. Hill, "Professor Simon Newcomb as an Astronomer," *Science* (September 17, 1909), 353–357.

[4] Arthur L. Norberg, "Simon Newcomb and Nineteenth-Century Positional Astronomy," Ph. D. dissertation, University of Wisconsin-Madison, 1974; Norberg, "Simon Newcomb's Early Astronomical Career," *Isis*, **69** (1978), 209–225; and "Simon Newcomb's Role in the Astronomical Revolution of the Early Nineteen Hundreds," in SOJ, 74–88. The broader study is Albert E. Moyer, *A Scientist's Voice in American Culture: Simon Newcomb and the Rhetoric of Scientific Method* (University of California Press: Berkeley, 1992), shorter versions of which appear in "Simon Newcomb: Astronomer with an Attitude," *Scientific American*, **279** (October, 1998), 88–93, and "Simon Newcomb at the Nautical Almanac Office," in NAOSS, 129–145.

to be a quack doctor at Moncton, Nova Scotia, an episode that made a sufficient impression for Newcomb to devote an entire chapter of his *Reminiscences* to it. Departing his native country in September, 1853, he made his way to a teaching post at a country school at Massey's Cross Roads in Kent County, Maryland, where his father had settled.[5] The following year he moved on to a small school in Sudlersville, Maryland, and finally (in 1856) to a tutoring position some 20 miles from the City of Washington. During this period Newcomb frequented the library of the Smithsonian Institution, met its Secretary, Joseph Henry, by chance in the library, and was recommended by him to the Coast Survey Office. J. E. Hilgard at the Coast Survey in turn recommended him to Joseph Winlock at the Nautical Almanac Office in Cambridge, Massachusetts, where Newcomb arrived in January, 1857. It is remarkable that, up to this point, Newcomb was entirely self-taught in mathematics and astronomy, and, though he studied under Benjamin Peirce at the Lawrence Scientific School of Harvard in 1857–58 (at the same time as Asaph Hall), he remained largely self-taught for the rest of his life.[6]

Upon arriving at Cambridge, Newcomb entered the happy ambience that we have already described in chapter 3. He took well to the life of a "computer" (which paid $30 per month), and wrote several early papers on the orbits of the minor planets, resulting in his first substantial investigation – showing that the orbits of the known minor planets could not have originated from the explosion of a single planet, thus disproving a theory of Olbers.[7] Fifty years later the celestial mechanician E. W. Brown still called this one of Newcomb's most important papers. It was only the beginning of Newcomb's prolific production of work, which would eventually include hundreds of papers and many books.[8]

In October, 1861, with the departure of several Professors of Mathematics (as well as Superintendent Matthew Maury) to the South during the Civil War, Newcomb (Figure 8.1) obtained a position at the Naval Observatory. Among Newcomb's early

[5] On Newcomb's father see Moyer (ref. 4), pp. 20–24, and Sara Newcomb Merrick, "John and Simon Newcomb: The Story of a Father and Son," *McClure's Magazine*, 35 (October, 1910), 677–687. Sara was Simon Newcomb's sister.

[6] *Reminiscences*, chapters 1 and 2; Norberg, "Simon Newcomb's Early Astronomical Career" (ref. 4), and Norberg, "Simon Newcomb and Nineteenth Century Positional Astronomy," chapter 2. Among the obituaries, memorials, and eulogies of Newcomb, in order of usefulness, are W. W. Campbell, "Simon Newcomb," BMNAS, 17 (1916), 1–18; E. W. Brown, "Simon Newcomb," Bull. *American Mathematical Society*, 16 (1910), 341–355; G. W. Hill, "Simon Newcomb as an Astronomer," *Science*, 30 (1909), 353–357; and T. J. J. See, "An Outline of the Career of Professor Newcomb," PA, 17 (1909), 465–481. See also Brian G. Marsden, "Newcomb," DSB, 33–36. In the centennial year of Newcomb's birth, a cairn erected by the Historic Sites and Monuments Board of Canada was unveiled on August 30, 1935 by Newcomb's daughter, Mrs Joseph Whitney. The memorial is located at Wallace Bridge, Nova Scotia, within a few feet of the site of the house in which Newcomb was born. Woolard, JWAS, 26 (April 15, 1936), 139–150.

[7] Newcomb, "On the Secular Variations and Mutual Relations of the Orbits of the Asteroids," *Memoirs of the American Academy of Arts and Sciences*, new series 5 (1860), 123–152. Newcomb continued this work in "Determination of the Law of Distribution of the Nodes and Perihelia of the Small Planets between Mars and Jupiter," AN, 58 (September 25, 1862), columns 209–220.

[8] For a list of Newcomb's works, see Raymond C. Archibald, "Bibliography of the Life and Works of Simon Newcomb," *Transactions of the Royal Society of Canada*, 11, section 3 (1905), 79–110.

Figure 8.1. Newcomb at the time he came
to the Naval Observatory in 1861.

work here was the determination of the Sun's distance from the Earth (the solar parallax) based on observations of Mars made in 1862. More important than that, and
indicative of Newcomb's preoccupation with accuracy and standards, was his small
catalog of stars and its corrections for freeing their positions from systematic errors
in right ascension. Newcomb also reformed the observing procedure such that right
ascensions and declinations of stars were more systematically observed, and initiated
the program of fundamental observations with the 8.5-inch transit circle.[9]

Newcomb conceived a research program based on the work of many observatories, with the aim of producing new tables of the positions of all the planets, including new astronomical constants on which they were based. As Norberg has pointed
out, this was carried out in three stages. First, already by 1866 he had produced tables
of Neptune, and by 1873 tables of Uranus.[10] Secondly, he revised Hansen's theory of
the Moon and its associated tables, publishing the results in 1878. Finally, as Director
of the Nautical Almanac Office beginning in 1877, he supervised the generation of

---

[9] Newcomb, "Investigation of the Distance of the Sun and of the Elements which depend upon it,
from the Observations of Mars made during the Opposition of 1862, and from other Sources,"
*WO for 1865*, Appendix II; "On the Right Ascensions of the Equatoreal fundamental Stars and the
Corrections necessary to reduce the Right Ascensions of different Catalogues to a mean
Homogeneous System," *WO for 1870*, Appendix III. On the Professors of Mathematics, see
Newcomb, *Reminiscences*, pp. 99–101.
[10] "An Investigation of the Orbit of Neptune, with General Tables of its Motion," *Smithsonian
Contributions to Knowledge*, **15**, article 2; "An Investigation of the Orbit of Uranus, with General
Tables of its Motion," *Smithsonian Contributions to Knowledge*, 9, article 4 (1874), published
separately in 1873.

new tables for all the planets. The astronomical constants were adopted in 1896 at the International Astronomical Congress held in Paris, and were adopted in the major ephemerides in 1901 – fulfilling Newcomb's lifelong ambition that twentieth-century astronomy begin on a new and more accurate basis.[11]

Most importantly for his subsequent work, Newcomb showed an almost obsessive interest in the motion of the Moon. That motion had been one of the main reasons for the founding of the Royal Observatory at Greenwich, with the goal to make practical the not-yet-proven method of lunar distances for determining longitudes at sea. However, as Newcomb noted, while Greenwich astronomers made many observations, the tables of the Moon's motion were left to foreigners to construct. Specifically, the German astronomer Peter Hansen, in Newcomb's opinion "the greatest master of mathematical astronomy during the middle of the century," had derived tables of the Moon's motion based on Greenwich observations made during 1750–1850, and published by the British government in 1857. By the time Newcomb began his career at the Naval Observatory, a question of great interest was the accuracy of Hansen's tables, not only for practical but also for theoretical reasons. In Newcomb's view the Moon's complicated motions were the most severe test for the theory of gravitation; indeed, near the end of his life he believed he had evidence that the theory did not account for all its motion, just as Leverrier had found an unexplained motion in the perihelion of Mercury.

Having decided on the importance of the Moon's motion for testing the theory of gravitation, Newcomb characteristically sought a way to focus all of his energies on the problem. Already in the fall of 1869, having completed his work on the transit circle and dissatisfied with its performance, Newcomb proposed his transfer from the Observatory to the Nautical Almanac Office, in order to study the Moon's motion, "a much more important work than making observations." But Superintendent B. F. Sands argued that Newcomb could do the lunar-orbit work at the Naval Observatory, where the building was better, the growing library was at hand, and Newcomb was free to participate in the early work with the 26-inch refractor and the transit-of-Venus expeditions. Newcomb remained, but undoubtedly kept a close eye on the Nautical Almanac Office, where he knew he could carry out his own program of research, especially if he was Director of the office.[12]

In the meantime, Newcomb made do as best he could on his special-interest problem. One way to test Hansen's tables was to see how well they agreed with observations prior to 1750. In 1871 Newcomb decided to further his work on the Moon by obtaining high-quality historical observations made in Europe. Taking advantage of

---

[11] Norberg (1974) ref. 4 above, p. 152. The effect of Newcomb's new tables may be seen in context in Table 12.5 of this work.

[12] *Reminiscences*, pp. 114–115 and 202–203. For further background, and Newcomb's desire to leave the Observatory by 1868, see Norberg, "Simon Newcomb's Early Astronomical Career," 220–225, and Norberg, "Simon Newcomb and Nineteenth Century Positional Astronomy," chapter 2.

an extended stay in Europe following the Naval Observatory expedition to observe the solar eclipse in Spain in December, 1870, Newcomb traveled to the Paris Observatory in March of the following year, where its Director, Charles Delaunay, placed its archives at Newcomb's disposal. The data he needed were found not in meridian observations of the Moon, which were subject to many errors at that time, but in occultation observations, which extended knowledge of the Moon's motion back to 1675. Although Newcomb also used historical observations of eclipses of the Sun and Moon to reach his conclusions, the Paris occultations were in a class by themselves and gave originality and authority to his results. "The material I carried away proved the greatest find I ever made. Three or four years were spent in making all the calculations," Newcomb wrote. He published his calculations in 1878; in a single stroke 75 years were added to the history of the Moon's motion.[13]

Already by 1874 it was clear to all that Newcomb was on a fast course of an extraordinary career. In February of that year he received the Gold Medal of the Royal Astronomical Society of Great Britain "for his Researches on the Orbits of Neptune and Uranus, and for his other contributions to mathematical astronomy," a high honor bestowed on few Americans.[14] At the same time, he had just been placed in charge of the largest telescope in the world, the great 26-inch refractor, which may have inspired James Lick to construct the 36-inch telescope in California, a project in which Newcomb played a considerable consulting role.[15] As we have seen, Newcomb played an important role as Secretary of the Transit of Venus Commission, whose work would be carried out in December by eight far-flung American expeditions. Also his early work on the Moon, which was already appreciable, promised much more.

So great was Newcomb's reputation by 1875 that he was approached in that year by Harvard University President Charles W. Eliot to succeed Joseph Winlock as Director of the Harvard College Observatory. Newcomb was thus at a crucial crossroads in his career. "I thus had to choose between two courses," he recalled. "One led immediately to a professorship in Harvard University, with all the distinction and worldly advantages associated with it, including complete freedom of action, an independent position, and the opportunity of doing such work as I deemed best with the limited resources at the disposal of the [Harvard] observatory. On the other hand was a position to which the official world attached no importance, and which brought with it no worldly advantages whatever." Although Secretary of the Navy George Robeson counseled him to leave government service for Harvard, Newcomb decided to stay. Bound to Washington by strong family ties, optimistic that the relationship between the "scientific and literary classes" and the politicians would improve, and feeling that

---

[13] Newcomb, "Researches on the Motion of the Moon. Part I: Reduction and Discussion of Observations of the Moon Before 1750," *WO for 1875* (1878), Appendix II. The work on the Paris observations is described on pp. 116–189. Newcomb's conclusions are given on pp. 274–280.

[14] "Address Delivered by the President, Professor Cayley, on presenting the Gold Medal of the Society to Professor Simon Newcomb," *MNRAS*, **34** (1873–74), 224–233.

[15] Campbell, "Simon Newcomb" (ref. 6), 7.

Figure 8.2. Newcomb as a Professor of Mathematics, U. S. N., with his wife, Mary Caroline (Hassler) Newcomb, in 1864. Here he is in the uniform of a Lt Commander; he rose to the rank of Captain in 1897 and Rear Admiral (retired) in 1906. Professors of Mathematics were staff officers, rather than line officers, and could be recognized as such from the olive-green cloth between the gold stripes.

Nature had "best fitted" him for the mathematical astronomy he was doing at the Naval Observatory, Newcomb remained in Washington.[16]

His decision final, Newcomb made the very best of his Washington years. In 1878 he published his *Popular Astronomy*, which went through eight editions by 1896, saw smaller editions through 1918, and was translated into German (1881), Norwegian (1887), and Russian (1896). Beyond astronomy he published on a wide variety of subjects, including economics, social science, and non-Euclidean geometry.[17]

In addition to his professional duties, Newcomb (who became a naturalized U. S. citizen in 1864) had a busy personal life. On August 4, 1863, two years after coming to the Observatory, he had married Mary Caroline Hassler (Figure 8.2), daughter of Charles A. Hassler and granddaughter of Ferdinand R. Hassler, the first Superintendent of the Coast Survey. Newcomb recorded in his diary "Was this evening married to the woman of my choice. Ferdinand A. Hassler, brother of the bride, and Dr. Wm. Harkness of the observatory waited on me." Their life, W. W. Campbell recalled, was happy in all respects. "Mrs. Newcomb was able and constant in thoughtfulness for his comfort, health, and happiness, and the remarkably strong individuality of each was thoroughly respected by the other." Together, they raised three daughters: Anita (born 1864), Emily Kate (born 1869), and Anna Joseph (born 1871)

---

[16] *Reminiscences*, pp. 213–214.
[17] See Moyer (ref. 4, 1992) on Newcomb's broader work.

Figure 8.3. Newcomb with his wife and young family at the Naval Observatory, 1874.

(Figure 8.3).[18] At the same time, Newcomb was building personal qualities that would long be remembered; more than one colleague recalled his "noble head, piercing eyes, a look full of strength – steady, direct, penetrating."[19]

## 8.2 Superintendent of the Nautical Almanac Office, 1877–97

On the departure of Joseph Winlock in 1866, Newcomb must have watched with interest as J. H. C. Coffin (Figure 8.4) was made Superintendent of the Nautical Almanac Office. Coffin, one of Maury's earliest recruits to the Naval Observatory in 1845 as a Professor of Mathematics, had gone on to head the Department of Mathematics at the Naval Academy in 1855, and, upon William Chauvenet's departure in 1860, also became head of the Department of Navigation and Astronomy. There was no question at this juncture of the young Newcomb taking the job for which 12 years

---

[18] Newcomb's naturalization papers are "Supreme Court of the District of Columbia, Naturalization of Simon Newcomb," June 16, 1864, copy in USNO Library Archives. Newcomb had applied on June 10, 1858 in Massachusetts. In becoming a U. S. citizen, Newcomb specifically renounced all sovereignty to the "Kingdom of Great Britain and Ireland." On the Newcomb diary entry on his wedding day see Norberg, p. 186.

[19] "Address of Rt. Hon. James Bryce, Ambassador from Great Britain," in "Simon Newcomb, Memorial addresses," p. 136.

Figure 8.4. Professor of Mathematics J. H. C. Coffin, U. S. N., Newcomb's predecessor as Superintendent of the Nautical Almanac Office. He is seen here with the rank of Commander.

later he would clearly be the frontrunner; at the age of 30 he had only eight years of experience and had not yet made a reputation. Thus it was Coffin who would continue the work of Davis and Winlock at the Nautical Almanac Office. Upon his arrival, the work of the Office was already highly valued; in 1866 the Chief of the Bureau of Navigation wrote to the Secretary of the Navy that the *American Ephemeris and Nautical Almanac* "has fairly taken the place of foreign publications of the kind on board of our merchant vessels, and it is used exclusively on board the vessels of the navy, and by astronomers and surveyors generally throughout the country. As a national work its value cannot well be over-estimated." Coffin was content at shepherding this work over the next 12 years. By one account, as evidenced in the volumes of the *Almanac* from the period 1869–80, Coffin's influence "although appreciable, cannot be called great. New positions of the standard stars were introduced on more than one occasion and changes of detail have from time to time been introduced into the work, but the general plan has remained unaltered."[20]

The most remarkable event during Coffin's tenure was not in the *Almanac* itself, but in the Office, which was moved from Cambridge to Washington in 1866. The reasons, which had little to do with Coffin, were undoubtedly due to several factors. The original defining work of Benjamin Peirce was finished, and the following year Peirce himself would succeed Bache as Superintendent of the Coast Survey. Davis, the first Superintendent of the Almanac Office, was now head of the Naval Observatory, and he perhaps persuaded the head of the Bureau of Navigation to relocate the

[20] George C. Comstock, "John Huntington Crane Coffin," BMNAS, **8** (1916), 6. Thornton A. Jenkins to SecNav, in *Report of the Secretary of the Navy* (December 3, 1866), p. 134.

Figure 8.5. Newcomb in 1877, when he became Superintendent of the Nautical Almanac Office.

Nautical Almanac Office to Washington. The result was that the Office was now located in the same building as the Hydrographic Office "within a short distance of the [Navy] Department, and consequently more strictly under its immediate supervision than heretofore."[21] Although he had yet to be joined to the Naval Observatory, Newcomb undoubtedly took the opportunity of its new proximity to visit the office he would one day head.

On Coffin's mandatory retirement from the Navy, on September 15, 1877, Newcomb took charge of the Nautical Almanac Office (Figure 8.5). The change, he later recalled, was "one of the happiest of my life." He was now in a position of "recognized responsibility," and, because he had complete control of the Office, he could now plan and carry out the research he desired. The Nautical Almanac Office at the time had been moved and its quarters was "a rather dilapidated old dwelling-house, about half a mile or less from the observatory, in one of those doubtful regions on the border line between a slum and the lowest order of respectability." The permanent occupants of the Office were Newcomb, his senior assistant Mr Loomis, a proof-reader, and a messenger. All of the computers worked at their homes.[22]

One of Newcomb's first steps was to secure office space on the top of the new Corcoran Building. He also centralized all work in Washington. At the height of his work, the computer staff was, according to Newcomb, equal to the three or four greatest observatories of the world. "The programme of work which I mapped out, involved, as one branch of it, a discussion of all the observations of value on the positions of the sun, Moon, and planets, and incidentally, in the bright fixed stars, made at the leading observatories of the world since 1750. One might almost say it involved

---

[21] Jenkins (ref. 20), p. 134. AR (1866), 139.    [22] Newcomb, *Reminiscences*, p. 214.

repeating, in a space of ten or fifteen years, an important part of the world's work in astronomy for more than a century past."[23]

Newcomb's ultimate goal was to predict as precisely as possible the positions of the planets, one of the long-term objectives of astronomy. We recall that, for the *American Ephemeris*, Winlock had constructed new tables of Mercury based on the formulae of Leverrier, and in 1872 Hill had constructed new tables for Venus. Old tables, however, were still being used for Mars, Jupiter, and Saturn. Newcomb's goal, then, was to be able to compute ephemerides from a single uniform and consistent set of data. Just as a single observatory such as Greenwich adopted consistent methods for observation, Newcomb wished to bring uniformity to the computed positions based on observation. This meant, for example, a uniform set of planetary masses, each determined as accurately as possible, and each used in an adopted best theory.[24]

As a vehicle for documenting this work Newcomb founded the *Astronomical Papers Prepared for the Use of the American Ephemeris and Nautical Almanac*, which Campbell ranked "among the priceless treasures of astronomical literature." In the first volume, published in 1882, Newcomb explicitly stated the purpose of this series of papers as "a systematic determination of the constants of astronomy from the best existing data, a re-investigation of the theories of the celestial motions, and the preparation of tables, formulae, and precepts for the construction of ephemerides, and for other applications of the results." In the introduction to this volume, Newcomb made the first public announcement of his program. Even though he had it in mind when taking over the Superintendency of the Office in 1877, only now, after Congress and the Navy Department had supplied all the assistance asked for, including a force of eight to 12 computers, did Newcomb feel confident of carrying the program through. At the same time, he set forth the unpublished work now in progress and the program for its continuance, and called for cooperation of astronomers around the world.[25] The first volume, in which four of the six papers were authored by Newcomb himself, demonstrated the variety of topics that would be relevant to Newcomb's program. Newcomb discussed solar eclipses and transit-of-Mercury observations, compared Hansen's and Delaunay's theories of the Moon, and published his catalog of 1,098 standard reference stars. Albert A. Michelson discussed his experimental determinations of the velocity of light, and G. W. Hill calculated perturbations of Venus on Mercury. By Newcomb's death seven volumes had been published, with most of the papers by Newcomb, fully justifying Campbell's characterization of the volumes collectively as one of the great treasures of positional astronomy.

What were the raft of computers doing? One group was occupied with tables of planetary motion. By 1882 Leverrier's tables of the Sun and the inner planets had been

---

[23] *Reminiscences*, p. 217.

[24] Newcomb, "The Astronomical Ephemeris and Nautical Almanac," in *Side-lights on Astronomy* (Harper and Brothers: New York and London, 1906), pp. 191–215.

[25] Newcomb, *Astronomical Papers* (hereafter APAE), volume 1, Prefatory Note, Introduction, pp. x–xi. Campbell (ref. 6), p. 10.

Figure 8.6. G. W. Hill, master of celestial mechanics, from the frontispiece of his *Collected Works*.

partially reconstructed in a more convenient form. Meridian observations of Mercury made at Greenwich, Paris, and Washington were being compared with the tables, and similar comparisons of observed and computed values for the Sun and other inner planets were being contemplated. Other computers were putting on a uniform system Greenwich observations of solar-system objects made prior to 1830. Work on the perturbations of the outer planets was begun, and one of the most tedious parts of the work was the preparation of formulae and tables for computing the general perturbations of the planets. Through it all, Newcomb continued his work on the motion of the Moon.

Newcomb readily conceded that he never could have carried out his life work had it not been for his able assistants. Among them were John Meier, a Swiss student of Rudolph Wolf at Zürich, whom Newcomb described as "the most perfect example of a mathematical machine that I ever had at command." Another was Cleveland Keith (son of early Naval Observatory employee Ruel Keith), who supervised the reduction work and the construction of the tables of the planets. "Without his help," Newcomb wrote, "I fear I should never have brought the tables to a conclusion." The price may have been high, however; Keith died in 1896 just as the final results were being assembled.[26]

Best known among Newcomb's assistants, however, was George W. Hill (1838–1914), whom Newcomb called "the greatest master of mathematical astronomy during the last quarter of the nineteenth century."[27] Hill (Figure 8.6) had worked in

---

[26] Newcomb discusses his assistants in *Reminiscences*, pp. 217–225.
[27] *Reminiscences*, p. 219. For Hill see E. W. Brown, "Biographical Memoir of George William Hill (1838–1914)", BMNAS, **8** (1916), 275–309; and Carolyn Eisele, "Hill, George William," DSB, 398–400.

the Nautical Almanac Office in Cambridge beginning in 1861, and now Newcomb assigned him the most challenging job of all, the theory of motions of Jupiter and Saturn, made difficult because their great masses and relative proximity caused perturbations larger than those affecting the other planets. In 1881 Newcomb convinced Hill to move from his home in West Nyack, New York, to Washington, where Hill undertook the assigned tasks without the aid of assistants Newcomb offered. Eight years later, he produced his results in volume 4 of the *Astronomical Papers*. Newcomb pointed to the "eminently practical character" of Hill's research, in which he concentrated not so much on elegant formulae but rather on the utmost precision in determination of astronomical quantities. Hill constructed tables of the motions of Jupiter and Saturn; by the time of their publication in volume 7 of the *Astronomical Papers*, he had already returned home to West Nyack.[28] "During the fifteen years of our connection," Newcomb wrote, "there was never the slightest dissention or friction between us."[29] A rare photograph of the Office shortly after Hill's arrival is shown in Figure 8.7.

In addition to all these activities, Newcomb used every opportunity to make new measurements that would improve the fundamental constants. Among the constants Newcomb wished to improve was the speed of light. This was in part because he felt that an accurate value for the speed of light combined with the constant of aberration could give a better value for the solar parallax than the transits of Venus, a fact he had emphasized already in his "Investigation of the Distance of the Sun" in 1867. Newcomb argued that Jean Foucault's experiment of 1862, which made use of a rotating and fixed mirror, could be improved by using a much larger scale, with a fixed mirror placed at a distance of 3 or 4 kilometers. In April, 1878 Newcomb brought the matter before the National Academy of Sciences, and a favorable report of the National Academy resulted in Congress appropriating $5,000 on March 3, 1879 for Newcomb's proposed experiments. In the meantime, however, a young ensign and instructor at the U. S. Naval Academy in Annapolis, Albert A. Michelson (1852–1931), had independently proposed to undertake similar experiments. By March, 1878 Newcomb knew of Michelson's work, and on April 26 Michelson began correspondence with Newcomb.[30] Shortly after Newcomb had obtained his appropriation, Michelson began his experiments at his own expense, with the first results published in April, 1879, Newcomb published the results of Michelson's series of summer experiments in the first volume of his *Astronomical Papers*. For the Congressionally funded speed-of-light experiments, Michelson was detailed at Newcomb's request to

---

[28] G. W. Hill, "Tables of Jupiter," APAE, volume 7, part 1 (1895), pp. 1–144; followed by "Tables of Saturn," pp. 147–285. As mentioned in chapter 6, Hill stayed with the senior Asaph Hall for several years after his move to Washington.

[29] *Reminiscences*, p. 222.

[30] Some of the Newcomb–Michelson correspondence is reprinted in Nathan Reingold, ed., *Science in Nineteenth Century America* (Hill and Wang: New York, 1964), pp. 275–306. Reingold points out that the relation of Newcomb to Michelson began as master and apprentice, but that in the area of the speed of light, the apprentice soon passed the master. For context see Dorothy Livingston, *The Master of Light* (Scribner's: New York, 1973), pp. 45–66; for Newcomb's 1867 paper see WO for 1865, Appendix II.

Figure 8.7. Nautical Almanac Office personnel, 1883. The photograph was taken on the loggia of the east wing of the State, War and Navy Building, later known as the Old Executive Office Building and the Eisenhower Building. The famous mathematician G. W. Hill is seated second right. Seated in the center is Lt Edward W. Sturdy, in charge of the Office while Newcomb was visiting European observatories after his observations of the transit of Venus in 1882. Seated second from the left is Eben Jenks Loomis. Others have not been identified with certainty, but members of the Office at the time were E. Austin, Roberdeau Buchanan, S. Carrigan, George Eastwood, John Meier, W. F. Mck. Ritter, John Runkle, and John Van Vleck. On the location of the photo see p. 508, ref. 10.

the Nautical Almanac Office; he remained as Newcomb's assistant in this work from the fall of 1879 until September 1880. Using a refined Foucault method, in summer 1880 Newcomb and Michelson jointly conducted experiments with a fixed mirror on the grounds of the Naval Observatory, and an observing station at Ft Myer across the Potomac River (Figure 8.8). The following summer the fixed mirror was placed at the foot of the Washington Monument and, Michelson having departed, ensign J. H. L. Holcombe assisted Newcomb. The resulting velocity of light based on these experiments was $299,860 \pm 30$ km s$^{-1}$.[31] Michelson went on to refine these results even more in the work for which he became famous. Although this resulted in the standard value used for the speed of light, for Michelson this was only a prelude to his work with Morley in 1887, which yielded negative results for the relative motion of the Earth in the supposed ether, a result that brought into question the basic assumptions of

[31] Newcomb, "Measures of the Velocity of Light made under the Direction of the Secretary of the Navy during the Years 1880–'82," APAE, volume 2 (1891), pp. 107–230; Newcomb's value for the speed of light is given on p. 202. On the circumstances of Michelson's detail to the Almanac Office, see Livingston (ref. 30), p. 63.

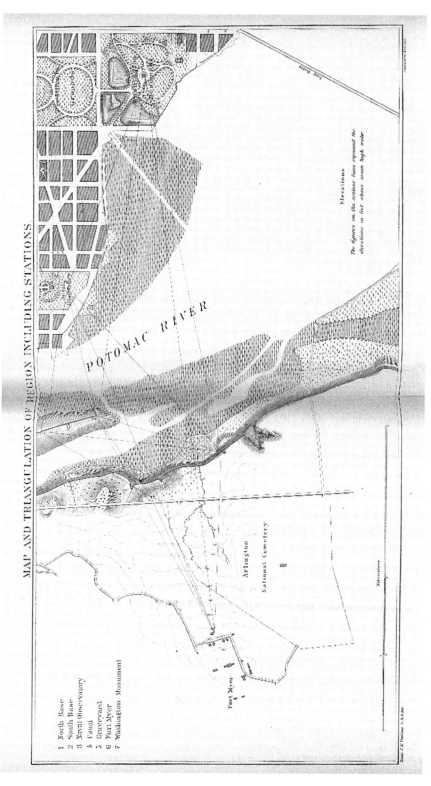

Figure 8.8. Newcomb and Michelson's speed-of-light experiments in 1880–81 used the Naval Observatory (top), Fort Myer (left), and Washington Monument (right) as stations for the mirrors at distances apart of several kilometers. From *Astronomical Papers of the American Ephemeris*, volume 2 (1891), ref. 31.

propagation of waves in classical physics.[32] For Newcomb, despite the new determinations of the speed of light, uncertain values for the aberration constant kept the elusive solar parallax uncertain.

The patronage of the Navy and the nation for Newcomb's work is in some ways surprising. Not only was the Nautical Almanac Office force greatly increased in order to undertake Newcomb's program, but also the *Astronomical Papers* were published by the Navy's Bureau of Navigation. From the outset Newcomb frankly admitted the limited immediate value of his investigations for practical applications. Existing tables of the planets, he wrote, were "not unsatisfactory" for current purposes; with the exception of the Moon, he saw "every reason to suppose that the tabular positions will serve the purposes for which they are immediately required in navigation and practical astronomy." Newcomb, however, was not satisfied with such a narrow victory over Nature, insisting that "when we take a wider view and consider the general wants of science both now and in the future, we find that in the increasing discordance between theory and observation there is a field which greatly needs to be investigated." In this Newcomb was entirely consistent with the Office's first Superintendent, Lt C. H. Davis.[33] The liberality of the Navy with regard to Newcomb's work would appear to be evidence of its interest in advancing pure science as well as its practical applications.

In 1895 Newcomb's preliminary results were finally published as *The Elements of the Four Inner Planets and the Fundamental Constants of Astronomy*, completed in 1899 with his publication of the tables of Uranus and Neptune. In Brown's estimation, "this volume gathers together Newcomb's life-work and constitutes his most enduring memorial."[34]

In 1896 occurred what Newcomb described as "the most important event in my whole plan," implementing the new system of astronomical constants. This occurred at an International Congress held in Paris. David Gill had first suggested in 1894 that such a conference be held to stimulate cooperation among the principal almanac offices, and Arthur M. W. Downing, Director of the British Nautical Almanac Office, took the initiative to put together the Paris conference in May 1896. Represented at this meeting were the American, British, German, and French almanac offices. If agreement could not be reached here, much of Newcomb's life work would have been bypassed. Although almanac offices officially change astronomical constants only with great reluctance due to the amount of work involved and the ripple effect on the reduction of astronomical observations, in the end they agreed that, beginning in 1901, Newcomb's constants would be used. This agreement, which assured Newcomb's reputation well into the future, nevertheless met with staunch resistance among many American astronomers, especially Lewis Boss and the editor of the

---

[32] On the Michelson–Morley experiment see Loyd Swenson, *The Ethereal Aether: A History of the Michelson–Morley–Miller Aether-Drift Experiments, 1880–1930* (University of Texas Press: Austin, 1972).    [33] Newcomb, APAE, volume 1, Introduction, p. vii.    [34] Brown (ref. 6), p. 347.

Figure 8.9. Newcomb and his family, mid-1890s, courtesy of Alan Wilson, Newcomb's great grandson.

*Astronomical Journal,* S. C. Chandler.[35] However, as we shall see in chapter 12, because Newcomb retired in 1897, it would be left to Newcomb's successor, William Harkness, to implement the controversial policy.

### 8.3 Newcomb's legacy

On March 12, 1897, Newcomb retired, severing official ties with the Nautical Almanac Office he had entered 40 years before. Although he was surrounded by family (Figure 8.9), Newcomb's idea of retirement was not rest and recreation. Nor was it the end of his involvement with the Nautical Almanac Office or the Naval Observatory. In connection with the former, Secretary of the Navy Herbert secured an appropriation for Newcomb to complete work left unfinished at his retirement, and likewise Carnegie Institution funding allowed Newcomb to continue his studies on the motion of the Moon. As we shall see in the next chapter, in connection with the Observatory, Newcomb was a strong proponent of a civilian Director.

Honors and awards were showered on the aging Newcomb, at least one of them with mixed feelings. Although three years before his death a special Act of Congress granted Newcomb the rank of Rear Admiral, Newcomb stated that "So far

---

[35] This controversy has been described in Arthur L. Norberg, *Simon Newcomb* (1974), pp. 381–402, and Norberg, "Simon Newcomb's Role in the Astronomical Revolution of the Early Nineteen Hundreds," in SOJ, 75–88.

as my views were concerned, the rank was merely a pro forma matter, as I never could see any sound reason for a man pursuing astronomical duties caring to have military rank." Even in the wake of Congressional action intended to honor him, Newcomb maintained an air of superiority; as evidenced by the quotation at the head of this chapter he believed himself to be in "an aristocracy ranking above all others in the scale of being," including members of Congress. Nevertheless, in addition to his other accolades, at the end of his life Newcomb could officially be addressed not only as "Professor Newcomb," but also as "Admiral Newcomb." Distinguished scientist though he was, Newcomb was not perfect. One of his most notorious failings was his adamant claim that heavier-than-air flight was impossible, a claim all the more trenchant because it was made as the Wright Brothers were beginning their experiments.[36]

Newcomb remained vigorous until the last year of his life. As late as 1907 he journeyed through his beloved Alps, and was complimented by the German Emperor on his youthful appearance. In September, 1908, he unexpectedly became ill at a meeting of the Board of Overseers of Harvard College, and early in 1909 he was diagnosed as having an inoperable cancer. Aware that his days were numbered, Newcomb rushed to complete his work on the motion of the Moon, and dictated the last of this memoir only three weeks before his death. One colleague recalled "At times opiates had to be administered to deaden his nerve-racking pain, but heroically, with indomitable will, he struggled on. I sat at his bedside in April 1909; three months had greatly sapped his energies, his voice was much less vibrant, his eye less keen. But on he forged – glowing spirit conquering frail flesh. On June the fifteenth his last work was completed and the preface dated and signed."[37] A Naval surgeon, kept on after mandatory retirement by order of the Secretary of the Navy, attended the great astronomer "so long as Professor Newcomb needed him." But there was no hope, and Newcomb died about 3 o'clock on the morning of July 11, 1909.

The Moon memoir was published in 1913 as "Researches on the Motion of the Moon, Part II" in the *Astronomical Papers*, 35 years after Part I had been published in 1878.[38] The study, which Newcomb completed with the aid of his assistant Frank E. Ross, spanned 2,600 years of observations, beginning with observations of lunar eclipses from 720 BC to 134 AD found in Ptolemy's *Almagest*, continuing with observations by Arabian astronomers from 829 to 1004, seventeenth-century eclipse and occultation observations, and lunar occultations since 1670. In the end, Newcomb lived to realize that his work, taken together with the research of E. W. Brown, did "prove beyond serious doubt the actuality of the large unexplained fluctuations in the

[36] Newcomb, "Is the Airship Coming?" *McClure's Magazine*, **17** (1901), 432–435; "The Outlook for the Flying Machine," *Independent*, **55** (1903), 2,509–2,512, reprinted in Newcomb, *Side-lights on Astronomy*, pp. 330–345.

[37] Raymond Claire Archibald, "Simon Newcomb, 1835–1909: The Unveiling of a Monument erected to his Memory by the Canadian Government," *Scripta Mathematica*, **4** (1936), 51–56.

[38] "Researches on the Motion of the Moon, Part II," *APAE*, volume IX, Part I (1912), pp. 1–249.

Moon's mean motion to which I have called attention at various times during the past 40 years."[39] Brown, Innes, Spencer Jones, de Sitter, and others later showed that the cause of the fluctuation was the irregular rotation of the Earth, a problem we shall take up in chapter 11 in connection with the determination of time, and in chapter 12 in connection with the determination of astronomical constants.

On July 14 the funeral service was held in the Church of the Covenant. President Taft, the Secretary of the Navy, many ambassadors, and other high government officials were among those in attendance. Although Newcomb in life had never considered himself a military man, he was buried with the military honors of a Rear Admiral in Arlington National Cemetery. Representatives of two foreign countries were among the honorary pallbearers, and the body was escorted to Arlington by several companies of marines and their band.[40] A few months later Newcomb's library of some 4,500 books and 7,000 pamphlets was purchased for the library of the College of the City of New York.[41]

Newcomb's great achievement, in the opinion of the eminent Yale astronomer E. W. Brown (who followed up on Newcomb's work by producing tables of the Moon), was not in purely theoretical mathematical investigations, nor in observational astronomy, but in the combination of the two, the comparison of theory and observation. "He was a master, perhaps as great as any that the world has known," Brown wrote, "in deducing from large masses of observations the results which he needed and which would form a basis for comparison with theory." However, Brown noted, Newcomb was not at home in the purely mathematical side of celestial mechanics, where he produced no new methods for dealing with the motions of solar-system bodies.[42] Nevertheless, through his science, politics, and the sheer force of his character, Newcomb left an indelible mark on the history of the Naval Observatory and American astronomy.

---

[39] Newcomb describes this work in his final paper on the Moon 'Fluctuations in the Moon's Mean Motion," MNRAS, **69** (1909), pp. 164–169.

[40] Newcomb, *Reminiscences*, p. 233; "Introductory Remarks of Mr. Charles K. Wead" and "Simon Newcomb," *Memorial Addresses at a meeting of the Philosophical Society of Washington*, held December 4, 1909, p. 135; H. H. Turner, MNRAS, **70**, 310; the circumstances of death and the funeral are given in T. J. J. See, "An Outline of the Career of Professor Newcomb," PA (October, 1908), 465–481.

[41] "Presentation of the Newcomb Library," *The City College Quarterly*, **6** (1910), 233–243. Milton Updegraff, Newcomb's successor as Director of the Nautical Almanac Office, spoke at the dedication of the "Newcomb Library."   [42] Brown (ref. 6), p. 353.

# Part III
# The twentieth century

# 9 Observatory Circle: A new site and administrative challenges for the twentieth century

> . . . the warrant of King Charles for the erection of Greenwich Observatory was wise. The gentlemen who advised him builded better than they knew. Can our government afford to be less enlightened than his of two hundred years ago? Must not, rather, proper steps be taken to put this great national institution upon a healthy basis, as well of site as of finance?
>
> Rear Admiral John Rodgers, 1877[1]

> To make such astronomical observations, derive and publish such data, and arrange for the supply of such nautical instruments as will afford to United States naval vessels and aircraft, as well as to all availing themselves thereof, means for safe navigation, including the provision of accurate time, and, while pursuing this primary function, to contribute material, within the capabilities of the available astronomical equipment, to the general advancement of navigation and astronomy.
>
> U. S. Naval Observatory Mission, 1927[2]

As early as 1877 two administrative issues began to pervade the work of the Naval Observatory and claim the attention of its astronomers and management. One was the feeling that the Observatory needed a new location, a claim advanced by most of the Observatory staff. This belief, however, became inextricably intertwined with the more contentious claim that the Observatory might also need a new organization and administration. From about 1877 until the new Observatory was finally completed in 1893, these issues were never far from the minds of Naval Observatory astronomers in their daily work, and certainly preoccupied their naval administrators. Both issues required the attention of Congress, and although the first issue was definitively

---

[1] John Rodgers, *Reports on the Removal of the United States Naval Observatory* (Government Printing Office: Washington, 1877), p. 6.

[2] The mission statement was first formulated in a memorandum from the Secretary of the Navy to the Superintendent, December 23, 1927, USNOA, AF, "1927 Organizational Formula"; it was published in AR (1928), 5. The memorandum superseded the "Regulations for the Government of the U. S. Naval Observatory," approved January 25, 1904, which can be found in the USNO Library. By comparison the current mission statement reads "Make such observations of celestial bodies, natural and artificial, derive and publish such data as will afford to United States Naval vessels and aircraft, as well as to all availing themselves thereof, means for safe navigation, including the provision of accurate time; and while pursuing this primary function, contribute material to the general advancement of navigation and astronomy."

answered with the abandonment of the old site in 1893, the question of administration proved much more difficult and spilled over well into the new century.

With these crises weathered, the Observatory, like the rest of the country, was plunged into war and depression. War placed special burdens on the Observatory, but, ironically, it was during the Great Depression that an important modernization program was undertaken. In the area of organization, the administrative structure has remained remarkably stable since World War II, with the Observatory not only remaining under the Navy Department, but under the same office within the Navy, the Chief of Naval Operations. Moreover, despite the gain and loss of some functions, the Observatory's mission also remained remarkably stable throughout most of the twentieth century. Since 1927, the mission statement at the head of this chapter has been the guideline within which all administrative actions are interpreted. As with other institutions, the issues of personnel, budgets, and organization preoccupied the administration. While the final three chapters of this study will concentrate on scientific accomplishments, in this chapter we highlight the move to the new site and the administrative issues that would determine the shape of the Observatory throughout the twentieth century.

### 9.1    A new site: Origins and development of Observatory Circle

Less than three decades after the Observatory's completion in 1844 at the E Street site, serious suggestions were being made that it be moved. It is remarkable that the first substantial suggestions for the removal came not from within the Observatory, whose occupants had to endure health conditions that Maury had already declared harmful, but from the outside. As early as 1872 a Board of Survey, constituted by Congress for the improvement of the harbor of Washington and Georgetown, had suggested that the high hill on which the Observatory stood should be leveled, and the material forming the hill used to fill in low ground and reclaimed land as part of the general improvements for the city. The Board emphatically stated that no harm should come to the Observatory, but that it should be moved to a better site in the District. Having brought the matter before the Secretary of the Navy, the Board noted that the Navy Department did not object to moving the Observatory to a healthier site. The Board even suggested that the sale of the old site, still known as "Reservation No. 4" on many maps, might result in sufficient funds to purchase a new site, where the institution could be developed "on a scale which its merits and importance demand."[3]

This suggestion was not acted on at the time, but neither was the idea completely lost. Indeed, it appears to have played an important role five years later. When,

---

[3]  Extract from "Report of the Board of Survey, ordered by Congress, on the improvement of the Harbor of Washington and Georgetown, District of Columbia, 1872," in Rodgers, *Reports* (ref. 1), p. 11. The Board included General A. A. Humphreys (Chief of Engineers) as its President, Benjamin Peirce (Superintendent of the Coast Survey), General O. E. Babcock (Commissioner of Public Buildings and Grounds), Henry D. Cooke (Governor of the District of Columbia), Alexander R. Shepherd (Board of Public Works, District of Columbia), and C. P. Patterson (Hydrographic Inspector of the Coast Survey).

on May 1, 1877, Admiral John Rodgers became Superintendent of the Naval Observatory, one of his chief aims was to remove the Observatory to a better site.[4] At the height of a distinguished Naval career, the senior Rear Admiral on the Navy's active list, an incorporator of the National Academy of Sciences and member of one of the most prominent naval families in the United States, Rodgers (seen in Figure 5.2) knew how to get things done. It is undoubtedly no coincidence that, a few months after his appointment, the District of Columbia's Board of Health referred the area's health problem to its Sanitary Committee, which reported in September, 1877, that the intolerable conditions around the site of the Observatory continued unabated, and recommended action by Congress. The committee's report included a letter from engineer and Quartermaster General M. C. Meigs urging that the high ground around the Observatory should be spread over the surface to cover the mud of the flats, and further recommending "that the U. S. Naval Observatory should be removed to the high grounds north of the city, and thus be lifted above the fogs and malaria, which now interfere with astronomical observations by obstructing vision and shattering the nerves of the observers."[5]

Within days, Rodgers had swung into action. "I found upon taking charge of the Observatory," he wrote the Secretary of the Navy, "that the malarious influences surrounding it were notorious, and that from May to about the middle of October the officers whose services were necessarily in the Observatory at night, paid the penalty in impaired health and in diminished efficiency. The fogs which arise from the river, driven by the prevailing winds, float above the instruments and lessen their useful-ness." Moreover, Rodgers argued, the present Observatory was in a "very dilapidated condition" requiring almost $30,000 in repairs, an expense that might be avoided by the removal to a new site. For these reasons Rodgers recommended "that a suitable site, north of the city and inside the District of Columbia, be procured for a new Observatory." The area allotted "need not necessarily be more than twenty-five or thirty acres in extent; but as much as this is needed, since, if surrounded by dwellings or factories, the smoke would obscure the clearness of vision, the traffic would shake the instruments, and some high structure, if placed upon the meridian near our instruments, might hide a useful part of the heavens," Rodgers wrote. For this pro-curement, and for plans and specifications (but not actual construction) for a new Observatory, Rodgers asked $100,000.[6] Considering that Congress had appropriated

---

[4] An article in the *Journal of Commerce* for January 16, 1878 (USNOA, AF, "Newspaper articles on removal of USNO, 1878–1880") says that Rodgers had determined to move the USNO almost immediately upon entering his duties in May, and that others had memorialized Congress, including most recently Robert C. Winthrop of the Academy of Arts and Sciences, Boston.

[5] Letter from Quartermaster-General M.C. Meigs to the Sanitary Committee of the Board of Health of the District of Columbia, August 16, 1877, in Rodgers, *Reports* (ref. 1), pp. 11–12. This was reported in the *National Republican* newspaper, September 6, 1877, as part of the coverage of the Board of Health report. USNOA, AF, folder marked "Newspaper articles on removal of USNO, 1878–1880."

[6] John Rodgers to SecNav R. W. Thompson, September 15, 1877, in Rodgers, "Report of the Superintendent on the Removal of the Naval Observatory," in *Reports* (ref. 1), p. 3.

$170,000 a few years earlier for the transit-of-Venus expeditions, such a request on a matter affecting the future of the Observatory was not unreasonable.

Both from the medical profession and from his staff Rodgers sought written testimony to the twin evils of malaria and fog at the Observatory. From two medical Directors of the Navy who had served most of the period between 1855 and 1873, from the Navy Medical Inspector, and from three doctors in the area Rodgers obtained ample support of the reality of the malaria problem. One Medical Director forwarded the death certificate of Gilliss, which noted the "insalubrity" of the Naval Observatory's location; the other noted frequent cases of malaria among the Observatory staff, caused by "miasmata" from the adjacent marshes, and suggested that the removal of the Observatory would allow city authorities to carry out the previously suggested improvements to the area. Another of the doctors testified that a principal cause of Admiral Davis's death was his residence at the Observatory.[7]

Among the Observatory staff, astronomer Mordecai Yarnall especially gave poignant testimony of the impact on individuals of the current site. Maury's family, he noted, had been sick almost every summer; Captain Gilliss had to use quinine frequently; Admiral Sands removed himself and his family to the country, but was "reduced to the edge of the grave" when he stayed a part of one summer; Admiral Davis and his wife both suffered from malaria; Rev Moses Springer died at an early age from causes directly traced to living near the Observatory; Professor Hubbard slept at the Observatory and "perished a martyr to malaria;" and Ferguson died of the same cause. "We now cease to sleep at the Observatory . . . at the expense of our efficiency," Yarnall concluded. Hall, Eastman, Harkness, and Holden gave similar opinions on the foggy conditions, and several referred to the better conditions in "the hills to the north of the city." Newcomb – already the crafty pleader – drew attention to the $440,000 cost of the Russian observatory at Pulkovo and the similar scale of one currently being built in Austria. He estimated that the working capacity of the observers would increase by a third if they could work and sleep near the Observatory.[8]

By way of additional justification for his request, Rodgers included a report on "the work of the observatory and its appreciation abroad," in which he clearly emphasized its practical value. Comparing the Naval Observatory to the Royal Observatory at Greenwich, Rodgers pointed to the origins of the American observatory in navigation and the fact that "the larger part of its work is in fields of immediate practical value." All longitudes in the country are "determined by means of or referred to this Observatory," he emphasized, the chronometers are kept and rated, a time ball is dropped both at the Observatory and via telegraph in New York, and essential aid is given to those responsible for the *American Ephemeris and Nautical Almanac*. Although Rodgers remarked that the search for new objects had never been a part of the regular work of the Observatory "because it has generally been felt that an institution sup-

---

[7] Rodgers, *Reports* (ref. 1), pp. 6–7.
[8] Ibid., pp. 7–10. The manuscript letters of some of these are in USNOA, AF.

ported at the expense of the nation should confine its energies to fields known to be remunerative," he nevertheless bragged about Ferguson's previous discovery of new asteroids and Hall's recent discovery of the moons of Mars. In fact, the moons of Mars put the Naval Observatory into the minds of Congress and the public as never before, and the politically astute Rodgers undoubtedly saw an opportunity to take advantage of the publicity. Finally, as evidence of its international reputation, Rodgers noted that the work of the Naval Observatory, as reviewed in the German Astronomical Society's *Vierteljahreschrift*, showed it second only to Pulkovo observatory. The Naval Observatory amply returns the sums expended on it, Rodgers concluded, and should not be hampered by fog and malaria.[9]

The eventual results of Rodgers's persistent efforts are simply stated: in 1881 the Navy purchased a new site, throughout the period between 1886 and 1891 Congress appropriated funds for new buildings and expenses related to moving, and in 1893 the Observatory occupied its new site.

However, the 15 years between concept and completion were filled with politics and intrigue, commission recommendations and rejections, lofty questions about the role of government in science, and petty concerns. Moreover, the request for a new observatory site became embroiled in the question of whether the Observatory could not be improved "by putting it under a scientific head, and by creating it a national observatory under some other department of the government rather than under that of the Navy, as at present." Writing to Secretary of the Navy Thompson on November 26, 1877, Rodgers argued strongly against civilian control, much less taking the institution away from the Navy. Professors of Mathematics Yarnall, Hall, Harkness, Nourse and Eastman concurred in keeping the Observatory as it was, to the extent that they all signed a single letter to Rodgers on the subject. However, Newcomb, who had been Superintendent of the Nautical Almanac Office for less than three months, and his protégé Holden, parted company with their colleagues on this subject; both felt the institution should remain under the Navy, but with a civilian head.[10] It was only the beginning of the controversy.

Meanwhile, Rodgers had set the necessary wheels in motion for the new observatory. In January of 1878 a bill was introduced in the Senate calling for the removal of the Naval Observatory to a still undetermined site. The bill, known as S. 493, was referred to the Committee on Naval Affairs, which submitted a detailed report clearly based on Rodgers's previous reports to the Secretary of the Navy, and recommended

---

[9] Ibid., pp. 12–15. More than half of the work cited by way of comparison with Pulkovo was due to Newcomb alone!

[10] Ibid., Rodgers to SecNav Thompson, November 26, 1877, and letters from the Professors of Mathematics to Rodgers, November 24, 1877, pp. 16–20. A one-sided retrospective look at the civilian controversy at the time of Rodgers's appointment is given in a report accompanying a bill to change the management of the Observatory in 1892; see "Naval Observatory," House Report No. 926 (Adverse Report of March 30, 1892 to accompany H.R. 3996), 52nd Congress, first session, pp. 2–3.

passage of the bill.[11] For undoubtedly political reasons, the thrust of the bill was changed to appoint a commission to ascertain the cost of removing the Observatory. This bill passed the Senate on January 24; significantly, nothing was said about a reorganization in the bill or the report. However, the reorganization issue had not died, and the bill still had to navigate the House side of Congress. On the same day that the Senate bill passed, Yarnall, Hall, Harkness, Nourse, Eastman, and Holden sent a letter to the National Academy of Sciences, objecting to a petition being circulated among members of the Academy and addressed to Congress. The writers of the petition held it a "grave error" to use public funds for a new National Observatory "without assigning to professional astronomers any part in designing or directing it." The Academy petition recommended a scientific commission to determine what direction a new Naval Observatory should take, as well as the ablest superintendent (Navy or civilian) and a board of visitors to periodically examine the institution. Yarnall and his colleagues urged Academy members not to sign the petition, however much they might approve of certain parts of it. A week later Rodgers himself, at the urging of the Professors, entered the fray "with reluctance," arguing that the Naval Observatory was a creditable national institution in which the astronomers already had a say.[12]

An even more fundamental question was raised in the Senate in February: What was the benefit of an observatory to the government? This inspired Rodgers's "Report on the Usefulness of Government Observatories," an expansion of his earlier report on the work of the Observatory.[13] Again it emphasized practicalities of navigation: Through the *Nautical Almanac* and the lessening of ship insurance, "astronomy enters into the price of every pound of sugar, every cup of coffee, every spoonful of tea." Rodgers also distinguished between private and government observatories: "The government observatory works on old themes; the private observatory devises new ones. Both are necessary in the world, and neither can be spared. Very fortunately, only one claims government aid." How much astronomy would change in the next century!

In late March of 1878 the House considered its own version of the bill, H. R.

---

[11] "A Bill to provide for the Removal of the Naval Observatory," printed as S. 493, 45th Congress, second session, read twice by Mr Sargent and referred to Committee on Naval Affairs, January 16, 1878. The Committee on Naval Affairs produced Report No. 33 to accompany S. 493, and recommended passage. USNOA, AF, folder marked "New Naval Observatory, Congressional Bills and Reports."

[12] The modified S. 493, "A Bill to appoint a commission to ascertain the cost of removing the Naval Observatory," is in USNOA, AF, "New Naval Observatory, Congressional Bills and Reports" folder. On the Academy incident see Yarnall *et al.*, "To the Members of the National Academy of Sciences," printed letter dated January 24, 1878, USNOA, AF. This letter reprints the three paragraphs of the original Academy circular, and objects succinctly in a single paragraph. Rodgers's printed letter dated January 30, 1878, located in the USNO Library, responds in more detail sentence-by-sentence. At the same time this was happening, a rival joint resolution was introduced in the Senate for a commission of scientists to investigate the establishment and location of a national observatory at high altitude on the Western plains, Senate Resolution 16, 45th Congress, second session, and Senate Miscellaneous document No. 25.

[13] John Rodgers, *Report on the Usefulness of Government Observatories* (Government Printing Office: Washington, 1877). That this document was transmitted to the Senate on the same day as their request (February 15, 1878) indicates that Rodgers had prepared it already as a result of questions raised at the end of 1877.

4129, to appoint a commission. The subsequent Act of Congress, approved by both houses and signed by the President on June 20, 1878, provided for a commission of three to invite sealed proposals or offers of sale of an appropriate site; to report on the propriety and expediency of disposing of the old Observatory grounds and buildings, and to examine any land in the District not offered for sale by its owners.[14] The Commission on Site, appointed by the President, chaired by Rear Admiral Daniel Ammen, and including Brevet Major General J. G. Barnard and Leonard Whitney, first met at the Navy Department July 15. It drew up an advertisement that ran weekly in the local papers over 30 days, soliciting proposals for a site from landowners. The solicitation ended on August 28, 1878, by which time some 79 bids had been received. At noon of that date, "in the presence of a large number of parties interested," the sealed bids were opened at the Navy Department. Over the next six weeks Ammen and the other local resident of the Commission examined all the sites. On October 14 the Commissioners, along with Admiral Rodgers and Asaph Hall, revisited all the eligible sites. Their choice was a site known as Clifton, a 45-acre site owned by "Mr. Elverson" of Philadelphia, who offered the land for sale at $667 per acre, for a total of $30,015.[15]

The Commission had to report also on the cost of removal of the Observatory, which it took to mean requiring new construction "on new and improved plans." For this Rodgers was ready with a plan of the new Observatory, as a result of consulting "the executive officers of the Observatory," and eminent astronomers in the U. S., who had been consulted by circular letter. Although the Commission had not been given authority to propose a plan for the Observatory, some provisional plan was necessary in order to ensure that analyses of possible locations were adequate. In carrying out this duty, the final report of the Commission, dated December 7, 1878, proposed "that in constructing these buildings, a National Observatory should be provided, which, while satisfying the practical astronomical exigencies of the military and commercial marine of the United States, shall also meet the higher and more universal demands of science, by equality in all its material means with other great national observatories."[16] The total cost, including $30,000 for the new land, $85,000 for the Superintendent and Professor's quarters, and $161,000 for the Observatory building, was $350,000. The Commissioners recommended that the old site not be disposed of due to depressed real-estate values, and that instead it be graded and the dirt used to fill in those areas along the original plans of the Board of Survey in 1872.

---

[14] "A Bill to appoint a Commission to ascertain the cost of removing the Naval Observatory," March 29, 1878. Mr Benjamin Harris from the Committee on Naval Affairs (CNA) reported H.R. 4129 as a substitute for S. 493, which was read twice and recommitted to the CNA. The CNA produced House Report No. 454, which details the differences from S. 493. The final Act of Congress, approved June 20, 1878, is U. S. Statutes at Large, volume 20, p. 241, reprinted in Ammen et al., Report of the Commission on Site for the Naval Observatory (Government Printing Office: Washington, 1879), p. 3.

[15] Ammen et al., Report (ref. 14), pp. 3–6. Details on the 79 offers of land are given in Appendix B of this report. Other appendices give a general description and specifications for the proposed Observatory building.    [16] Ammen et al., Report, p. 6.

So politically charged was the site selection process that Congress ignored the Ammen report, ostensibly for reasons of inaccessibility of the sites, and so its recommendation for Clifton fell stillborn. However, the Observatory's Professors of Mathematics also had qualms about the Clifton site. An undated and unsigned letter in the Naval Observatory Archives indicates that the Ammen Commission had chosen Clifton because it was less expensive than the others, even though not as good for astronomical purposes in terms of room for buildings and commanding a view of the horizon. Pending Congressional approval of Clifton, all five Professors visited the superior sites; three agreed with the Commission's choice on the basis of the issue of funding, and two argued that the Congressional mandate was only to find the best site, which they considered to be the adjoining Barber estate, which was also offered for sale to the Ammen Commission. An anonymous letter writer (probably Rodgers) wrote that he "should consider it a great misfortune to astronomical science in this country to place the only government observatory at Clifton when its southern horizon could be cut off by the erection of ordinary dwellings and a clear view obstructed by the smoke and heat from the chimneys."[17]

The Barber estate now being the favorite, undoubtedly because of the lobbying of Rodgers, on February 4, 1880 Congress voted to have a second commission to consider and resolve site selection, and also included $75,000 for site purchase.[18] This Congressional Commission, composed of Senator William P. Whyte (Maryland), Representative Leopold Morse (Boston), and Superintendent Rodgers, was empowered to make the final decision. According to contemporary newspaper reports, by April the Observatory Professors still oscillated between the initial majority favoring Clifton, and the Barber estate where there would be ample space for their residences; they were accused of meddling in the work of the Commission.[19] Unable to decide

---

[17] USNOA, AF, folder marked "Removal of U. S. Naval Observatory to Present Site, II."

[18] "An Act to locate and purchase a new site for the United States Naval Observatory," Approved February 4, 1880. U. S. *Statutes at Large*, volume 21, pp. 64–65, printed in RSN, November 28, 1881 (Washington, 1881), p. 738.

[19] "The Proposed Naval Observatory," newspaper article of unknown origin, dated April 17, 1880, USNOA, AF, folder marked "Newspaper articles on Removal of USNO, 1877–1880." On May 26, 1880 the *New York Herald* (USNOA, ibid.) reported friction between the two Congressman on the one hand and Rodgers on the other as to choice of site and architect. Senator Whyte and Congressman Morse wanted to accept a site near the Soldiers' Home, which suffered from tremors, and were again threatening to outvote Rodgers. Responding to this charge for *The Evening Star* of Washington of May 29, 1880, Whyte admitted that he and Morse had views opposed to Rodgers. The congressman wanted the new Observatory near the city (one of the sites near the Soldiers Home), so that "members of Congress and visitors at Washington may enjoy going there and looking through a telescope. The people must pay for the observatory, and we want it where it will do them some good." The Professors and others, Whyte complained, wanted the Observatory "in some obscure spot way out the other side of Georgetown, because it will be convenient to their residences," and were making excuses about tremors from nearby roads in order to reject his choice. Whyte did not accept the idea that a wagon would shake a large building, but was willing to see more evidence. In correspondence with Rodgers dated June 29, 1880, Morse remarked that the Barber site "is most favored by the Professors." USNOA, AF, folder marked "Removal of U. S. Naval Observatory to Present Site."

among themselves, in January, 1881 the Commission delegated to a blue-ribbon and technical subcommission composed of F. A. P. Barnard (President of Columbia College in New York), Henry A. Rowland (Professor of Physics at Johns Hopkins), and Charles S. Hastings (Assistant Professor of Physics at Johns Hopkins University) the task of examining four sites that the Commission had selected.[20] These four properties, including the Barber, Middleton, Whitney, and Petworth sites, were to be examined with a view toward any objections as to vibrations, obstructions, proposed improvements, and difficulty of access. The Commission also desired to know what effect an already proposed railway down Rock Creek Valley might have on the sites.

Although these were near the Clifton site, the first two just north of Georgetown and the latter two near the Soldier Home, Clifton was now out of the running. Citing the need for a steady atmosphere, the Subcommission agreed to avoid factories, clustered dwellings, large areas without vegetation, and highways. The latter especially were objectionable due to tremors. "National observatories have not been founded by governments simply out of the generous desire to promote the advancement of science," the Subcommission members wrote. "The notion which has prompted their erection has been far less disinterested than this, and is found in the encouragement such institutions afford to commerce by the security with which they surround navigation. It is probable that no appropriations from the public treasury of any commercial nation have ever been repaid so many thousand fold as those which been made for the improvement of our knowledge of the positions and movements of the heavenly bodies." Long periods of patient observation are needed in order to secure these benefits, the Commission emphasized, and so seclusion is an advantage rather than an objection.[21]

Carrying out their duties during a deep snow, a fact that, as they noted, impaired geological investigations, the Subcommission members made the rounds of each site. Crucial to their final decision were the tests for tremors made at each site by a team of observers headed by assistant astronomer H. M. Paul. One by one the alternatives were eliminated. The Middleton place was too subject to tremors from wagons on the nearby road, and in any case had not enough room at the summit for the buildings; the Whitney place was also subject to road tremors from traffic; and Petworth

---

[20] "Memo Addressed by Senator Whyte to Professors Barnard and Rowland and Dr. Hastings, giving the considerations which are to govern the selection of a new site for the Naval Observatory," USNOA, AF. The memo itself is undated, but a letter from Senator Whyte "in the Senate chambers" to Admiral Rodgers dated January 10, 1881 notes that Whyte forwarded the memo to Daniel Gilman, President of Johns Hopkins University, and Gilman replied on January 8 that the two Hopkins Professors would meet in Washington with President Barnard on the following Tuesday. Several "officers of instruction and government" at Johns Hopkins had written Whyte urging him to consult scientists in the site selection, and this probably was the origin of the Hopkins'consultants.

[21] F. A. P. Barnard, Henry A. Rowland, and Charles S. Hastings, "Results of the Examination of Certain Sites Suggested as Suitable for The Naval Observatory, made at the request of The Commission on the Observatory," RSN, November 28, 1881 (Washington, 1881), pp. 738–745.

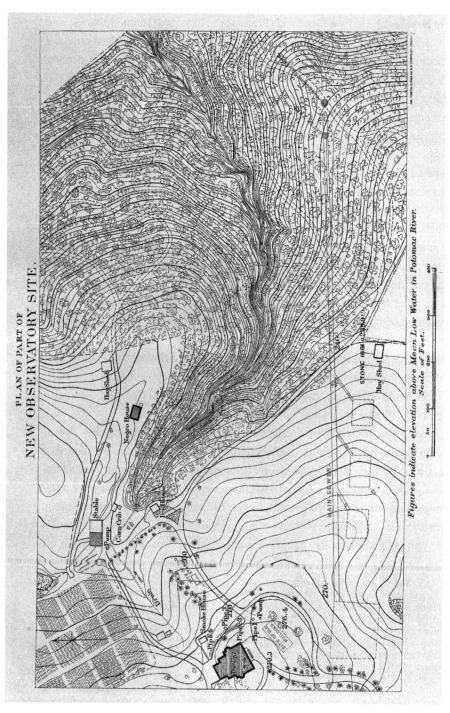

Figure 9.1. This map, based on a survey of the Barber site by the U. S. Coast and Geodetic Survey in 1881, includes a part of the new Naval Observatory site, showing original structures on the Barber estate. From the *Annual Report* (1882), 134.

was liable to tremors caused by the trains on the Baltimore & Ohio railroad tracks. In the end the Subcommission found least objection to the Barber estate, and thus was the location for the new Observatory finally settled.

The Barber estate was a property purchased by Cornelius and Margaret Barber in 1834, and part of a tract of land called Pretty Prospect. Northview, as the Barbers called it, was now run by the widowed Margaret Barber, who lived there with 15 slaves and 20 bondsmen and their children. Its 70-acre topography, which can be seen in a contemporary map (Figure 9.1), was very favorable for the construction of an observatory, with a clear horizon in all directions. The Subcommission was especially impressed that "it is remote from any public road, being approachable only by a private carriage path, so that it enjoys the inestimable advantages of seclusion, quiet, and freedom from disturbance either by frequent visitors or by passing vehicles." Surrounding the site were deep ravines, which "are likely long and perhaps forever to prevent its being hemmed in by buildings, whatever may be the growth of the city." From the present conditions, then, the Subcommission found no objections to the site. It did, however, worry about the possibility that a branch of the Baltimore & Ohio Railroad might be brought down Rock Creek valley. However, because the Commissioners did not think the tortuous path of Rock Creek was a likely railroad route, especially since Congress would have to approve such a route, they did not believe this to be a serious objection. The Commissioners accepted the report of their Subcommission, and, on March 8, 1881, the Barber estate was purchased for $63,000, more than twice the price for which the Clifton site could have been purchased, but well within the $75,000 allowed. The estimated cost for the new Observatory and all associated expenses was $586,138; an 1881 map (Figure 9.2) shows the proposed buildings. Another contemporary map shows the locations for the old and new sites, separated by only about 2 miles (Figure 9.3).[22]

John Rodgers died on May 5, 1882. He had carried the Observatory far, but his untimely death deprived the new Observatory of its most influential advocate. With the site purchased but no appropriation for buildings, the movement for a new Observatory came to a standstill. Although the Secretary of the Navy requested appropriations for the new Observatory in 1883 and 1884, there the matter rested.[23]

---

[22] Barnard, Rowland, and Hastings, ibid., pp. 13 and 737. Paul's report on tremors at the site exists only in manuscript, USNOA, AF, folder marked "Removal of U. S. Naval Observatory to Present Site." For more background on the Barbers and their estate, see Gail S. Cleere, The House on Observatory Hill (Government Printing Office: Washington, 1989), pp. 2–6.

[23] One of Rodgers's successors wrote that Rodgers's death "caused a cessation of active efforts to obtain the necessary appropriation." George Belknap to F. A. P. Barnard, in Barnard report (ref. 21), p. 25. In the Annual Report (AR) for 1883, pp. 268–269, Chief of the Bureau of Navigation J. G. Walker endorsed Superintendent R. W. Shufeldt's request for $586,138 for the new Observatory, and forwarded it to Secretary of the Navy William Chandler, who recommended approval, but to no avail (RSN (1883), 25). The following year (AR (1884), 4) Shufeldt's successor S. R. Franklin resorted to requesting at least a portion of the funds, and, although the same Secretary of the Navy recommended approval of the funds in his Report to the President dated December 1, 1884, it was again to no avail, RSN (1884), 29.

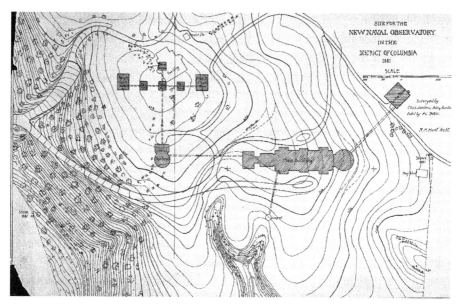

Figure 9.2. The site for the New Naval Observatory, showing the proposed layout of buildings according to the plan in which they were eventually constructed. The residence for the Barber estate is shown in outline, marked "Dwelling" both here and in Figure 9.1.

Meanwhile, larger events in organization of Federal science were taking place. In 1884 a joint Commission of Congress, chaired by Senator W. B. Allison of Iowa, was established to "secure greater efficiency and economy of administration" of the Geological Survey, the Coast and Geodetic Survey, the Army's Signal Service, and the Navy's Hydrographic Office. Among the suggestions, a committee of the National Academy of Sciences proposed to the Commission that most science activities be placed under a Department of Science. Geological Survey Director John Wesley Powell suggested that instead the scientific bureaus be placed under the Smithsonian Institution. The Naval Observatory became peripherally involved in these hearings when Powell suggested removing the Observatory from the Navy, consolidating it with the Coast and Geodetic Survey in the Treasury Department, and transferring to the Navy the hydrographic portion of the Coast Survey's work in return. The Allison Commission, however, took no action on either the Observatory transfer or any of the other proposals, thus lending tacit agreement to the *status quo*.[24]

[24] A. Hunter Dupree, "The Allison Commission and the Department of Science, 1884–1886," chapter 11 of *Science in the Federal Government* (Harvard University Press: Cambridge, Massachusetts, 1957), pp. 215–231. The Allison report is "Testimony before the Joint Commission to Consider the Present Organizations of the Signal Service, Geological Survey, Coast and Geodetic Survey, and the Hydrographic Office of the Navy Department, with a view to secure greater efficiency and economy of administration of the public service in said bureaus, authorized by the Sundry Civil Act approved July 7, 1884, and continued by the Sundry Civil Act approved March 3, 1885," 49th Congress, first session, Senate Miscellaneous Document 82. This 1886 Senate document has been reprinted by Arno Press (New York, 1980). The Navy Department did not allow Newcomb to serve on the National Academy committee because of a possible conflict of interest.

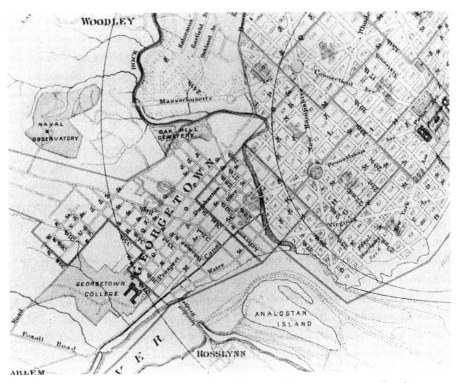

Figure 9.3.  A map showing the site of the Naval Observatory at 23rd and E. Streets (bottom right), and the new site (top left). The distance between the two sites is about two miles. The "President's House" is on the far right. Massachusetts Avenue has not yet been extended.

The process for removal of the Naval Observatory was restarted in an unexpected way on April 22, 1885. On that date Secretary of the Navy William C. Whitney sent a letter to the President of the National Academy of Sciences asking for advice on changing the time of the beginning of the astronomical day. This had been one of the issues recommended by the International Meridian Conference in October 1884; it was now coming before Congress and Whitney was acting in response to a Senate resolution on the subject dated February 2. Whitney took the opportunity to request the National Academy of Sciences to consider not only this question, and the advisability of Congressional funding for the solar eclipse of August 1886, but also "as to the advisability of proceeding promptly with the erection of a new Naval Observatory upon the site purchased in 1880" (actually 1881).[25] The President of the Academy responded by naming a distinguished Committee of the National Academy chaired by F. A. P.

<hr />

[25] Whitney to O. C. Marsh, President of National Academy, April 22, 1885, in "Letter from the Secretary of the Navy, transmitting in compliance with Senate resolution, February 2, 1886, [F. A. P. Barnard's] report of National Academy of Sciences upon the proposed new Naval Observatory," Senate Executive Document No. 67, 49th Congress, first session, sent by SecNav Whitney to the Senate February 9, 1886 and referred to the Committee on Naval Affairs on February 10, pp. 20–21.

Barnard, the President of Columbia College, who had also been a member of the Subcommission on site selection. Other members included Alexander Graham Bell (a friend of Newcomb and famous for his newly invented telephone), geologist J. D. Dana of Yale, solar astronomer S. P. Langley at Allegheny Observatory in Pennsylvania, scientist and Congressman Theodore Lyman of Massachusetts (a guiding force on the Allison Commission), astronomer E. C. Pickering at Harvard, and astronomer C. A. Young at Princeton.[26]

The work of this committee quickly became embroiled in the question of the identity of the Naval Observatory. The advisability of proceeding with a new Observatory "must very much depend upon the functions which it is desired or designed that the Observatory shall in the future fulfill," Barnard wrote the Secretary of the Navy. Although the Naval Observatory began with purely practical duties, he continued, it now contributes as much to the advancement of astronomy as to its practical services to the Navy, "and indeed, in the view of the Academy, much more." Otherwise, he pointed out, Congress would not have appropriated funds for a 26-inch telescope. If the Observatory were to conduct both practical and pure astronomy in the future, Barnard wrote, then the decision to proceed with the construction would be simpler than if it were only to undertake practical astronomy. However, this raised the further related questions regarding "the organization of the working astronomical corps, and the direction of the operations of the Observatory." Although the Observatory had been served admirably by competent Navy officers with Maury, Gilliss, Davis, and Rodgers, Barnard argued, in the future it was possible that "the service should not be able to furnish a man whose qualifications to act as director of the Observatory should be fully recognized in the scientific world." In that case the Secretary of the Navy should have the option to choose a military or civilian director as he sees fit. This was no half-baked idea; rather, on this provision hinged the committee's decision: "supposing that provision secured, there can be no doubt that the committee and the entire Academy will, with one voice, advise the erection of the proposed new building. In the absence of such a provision I think they would prefer to be excused from offering any advice."[27]

Barnard's letter went unanswered, and only at the beginning of November, 1885, did Superintendent George Belknap forward further information about the Naval Observatory. Clearly chafing at a series of nine questions posed by Barnard, Commodore Belknap argued that "It is first of all a naval institution, its astronomical work being, so far as the naval service proper is concerned, of a purely secondary consideration. Its officers, with the exception of three observers and one computer, are commissioned in the Navy, and its work and reputation are the property of the service." After explaining why this was so, Belknap concluded "It is the creation of the line of the Navy, its founder having been Gilliss. If the time has come when the purely

[26] O. C. Marsh to F. A. P. Barnard, April 28, 1885, in *ibid.*, p. 21.
[27] F. A. P. Barnard to SecNav Whitney, June 11, 1885, in *ibid.*, pp. 21–22.

scientific side of the institution has outgrown the needs of the naval service, the converse is true, namely, that the Navy has no need of it, or of the scientific staff. If the so-called scientific men of the country think that the time has come to apply to Congress for money to build a national observatory, the Navy will not stand in its way; only it will take no responsibility for it, and will be glad to see it go to another Department of the Government, and to be under purely civilian control, including professors with civilian appointments instead of naval commissions."[28] The battle lines were clearly drawn.

By November 30, 1885, the National Academy Committee had submitted its report, which recommended that the beginning of the astronomical day be changed from noon to midnight, that there be no funding for the eclipse due to the lack of time to prepare, and that the Observatory be moved, but under a civilian head: "Let all those instruments in the present observatory which can be of service to the Navy, with so many of the astronomical staff as may be needed to use them, be transferred to Annapolis, and let the observatory of the Naval Academy, strengthened by these accessions, be styled the Naval Observatory. Let the remaining instruments, which probably will be the 26-inch telescope, the transit circle, and the prime vertical transit, be reserved to be installed in a new edifice to be erected upon the site purchased in 1880 [sic] . . . and let this new structure be styled the National Observatory of the United States."[29]

On February 10, 1886, the Navy transmitted the National Academy report to Congress, where it was referred to the Committee on Naval Affairs. That the committee did not accept any of the Academy's recommendations is evident in the Congressional appropriation made during this session for construction of the new Observatory at the Barber site. The Naval Appropriations Act of July 26, 1886 provided $50,000 "for commencing the erection of the new Naval Observatory," with the proviso "that the construction of no building shall be commenced except an observatory proper, with necessary offices for observers and computers." By October the Superintendent reported "the plans are undergoing a final revision, and steps are being taken to carry them into execution." The Naval Appropriations Act of March 3, 1887, provided a further $60,000 (including $10,000 for a new meridian circle), and specified that the total cost of the new Observatory should not exceed $400,000. This limit was reached with the further appropriations of $50,000 for continuing, and $240,000 for completing the Observatory, as provideed in the Naval Appropriations Acts of September 7, 1888 and March 2, 1889 (see Appendix 5).[30]

---

[28] Belknap to Barnard, November 10, 1885, ibid., pp. 27–28.
[29] Report of the National Academy of Sciences upon the proposed new Naval Observatory, 19, in Letter (ref. 25). Despite the recommendation on the astronomical day, which Newcomb opposed, the beginning of the astronomical day was not changed to midnight until 1925. See chapter 11, ref. 8.
[30] These provisions of the Naval Appropriations Acts are to be found in *Statutes at Large*, volume 24, pp. 156 and 585; *Statutes at Large*, volume 25, pp. 463 and 814. See also AR (1886), 12.

Having secured the appropriations, the question now was how to proceed with construction and what instruments were to be purchased. In expectation of this need, already in 1883 Simon Newcomb had made a trip to Europe for just this reason. Newcomb, now well entrenched as Superintendent of the Nautical Almanac Office and still agitating for a civilian to head the new Observatory, visited the observatories of Paris, Neuchâtel, Geneva, Vienna, Berlin, Potsdam, Leyden, and Strassburg, as well as the workshop of Repsold in Hamburg. His subsequent report provided details not only on the telescopes, transit circles, and clocks of European observatories, but also on their buildings. Although the Secretary of the Navy duly transmitted Newcomb's report to President Chester Arthur, and although the President commended it to the consideration of Congress, the report seems to have had little practical effect on the new Naval Observatory.[31]

That a need was still felt for more information in the wake of the 1886 appropriation for the new Naval Observatory is evident in the fact that, when, in March, 1887, Lt Albert G. Winterhalter was sent as the Naval Observatory representative to the International Astrophotographic Congress in Paris, he was also sent on an extensive visit to European observatories. His itinerary included England, France, Germany, Austria, Holland, and Italy. Since the new Observatory would clearly become a reality, Winterhalter's mission was now more urgent, and the instructions more specific, than Newcomb's. Superintendent R. L. Phythian particularly instructed him to "ascertain as much as possible concerning the recently constructed dome of the Nice Observatory – its cost and its adaptability to the purpose for which it was designed. You will also ascertain what would be the price for complete plans and specifications of such a dome, to be 41 feet in diameter. You will forward to me a written report concerning this at the earliest practicable moment, and also embrace in it any other items of information you may think it desirable for the Observatory to possess in connection with large domes."[32] Beyond that Winterhalter saw it as his duty "to consider especially the building and equipment of observatories, taking cognizance of modern improvements, and bearing in mind the necessities of the New Naval Observatory now in process of erection."[33] Aside from any impact of Winterhalter's trip on the new Naval Observatory, and the interesting account of the Astrophotographic Congress, his detailed report in 1887 stands as a perceptive,

---

[31] Newcomb's full report is found in "Message from the President of the United States, Transmitting a Communication from the Secretary of the Navy, including a report on recent improvements in astronomical instruments," Senate Executive Document No. 96, 48th Congress, first session, pp. 1–20, in *The Executive Documents printed by order of the Senate of the United States for the First Session of the Forth-Eighth Congress* (Government Printing Office: Washington, 1884).

[32] Albert G. Winterhalter, *The International Astrophotographic Congress and A Visit to Certain European Observatories and Other Institutions* (Government Printing Office: Washington, 1889), also Appendix I to *WO for 1885*, p. 327. On June 14 Phythian received Winterhalter's report and a printed pamphlet on "The Great Dome at Nice," followed by three more reports on the same subject on June 29; pp. 285–295. Winterhalter's general report on his visit to Mr Bischoffsheim's Observatory on Mont-Gros, Nice, is on pp. 75–95.      [33] *Ibid.*, p. 73.

Figure 9.4. Richard Morris Hunt, the "dean of American architects" in the late nineteenth century, was the architect of buildings for the new Naval Observatory.

detailed, and well-illustrated description of astronomical observatories and instruments at the time.

Of even more immediate concern was the choice of an architect for the Observatory buildings. Although in its 1881 report the Commission on Site had accepted the plans of architect C. A. Didden, in the end Richard Morris Hunt (1827–95), who has been called "a key figure of America's Gilded Age," was chosen as architect for the new Naval Observatory buildings. Hunt (Figure 9.4), widely recognized as the dean of American architects in the late nineteenth century, designed many mansions for his wealthy clients, as well as commercial and public buildings. Only three years before his Naval Observatory commission, the 89-foot pedestal he had designed for the Statue of Liberty was erected in New York harbor. "During his lifetime," his biographer has written, "he was one of the most widely honored of nineteenth-century Americans, both at home and abroad."[34] Aside from Hunt's secondary role in the extension of the Capitol, the Naval observatory was his first work for the federal government, followed in 1889 by his design of several buildings at the United

[34] Paul R. Baker, *Richard Morris Hunt* (MIT Press: Cambridge, Massachusetts, 1980), p. xiv; on the Naval Observatory pp. 380–381. The debate as to choice of architect went back at least to 1880, as reported in the *New York Herald* for May 26, 1880 (ref. 19). The Congressmen wanted architects from their own districts, whereas Rodgers wanted one of some reputation.

Figure 9.5. An artist's view of the physical layout of the Observatory, made in 1894 shortly after its construction had been completed. The view is toward the north, and features the main building (1), the Superintendent's residence behind it, the cluster of transit-circle buildings and "clock house" and the dome for the 26-inch refractor on the left. The clock house marked the exact center of Observatory Circle, just then being demarcated by act of Congress, and defined by a 1,000-foot radius from the center.

States Military Academy. Before his death in 1895, Hunt would design the Administration Building of the World Columbian Exposition in Chicago (1893), the last major achievement of his distinguished career.

By October, 1887, Superintendent Phythian reported that Hunt's plans were "nearly completed" and would soon be submitted for approval to the Navy. It was, however, October 2, 1888 before the Navy Department awarded a contract to P. H. McLaughlin & Company for the erection of nine buildings at the new site, stipulating that the work should be completed by April 2, 1890. A view of the physical layout of the nine buildings can be seen in Figure 9.5. The facility was dominated by the main building and the giant dome of the 26-inch telescope, but also included the cluster of five structures associated with the transit-circle buildings and the central clock house, the prime-vertical building (nearby to the north), and the boiler house. Although work

Figure 9.6.  The main building under construction, April 5, 1890. The view is toward the northeast.

began immediately, little did anyone know that almost five years would pass before any of the buildings would be occupied. By September, 1890 the Superintendent stated in his *Annual Report* that the walls of the main building were nearly finished, the roofing ready to put in place, and the building scheduled to be under cover by November (Figure 9.6). The same report noted that the building for the 26-inch telescope was almost completed (Figure 9.7), the clock and observers' rooms already under cover, the transit-circle rooms under construction (Figure 9.8), and the boiler house completed. The Barber mansion was in the process of being demolished in order to make room for the construction of the prime-vertical building. However, then construction slowed, the contract was extended a number of times, and was forfeited September 8, 1891. A new contract was entered into on February 16, 1892, the government accepted the buildings on February 11, 1893, and the old Naval Observatory was abandoned as an observatory on May 15, 1893.[35] A view of the completed main building is shown in Figure 9.9.

Despite the European reports of Newcomb and Winterhalter, in the end the contract for the two equatorial-telescope domes went to Warner and Swasey, who proposed to build them for $20,000 on principles of construction identical to those of "domes we have built for many universities and Colleges throughout the country,

[35] AR (1887), 8; AR (1888), 7–8; AR (1890), 100; AR (1893), 3.

Figure 9.7. The building for the 26-inch telescope under construction, looking east.

Figure 9.8. The Transit House complex under construction with the Barber mansion at the center of what would become Observatory Circle, 1890. The east and west transit-telescope buildings are still under construction. Completed are the central "clock house," where the most sensitive clocks would be placed, and the observers' rooms. The Barber mansion was torn down soon after this photograph was taken.

*U.S. Naval Observatory, circa 1910*

Figure 9.9. A genteel era, about 1902, showing the main building with the dome for the 12-inch telescope and tennis on the front lawn. The Library is on the left. The one-legged man with crutches is Lt-Cdr E. E. Hayden, in charge of Time Service.

among which we may mention Johns Hopkins University, at Baltimore, and Smith College, at Northhampton, Massachusetts."[36]

The complexity of the overall undertaking of moving an observatory, quite aside from the construction of the buildings, was impressive. As Superintendent F. V. McNair reported a few months after the move, contracts had also to be made for enclosing the grounds; for grading, "macadamizing," and guttering the roads around the Observatory buildings; for bookcases, railing, and shelving; for transporting instruments, piers, and books; for repairing and remounting the 26-inch equatorial, meridian circle, and prime vertical, and also for a new 6-inch meridian circle; for furniture and gas or electric fixtures for the main building, as well as the equatorial, transit-telescope, clock, and magnetic buildings and the boiler house; for construction of a well tank, cistern, pump, boiler and pipes for water supply; for installation of electric plant; and for the standard clock and clock room fittings.[37] The Naval

---

[36] Warner and Swasey to R. L. Phythian, June 5, 1888, USNOA, AF.

[37] Naval Appropriation Act, March 2, 1891, U. S. *Statutes at Large*, volume 26, p. 806, provides the detailed cost breakdown. An indication of the progress of construction may be had from the specifications, beginning with the "Circular Relating to the Construction of a New Naval Observatory," including Advertisement of April 10, 1888 (Navy Department, 1888). In chronological order, specifications were issued for repairing and remounting instruments, 24 pages (May 15, 1891); for material for the Observatory (November 24, 1891); for engine pumps, pump house, etc., 24 pages (1891); for an extension of the boiler house for the electric plant (1891); for conduits (tunnels) between buildings (undated); for a passenger elevator and a small

Figure 9.10. The Superintendent's residence seen here *circa* 1895 before the roads were paved. It became the home of the Chief of Naval Operations in 1928, then beginning in the 1970s, the home of the Vice President of the United States. Vice President Walter Mondale was the first to actually live in the house.

Appropriations Act of March 2, 1891, provided another $136,689 for these activities, capping total appropriations of $611,000 since 1880, including $75,000 for purchase of land and $400,000 for construction of buildings.

The last appropriation included $20,000 for a residence for the Superintendent (Figure 9.10), which was designed by local architect Leon Dessez (1858–1919). Construction began in 1891 and was completed in 1893. Its first resident was Captain Frederick V. McNair, followed soon after by Commodore Robert L. Phythian. This building was destined to have a storied history, later becoming the residence for the Vice President of the United States.[38]

The site environment was an important concern, both for access and for

(footnote 37 cont.)
freight elevator (undated); for bookcases and other library fittings (undated); for carpets, linoleum, and office furniture (undated); for a brick residence for the Superintendent (undated); for installation of an electric lighting plant (1892); for construction of a meridian circle and its accessories (1893); and for two brick dwellings (1895 and 1896). All of these are bound in a single volume, "U. S. Naval Observatory Specifications, 1888–1896," in the USNO Library. In addition there is a Proposal for the purchase of Naval Observatory Lands, with map of plots to be sold, and Sale of USNO Lands, dated February 11, 1895.

[38] On the construction of this residence see Gail S. Cleere *The House on Observatory Hill* (Government Printing Office: Washington, 1989), pp. 12–17. McNair also has another distinction that has become a part of Observatory legend: a three-dimensional image of his face appears inside the pier of the 12-inch telescope.

appropriate observing conditions. Massachusetts Avenue not yet having been extended as far west as the Observatory, the main entrance was that of the old Barber estate off Tennallytown Road. However, a new entrance was already on the horizon when the Superintendent reported in late 1893 that "Observatory lane, the entrance to the Observatory grounds from the Tennallytown road, has been graded, macadamized, lighted, and a new entrance gate erected. Roads have been graded and macadamized through a large portion of the grounds, and the valley contiguous to the boiler house practically filled in; the grounds have been inclosed as authorized, and the paths partially completed. When Massachusetts avenue shall have been graded, the principal entrance to the Naval Observatory will be on that thoroughfare." It would be 1904 before the grading of Massachusetts Avenue was complete, and a road about 550 feet long constructed and paved leading to the buildings at the center of the circle.[39]

Even more important than better access was the converse problem: the protection of the Observatory from surrounding influences. A bill to establish an extensive circular perimeter as a buffer to guard against traffic vibrations was already submitted in March, 1894, and passed in August. Thus Observatory Circle came into being. It was surveyed over the next few years by the U. S. Coast & Geodetic Survey (Figure 9.11).[40] In 1901–02 Navy appropriations were used for the purchase of all of the lands lying within Observatory Circle and not included in the original purchase, with the exception of two plots whose owners refused to accept the government appraisal and appealed to the courts. By 1904 some land still needed to be acquired (Figure 9.12), and a plan was proposed for Congressional consideration. An aerial view taken in 1919 (Figure 9.13) shows some non-U. S.-government facilities, including the Industrial Home School, still lying within the embryonic Observatory Circle. Even with the circle formed, until World War II, no fence existed around the circle, and public access was unimpeded.[41]

[39] AR (1893), 4. The grading of Massachusetts Avenue was in progress in 1902 and finished in 1904. "When Massachusetts avenue becomes a thoroughfare the Observatory grounds will form one of the most beautiful parks in the city of Washington, second only to the Soldiers' Home in extent and in picturesque beauty." AR (1902), 8. On Observatory Lane, the original entrance for the Barber estate and the Observatory, see AR (1904), 22.

[40] "A Bill to establish an observatory circle as a provision for guarding the delicate astronomical instruments at the United States Naval Observatory against smoke or currents of heated air in their neighborhood and undue vibrations from traffic upon the extension of public thoroughfares in the vicinity, and for other purposes," S. 1769, 53rd Congress, second session, March 14, 1894. The bill was passed as a joint resolution of Congress (Public Resolution No. 36) August 1, 1894; USNOA, AF, "Observatory Circle" folder. On the survey of the circle see E. D. Preston, "Establishment of the United States Naval Observatory Circle, and the Determination of the Geographical Position of the Center of the Clock Room," *U. S. Coast and Geodetic Survey Report of 1896*, Appendix 6, pp. 285–291.

[41] AR (1901), 9; AR (1902), 7; AR (1903), 11; AR (1904), 17–22, including map. The two plots of land in question were those of Young and Normanstone. In addition the Observatory was attempting to purchase the Industrial Home School: "The school is not in itself detrimental or objectionable to the Observatory; but at the same time it would be highly advantageous to own the land on which it stands." AR (1902), 8. The completion of the circle was still an issue in AR (1914), 19–20, and AR (1915), 19. Because of the crucial nature of the time functions and the Instrument Repair Shop, Superintendent J. F. Hellweg recommended that the entire perimeter be fenced during World War II. AR (1941), 31.

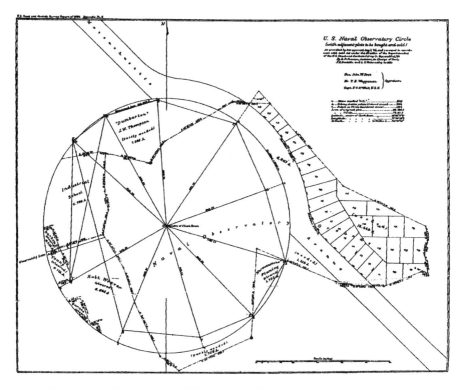

Figure 9.11. Coast Survey of Observatory Circle, 1896. From *Report of the Superintendent of the U. S. Coast and Geodetic Survey . . . ending June, 1896* (Government Printing Office: Washington, 1897), Appendix 6 (ref. 40). The radius of the circle is 1,000 feet, with the clock house at its center.

On May 15, 1893, the last personnel having left, and the last instruments having been removed, the old site of the Naval Observatory was formally abandoned for astronomical purposes, though it continued to be used for a variety of naval purposes.[42] Here Maury had ruled before his fateful decision in 1861; here too Gilliss and Davis had given their best and spent the last hours of their lives, here many astronomers had labored their entire careers in the service of applied astronomy. Now it was time to move on. Although the problem of the site had been resolved, the matter of administration lingered on.

## 9.2    The battle for civilian control

Even as the new Naval Observatory was under construction, the issue of its administration was reaching a boiling point. American astronomers realized that the situation was now more fluid than it had ever been, or was likely to be in the near

---

[42]  On the subsequent use of the Foggy Bottom site and buildings for medical purposes, see Jan K. Herman, *A Hilltop in Foggy Bottom: Home of the Old Naval Observatory and the Navy Medical Department* (reprinted from *U. S. Navy Medicine*, 1984).

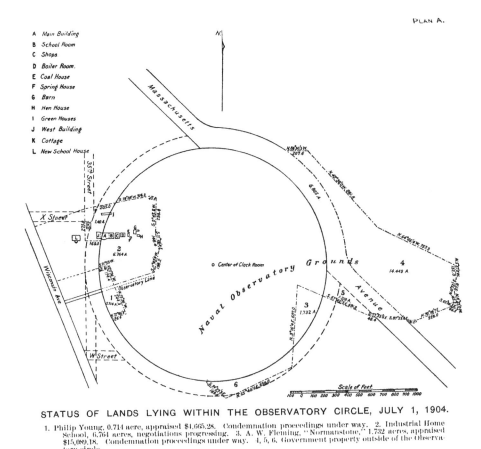

STATUS OF LANDS LYING WITHIN THE OBSERVATORY CIRCLE, JULY 1, 1904.

1. Philip Young, 0.714 acre, appraised $1,665.28.  Condemnation proceedings under way.  2. Industrial Home School, 6.764 acres, negotiations progressing.  3. A. W. Fleming, "Normanstone," 1.732 acres, appraised $15,089.18.  Condemnation proceedings under way.  4, 5, 6, Government property outside of the Observatory circle.

Figure 9.12.  Status of lands lying within Observatory Circle, 1904. From the *Annual Report* (1904), opposite p. 20.

future, and that organizational decisions made during the transition to a new site would probably quickly harden into fixed rules over the long term. In short, never had the moment been better to seize the initiative on the question of administration. For more than a decade after the move to the new Observatory in 1893, the issues of organization and a civilian director were constantly at the forefront. For decades after that, such administrative issues persisted in many forms, including attempts at transfer to the Commerce Department, the Hydrographic Office, and the Smithsonian. These developments also need to be seen in the context of the expansion and reorganization of American science.

### Attacks and defenses

The issue of administration was especially agitated by Lewis Boss (1846–1912), Director of the Dudley Observatory since 1876. Boss, a graduate of Dartmouth (1870),

Figure 9.13. An aerial view of the Observatory, 1919. The instruments and administration building are clustered around the center of the circle, which is the "clock house" just north of the dome for the 26-inch telescope, and between the two transit-circle buildings. The Superintendent's residence is at the top, and at the bottom are the photographic building and others. The Industrial Home school remains within Observatory Circle (upper left). See Table 10.1 for relative distances between major buildings.

had no formal training in astronomy, but had frequented the college observatory during his student days. While working as a clerk at the Census Office and the Land Office in Washington following graduation, he had also frequented the Naval Observatory, where he borrowed a chronometer and sextant and made careful observations. The Naval Observatory staff thought enough of his observations that, when the U. S. Northern Boundary Commission asked for the services of an astronomer in surveying the 49th parallel, Boss was recommended. Boss's success is evident in the fact that, when that effort ended in 1876, he was offered and accepted the Directorship of Dudley Observatory in Albany, New York. There he remained the rest of his life and built a reputation for Dudley in positional astronomy, culminating in the famous *Preliminary General Catalogue* (1910), which was published two years before his death.[43]

[43] On Boss see Benjamin Boss, "Biographical Memoir of Lewis Boss, 1846–1912," BMNAS, **IX** (Washington, 1920), 239–260. In 1906 the Carnegie Institution established a Department of Meridian Astronomy, under the direction of Boss.

His son Benjamin would carry on Lewis's work, including the famous *General Catalogue* (1937).

Several factors prompted Boss to enter the Naval Observatory controversy in 1891, including a sincere desire to see an improved National Observatory that might help establish an American Astronomical Society. Moreover, Boss was just at this time in a state of flux, negotiating to move his observatory to a more favorable site, where he could better undertake an ambitious new program of positional observations that would lead to the 1910 *Preliminary General Catalogue.* He knew that such a massive program could be much better undertaken with the facilities and resources of the Naval Observatory, and was considering applying for a position as Professor of Mathematics as a stepping stone to Superintendent; Newcomb had indicated in correspondence that Boss was his personal favorite for the latter position.[44] This was impossible under the current setup, but Boss was well connected politically and ambitious enough that he knew this might be changed.[45] Certainly the attention he gave to the subject shows much more than a passing interest in the fate of the Naval Observatory.

Boss was aware of the long history of the administration question at the Naval Observatory. He wrote in 1891 that "During the last decade there has been a distinct retrogradation at the Naval Observatory in the output of results as well as in morale. There is probably not a practical astronomer in the land who does not realize this fact. During the last year or two, there have been increasing indications that the old, irrepressible conflict is about to break out again with renewed force."[46] Boss was also aware of the broader context, emphasizing Congressional action in the controversy between civilian and Army surveys for geographical and geological purposes, and the transfer of the Weather Service from the Army to the Agriculture Department. Congressional action, he suggested, was once again appropriate to settle the analogous question at the Naval Observatory. By the end of 1891 Boss had elaborated his arguments to the extent that, when they were printed in a separate brochure (through the contributions of American astronomers), it came to 68 closely printed pages. Boss discussed in great detail the distinction between naval and astronomical

---

[44] Howard Plotkin, "Astronomers vs. the Navy: The Revolt of American Astronomers over the Management of the United States Naval Observatory, 1877–1902," *Proc. APS,* **122** (December 18, 1978), 385–399: 389.

[45] Benjamin Boss (ref. 43), p. 244 notes that Boss "strongly espoused the candidacy of James G. Blaine for President in the campaign of 1884, when Grover Cleveland was elected, though his personal relationship to Mr. Cleveland was most friendly. He also devoted much time and energy to the senatorial campaign of William M. Evarts. Senator Evarts attributed much of the success of his campaign to the efforts of Professor Boss." Cleveland was President 1885–89 and 1893–97, with Benjamin Harrison being President in the intervening period. Boss clearly had connections in Congress that he did not hesitate to use.

[46] Lewis Boss, "An Irrepressible Conflict," *The Sidereal Messenger,* **10**, No. 4 (April, 1891), 161–168: 166. The copy in the Naval Observatory Library is well annotated, including one note calling Boss "Brutus." The phrase "an irrepressible conflict" is an allusion to a statement of Lincoln's Secretary of State, William Seward, regarding the Civil War. For Boss's connection to the "time wars" then going on regarding dissemination of time, see Bartky, *Selling the True Time,* pp. 199–200 and references.

observatories in other countries, the record of the Naval Observatory, and why an astronomer rather than a naval officer should direct it. By 1892, however, Boss had received approval for his new site at Dudley, and ceased his leading role in the Naval Observatory issue.[47]

In his report for 1891 the Secretary of the Navy Benjamin F. Tracy recommended legislation to enable the President to appoint, from inside or outside the naval service, "the ablest and most accomplished astronomer who can be found for the position of superintendent." Seizing on this recommendation, B. A. Gould and E. C. Pickering led a large number of American astronomers in petitioning Congress for legislation to place a "practical astronomer" at the head of the Naval Observatory. "A body of the ablest astronomers, thrown together without adequate guidance, without any problem to solve, and without a concerted plan of work, can no more achieve success than an army can march and fight without a general. All experience shows that the success or failure of a scientific as of a business institution depends upon its direct- ing head," they wrote. These calls for action were included in a bill introduced in the Senate in January, 1892.[48] However, the Committee on Naval Affairs of the House of Representatives, reaching back to 1874 for Joseph Henry's favorable review of Admiral Sands's superintendency, citing Benjamin Peirce and other scientists in favor of Admiral Rodgers, and noting the disastrous reign of Leverrier at the Paris Observatory and his replacement by a more efficient naval administration, produced an adverse report on the House version of the same bill, halting further action.[49]

Even as the move to the new site was inexorably moving forward, in the matter of administration Superintendent McNair was leaving nothing to chance. McNair dis- tributed new regulations on a trial basis in August of 1892, and, although by these reg- ulations Harkness was appointed "chief astronomical assistant" to the Superintendent on October 21, when the new regulations for the Naval Observatory were formally promulgated in April, 1894 by Secretary of the Navy H. A. Herbert, Boss and other American astronomers could not have been happy. The superintendent – an officer of the Navy – remained the head of the Observatory. The "chief astronomical assistant" and the "chief nautical assistant" to the superintendent, by their very titles as well as their job descriptions, remained entirely subordinate to him. An

[47] Lewis Boss, *A Statement in Respect to the United States Naval Observatory and its Organization* (C. van Benthuysen & Sons: Albany, New York, 1891); Plotkin (ref. 44), 389.

[48] "Memorial for a Change of Management of the United States Naval Observatory," U. S. Senate Miscellaneous Document No. 40, 52nd Congress, first session, presented by Mr Morrill and referred to Committee on Naval Affairs January 14, 1892. The bill includes a lengthy discussion of why the director of the Naval Observatory should be a practical astronomer. There is no doubt that Lewis Boss was the author of this discussion, which presents similar arguments and phrases like "astronomical mob" that Boss had used earlier. For more details, see Plotkin (ref. 44), 389–390.

[49] "Naval Observatory . . . Adverse Report to accompany H. R. 3996," House of Representatives, Report No. 926, 52nd Congress, first session, submitted by Mr Lodge and laid on the table March 30, 1892. This document gives a history going back to 1870 of attempts to place the USNO under civilian control. Henry Cabot Lodge was a Massachusetts Republican; Captain C. H. Davis II was his brother-in-law.

"Observatory Board" consisting of the Department chiefs, was to "discuss and recommend" the scope and character of all work done at the Observatory.[50]

Six months later, however, a new set of regulations was issued, making provision for the positions of Astronomical Director and Director of the Nautical Almanac.[51] "After the removal to the new site and after another unsuccessful attempt to influence legislation," recalled the Superintendent some years later, "such pressure from within and without was brought on the Secretary of the Navy, as to induce him to make a concession by the establishment of the non-descript office of 'Astronomical Director.' The order establishing this office virtually removed the immediate control of the astronomical department from the Superintendent to the head of that department. While responsible for the extent and scope of the astronomical work, the 'Astronomical Director' was powerless in matters of discipline and finance. The office was, in fact, a humbug and a sham, and it was a mistaken concession, which only strengthened the hands of intrigue, and had no direct bearing on the difficulties of the question. The office was refused by those who had been most industrious in fomenting discord. It was accepted, in perfect good faith, by Professor Wm. Harkness, who retained it until his retirement in December, 1899, and it must be understood that the office was respected and strengthened and the provisions of the order carried out in absolute good faith by the Chief of the Bureau of Equipment and the Superintendent."[52] Harkness was appointed Astronomical Director on September 21, 1894 and took on additional duties as Director of the Nautical Almanac Office on June 30, 1897. The same document described Harkness's relation with the Superintendent as "cordial and intimate," "a common interest and determination on both sides to make the best of the case, made an otherwise impossible situation tenable for the time being."[53] It would not remain so after Harkness, as we shall see.

That the move to the new site and the promulgation of new rules hardly settled

---

[50] "Regulations of the U. S. Naval Observatory," April 21, 1894. This 14-page text contains as its opening the first statement of the Observatory's mission, including primarily work related to navigation, and secondarily "astronomical investigations of general or special scientific interest." These were essentially the same as the "Initiatory Regulations of the U. S. Naval Observatory," dated August 1, 1892, which were written by Superintendent F. V. McNair on a trial basis before being recommended to the Navy Department for approval.

[51] The revised regulations are dated September 20, 1894.

[52] Undated and unsigned document in USNOA, AF, "Davis" folder, p. 10. Internal and external evidence indicates that it was most probably written by C. H. Davis II around 1901. Another 13 page memorandum known to have been written in 1902 by Davis for his successor Colby M. Chester contains similar arguments and phraseology. In the latter memo Davis wrote "The office of astronomical director was a miserable subterfuge, adopted at a time when the affairs of the Observatory were in a state little short of chaos. It was a nondescript office, carrying responsibility without power. It gave a pretext to the enemies of the Observatory to speak of the 'dual head' always to the disparagement of the superintendent, whose final authority was never affected or changed, but who bore the odium of every criticism and complaint that was made against the Observatory," p. 2.

[53] *Ibid.*, p. 11. Harkness was appointed Director of the Nautical Almanac because other Professors of Mathematics at the Observatory were retiring in rapid succession, and outside appointments were political rather than meritorious. This caused a problem with other vacancies also. The author of the document wrote "The work was too much for one man. Professor Harkness held up until his retirement, when his health failed from overwork."

Figure 9.14. Professor of Mathematics Stimson J. Brown served as Astronomical Director and Director of the Nautical Almanac Office before his infamous demise as part of the controversy over civilian control over the Observatory.

the matter is evident in the fact that, in July, 1895, Secretary of the Navy Hilary A. Herbert ordered Naval Observatory astronomer Stimson J. Brown to visit Greenwich, Paris, and Berlin "for the purpose of observing and reporting upon the organization and methods of work of the national observatories in those cities." Brown (Figure 9.14), a graduate of the Naval Academy (1876), had been at the Observatory since 1881. During 1885–87 he used the 4-inch Repsold meridian circle at the Naval Academy for observations of the southern zones of the German Astronomical Society, and from 1887 to 1890 was at Washburn Observatory observing a similar program.[54] In submitting his report to the Secretary of the Navy in early 1896, Brown remarked on the contrast between the organization and methods of the Europeans and the Naval Observatory: "In all of them there is recognized the right of astronomers outside of the observatories to a definite interest in the work of the Government institutions, and the necessity of a strict legal control of the character and continuity of the observations and the instrumental equipment for their prosecution. There is no doubt that their successful work has been largely due to the recognition of these principles. The careful study of these matters has led me to the conviction that the future success of our observatory will depend upon the recognition of the vital points which appear in the organization of the European observatories."[55]

[54] On Brown see *WWS*, and A. N. Skinner, *Science* (ref. 61), 10–11.
[55] "Report of Prof. S. J. Brown, U. S. N., Upon the Organization and Methods of Work of European Observatories," February 19, 1896. Printed in *Report of the Board of Visitors . . .*, Exhibit N, pp. 61–77: 61.

At least in part the problem was also due to the agitation of Simon Newcomb, now nearing mandatory retirement as Superintendent of the Nautical Almanac Office. Newcomb had instigated official Navy inquiries about the Naval Observatory on three distinct occasions: December, 1893, March 12, 1896, and April 6, 1897. By the third attack Harkness, who as Astronomical Director had to answer them, was becoming impatient: "This is the third time I have been called on to review Professor Newcomb's attacks on the Naval Observatory under the plea that it was not furnishing a sufficient number of observations for the use of the Nautical Almanac Office . . . I hope that the answer I have now made will tend to bring this vexatious controversy to a close."[56] But the controversy was only beginning.

### The Board of Visitors

More serious than Brown's conclusions and Newcomb's attacks was the continuing issue of an outside Board of Visitors. Such a Board, which had already been recommended by Secretary Tracy in 1891 and repeatedly urged by Superintendents, was a feature of most European observatories, as Brown had not failed to point out.[57] The issue began to heat up again in 1897. In response to a letter from the Secretary of the Navy in September of that year, Superintendent Charles Davis II (Figure 9.15) and his superior at the Bureau of Equipment submitted suggestions for the organization of the Observatory. Their plan included the Superintendent and his assistant (both line officers), four astronomers who would be Professors of Mathematics and head the "astronomical departments," assistant astronomers not to exceed six, computers, the head of the Nautical Almanac (its organization to remain as at present), and a line officer to head each of the Division of chronometers and time service and the Division of magnetism and meteorology. The Superintendent and the four senior astronomers would form the "Observatory Council," which would coordinate the work set out by a Board of Visitors, appointed by the Secretary of the Navy on recommendations from the National Academy of Sciences. The Board would meet once per year. Notably absent from the plan was any mention of an Astronomical Director.[58]

The issue of organization, and in particular of a Board of Visitors, took on

---

[56] Harkness letter dated April 22, 1897, quoted in "Memorandum for the Chief of the Bureau of Equipment," c. November, 1902, USNOA, AF, folder labeled "Attacks and Defenses, 1902–1904." After the first attack Asaph Hall, now retired, wrote to his son Asaph Hall IV "Newcomb has begun a fight at the Naval Obsy by writing a letter to the Secy of the Navy complaining of the conduct of the Obsy. It is said he alludes to a professor there who thinks he knows about theory. It is assumed this refers to Harkness, who is very mad and is undertaking to annihilate Newcomb in a long reply." Asaph Hall III to Asaph Hall IV, April 13, 1894, AHP, Madison Building, Library of Congress. Hall also noted in another letter to his son dated July 29, 1894 that Newcomb and Superintendent McNair were meeting SecNav at the latter's house to discuss the issue.

[57] AR ending June 30, 1885, 10, renews the suggestion for a Board of Visitors. See also the Board of Visitors report (ref. 58), p. 8.

[58] "Proposed Organization of the Observatory," F. E. Chadwick and C. H. Davis to SecNav, September 7, 1897, reprinted in *Report of the Board of Visitors to the United States Naval Observatory* (Washington, 1899), pp. 38–40. On the term "line officers," see chapter 5, ref. 11.

Figure 9.15. Superintendent Charles Davis II, son of the first Superintendent of the Nautical Almanac Office, was embroiled in the controversy over Navy versus civilian control of the Observatory.

renewed life with the active role of American astronomers. Just at this time they were organizing a professional astronomical society on a national basis (with Simon Newcomb to be its first President), and one of their first concerns was the situation at the Naval Observatory. Already at the second meeting of the Astronomical and Astrophysical Society of America, held at Harvard in August, 1898, an entire morning was devoted to the condition of the Naval Observatory. The opinion was "nearly unanimous" that the scientific output of the Observatory was far below what it should be for its appropriations. A committee, consisting of Pickering (Harvard), Hale (Yerkes), and Comstock (Washburn) was appointed to bring the matter before the Secretary of the Navy. The following week the AAAS met in Boston and a committee consisting of Pickering, T. C. Mendenhall (Worcester Polytechnic Institute), and R. S. Woodward (Columbia) was appointed to cooperate with the committee of astronomers. The latter committee sent a questionnaire on the Naval Observatory to all American astronomers. The result was 51 to two that a change in organization was desirable, 51 to three that the direction should be given to an astronomer, and 43 to six that a visiting committee should be appointed. However, opinion was evenly divided 19 to 19 on whether the Observatory should be transferred to another branch of government.[59]

When the Astronomical and Astrophysical Society met in Washington in February, 1899, its Committee submitted a memorandum on the Naval Observatory to Secretary of the Navy Long; in March he agreed to appoint the Board of Visitors. Under these circumstances there was little Superintendent Davis could do but agree that the

---

[59] *Ibid.*, pp. 9–10, and appendices E, F, and G (pp. 42–48) for the questionnaire, answers, and related documents, including the names of the respondees. George E. Hale, Chairman of the astronomy meeting, had informed Secretary Long on August 22, 1898 of the formation of the Committee, USNOA, AF. On the formation of the Astronomical and Astrophysical Society of America (now the American Astronomical Society) and Newcomb's role see *The American Astronomical Society's First Century*, ed. David DeVorkin (American Institute of Physics: Washington, 1999).

time was ripe for a Board of Visitors.[60] *Science*, the official publication of the AAAS, which included on its Editorial Committee Newcomb, Woodward, Pickering, and Mendenhall – all sympathetic to civilian control – lost no opportunity to focus on the Naval Observatory. Three times in 1899 the Naval Observatory appeared prominently on its pages. In the first issue of 1899 Observatory astronomer A. N. Skinner wrote a balanced article in which he gave the history of the Observatory and its present operation, carefully avoiding any opinion on its administration but concluding that, despite the naval limitations on the scope of its operations, it had made other contributions to science. In March *Science* published 14 letters from astronomers and other scientists in response to a questionnaire. It is clear from these letters that the AAAS committee itself favored transfer from the Navy. By June the magazine was applauding the recent appointment of the Board of Visitors, which it said had a "difficult and delicate task" of far-reaching importance, "perhaps determining the character of our government astronomy for fifty years to come." The magazine hoped that the Board would not confine itself to details, but take into account the whole history of the institution: "What we are concerned with are the work and results of the most richly endowed and liberally supported astronomical observatory in the world." Emphasizing that taxpayers' money was at stake, the article went on: "For more than fifty years we have been trying an experiment in astronomical administration which no other nation ever thought of trying and which we ourselves have never tried in any other field, that of managing a great national observatory like a naval station." When a critical examination of the output of the institution is done, "it is a serious question whether any other than an adverse conclusion can be reached."[61]

In his own proposal Superintendent Davis had suggested that the Board of Visitors consist of one member from both the House and Senate Naval Committees and three eminent astronomers. So it was when, on June 30, 1899, the Board of Visitors, consisting of Senator William E. Chandler, Representative Alston G. Dayton, E. C. Pickering of Harvard, George C. Comstock of Washburn Observatory, and George E. Hale of Yerkes Observatory, met at the Naval Observatory.[62] Davis submitted documents giving the background of the formation of the Board. At the request of the two Congressmen, the three astronomers submitted the results of discussions and

---

[60] A "my dear Harry" letter from Senator H. C. Lodge to Davis on September 28, 1898 – a month before Davis took over as Superintendent, showed Lodge to be a staunch ally: "If there is any further attack on the Observatory, you may rely on me always." USNOA, AF, Davis 1898 file. This support is hardly surprising considering that C. H. Davis had been Lodge's father-in-law, and C. H. Davis II was his brother-in-law.

[61] A. N. Skinner, "The United States Naval Observatory," *Science*, **9** (January 6, 1899), 1–16 and 859; "Discussion of a National Observatory," *Science*, **9** (March 31, 1899), 467–476, where the respondents included Simon Newcomb, Asaph Hall, C. A. Young, T. C. Mendenhall, R. S. Woodward, C. L. Doolittle, W. H. Pickering, Arthur Searle, Frank W. Very, David P. Todd, G. W. Hyers, E. A Fuertes, W. L. Elkin, and James E. Keeler; and the editorial "United States Naval Observatory," *Science*, **9** (June 23, 1899), 857–859. It is likely that Newcomb himself wrote the editorial.

[62] *Report of the Board* (ref. 58), p. 5. SecNav Long wrote the Board on June 28, 1899 a "Memorandum respecting the Functions of the proposed board of visitors to the Naval Observatory," pp. 40–41.

correspondence centering around the second meeting of the Astronomical and Astrophysical Society of America and the AAAS. The Board met in seven sessions over the next three months. It found that, in respect to buildings, instruments, and equipment (not scientific output), the Naval Observatory had become "an astronomical observatory of the first rank," most of whose equipment, however, had little to do with the direct requirements of the naval service. "In view, however, of the absence of a national university, a department of science and industries, or other department or bureau of the Government especially suited to the conduct of scientific work, and in view of the diversity of opinion among American astronomers upon the question to which existing Department the Observatory could be widely transferred, we believe it to be inexpedient for us at the present time to further consider the subject of such transfer."[63]

However, the Board had plenty to say about the organization of the Observatory under the Navy. It disapproved of the Davis plan of 1897, mainly because "the proposed transfer of duties and responsibilities from a single director to a committee of five appears to us a step in the wrong direction," especially when the Superintendent held the power of veto. It held that an astronomer should head the scientific work: "In the history of observatories we have been unable to find a case of successful administration without a competent astronomer in immediate supervision of the work, and we believe that the ideal conditions for the successful administration of an astronomical observatory are most nearly realized when a professional astronomer is made the responsible director of the work." If a line officer is to remain the head of the Observatory, the Board concluded, an astronomical director should direct its scientific work, but not also be the head of the Nautical Almanac Office, as Harkness currently was. The Board recommended gradually abolishing the position of Professor of Mathematics, in accordance with Secretary Long's suggestion, and even gave a list of salaries needed to attract astronomers who would replace the Professors.[64] Finally the Board recommended the establishment of a permanent Board of Visitors.

One may only wonder what William Harkness, in the closing few months of his 37-year career at the Naval Observatory, thought of the Board report. Of Superintendent Davis's reaction there is no doubt. Davis considered the Board's report "an attack on the Observatory," and complained that the Board had proposed sweeping reorganization, while not addressing the issues it had been summoned to address. Comparing the manpower at Greenwich, Harvard, and the Naval Observatory, Davis asserted that the delay in publication stressed by the Board as showing bad management was entirely reasonable because of the several years taken up with the move of the Observatory. He also chided the Board for charging the

---

[63] Ibid., pp. 5–6.
[64] Ibid., p. 8 and Appendix L. A list of currently active Professors of Mathematics is given in Appendix C. The Board met on June 30 (two sessions), July 1 and 3, and September 26, 27, and 28.

Observatory with having two directors, and Davis left no doubt that he felt there was but one person at the helm: "The Observatory, like other naval establishments, has but one 'head,' ... the astronomical director of the Observatory is the head of the astronomical department, just as a naval officer is the head of the department of nautical instruments. To charge two 'heads' to the Observatory in order to increase the apparent extravagance of maintenance is clearly a perversion of fact. If the Observatory has two heads then it has eight heads, one for each department, and including the superintendent."[65] Davis, not keen to repeat the "experiment" of a Board of Visitors, nevertheless recommended that such a Board might be appointed from time to time at the convenience of the Navy, but with a more constructive outlook than the group of outsiders just announced.

Harkness retired on December 17, 1899, and on the same day was promoted from Captain to Rear Admiral. His statement of the needs of the Observatory at that time showed his frustration with the controversy. "Complaints are frequently made that we do less work than other great national observatories," he wrote, "but some of these institutions have forces two or three times larger than our own, and until that disparity is at least partially remedied we can not hope to compete with them." From a technical point of view Harkness recorded his "deep sense" of the need for two more (human) computers, together with a 13-inch photographic telescope and an 8-inch heliometer. As other observatories advanced in the photographic art as applied to astronomy, the Observatory never was provided with the instruments requested. Even so, Harkness undoubtedly found the administrative issue more serious than the lack of two instruments.[66]

### The Brown affair

Upon the retirement of Harkness, S. J. Brown, Professor of Mathematics since 1883 and next in seniority to Harkness in that position, became Astronomical Director and Director of the Nautical Almanac Office. He would play a new, and tragic, role in the story, for the Board of Visitors' report was about to have its ultimate test in Congress. On April 20, 1900, a bill to reorganize the Naval Observatory, incorporating the Board's recommendations and those of the Astronomical and Astrophysical Society of America, was introduced in the Senate and referred to the Committee on Naval Affairs. Although its 34-page report details the status of American observatories at the turn of the century, the bill subsequently died. Nonetheless, it provided the backdrop for the most serious administrative crisis in the Observatory's history.[67]

Soon after the Senate bill's introduction, Davis's response to the Board of Visitors' report came under severe criticism, as did Brown's first *Annual Report*, setting

---

[65] "Board of Visitors," in AR (1900), 6–12: 10.    [66] AR (1899), 18.

[67] 56th Congress, first session, Senate Report No. 1,043, "Reorganization of the Naval Observatory," Report to accompany S. 2019, submitted by Mr Chandler from the Committee on Naval Affairs, April 20, 1900. USNOA, AF, folder marked "Reorganization of the Naval Observatory, 1900."

off another round of political repercussions. The reaction of Davis to the Board of Visitors' report was criticized in two editorials in *Science*, which reprinted in its pages much of the Observatory's *Annual Report* for 1900.[68] In a private letter to Davis a week later referencing both the *Annual Report* and the *Science* editorials, Harkness severely criticized Brown's administration and work during his first year.[69] Apparently upset with his successor's actions regarding personnel and instruments, he charged that Brown was neglecting "the first and most important business of the Naval Observatory," namely observations on the Sun, Moon, planets, and nautical-almanac stars. He also charged him with not understanding some of the fundamental principles of the instruments he was proposing to change. While warning Davis that the enemies of the Observatory were sure to seize on these points, Harkness assured him that "I am more than willing to aid you in defending the interests of the Observatory in any way in my power."

Davis would need all the help he could get; *Science* alone ran nine items on the Naval Observatory in 1901. Furthermore, on January 21 a bill was introduced in the Senate "To organize the National Observatory of the United States," by renaming it "the National Observatory of the United States, placing at its head an eminent astronomer appointed by the President, and making the Board of Visitors a permanent body.[70] Moreover, important events were occurring at another level, as detailed in a letter from Brown to Lick astronomer W. W. Campbell. Brown stated that he and Newcomb had been called before Secretary of the Navy Long in mid-January as a result of a query of

---

[68] "The Naval Observatory Report," *Science*, **13** (January 4, 1901), 1–14; and "A Notable Official Report," *Science*, **13** (January 11, 1901), 41–45. The author of these editorials was again probably Newcomb (see the following note). Reaction to the editorials came in "The U. S. Naval Observatory," *Science*, **13** (January 18, 1901), 113–114, and (January 25, 1901), 150–151. The first was signed "M," the second was from Frank Bigelow. M noted that "A distinguished statesman whose loyalty to the interests of the Navy Department has long been known was made chairman of the Board and it is an open secret that he peremptorily cut off all suggestions looking to the real emancipation of this great institution." Both the first editorial and Bigelow's article are interesting for the issue of the cessation of magnetic observations. Bigelow favorably commented on the freedom of astronomers at the Naval Observatory compared with Harvard, and on how the current setup was good, insofar as the scientists were free to do their work while the naval officer administrates: "A scientist can take up these duties only by abandoning his researches, and it is little more than a dream to suppose that one can carry both along together." Also, he remarked, the Naval Observatory owes its prosperity to the need for practical results; Congress will not support it otherwise. Responding in the February 1 issue of *Science* (pp. 195–196), Newcomb called the issue "probably the most important subject at present before American men of science."

[69] Harkness to Davis, January 15, 1901. Harkness was writing from Jersey City. Brown's report as Astronomical Director is in AR (1900), 13–25. Harkness refers to the article in *Science* of January 4, 1901 and agrees with Davis that Newcomb is probably the author.

[70] U. S. Senate, 56th Congress, second session, S. 5671, "A Bill to organize the National Observatory of the United States," introduced January 21, 1901 by Senator Morgan and referred to the Committee on Naval Affairs. In its February 15 issue, pp. 276–277, *Science* reprinted the text of the bill and urged scientists to support the bill through letters to Congressmen and the Naval Affairs Committee. In its February 22 issue, p. 314, *Science* noted that the Senate had passed the Naval appropriations bill providing for a Board of Visitors, but still with a line officer as head. So *Science* urged passage of Senator Morgan's bill. However, by March 15 *Science* reported the appropriations bill passed in the House "after stout and repeated resistance by the House conferees," with the amendment that further legislation might be considered by Congress.

Newcomb about the tables of the Moon. Being asked at the meeting for his frank assessment of the situation at the Observatory, Brown gave his opinion, with the approval of Newcomb. "In the interview," Brown continued in his letter to Campbell, "at which Prof. Newcomb was present, he [Secretary Long] intimated clearly that he would be inclined, in the absence of remedial legislation at this session of Congress, to exercise the authority which he possesses of placing a Professor of Mathematics at the head of the Observatory. The following week he sent for me, and told me that he had considered the matter carefully, and had decided to order me as Superintendent of the Observatory in case Congress failed to act." Brown was now writing Campbell asking him to write a personal letter to the Secretary of the Navy assuring him that such a move would be viewed favorably by American astronomers. Secretary of the Navy Long's motives, according to Brown, were "to free the Observatory from the continuous criticisms to which it has been subject in the past, and bring it in harmony and cooperation with the astronomical sentiment of the country.[71]

Brown's ambitions were dashed a month later when he was dismissed in spectacular fashion, and with him was abolished the position of Astronomical Director. According to a document probably written by Davis, "Professor Brown's administration of these offices was simply scandalous. His one object was to turn a difficult and embarrassing situation to his own personal advantage. To accomplish this end, while openly professing loyalty to the naval administration and a desire to harmonize conflicting interest, he was in secret stirring up dissension and discord, conspiring with those within and intriguing with those without, and even practicing trickery and deceit toward his own confederates. An investigation of Professor Brown's conduct and secret correspondence revealed the most disgraceful conditions that had ever existed at the Observatory. In his stupid contempt of the Superintendent he had ventured almost openly upon the most flagrant violation of law and regulation. The situation was too acute to last. In March 1901 the facts of the case being fully revealed, Professor Brown was relieved from duty by the Superintendent [Davis] and charges were preferred against him of conspiracy, falsehood, disrespect and insubordination, attempt at intimidation, and gross neglect of duty. In order to escape an investigation of these charges Professor Brown accepted without protest his detachment from the Observatory, tantamount to a disgraceful dismissal, with the most scathing letter of rebuke ever addressed by the Navy Department to a commissioned officer."[72] While

---

[71] S. J. Brown to W. W. Campbell, February 6, 1901, USNOA, AF. Brown states that he had confidentially related in detail the situation to Professors of Mathematics Skinner and Eichelberger, who could answer further questions at a meeting in San Francisco. Brown also noted that he had already received a very strong letter of endorsement from Columbia's Barnard.

[72] Davis file, USNOA, AF, p. 13. See also the Davis memorandum to his successor, Colby M. Chester, 1902, p. 6, indicating that others were raising problems for Brown at the Naval Academy. This memo also indicates that Secretary Long was the principal witness against Brown. The date of Brown's detachment is given in the *Annual Report* ending June 30, 1901 as March 25 (p. 10). On March 28 Professor W. S. Harshman filled Brown's place as Director of the Nautical Almanac.

Professors of Mathematics were staff officers rather than line officers, they were still subject to the code of military justice. *Science* reported the charges against Brown: "that the accused resorted to intriguing methods to bring about the administration of affairs which he desired; that he made statements as coming from Captain Davis which that officer controverts; that he threatened the superintendent with attacks upon the floor of Congress, and neglect of duty." The papers, *Science* noted, were transmitted to Admiral Bradford at the Bureau of Equipment, and forwarded to Secretary Long, but were not made public: "owing to the personal nature of the controversy, the officials have surrounded the matter with the greatest secrecy."[73] By April 5, *Science* had reprinted a *New York Evening Post* report that Brown had been detached from the Observatory, commenting "It would seem from this that Secretary Long shares with Capt. Davis . . . the belief that Professor Brown transgressed the naval regulations in his efforts to have Congress pass the legislation needed to make the institution a great national one, and not a mere adjunct to the navy."[74]

Brown returned to the Naval Academy as an instructor in mathematics, and in 1906 became head of the Department of Mathematics, a post in which he served until 1910. Brown undoubtedly never forgot his Naval Observatory career, or its terrible culmination. Nor did the Navy. Writing to his relief in 1902, Davis stated that the office of Astronomical Director "was abolished on Brown's detachment in 1901. It should never be revived."[75] Nor did the reorganization bill pass the current session of Congress. The Navy had survived the most serious challenge to its control over the Observatory. When new regulations were issued on January 25, 1904, an Astronomical Director was not part of the plan.

### Continuing challenges

Despite his call for an occasional Board of Visitors, it was nevertheless undoubtedly with some chagrin that Davis was forced by Congressional legislation to entertain another Board of Visitors in 1901. The same legislation provided that "the Superintendent of the Naval Observatory shall be, until further legislation by Congress, a line officer of the Navy not below that of Captain." The Board, chaired by Princeton astronomer C. A. Young, consisted also of Charles F. Chandler, Asaph Hall Jr, Ormond Stone, William R. Harper – and once again Harvard nemesis E. C. Pickering. Its much shorter report recommended that no Astronomical Director be appointed, not because the Board objected to such a civilian director as the sole head of the Observatory, but because a dual head "has been found to work unsatisfactorily." At the same time the Board recorded its "unanimous judgment" that legislation should be introduced to have a civilian astronomical director as the sole Director of

---

[73] *Science*, **13** (March 15, 1901), 437–438. See also PA, **9** (1901), 216–217. On the distinction between staff officers and line officers in the U. S. Navy, see chapter 5, ref. 11. For Newcomb's comments on this distinction with regard to Professors of Mathematics, see his *Reminiscences* (Houghton & Mifflin: Boston and New York, 1903), pp. 99–101.     [74] *Science*, **13** (April 5, 1901), 550.
[75] Memorandum of Superintendent Davis to his relief, 1902, p. 3, USNOA, AF.

the Observatory. It also recommended that vacancies for assistant astronomer be filled by examination.[76] Not surprisingly, Davis once again saw the Board's report as an attack, and it was therefore with relief that he reported that Congress had refused to fund the Board for 1902 and 1903.[77] In fact, it was the last "Board of Visitors," though not the last investigating Board, the Observatory would see for many years.

Captain Davis had overseen one of the most tumultuous periods in the Observatory's history. At his departure on November 1, 1902, he left a detailed summary for his successor Rear Admiral Colby Chester, including the status of the instruments, grounds, and other projects. He was able to report that the "Observatory is in a thoroughly efficient and harmonious condition at the present time, so that you need have no misgivings or anxiety in taking charge that anybody within the Observatory is plotting to supplant you or is not loyal. On the contrary, I have no hesitation in saying that every individual now on duty within the Observatory limits will be absolutely loyal to you and is absolutely loyal to the naval administration." But he could not resist one last look back at the astonishing events of his tenure. In the settlement of the trouble at the Observatory, Davis wrote, it was necessary to detach four Professors of Mathematics: Henry Paul, Brown, T. J. J. See, and Milton Updegraff. Paul was detached "for general indolence and inefficiency. He is a harmless individual without much force of character, or of rather a cheerful disposition, but with a constitutional objection to labor of any kind. His presence here got to be so scandalous that I demanded his detachment, but it was not on account of any overt act on his own part."[78] T. J. J. See was transferred to the Naval Academy in September, 1902.[79] Davis's memorandum to his relief states that "See was a participant in the conspiracy of which Brown was the chief, and was guilty in an equal degree. If Brown had been brought to trial See would have been convicted also. The detachment of See and Updegraff was not procured by me. It came quite accidentally, but I was very glad when it did come, because those two men were a disturbing element and they were the only ones who were not loyal to the Superintendent."[80]

Davis knew, however, that this was not the end of the fight over control of the

---

[76] *Report of the Board of Visitors to the Naval Observatory for the year 1901* (Government Printing Office: Washington, 1901). The Board met for several days beginning April 9 and October 29, 1901. The Board was established by the Naval appropriations act of March 3, 1901, which went into effect July 1. *Statutes at Large*, volume 31, p. 1,122.     [77] AR (1902), "Board of Visitors," pp. 9–11.

[78] Davis memorandum to Colby Chester, 1902, p. 5. H. M. Paul was detached March 3, 1899.

[79] T. J. J. See, "Memorandum about Orders to the Naval Academy of Professors of Mathematics, September, 1902, and about the General Policies of the Naval Observatory" is referred to in an untitled document in USNOA, AF; the See document itself has not been found. See was detached and ordered to the U. S. Naval Academy on September 20, 1902. Updegraff received the same orders on the same day. See, however, would return to man the Observatory's Mare Island station from 1903 until his retirement in 1930.

[80] Davis Memorandum (ref. 78), p. 7. Davis went on to say that "See is an extremely curious character. I have often thought, and those associated with him here have thought that he is really not mentally responsible. There is nothing on record against See. I never proved anything against him; although I am fully assured of misconduct on his part, I never had any tangible evidence."

Observatory. "I must turn this fight over to you," he wrote to Chester, "but as it is perennial and never ceases I could not do otherwise than to leave it in an unsettled condition." He particularly worried about legislation affecting the Observatory being attached as a rider to an appropriations bill, and he urged great vigilance on the part of Chester. In fact, Chester reported in his *Annual Report* for 1903 that most of the time of the Superintendent and staff during the latter half of the year was given to answering questions from three investigating boards, two of them originating in the Navy, and one ordered by the President of the United States as part of a larger effort in the administration of science.[81]

The source of the first Navy investigation was once again the persistent Simon Newcomb, now retired, who asked the Secretary of the Navy to examine the work of the Naval Observatory in comparison with that of the Royal Observatory at Greenwich.[82] As a result, in early 1903 Secretary of the Navy William H. Moody constituted a Board to look into eliminating or transferring any of the work at the Naval Observatory elsewhere. After receiving a 50-page report on the subject from Superintendent Chester, the Board concluded "the regular work of the Naval Observatory is essential to the Navy; it can be systematically and successfully accomplished only under Government control; and no portion of it should be discontinued or transferred to other than the control of the Navy Department."[83] The General Board of the Navy, of which Admiral of the Navy George Dewey was president, discussed the same issue and rendered a similar decision.

Perhaps more serious was the Presidential Board, which had a much broader mandate: to centralize the scientific bureaus springing up throughout the Executive. In 1903 President Theodore Roosevelt appointed a Committee on the Organization of Government Scientific Work, chaired by Charles D. Walcott of the Geological Survey. Roosevelt seems to have taken a personal interest in having both the Naval Observatory and the Hydrographic Office transferred from the Navy Department to the newly created Department of Commerce. Prominent scientists, including W. W. Campbell, wrote in favor of doing just that.[84] The Committee met for four months, gathering reports from the scientific bureaus, including the Naval Observatory, but

---

[81] Ibid., p. 9. AR (1903), 6.

[82] Newcomb to SecNav Moody, November 6, 1902 USNOA, AF; the return address is Newcomb's P Street home. The reply to Newcomb's charges, possibly from Bureau of Equipment head Bradford, drew attention to the three previous attacks by Newcomb on the Observatory's management and direction.

[83] AR (1903), 6–7. The Board, constituted February 3, 1903, consisted of RADM (retired) Francis M. Ramsay (chair), Captain John E. Pillsbury and Cdr Charles J. Badger. Its recommendations, dated April 3, 1903, and a copy of the report are in USNOA, AF, folder marked "Attacks and Defenses, 1902–1904." The Board did recommend that the magnetic work, which had been suspended since June 1898 when the proximity of electric car lines prevented the obtaining of correct results, be permanently discontinued since it was being done at the Coast and Geodetic Survey.

[84] W. W. Campbell to Gifford Pinchot, May 20, 1903, and W. W. Campbell to Charles D. Walcott, May 25, 1903, USNOA, AF. Other members of the Committee were Gifford Pinchot (secretary, Bureau of Forestry), Garfield, Army General Crozier, and Chief Constructor Admiral Bowles of the Navy.

the reports were never published, and the Committee had no effect on scientific organization.[85]

Although none of these investigations resulted in action, a compromise of sorts in the issue of military versus civilian control of the Naval Observatory was reached with the formation of the Astronomical Council by order of the Superintendent July 29, 1908, soon approved by Secretary of the Navy Newberry. The Council was composed of the Superintendent and his Assistant, such assistants in charge of the astronomical divisions as the Superintendent may designate, and the Director of the Nautical Almanac. The Council was to be "guided by the fact that the most important astronomical duty of the Government is the publication of a nautical almanac, and as that is intended not only for the use of navigators, but also of astronomers in the most delicate investigations known to their science, it should be kept up to the highest attainable pitch of accuracy." To that end, it noted, fundamental meridian observations were essential.[86]

In spite of the formation of the Astronomical Council, the Observatory endured more attacks over the next decade. All ended in failure for the proponents of civilian control, but they demonstrate their dogged persistence. The first attack was a result of Simon Newcomb's last effort; E. C. Pickering said of Newcomb that "When dying, one of his last efforts was to see the President of the United States with the result . . . of the recommendation by the President in his annual message to Congress of November, 1910, that the administration [of the Naval Observatory] should be changed." In that annual message President Taft indeed proposed keeping the Observatory under the Navy, but with a civilian head, analogous to, and with the same salary as, the Superintendent of the Coast Survey and the Director of the Geological Survey. Legislation was submitted for that purpose, the House Committee on Naval Affairs recommended passage, but in the end again there was no action.[87]

---

[85] In March Pinchot wrote SecNav requesting that he inform the Committee what scientific work was now being carried out in the Navy. Gifford Pinchot to SecNav, March 27, 1903, USNOA, AF. Chester recommended no change, except that the magnetic work be transferred to the U. S. Coast and Geodetic Survey. Chester recounts the situation in an eight-page typescript, Rear Admiral Colby M. Chester to Rear Admiral L. H. Chandler, April 5, 1921, USNOA, AF. On the general program for science of Roosevelt, see Dupree (ref. 24), pp. 294–295. See also "The Endowment of Astronomical Research," *Science*, May 8, 1903, possibly by Newcomb.

[86] The order creating the Astronomical Council is printed in AR (1908), 5–6.

[87] The Pickering quote is from "Memorandum" dated February 4, 1914, probably written by Asaph Hall Jr, USNOA, AF, "Pickering, 1914" file. The Bill is House of Representatives, 61st Congress, second session, H.R. 22685, "A Bill to establish a naval observatory and define its duties, and for other purposes," introduced by Mr Dawson and referred to the Committee on Naval Affairs March 10, 1910. House of Representatives, 61st Congress, second session, Report No. 781, "To Establish a Naval Observatory, Define its Duties, etc." Report to accompany H.R. 22685, Committed to the Committee of the Whole House on the State of the Union March 16, 1910. The report contains the two long paragraphs from President Taft's message. The report was referred to the Senate Committee on Naval Affairs on March 31, and was reported out of the committee without amendment on April 20. Why it did not pass is not known. Taft's address was announced in a note by S. D. Townley, "The Naval Observatory," PASP, **22** (1910), 38–39, where Astronomical Society of the Pacific members were encouraged to write their Congressmen in favor of the bill. Taft was President 1909–1913, followed by Woodrow Wilson.

In 1911–12 the Director of the Hydrographic Office proposed merging with the Naval Observatory. Superintendent J. L. Jayne characterized it as "a case of the child adopting its parent," and, although much was written about the subject, perceived administrative difficulties prevented it.[88] In 1912 Charles Walcott, now head of the Smithsonian Institution, proposed that the Observatory be placed under his organization. When William Eichelberger (Director of the Nautical Almanac) was called upon by the chairman of the House Committee on Appropriations for an opinion on the matter, he wrote that an astronomical observatory would secure the best results in the hands of a professional astronomer.[89] In 1918 another attempt was made to transfer the Observatory to the control of the Smithsonian's Board of Regents and to have it renamed the "United States National Observatory." The matter dragged on until 1921, when Walcott called the plan to place government scientific bureaus, including the Observatory, under the Smithsonian "entirely impracticable."[90]

Meanwhile, trouble continued to flare from astronomers. In 1913, Pickering, in his Presidential address for the AAAS, severely criticized the Observatory, and suggested that it be placed under another government department, especially the Smithsonian, with a civilian head.[91] In 1915 Yale astronomer E. W. Brown wrote President Woodrow Wilson, suggesting that the Observatory be made a separate department outside the Navy, again to no avail.[92]

Although none of these attempts was ultimately successful, another Congressional action had more effect on the Observatory. In the face of World War I, the administrative structure of the Navy was reorganized, and, by amendment 89 of the Naval Appropriations Act of 1916, Congress forbade further appointments to the Corps of Professors of Mathematics of the U. S. Navy. The 16 Professor billets then on Navy rolls were gradually phased out as their occupants retired, bringing the Corps to a close some 20 years later. On the retirements of Asaph Hall Jr and William S.

[88] J. L. Jayne, "Memorandum for Chief of Bureau of Navigation," USNOA, AF, "Hydrographic Office Merger, 1911–12" folder. The Memorandum is accompanied by lengthy statements from Professors Hall, Updegraaf, and Littell, and astronomer G. A. Hill.

[89] Hearings, Sundry Civil Appropriation Bill for 1913, Part I, pp. 558–579. W. S. Eichelberger to Hon John J. Fitzgerald, Chairman House Committee on Appropriations, March 30, 1912, in folder "Bill to Place USNO under Smithsonian," USNOA, AF.

[90] House of Representatives, 65th Congress, second session, H.R. 10954, "A Bill to change the name of the United States Naval Observatory, at Washington, District of Columbia, to transfer the same to the Smithsonian Institution, and for other purposes," introduced by Mr Humphreys and referred to the Committee on Naval Affairs March 22, 1918. See also "The Proposed Transfer of the United States Naval Observatory to the Smithsonian Institution," *Science*, **47** (April 19, 1918), 383, which reprints a letter from SecNav Daniels to Chairman Padgett of the House Committee on Naval Affairs emphatically disapproving of H.R. 10954, and citing the Observatory's involvement in war work.

[91] Pickering's address "The Study of the Stars" is printed in *Science*, **39** (January 2, 1914), 1–9; the discussion on the Observatory is on p. 5. Pickering's address set off a reaction at the Naval Observatory, led by Asaph Hall Jr, USNOA, AF, "Pickering, 1914" file.

[92] E. W. Brown to President Woodrow Wilson, January 7, 1915, USNOA, AF, folder "E. W. Brown to President, 1915." This folder also contains the Navy Department's reply.

Eichelberger in 1929, and Frank B. Littell in 1933, this breed of employees exits the stage of Naval Observatory history.[93]

## 9.3    War, depression, and modernization

World War I brought renewed attention to the Naval Observatory, and a chance to prove its practical use to the country. As European observatories were crippled by war conditions, additional responsibilities were thrown on the Observatory, especially in regard to its navigational duties.[94] The war most directly affected the Observatory's "Material Department" and Nautical Instrument Repair Facility. The latter was established in 1913, followed in 1917 by an Aviation section, whose purpose was to facilitate work on the design, development, standardization, procurement, and issue of aeronautical instruments. The Material Department was responsible for everything from the testing of compasses (including new models of gyrocompass), to purchase and inspection of aeronautical instruments. It even handled requisitions for purchase of feed and supplies for homing pigeons at air stations. "Up to the date of the armistice, in November, 1918," Superintendent Rear Admiral J. A. Hoogewerff wrote, "the personnel of the Naval Observatory was concentrated chiefly on supplying the compasses and compass equipment, navigational instruments, instruments for aviation, nautical almanacs, and time service, not only for a greatly increased Navy, but for the Shipping Board to supply its vessels with navigational equipment." The capacity of companies manufacturing nautical instruments, particularly compasses and binnacles, was entirely inadequate to supply the demand, so that other companies, counting on the technical advice of the Observatory, had to be induced to undertake the work.[95] Staff at the Observatory were in short supply too; retired officers with previous technical duty returned to the Observatory to handle the increased workload. By drawing on U. S. Coast and Geodetic Survey personnel, the Nautical Almanac Office was prepared to duplicate computations normally furnished during peacetime by foreign governments.

Because of dependence on foreign countries for optical glass, one unusual difficulty during the war was a shortage of binoculars and small telescopes for submarine lookouts. Accordingly, under a program termed "Eyes for the Navy", in 1918–19 Assistant Secretary of the Navy Franklin D. Roosevelt appealed to "the patriotic citizens of the country" to lend their glasses to the Navy for the duration of the war at a nominal rental (Figure 9.16). The result was that 52,000 binoculars and spyglasses were so loaned, of which 32,000 were suitable. Citizens also loaned their

---

[93] On the Navy's Professors of Mathematics see C. J. Peterson, "The United States Navy Corps of Professors of Mathematics," *Griffith Observer* (Februrary, 1990), 2–14.

[94] AR (1917), 3–4 and 23.

[95] AR (1919), 3–13. On the establishment of the Nautical Instrument Repair Shop see G. A. Weber, *The Naval Observatory* (Johns Hopkins University Press: Baltimore, 1926), p. 33; and on the work of the Material Department, ibid., pp. 64–65.

Figure 9.16. An "Eyes for the Navy" poster to encourage citizens to loan binoculars and spyglasses to the government during World War I.

personal chronometers, sextants and other navigational instruments for the war effort.[96]

### Modernization

By the 1920s there was a strong feeling that the astronomical functions of the Naval Observatory were falling behind the times. The declining situation was even perceived from within the institution. When, during the 1929 Appropriations hearings, Congressman Burton L. French remarked to Naval Observatory astronomer James Robertson "Your thought is that in the last 30 years other nations have, through the national observatories, been going forward, while we have been doing the practical or necessary work, and have been neglecting that which was purely scientific," Robertson replied "Yes, sir; that is it exactly. It has been neglected entirely." Superintendent Charles S. Freeman agreed: "As Mr. Robertson has said, we have rather been marking time at the Naval Observatory for almost a generation."[97]

Although one must always be cautious of such statements, especially when they are made by incoming institutional heads in the context of requests for funding, nevertheless there was more than a little truth in the perception by the Navy and civil-

---

[96] AR (1919), 3–4 and 12. "Our Navy needs your Binoculars," *The American Horologist*, **9**, no. 9, p. 49.

[97] "Hearing before Subcommittee of House Committee on Appropriations," consisting of Messrs Burton L. French (Chairman), Guy U. Hardy, John Taber, William A. Ayres, and William B. Oliver, Navy Department Appropriation Bill for 1930, held Friday, January 11, 1929 (Government Printing Office: Washington, 1929), pp. 806 and 808. A. James Robertson was about to become Director of the NAO, and Freeman was Superintendent of the Observatory.

ians alike that the Naval Observatory was no longer a first-rate institution in comparison with other observatories. Spectroscopy and photography, for example, formed no major part of its work, despite constant requests by staff for funding for both. Nonetheless, from small beginnings, and against considerable odds compounded by the Great Depression, a new era gradually emerged at the Observatory, which may be seen in historical context (and in fact was seen at the time) as an era of modest modernization. The modernization program begun in 1927, spearheaded by a new and ambitious Superintendent (Freeman), implemented with the help of political connections by a prominent staff member (Robertson), and encouraged by outside astronomers who had long expressed the belief that the Naval Observatory had become outmoded, would greatly affect the next generation of astronomers at the Observatory. Although a new dark-sky site was no part of the original modernization plan, in retrospect the establishment of the Flagstaff station may be seen as a culmination of this era, especially since the very reason for its establishment was to acquire a better site for the 40-inch reflector that would become the centerpiece of the modernization plan.

In September, 1927, little more than a year before the stock-market crash would bring on worldwide depression, Captain Charles S. Freeman became Superintendent of the Observatory. Immediately he saw the extent of the task before him. "A very exhaustive study of the practical needs of the naval observatory began shortly after I took charge of it," he wrote, "on the basis of finding out whether the observatory was performing its astronomical functions in a manner actually commensurate with the astronomical output of the various first-class powers." The answer was found to be "no," and, especially in the area of photographic astronomy, the Observatory was judged "woefully deficient."[98]

Undoubtedly at Freeman's instigation, a new organization and mission statement was promulgated by the Secretary of the Navy on December 23, 1927, only four months after Freeman's arrival. While emphasizing nautical instruments, astronomical observations for navigation, and accurate time, it also mentioned contributing "within the capabilities of the available astronomical equipment, to the general advancement of navigation and astronomy." Elaborating on the Observatory's mission in a detailed report dated July 2, 1928, Captain (later Admiral) Freeman wrote that the institution had "a dual character. Basically and primarily the reason for its existence derives from the needs of the Navy." However, it is also a national observatory, he emphasized, and sometimes pursues goals "based upon the intrinsic worth of knowledge for the sake of knowledge." Just as the Weather Bureau ministers to the whole nation, not just to the Agriculture Department of which it is a part, he opined, so the Observatory serves not only the navigator, but also the general astronomer, the

---

[98] Ibid., p. 807.

engineer, the surveyor, the geodesist, hydrographer, geophysicist, and meteorologist, and anyone requiring accurate time. Still, he remarked, walking a narrow plank, "in rendering such service, the observatory is not permitted to venture far into the fields of abstract research along astronomical lines," work more properly carried out by non-governmental observatories. Government work, in his estimation, "is habitually restricted to utilitarian objects or to heavy, continuous work which cannot find adequate support elsewhere."[99]

A statement of the "Field of Work" of the Observatory, and a detailed "General Plan" in Freeman's report made more concrete the requirements specified in the mission statement. It is striking how many of them on the positional astronomy side were tied directly to the *American Ephemeris*, the one product whose practicality no one could doubt. Observations of the stars, planetary satellites, and asteroids were all mentioned specifically with regard to data for the *Ephemeris*. The organization chart (Figure 9.17) showed the importance of the Nautical Branch, which included not only the equipage, compass and maintenance Divisions, but also the time service as a section of the equipage Division. The Nautical Branch employed fully a third of the Observatory's staff at this time. Their functions were distinct from those of the Astronomical Branch, which included all of the telescopes as well as the Ephemeris Department (the Nautical Almanac Office). The Time Service would not emerge as a Department equal to the observing and ephemeris Departments until the 1940s. However, a step forward was taken in 1928 when Paul Sollenberger was appointed the first civilian Director of Time Service under Freeman; a staff photograph at the time (Figure 9.18) included Sollenberger, among other Time Service employees. (The same photograph shows future Department Directors C. B. Watts and Gerald Clemence, as well as the Nautical Almanac Office Director at the time, James Robertson; but it does not show employees of the Material Department of the Nautical Branch.)

The man who would carry out the modernization plan was not Freeman, but Julius Frederick Hellweg, who would serve as Superintendent during 1930–46, longer than any other Superintendent except its first, Matthew F. Maury. Ironically, it was under Hellweg (Figure 9.19) in the midst of the Great Depression that the Observatory carried out its first modernization program in the twentieth century, a program that would result in two new telescopes, an innovative 40-inch reflector, and a 15-inch telescope. Although, because of the Depression, Hellweg returned $50,000 to the Treasury, instituted cutbacks in salary and personnel, and indefinitely postponed some items of the plan, the modernization program eventually bore fruit at the Observatory – with what effect we shall see in chapter 10.[100]

---

[99] AR, **28** (July 2, 1928), 3. The full 1927 mission statement, found at the head of this chapter, is to be found on p. 5 of this AR.

[100] AR, **34**, 3. Hellweg, who died at age 94 on March 10, 1973, was a graduate of the Naval Academy and saw service on destroyers, on battleships, and at ordnance plants prior to coming to the Observatory. Although he retired in 1935, he remained Superintendent, and rose to the rank of Commodore.

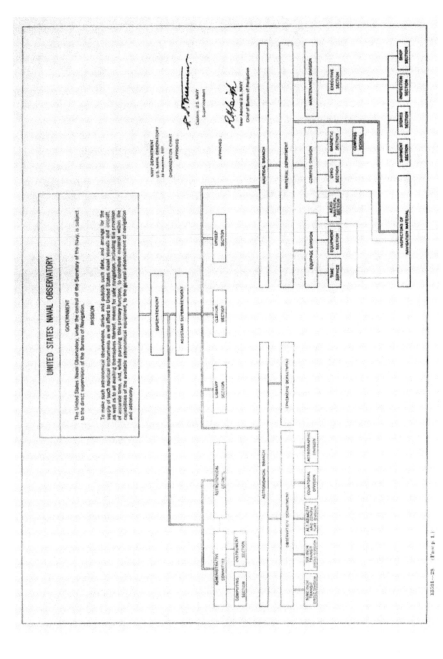

Figure 9.17. The 1927 organization chart, showing the division of work between the Nautical Branch and the Astronomical Branch. A third of the Observatory staff was in the Nautical Branch, including the Time Service, while the observation and ephemeris functions came under the Astronomical Branch.

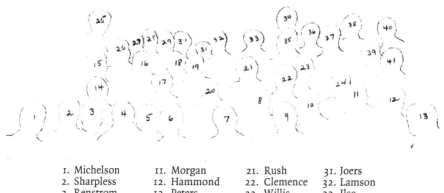

| | | | |
|---|---|---|---|
| 1. Michelson | 11. Morgan | 21. Rush | 31. Joers |
| 2. Sharpless | 12. Hammond | 22. Clemence | 32. Lamson |
| 3. Renstrom | 13. Peters | 23. Willis | 33. Ilse |
| 4. Watts | 14. Scott | 24. Fisher | 34. Burton |
| 5. Willis, Elsie | 15. Nordberg | 25. Sharnoff | 35. Seewald |
| 6. Hamilton | 16. Draper | 26. Adams | 36. Browne |
| 7. Lewis | 17. Sollenberger | 27. Mattes | 37. Phenix |
| 8. Savage | 18. Larrivee | 28. Raynsford | 38. Haupt |
| 9. Robertson | 19. Whittaker | 29. Bestul | 39. Pawling |
| 10. Hedrick | 20. Krampe | 30. Liferock | 40. Lyons |
| | | | 41. Snow |

July, 1932

Figure 9.18. Staff photo, July, 1932. See index for first names.

Figure 9.19. Superintendent J. F. Hellweg, who served as Superintendent of the Naval Observatory from 1930 until 1946, and saved it from termination. Hellweg is seen here looking into the clock vault shown in Figure 11.6. This photograph appeared in *National Geographic* for March, 1942.

### Presidential involvement

The Great Depression also saw national politics once again impinge on the Observatory when the Hoover administration attempted its transfer to another branch of government, and Franklin D. Roosevelt (a former Secretary of the Navy) tried to abolish it altogether in 1933. Seen from the outside, both were mere ripples of much larger political currents of the time. Seen from inside the Naval Observatory, the survival of a century-old institution was at stake.

The first presidential action began in 1929 with Herbert Hoover's plans for reorganization of the Departments of the Executive Branch upon taking office. However, the Wall Street crash occurred in his first year in office, and only on June 30, 1932, did Congress enact provisions to reorganize those Departments. In December Hoover sent a message to the Senate and House with specific recommendations for reorganization – to take effect within 60 days unless Congress acted.[101] Among the recommendations was that the Naval Observatory be transferred to the Commerce Department, along with the Navy's Hydrographic Office, where they would form a Merchant Marine Division with the Coast and Geodetic Survey and other nautical agencies.[102]

Congress acted quickly; a week later the House Committee on Naval Affairs convened hearings with a view toward making a recommendation to the full House. Testifying on December 17 were Chief of Naval Operations Admiral William V. Pratt, Rear Admiral Upham, Chief of the Bureau of Navigation, the Hydrographer of the Navy, and the Superintendent of the Naval Observatory. They were followed by a

---

[101] House Document 493, 72nd Congress, second session, "Message from the President of the United States [to the Senate and House of Representatives] transmitting a Message to Group, Coordinate, and Consolidate Executive and Administrative Agencies of the Government, as Nearly as May be, According to Major Purposes," December 9, 1932.    [102] Ibid., p. 5.

representative of the Coast Survey and the Bureau of the Budget. Their arguments were by now familiar, and on January 11, 1933, the House Committee recommended "strongly against" the transfer.[103]

This Hoover-led crisis had hardly been weathered when word came two weeks later that President-elect Roosevelt was considering abolishing certain agencies, including the Naval Observatory. *Science Today* reported that "From Warm Springs, Ga., where President-elect Roosevelt is preparing an 'economy' program for Uncle Sam after March 4, comes the suggestion that the U. S. Naval Observatory is a 'luxury' or at least is performing functions which private agencies are performing better. The grindstone is sounding upon the executioner's ax, we are given to understand."[104] *Science Today* went on to editorialize about the usefulness of the Naval Observatory, and characterized the Roosevelt plan as ill-advised.

By March the Observatory had further cause for worry in the form of the article "What Roosevelt Intends to Do" in *Collier's* magazine. Its author, George Creel, had been chairman of the Committee on Public Information during World War I, and worked with Roosevelt when he was Assistant Secretary of the Navy. The article left no doubt of his feelings: "Hidebound conservatives, fearful of change, may expect an unhappy time of it these next four years, for Franklin D. Roosevelt is going to take office with his face turned to the left. It is his fixed belief that America has come to the end of an era – the era of unplanned, unbridled and wasteful production – and that what we are now enduring is in no sense a 'slump,' but the breakdown of a system." What Roosevelt wanted, the article reported, was permanent retrenchment of the Federal establishment, cutting out scores of bureaus, boards, and commissions. In particular, "One of Franklin Roosevelt's own favorite illustrations of how money can be saved is the Naval Observatory, a moss-grown institution that costs $300,000 a year exclusive of an admiral, his aides and other Navy personnel . . . . At the Observatory they study stars with a telescope made in 1840, and the two major duties are to set a clock each day and get out the Naval Almanac, a task that is the last word in perfunctoriness." Similar statements were made in Congress.[105]

This article precipitated a negative reaction from many astronomers, some undoubtedly solicited by Hellweg and the leading astronomers at the Observatory.

---

[103] "Hearing [No. 836] on Proposed Transfer of the Hydrographic Office and the Naval Observatory from the Navy Department to the Department of Commerce," U. S. Congress. House Committee on Naval Affairs, December 17, 1932; Report [No. 846] of the Committee on Investigation of Budget's Recommendation to Transfer the Naval Observatory and the Hydrographic Office from the Navy Department to the Department of Commerce, House Committee on Naval Affairs, January 11, 1933.

[104] Reported in Watson Davis, "Uncle Sam's Observatory," *Science Today* (January 30, 1933) (Washington: Science Service) in Hellweg, "U. S. Naval Observatory, 1809–1948, First Draft Narrative" (hereafter, "Narrative"), typescript in USNO Library, Appendix, p. 91. Pages 91 ff. of the appendices deal with the Roosevelt issue.

[105] George Creel, "What Roosevelt Intends to Do," *Collier's: The National Weekly* for March 11, 1933, 7–9 and 34–36; on the Observatory see p. 34. Creel claimed that this opinion came "from the lips of the President." Hellweg, Narrative, Appendix 1, p. 2. On Roosevelt and the Navy see *FDR and the U. S. Navy*, Edward J. Marolda, ed. (St Martin's Press: New York, 1998).

S. A. Mitchell wrote to the Navy Secretary from the University of Virginia that the Naval Observatory's work in astrometry, time, and almanacs was "second to none," and suggested an impartial committee review its work. H. D. Curtis wrote from the University of Michigan that there would be a "riot" if the functions of the Observatory were dropped. Mt Wilson astronomer Seth Nicholson wrote Hellweg in support, as did the chief of the Division of Geodesy of the Coast and Geodetic Survey. Others wrote to H. E. Burton, and Harold Spencer Jones, Great Britain's Astronomer Royal, wrote James Robertson that the Observatory was doing work "of the greatest value to astronomy." At another level of society, James Stokely, on behalf of the Rittenhouse Astronomical Society in Pennsylvania, wrote directly to President Roosevelt.[106]

Considering the long history of the battle for civilian control, it is little surprise that not all astronomers were supportive. W. W. Campbell and Frank Schlesinger (the latter of whom had given advice on the Observatory's modernization program a few years earlier) were still trying to wrest the Observatory from the Navy, and discussed possible meetings with Roosevelt. When a Board of three was appointed to revamp all Federal activities, Hellweg swung into action. By his account, he provided data to the Board, but was not allowed to present it. Afraid that the Observatory was about to be abolished without having its day in court, Hellweg decided to go to the highest levels of government, with his superior's approval to do so. On hearing from Postmaster General Farley that the Observatory (and presumably other agencies) had been lost by decision of the Cabinet, he pleaded for a meeting with Roosevelt. The Observatory's drawn out near-death experience came to an end in the summer of 1933, when Hellweg accompanied the Secretary of the Navy to a meeting with President Roosevelt himself. Hellweg presented Roosevelt with "seven different statements" demonstrating that the expenditures at the Observatory had been vastly overstated. After the fifth statement, Hellweg recalled, "the President, very evidently much surprised, banged his fist on his desk with 'all right, Captain, you keep it.'"[107]

### World War II

Hellweg also saw the Naval Observatory through World War II. The Observatory proved indispensable in the war effort, especially with its chronometer function, time signals, *Almanac* activities, and other activities essential for navigation. With the entry of the United States into the War, the Observatory's nautical instrument sections (both compass and equipage) were placed under the Bureau of Ships, and the astronomical sections under the Bureau of Navigation; the latter were transferred to the Chief of Naval Operations the following year (April, 1942). The Instrument Repair Shop (expanded and placed in a new building as of 1942), although under the Bureau

---

[106] Copies of these letters are found in Hellweg, Narrative, Appendices, pp. 99–137.
[107] Campbell to Schlesinger, John Hall material, USNOA, BF; Hellweg, "Attacks and Defense," in Narrative, Appendix 1, pp. 2–3. This account is by Hellweg and has not been confirmed independently.

of Ships, remained directly under the control of the Observatory during the war. Two new classes of apprentice, instrument makers and chronometer repairers, were instituted. Shortages of personnel to carry out this work were a constant problem, with women being employed "in all lines of work where suitable."[108]

Before the United States's entry into the War, *Science* reported that the shortage of chronometers was unlikely to be as acute as during World War I. The difference was radio time signals, which by this time had increased both in accuracy and number to the extent that "a navigator could now operate satisfactorily with no timepiece but a dollar watch." With time signals transmitted 20 times a day at different frequencies from Arlington (Virginia), Mare Island (California), and elsewhere, even a relatively poor timepiece could be checked frequently and its error determined. Yet, in actual fact, the chronometer shortage during the war did become acute. Although a few American firms, including William Bond & Son of Boston, and Bliss and Negus (both in New York City), made finished chronometers, these were still assembled from imported parts. By the time the United States entered the war in December, 1941, neutral Switzerland had almost ceased supplying the United States with chronometers because the Axis powers did not permit exports outside the Axis countries. British makers of chronometers were unable to supply the United States because of the needs of their Admiralty. The Naval Observatory therefore had to turn to domestic watch makers, in what has been described as "one of the most remarkable stories in all the annals of chronometry." The Hamilton Watch Company and the Elgin Watch Company agreed to the challenge, but in the end only Hamilton met the Navy's requirements of an average daily rate of 1.55 seconds per day (Figure 9.20). For the first time ever a manufacturer undertook the mass production of marine chronometers. After two years of research, development, and testing its assembly line, Hamilton delivered its first two chronometers to the Observatory on February 27, 1942. During the war Hamilton produced 8,902 chronometers for the Navy, another 1,500 for the Maritime Commission for merchant navy ships, and 500 for the Army and Air Force, its production reaching a peak of 546 chronometers per month during October, 1944.[109]

The work of the Nautical Almanac Office, especially, was increased during the war. Hollwog reported already in July, 1942, that "The French astronomers very soon ceased productive effort; the Germans, of course, could not be depended on. All the

---

[108] AR (1942), 2. For a detailed chronology of the work of the Maintenance Division of the Materials Department, see Marvin E. Whitney, *Military Timepieces* (American Watchmakers' Institute Press: Cincinatti, 1992), pp. 255–260.

[109] James Stokley, "Chronometers for the Navy," *Science*, **92**, No. 2,380 (August 9, 1940), 9, Supplement; Marvin E. Whitney, *The Ship's Chronometer* (American Watchmaker's Institute Press: Cincinnati, 1985), pp. 3–4. The latter is the definitive work on chronometers in America; on testing chronometers during this period see pp. 46 ff. Whitney worked in the Nautical Instrument Shop, 1941–50, see SJD interview with Whitney, May 2, 1988, USNO Library. The "remarkable-story" quotation is from Anthony G. Randall, *The Time Museum Catalogue of Chronometers* (The Time Museum: Rockford, Illinois, 1992), p. 47.

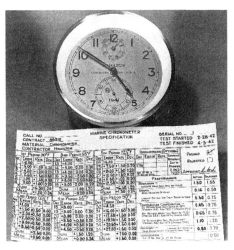

Figure 9.20. A Hamilton chronometer and its rating sheet, showing that the chronometer passed a series of tests between February 28 and April 5, 1942, meeting specifications for a daily rate below 1.55 seconds per day. The Observatory's annual reports prior to 1950 contain summaries of many such chronometer ratings.

other countries subjugated by the Germans are temporarily out. The British alone tried to continue their effort, but they, too, have been seriously handicapped as a result of the repeated bombings of London."[110] As we shall see in chapter 12, in addition to the increase in number of calculations required due to the lack of international cooperation, the Nautical Almanac Office proved important to the war effort in other ways as well.

The war years also sparked the first substantial construction of new buildings in many years, including new quarters for the instrument repair shop (the "Material Building," now known as Building 52), a new building for instrument storage (Building 56), and a cafeteria (Building 59) to accommodate the growing number of workers at the site. The chronometer shop alone expanded from ten to 80 employees during the war, necessitating its move to the "Material Building" as early as 1941. These buildings subsequently were used in part by the Observatory, as well as by other Navy and government tenants, including the Office of Naval Intelligence and the Weather Bureau.[111]

## 9.4  Post-war and Space-Age administrative developments

Despite the failed attempts to transfer or abolish the Naval Observatory, it was of course affected by internal Navy reorganizations, as well as by continued political events.

---

[110] AR (1942), 3. For more on the Observatory's war work see Whitney H. Treseder, "World War II at the U. S. Naval Observatory," manuscript dated August 12, 1997, USNO Library.

[111] AR (1943), 14. The Shop and other Divisions of the Material Department were moved from the "Boiler House" building to the new "Material Building" (Building 52) in October, 1941, and another new building (Building 56) was finished shortly after the war for instrument-storage boxes. However, with the war's end, it was turned over to the Weather Bureau. Whitney interview, pp. 5 and 41–43. The Translations Unit of the Office of Naval Intelligence occupied a part of Building 52 beginning in May, 1951, the Air Intelligence School from 1954 until 1957, the Technical Unit from January, 1955, the Naval Scientific and Technical Intelligence Center (NAVSTIC) from 1960, and the Acoustic Intelligence Analysis Facility from June, 1962; Capt Wyman H. Packard, A Century of U. S. Naval Intelligence (Department of the Navy: Washington, 1996), pp. 32, 138, 166, 191, and 376.

An important internal Navy change occurred during the war, when, on April 8, 1942, the Hydrographic Office and the Naval Observatory were transferred from the Bureau of Navigation to the Chief of Naval Operations, where the Observatory remains today.

Potentially more earth-shattering for the Observatory was a plan whose implementation would have moved the entire Naval Observatory from Washington, a story that also illustrates how the Observatory was affected by events beyond its control. In 1946 Congress authorized a hospital center to be established in the District of Columbia, the land to be provided by transfer from any Federal agency. In 1947 Secretary of the Navy Forrestal was informed that the Naval Observatory site had been chosen. The Secretary objected, saying that it would take eight years, ten million dollars, and a second observatory to continue vital work while such a move was under way. These arguments were to no avail; the plan was favored by Senator Millard Tydings of Maryland, who was Chairman of the Senate Military Affairs Committee, and whose wife was a member of the District of Columbia committee charged with selecting the site for the new "Washington Hospital Center."[112]

Word soon came that the Naval Observatory must move, within a month if possible. Indeed, there were good scientific reasons for moving; Sky and Telescope for July, 1947 reported that "Since its removal to the present site [in 1893], the seeing conditions have gradually deteriorated with the growth of the city until within recent years they have become so unsatisfactory that it has been evident for some time that a move to another location would eventually be necessary." Observatory astronomers, led by C. B. Watts, explained how the move could not be undertaken hurriedly due to the need for continuity in the observations; they needed a minimum of one year, preferably two. Finally the Secretary of the Navy agreed to the move, if the necessary time and funds could be made available.[113]

F. P. Scott, an astronomer in the 7-inch Transit Circle Division was put in charge of site testing, assisted by Alfred Mikesell, who had worked on the 40-inch reflector. Under the rushed circumstances they were given virtually unlimited resources for site testing. Mikesell and A. N. Adams, another transit-circle astronomer, went down the Appalachian range visiting local meteorologists, as far south as Asheville, North Carolina. Flagstaff, Arizona, the home of Lowell Observatory, was a name suggested by some. However, after consulting many astronomers, including Otto Struve, Gerald Clemence decided that the new site should be within 2 hours of Washington. Further site testing was done at four locations, two in the Lynchburg, Virginia, and two in the Charlottesville area, using Polaris telescopes borrowed from Palomar Observatory. On every clear night over 18 months, the telescopes observed

---

[112] Public Law 648 of the 79th Congress dated August 7, 1946 authorized the hospital center. MGEN P. B. Fleming (Administrator of the Federal Works Agency) to SecNav Forrestal, April 17, 1947 requested the transfer of land. "Brief of Proposed Move of U. S. Naval Observatory," typescript, USNOA, AF, and Mikesell OHI, pp. 98 ff.

[113] "Naval Observatory Seeks New Location," Sky and Telescope, 6, No. 69, 2. For more details see "Naval Observatory to be Moved from D. C.," The Washington Post, May 15, 1947, and other clippings in "USNO Proposed Move, 1947," USNOA, AF. See also AR (1947), and AJ, 53 (1948), 150.

Polaris, looking visually for image motion. In addition a 5-inch Alvan Clark refractor (one of the transit-of-Venus telescopes) was placed at each site to analyze seeing according to the so-called Pickering scale. Eventually one of the Charlottesville area sites, known as Piney Mountain, was selected. At the same time it was proposed "to establish one or more of the equatorials at a site near Flagstaff, Arizona." Congress passed legislation authorizing the relocation, and, in summer 1950, also authorized the sum of $7,000,000 to relocate the Observatory to Charlottesville; however, no further action was taken. Perhaps the deciding factor was the defeat of Senator Tydings in 1950, and within a few weeks the Observatory was told not to worry about moving.[114] Nonetheless, the idea of a Flagstaff site was not to be forgotten; in the mid-1950s much administrative energy was expended on the search for a new dark-sky site for the 40-inch telescope, which was indeed located in Flagstaff, Arizona in 1955.

Although the Observatory itself did not move, the transfer of the chronometer and nautical-instruments functions of the Observatory to the Navy Yard at Norfolk Virginia during 1950–51 did radically change the character of the Observatory, moving it in the direction that some of the reformers had recommended. However, the primary motivations were economic rather than internal to the Observatory: the reasons given in official correspondence were a potential savings of more than $700,000 over five years, reductions in Navy funding, decreasing workload in instrument repairs due to war surpluses and the inactivation of ships of the fleet, and the need for dispersing repair facilities for strategic considerations. On May 25, 1950 the Bureau of Ships recommended to the Chief of Naval Operations that the Navigational Instrument Repair Facility be disestablished in Washington, and transferred to a Naval shipyard. The date set for the disestablishment was January 1, 1951; only the repair functions for the navigational timepieces were transferred to Norfolk, the remaining functions being liquidated. The resulting move left some logistical problems for the Observatory; the Facility had also performed work for the Observatory, and on its departure the Superintendent requested four instrument makers in addition to the three already on board, and the retention of certain equipment of the Repair Facility. Although the Observatory had instrument makers throughout its history, this was the modern beginning of the Observatory's Instrument Shop, which has been located ever since in the same building as the Repair Facility, Building 52. In addition to the accommodation for instrument makers, compensation was also requested for supply and fiscal clerks and maintenance people, who had also been borrowed from the Repair Facility.[115]

---

[114] The $7 million is authorized in the military Construction Act of June 17, 1950, *Statutes at Large*, volume 64, p. 239. AR (1948), AJ, **54** (1949), 65–66; AR (1949), AJ, **54** (1949), 219–220; AR (1950), AJ, **55** (1951), 202; Mikesell OHI, p. 113. Duncombe, OHI, p. 6, relates that Senator Joe McCarthy may have been in part responsible for Tydings's defeat.

[115] "Navigational Instrument Repair Facility, Washington, D. C., Disestablishment of," Chief, Bureau of Ships to Chief of Naval Operations, May 25, 1950, USNOA, AF, folder "General Adm. Plan for NavObsy, 10–16–50". See also Whitney, *Ship's Chronometer*, 27–28; and Whitney OHI, pp. 36–38. Whitney states that the move was scheduled for December 31, 1950, but was not completed until as late as June, 1951.

Figure 9.21. Division Directors, March 10, 1953. Front row (From left to right) Gerald M. Clemence, Superintendent Captain F. A. Graf, Cdr R. E. Sinnott, Paul Sollenberger, and John S. Hall. Back row: E. W. Woolard, Chester B. Watts, William Markowitz, Francis P. Scott, and Gilbert J. Oates.

As a result of the removal of the nautical instruments function from the Observatory, fully half of the observatory's organizational structure was removed. Remaining were the Nautical Almanac Office (the "Ephemeris Department" in Figure 9.17), Six and Seven Inch Transit Circle Divisions, the Equatorial Division, and the Time Service, all under the Superintendent, who was a Navy Captain. The Directors of these Divisions in 1953, shortly after this reorganization, are shown in Figure 9.21.

As had been the case for the past 50 years since the notorious dismissal of Professor of Mathematics S. J. Brown, there was as yet no overall Astronomical Director. However, in 1958, the Brown affair having long been forgotten, the position of "Scientific Director" was established at the recommendation of the Superintendent. This action was no doubt in part because the Observatory's profile had been raised as the Space Age began. It was also a correction to an administrative difference that had festered since the late nineteenth century: The Director of the Nautical Almanac Office was a step above the other Division Directors in pay grade, a situation that had existed since the Nautical Almanac Office came to the Naval Observatory under Simon Newcomb in 1893. Increasingly, the Director of the

Figure 9.22. Scientific Directors Kaj Strand, Gart Westerhout, and Kenneth Johnston (left to right), in the lobby of the administration building, mid-1990s.

Almanac Office was consulted for crucial decisions, was initiating and supervising research throughout the Observatory, and was influencing appointments.[116] Thus, at his own instigation and after consultations with the Superintendent, in 1958 Gerald Clemence, Director of the Almanac Office, was placed in the newly created position of Scientific Director, and the Directors of the other Divisions were put on a par with the new Director of the Almanac Office, Edgar Woolard.[117] Only three others would serve as Scientific Director during the twentieth century (Figure 9.22): Kaj Strand (1963–77), Gart Westerhout (1977–93), and Kenneth Johnston (1993– ).

As we shall see in the following chapters, the Naval Observatory entered the Space Age haltingly. It might have been otherwise, had feelers succeeded in putting the Naval Observatory under the National Aeronautics and Space Administration (NASA), which was founded in October, 1958.[118] If the Observatory entered a relationship with NASA reluctantly, however, by the end of the twentieth century it had fully embraced NASA funding, with the award of $180 million for the Full-Sky Astrometric Mapping Explorer (FAME), with Scientific Director Kenneth Johnston as the Principal Investigator. By the end of the century, demands for an increase in astrometric accuracy demanded access to space.

As with all government institutions (and institutions in general), personnel, budget, and facilities were constant administrative concerns through the century. As

---

[116] Superintendent USNO (J. Maury Werth) to SecNav via CNO, August 23, 1970. The subject of this memo was "Abolition of Position of Scientific Director, U. S. Naval Observatory." This memo, in arguing for an abolition of the position, cited the reasons for initiating it in 1958. The 1958 initiating document has not been found.

[117] Duncombe OHI, 11 January, 1988, pp. 16–18. Clemence was in fact promoted from a GS-15 ($13,970) to a GS-17 ($15,500), and the other Division heads were eventually raised to be on a par with the Director of the NAO, rather than the NAO Director being lowered. (Clemence Service record; Clemence folder). This promotion, and the subsequent raising of grade levels of Division Directors and other staff, was undoubtedly another factor in the creation of the position of Scientific Director.

[118] The USNO apparently was invited to become a part of NASA, but Clemence turned this down. John Hall Interview # 2, p. 13; see also Duncombe OHI.

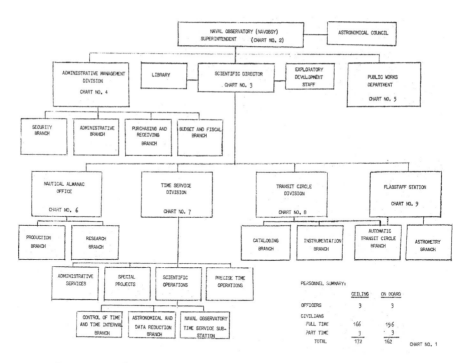

Figure 9.23. The 1976 organization chart, showing 162 civilian personnel on board, plus three officers.

of 1925 the Observatory had 106 employees, of whom a third were in scientific departments, one third in the Nautical Branch or Materials Department, and one third in administrative and support roles. At mid-century the staff consisted of 99 employees, of which half were in scientific departments. Because of the Space Age, by 1967 the staff had increased to 197. By the end of the century centralization and "regionalization" within the Navy and Department of Defense caused the number of support personnel to decrease at the Observatory, first in the area of human resources, then in facilities management and security. This resulted in total personnel numbering 113, of which a much higher proportion were scientific staff than at mid-century. The budget (about $300,000 in 1933 when Roosevelt tried to abolish the institution) was about 50 times that at the end of the century, exclusive of outside funding.[119]

The organization chart as of 1976 (Figure 9.23) shows few changes in terms of functions. The transit circle remained a predominant instrument for positional astronomy, and the Almanac Office, Time Service, and Flagstaff stations were still central organizational elements. By the end of the century, however, the organizational chart (Figure 9.24) showed some major changes, but without changing the mission. Two Transit Circle Divisions (the 6-inch and 7-inch Divisions), were

[119] For a breakdown of employees and their roles as of 1925 see Weber (ref. 95), pp. 67–69. For appropriations from 1875 to 1926 see Weber, pp. 87–89.

## U.S. Naval Observatory Organization Chart

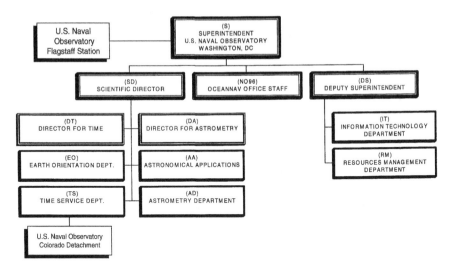

Figure 9.24. The 2000 organization chart. The Transit Circle Divisions have gone, The Nautical Almanac Office has been absorbed into The Astronomical Applications Department, and an Earth Orientation Department has been created.

combined into one Transit Circle Division in the reorganization of 1976, and in 1982 became the Astrometry Department, reflecting the advent of new astrometric technologies, even though the transit circles would see another two decades of service. The Nautical Almanac Office was gone from the chart as a major organizational element, having been subsumed in 1990 as only a small Division under the Astronomical Applications Department. This reflected the decrease in importance for celestial navigation in the wake of Global Positioning System (GPS) satellite navigation, as well as a broader concept of types of data, now computerized, necessary for a wider variety of users. Only the Time Service Department remained unchanged in name and function, but even here its Earth Orientation function was spun off into a new Department, reflecting new space geodetic techniques for accurately measuring the Earth's rotation. The Flagstaff station also remained and increased in importance, but in the mid-1990s the Time Service substation was moved from Richmond, Florida to Schriever Air Force Base in Colorado Springs, close to the GPS control station for which it provided the time. Both the Security and Public Works Departments, which formerly had been integral parts of the Observatory, were gone, casualties of regionalization.

Not shown on the current organization chart, but nevertheless an important part of the Observatory, is the Naval Observatory Library (Figure 9.25), a part of the Scientific Director's staff. Over 150 years the librarians (Appendix 1) have built its

Figure 9.25. The earliest photograph of the library, around 1900.

reputation as one of the best astronomy libraries in the world. Gilliss had acquired the nucleus of the collection during his trip to European observatories in 1842, some 700 volumes of astronomical works and 175 volumes of observatory publications. In addition to its annual purchases, the Library was enriched in 1866 by the donation of Gilliss's personal collection by his widow, by the acquisition of the Nautical Almanac Office library in 1909, and by the donation of the senior Asaph Hall's personal library in 1930. A rare book collection of over 800 pre-nineteenth-century volumes was also acquired, largely during the nineteenth century from funds set aside for that purpose. The Library's collection had reached 55,000 volumes by the beginning of the Space Age, and, reflecting the information explosion, more than 80,000 volumes by the end of the twentieth century.[120]

---

[120] The Gilliss donation is documented in *Report of the Secretary of the Navy* (Washington, 1867), p. 133. In 1930 Mrs Asaph Hall Jr presented 625 volumes of Asaph Hall Sr to the Library, AR (1930), 20. On the history of the Library to 1961 see Marjorie S. Clopine, "United States Naval Observatory Library: Resources and Treasures," *Special Libraries*, **52**, No. 2 (February, 1961), 78–81. Clopine was Librarian of the Naval Observatory at the time. The library collection as of 1975 is given in *Catalog of the Naval Observatory Library* (G. K. Hall, 1976). Astronomer E. S. Holden was the first to catalog the collection in 1879. The Library houses two busts, one of Gilliss, and one of Maury, the latter unveiled in 1931, AR (1931), 17. It also houses several scientific portraits, including those of George Saegmuller and Mordecai Yarnall (dating from the nineteenth century), as well as James Melville Gilliss, Asaph Hall III, Simon Newcomb, G. W. Hill, and Gerald Clemence (all commissioned in the 1990s).

Figure 9.26. A view of the Observatory toward the north, 1977. The administration building, dome for the 26-inch telescope, and transit-circle buildings are in the center. The original Superintendent's residence, now home to the Vice President of the United States, is just north of the administration building. At the bottom is "Building 56," which today houses parts of the Astrometry Department. The National Cathedral is at the top.

The physical site and buildings at Observatory Circle changed only gradually over the century. Those buildings occupied in 1893 still formed the core of the physical plant – the Administration building, the 26-inch-telescope building, the transit-circle-telescope buildings and the Superintendent's residence – all situated within Observatory Circle, gave the Observatory the physical layout it still maintains today (Figure 9.26). A few additional buildings had been added during the twentieth century, and some of the buildings changed function. Most notably, on February 2, 1962 the Simon Newcomb Laboratory – a substantial extension of the old Astrographic Building – was dedicated as the new facility for the Time Service Department and the Astrometry and Astrophysics Division (Figure 9.27). In connection with the opening of this facility, and in light of the close cooperation among the Royal Greenwich Observatory, the National Physical Laboratory (U. K.), and the Naval Observatory, Markowitz succeeded in bringing Harrison number 4 chronometer to the Naval Observatory for display in 1963–64, the only time the historic clock has ever

Figure 9.27. The Simon Newcomb Laboratory, which was dedicated in 1962, for the Time Service Department and Astrometry and Astrophysics Division, seen in a view looking southeast, October 19, 1961. The astrodome on the roof was built for an artificial-satellite tracker. The attached building on the right is the old astrographic building, and the detached building on the far right for many years held a 5-inch Alvan Clark refractor owned by the National Capital Astronomers.

left the U. K., aside from its sea trials (Figure 9.28).[121] The five living quarters, built for the heads of the observing Divisions and occupied by them until the 1960s, eventually became Admirals' residences.

Most remarkably in terms of usage, however, the Superintendent's residence would become the home for the Chief of Naval Operations in 1928, and the official residence for the Vice President of the United States in 1974. Vice President Nelson Rockefeller funded many upgrades to the residence, and every Vice President since Walter Mondale has lived there with his family. The bustling activities associated with having a Vice President on the Observatory grounds brought substantial changes to the ambience of Observatory Circle. In the tradition of John Quincy Adams and Abraham Lincoln, during their period of residence on the grounds Vice Presidents Mondale, George Bush, and Al Gore regularly visited the Observatory proper to view the heavens.[122]

[121] W. Markowitz, "Naval Observatory Exhibits Harrison No. 4 in US," *Horological Journal* (June, 1964); Markowitz, "Reminiscences", 21 October, 1986, USNO Library, pp. 2–7. The new Simon Newcomb Laboratory was dedicated on February 2, 1962. Dedication brochure is in USNOA, AF, Simon Newcomb Laboratory folder.
[122] The act authorizing the Secretary of the Navy to assign the Superintendent's quarters to the Chief of Naval Operations is *Statutes at Large*, volume 45, p. 1,018, December 10, 1928. Public Law 93–346, Joint Resolution dated July 12, 1974 "Designating the premises occupied by the Chief of Naval Operations as the official residence of the Vice President, effective upon termination of service of the incumbent Chief of Naval Operations." The resolution actually designates the

Figure 9.28. The famous Harrison No. 4 chronometer on display in the lobby of the Administration Building of the Naval Observatory in 1963. Astronomer James A. Hughes presides at one of the Observatory's regular public tours.

Through war, depression, attempted transfers and moves, internal and external reorganizations, and regionalization, the functions of the Observatory remained remarkably stable: space (positional astronomy), time, and navigation. The histories of these scientific functions are the subject of the final three chapters.

home as "the official temporary residence of the Vice President of the United States," USNOA, AF, folder marked Vice President's Residence. Furthermore, it designates the Secretary of the Navy to provide for staffing, care, maintenance, repair, improvement, alteration, and furnishing of the residence and grounds. Hearings on the subject were first held during the 89th Congress, February 24 and 25, 1966, USNOA, AF, Vice President's Residence. No funds were appropriated at this time. The discussions leading directly to the law of July 12, 1974 are described in the "Hearing Before the Subcommittee on General Legislation of the Committee on Armed Services," United States Senate, 93rd Congress, second session, on S[enate] J[oint] Res[olution] 202, "To Designate the Premises Occupied by the Chief of Naval Operations as the Official Residence of the Vice President, Effective upon the Termination of Service of the Incumbent Chief of Naval Operations." See Gail S. Cleere, *The House on Observatory Hill* (Government Printing Office: Washington, 1989) for the subsequent history of the Vice President's Residence.

# 10  Space: The astronomy of position and its uses

It is commonly to be expected that a new and vigorous form of scientific research will supersede that which is hoary with antiquity. But I am not willing to admit that such is the case with the old astronomy, if old we may call it. It is more pregnant with future discoveries to-day than it ever has been, and it is more disposed to welcome the spectroscope as a useful hand-maid, which may help it on to new fields, than it is to give way to it.

Simon Newcomb, 1897[1]

Astrometry is on the threshold of great changes due to the fact that this decade, alone, is witnessing an improvement of stellar positions equivalent to the total improvement of the previous two centuries.

E. Høg and P. K. Seidelmann, 1995[2]

We recall from section 5.2 the late-1930s opinion of Benjamin Boss that positional astronomy was languishing at the end of the nineteenth century compared with astrophysics, that the outlook for the field was discouraging unless new developments were forthcoming, and that it required new and unifying goals such as studying the structure of the universe to revitalize the field. At its new site, and even in the midst of its administrative problems, the mission of the Naval Observatory gave it both the motive and the means to lead the field of astrometry more than any other observatory in the United States. Yet, during the first half of the twentieth century, while it continued routine programs of positional astronomy, dabbled in the new and important technique of photographic determination of star positions, and occasionally showed a spark of innovation, the Observatory by its own admission did not live up to its promise as the national observatory for the United States. With few exceptions, neither new technology nor unifying goals drove its research program during these decades, despite new interests having been acquired after the move and an attempt at modernization in the midst of the Great Depression. The ever-present quest for accuracy remained, but with doubts about its usefulness for practical applications.

All of this changed gradually in the post-World War II era, and then radically

---

[1] Simon Newcomb, *The Problems of Astronomy: An Address by Simon Newcomb at the Dedication of the Flower Observatory of the University of Pennsylvania* (New Era Printing Co.: Lancaster, 1897), pp. 15–16.

[2] E. Høg and P. K. Seidelmann, eds., "Preface," in *Astronomical and Astrophysical Objectives of Sub-Milliarcsecond Optical Astrometry* (Kluwer: Dordrecht, 1995), p. xv.

with the beginning of the Space Age. New applications in science, navigation, and warfare brought the need for star positions with an accuracy undreamed of before. Thanks to Space-Age funding, the Observatory built its largest telescope, expanded its astrometric work to the Southern Hemisphere, and, by the end of the century, was using or testing state-of-the-art instrumentation and techniques, including interferometry, charge-coupled-device (CCD) detectors, and even an astrometric satellite known as the Full-Sky Astrometric Mapping Explorer (FAME). In this chapter we see how the old astronomy met the new in the first decades of the century, how the modernization program succeeded in providing only a modicum of progress, and how Space-Age needs returned the Naval Observatory to first-class status in positional astronomy. Although the century-long dominance of astrophysics would not allow the Naval Observatory to recapture the unique position it held in astronomy in the late nineteenth century, by the end of the twentieth century the renaissance of precision astrometry was a significant component for advances in astrophysics.

## 10.1    The old astronomy meets the new, 1893–1927

The period of the 1890s during which the Naval Observatory moved to its new site was a time of great transition in American astronomy. It had become clear that photography would play a major role not only in recording permanent images of the Sun, Moon, planets, comets, and nebulae, but also for charting the positions of the stars – the specialty of the Naval Observatory. The new directions for astronomy declared by S. P. Langley in his volume *The New Astronomy* (1884–87) were taking hold, and, although the composition of the celestial bodies that was the central focus of astrophysics formed no part of the original goals of the Naval Observatory, the new developments raised the question of whether the Observatory's goals should be expanded – at least if it wanted to maintain its self-styled status as a national observatory.

American astronomy changed in other ways, too, during the 1890s. In the last great binge of astronomical-lens technology, the world's largest refractors were installed at Lick and Yerkes Observatories, their 36-inch and 40-inch apertures far surpassing that of the "Great Refractor" that had placed the Naval Observatory at the forefront of American astronomy in 1873. More important than the size of their instruments, these new observatories were not bound by any practical mission such as the improvement of navigation; established by the generosity of individual benefactors who gave their names to the observatories, they could largely pursue whatever research programs they wished, subject only to the interests of the Director and a cadre of independent researchers. Finally, the 1890s also saw the professionalization of astronomy in America. The rise of the "Astronomical and Astrophysical Society of America" – its very name a result of negotiation between practitioners of the old and new astronomy – signaled not only an increasing number of astronomers in the United States, but also a more coherent identity and community. What part the Naval Observatory played in that identity is revealed not only by the politics and the setting described in the last chapter, but also by the science to be discussed in the present one.[3]

As the sole national observatory for the United States during the first half of the twentieth century, the Naval Observatory thus provides a unique opportunity to examine how an astronomical institution steeped in the old tradition of visual positional astronomy reacted to new developments in its field. Although it was bound by a mission, as most Federal agencies are, the Naval Observatory nevertheless had room for maneuvering in terms of new techniques and even new directions for research. How the institution handled this challenge and opportunity, in the context of meridian observations, the work of the equatorial telescopes, and attempts at astronomical photography and spectroscopy, is the story of this section. First, however, we must examine the driving force behind all this work.

### A crisis of purpose

A classic expression of the tensions between the old and new astronomy is found in an address by none other than J. R. Eastman – one of the stalwarts of the second generation of Naval Observatory astronomers – who addressed the Rochester meeting of the AAAS in August, 1892 on the eve of his retirement.[4] In Eastman's view, the problems of astrophysics, while resulting in "grand and important discoveries," were secondary to the "fundamental problems" of finding the positions and motions of the stars and the bodies of the solar system. The problems of positional astronomy, Eastman stated in an obviously defensive mode, "still lie at the foundation of the 'old' astronomy and cannot be relegated to the limbo of useless rubbish or to the museum of curious relics, not even to make room for the newborn astro-physics." The entire "astronomical superstructure," Eastman emphasized, rested on this foundation and the accuracy with which it is laid. That foundation was now in danger; despite the existence of some 250 observatories in the world, of which 60 were equipped with transit circles, a quarter of which were capable of good results in his estimation, Eastman reported that no "fundamental" work concerning star positions was in progress anywhere in the world. If only five of those institutions cooperated for a five-year observing program, he argued, a fundamental catalog might be produced to replace the Pulkovo catalog compiled three decades earlier. That the motive for such a new catalog apparently did not exist was the heart of Eastman's message to the AAAS, one undoubtedly based on his own experience.[5]

---

[3] Donald E. Osterbrock, *Yerkes Observatory, 1892–1950: the Birth, Near Death, and Resurrection of a Scientific Research Institution* (Chicago: University of Chicago Press, 1997); Osterbrock, John R. Gustafson, and W. J. Shiloh Unruh, *Eye on the Sky: Lick Observatory's First Century* (Berkeley: University of California Press, 1988); and Osterbrock, "The Quest for more Photons: How Reflectors Supplanted Refractors as the Monster Telescopes of the Future at the End of the Last Century," *Astronomy Quarterly*, 5 (1985), 87–95. On the formation and rise of the American Astronomical Society see *The American Astronomical Society's First Century*, ed. David DeVorkin (American Institute of Physics: Washington, 1999).

[4] J. R. Eastman, "The Neglected Field of Fundamental Astronomy," *Proc. AAAS*, 41 (Salem, 1892), 17–32. Eastman was at the time Vice President of the section on astronomy and mathematics.

[5] Ibid., p. 19. On the technical definition of "fundamental" star positions see chapter 5, pp. 175–176.

Nor, as one might have expected from someone completing three decades of service at the Naval Observatory, was Eastman's argument based on the utility of fundamental astronomy to navigation. Quite the contrary, Eastman made the following astonishing statement: "if every fixed observatory in the world were destroyed today, no interest of navigation or commerce would suffer for the next 50 years. The function of astronomy in promoting the development of navigation and in fostering the extension of commerce has been completed." The motive for the work of most observatories in the United States, Eastman said, was found in attempts to please the patrons of astronomy, whether they were individuals, corporations, or legislative bodies. Astronomy, in his opinion, had now reached the point at which the original goals had been fully achieved. The true purpose of astronomical study is that "it stimulates the highest form of intellectual activity, widens the already broad field of investigation and increases the sum of human knowledge. Whoever pleads the cause of astronomy on a lower plane discounts the intelligence of himself or of his audience. Why should the astronomer stoop to select a less noble theme, or consider it from a lower point of view?"[6] Newcomb's view (at the head of this chapter) of the spectroscope as a "useful hand-maid" to positional astronomy expresses a similar sentiment.

Ten years after Eastman's address it must have been clear even to the aging Newcomb that the status of the old astronomy was forever changed. Observatories scrambling for resources in many cases no longer did so with practical justifications, which would have pleased Eastman, but neither did they emphasize positional astronomy to the exclusion of astrophysics, as is evident in the report of the Advisory Committee on Astronomy of the newly founded Carnegie Institution. E. C. Pickering, George Ellery Hale, Lewis Boss, Simon Newcomb, and Samuel Langley agreed on general considerations such as the need for a Southern Hemisphere observatory and the need for more workers in the field and for cooperation in research. However, Langley and Hale pushed solar work and studies of stellar evolution relying on astrophysics; Pickering naturally focused on photometry and photography; and even Boss, while emphasizing meridian astronomy as well as parallax and astronomical constants, viewed them as a means toward determining the structure of the sidereal universe – a relatively new goal of astronomy. Only Newcomb pushed the problems of celestial mechanics, even as he must have seen the landscape changing all around him.[7]

---

[6] Ibid., p. 31.

[7] "Report of the Advisory Committee on Astronomy," *Carnegie Institution of Washington Year Book no. 1*, for 1902 (Carnegie Institution: Washington, January, 1903), pp. 87–160. Boss argued (pp. 117–122) that no observatory in the United States, including the Naval Observatory, made "fundamental" observations of star positions with the highest accuracy. As a result, for several decades the Carnegie Institution supported a Department of Meridian Astronomy, headed first by Lewis Boss and then his son Benjamin. As we shall see, perhaps partly in response to this criticism, the Naval Observatory soon initiated fundamental observations. For an assessment of the needs of astronomy some 15 years after the Carnegie Report see Henry Norris Russell, "Some Problems of Sidereal Astronomy," *Proc. NAS*, **5**, no. 10 (October, 1919) 391–416. This paper, communicated to the Academy on June 17, 1917, may be considered the equivalent of the Academy's modern "decade reviews" of astronomy.

How these broader events translated into practical problems at the Naval Observatory is encapsulated in the assessment of Lt Albert Winterhalter, a young sharp-minded Naval officer assigned to the Observatory, who reported his findings in the wake of a trip to European observatories at the time of the 1887 Astrophotographic Congress in Paris. Winterhalter was effusive in his praise of the most practical aspects of the Observatory's work. During his visits he "found nothing so well suited for distributing time-signals over large areas of territory as the Gardner system now and for several years past used at the Naval Observatory." The methods of chronometer rating at the Observatory he also found "most excellent," suggesting in his report only more studies on the effects of various influences on chronometers. The Observatory's current system of inspecting and verifying nautical instruments required only to be enlarged to encompass other instruments. However, regarding the more purely astronomical work of the Observatory, Winterhalter was more cautious, and sounded a warning note. While not even a large observatory can pursue all of modern astronomy, "there are certain lines of investigation which, in the interest of immediate progress, must be pursued." Among these was astrophysics, and Winterhalter noted in his 1887 report that "Nearly every great observatory has included more or less of physical observations in its routine work. Especially is this true in countries where no observatory exists specially devoted to such observations. An institution devoted solely to astrophysics is, of course, the highest expression of the cultivation of this branch of astronomy. In the absence of such an establishment, existing government institutions should be equipped to do the work."[8]

Celestial photography was another branch of the new astronomy. In Winterhalter's view, "it may easily be predicted that soon no observatory will be able to do without its aid in some form or another." Either the Naval Observatory should be equipped with appropriate instrumentation, or a separate institution should be founded, Winterhalter concluded. He indicated his preference by warning that "An observatory which takes no heed of or is not allowed to partake of the spirit of the age in which it works soon falls behind."[9]

At the turn of the twentieth century, therefore, the Naval Observatory faced a dual crisis of purpose. According to Eastman, it had satisfied the practical needs of the Navy for the next 50 years in terms of accurate star positions. At the same time, the new astronomy threatened to leave the Naval Observatory behind in any ambitions to be at the vanguard of American astronomy, as it had been in the last third of the nineteenth century. Yet, despite the internal sense of a new era brought about by a new site,

---

[8] Albert G. Winterhalter, *The International Astrophotographic Congress and A Visit to Certain European Observatories and Other Institutions* (Government Printing Office: Washington, 1889), p. 323. Also published as Appendix I to WO for 1885. Winterhalter does not mention the Smithsonian Astrophysical Observatory, founded in 1891 in Washington by Samuel P. Langley, the Institution's Secretary.

[9] Winterhalter, "General Conclusions," pp. 321–323. Winterhalter also noted that "the spheres of meteorology and astronomy have been all but separated in modern times. Each is best cultivated in quarters of its own."

and despite the pleas of outside astronomers for an expanded program, the mission of the Naval Observatory did not in fact change in any way with the abandonment of the old site in 1893. Nevertheless, as the century progressed, new technology drove the need for more accurate time and star positions, in ways that would have surprised even the forward-looking Eastman and Newcomb.

### Meridian instruments: Their work and their observers

For most of the twentieth century, classical astronomy was epitomized by the transit-circle telescope and related instruments for determining precise celestial positions. During the nineteenth century no respectable observatory was without these essential instruments in some form, and the Naval Observatory was among the best. Fewer and fewer observatories supported classical positional astronomy as the twentieth century advanced; thus the Naval Observatory remained the strongest supporter in the United States of this venerable tradition, just as Greenwich, Paris, and Pulkovo Observatories did in their countries. Indeed, as fewer universities supported the long-term research required for determining star positions, and as privately funded observatories turned toward more exciting projects, national observatories became virtually the sole upholders of traditional classical astronomy. They did this (despite Eastman's statement) nominally in support of celestial navigation. However, in a broader sense they carried out these programs also, as Eastman had suggested, in support of all astronomy, which requires reference stars for much of its work. As Superintendent C. H. Davis II stated to the Observatory's Board of Visitors in 1899, producing the *Nautical Almanac*, and all observations in support of it, was the most important duty of the Observatory. At the same time, he reiterated his father's dictum that the *Almanac* was intended not only for navigators, but also for astronomers.[10]

A glance at the publications of the Naval Observatory is sufficient to show that there was no abatement in meridian observations undertaken at the new site. The bulk of the labor at the Observatory remained in this area, and its commitment is evident in an increasing arsenal of instruments in the service of these goals. Even as Eastman published his landmark *Second Washington Catalogue* (1898) incorporating observations with the 8.5-inch transit circle from 1866–91, the Observatory was well into its new meridian observing program with the 9-inch transit circle. In addition to modifications to the old prime vertical, a new altazimuth instrument was purchased as well as a new Warner and Swasey 6-inch transit circle. The 9-inch and the 6-inch transit circles would be the workhorse instruments of the Observatory for positional astronomy during the first half of the century, beginning a pattern of having two separate transit-circle groups at the Observatory, observing (and in some ways competing) side-by-side. The same could be said well into the second half of the century, except that the 9-inch was replaced by a new 7-inch transit circle, which in the late 1960s

---

[10] Davis, AR (1900), 7.

Table 10.1. *Positions and characteristics of equatorial and meridian instruments at Observatory Circle*[a]

| Instrument aperture | Difference in latitude (feet) | Difference in longitude (feet) | Altitude above sea level (feet) | Maker | Focal length (inches) |
|---|---|---|---|---|---|
| 9-Inch transit circle | +7.47 | −82.00 | | Pistor & Martins | 107 |
| 6-Inch transit circle | +7.47 | +82.00 | | Warner & Swasey | 72 |
| 4.86-Inch prime-vertical transit | +52.40 | 0.00 | | Pistor & Martins | 77 |
| 26-Inch equatorial | −171.63 | 0.00 | 280 | Warner & Swasey | 389.66 |
| 12-Inch equatorial | −171.63 | −276.45 | 313 | Saegmuller | 180 |
| Meridian instrument of 40-foot photoheliograph | −354.22 | −103.16 | 265 | | |
| 10-Inch photographic equatorial (mounted 1911) | Near photoheliograph | | 250 | G. Peters | 113 |
| 5.3-Inch south transit | −171.63 | −225.38 | | Ertel & Son | 84 |
| Altazimuth | +269.72 | +156.40 | | Warner & Swasey | 50 |

Notes:

[a] With respect to the clock house, located at the center of Observatory Circle, longitude 5 h 8 m 15.78 s W of Greenwich, latitude +38 degrees, 55 minutes, 14.0 seconds. The relative positions of the instruments are pictured in the aerial view in Figure 9.13.

began a program of Southern Hemisphere observations. The two groups were combined in the late 1970s. The characteristics and relative positions of the meridian and equatorial instruments with respect to the clock house at the center of Observatory Circle are shown in Table 10.1, while Appendix 3 indicates graphically the years these instruments were in use.

The 9-inch transit circle (Figure 10.1) was not a new telescope, but a modification of the 8.5-inch Pistor and Martins transit circle that had been installed at the old site during 1866–91. Not only was its objective lens replaced by a new 9.14-inch Clark lens, but also the telescope tube was shortened by about 18 inches at each end to accommodate the 107-inch focal length, a task not carried out without problems.[11] This, it was hoped, would take care of some of the deficiencies that had been noted by Newcomb and others over its 30 years at the Foggy Bottom site. It was mounted in the fall of 1892 in the East transit house, and the published observations of the 9-inch transit circle would extend from 1894 to 1945. In addition to Aaron N. Skinner (who had been at the Observatory almost a quarter century when it moved to its new site),

[11] PUSNO, 1, vii–ix. The first new lens was of diameter 8.97 inches, but proved to be defective. More minute changes to the 8.5-inch lens are also described here. An eight-day Howard clock, placed in the clock room, was used as the standard sidereal clock, making use of the chronograph method. In 1933 the 9-inch transit circle house was rebuilt, giving it a roll-off roof, PUSNO, 15, part 5, 123.

Figure 10.1.  The 9-inch Transit Circle, about 1910. The instrument was completely remodeled from the 8.5-inch Pistor and Martins transit circle (Figure 5.6), purchased in 1865. The two cylindrical weights on either side of the telescope are part of a suspension system that allows very little weight to be placed on the pivots where the telescope turns. From *Publications of the U. S. Naval Observatory*, volume IX, part 1.

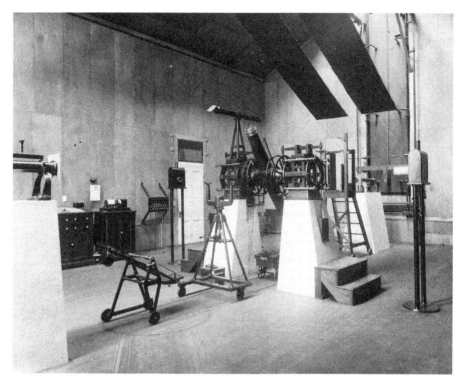

Figure 10.2. The 6-inch Warner and Swasey Transit Circle, which was first mounted in 1897, seen here in November, 1899. A device holding the mercury basin (used as a liquid mirror when the telescope is turned downward to determine the "constants" of level and nadir), and a carriage for reversing the instrument monthly, are seen in front of the instrument; the observer's chair is barely visible under the telescope. The piers to the north and south of the telescope hold the collimators (for determining the optical axis of the telescope) and the mark lenses, for determining shifts in azimuth. The constants of level, nadir, mark, and azimuth were taken every few hours so that shifts in the instrument could be taken into account in the observational data. A chronograph for recording timing data is against the far wall.

the telescope would occupy part of the careers of a new generation of astronomers at the Naval Observatory, including William S. Eichelberger, Frank B. Littell, and, most notably, Herbert R. Morgan.[12]

The six-inch transit circle (Figure 10.2), by contrast, was a brand new instrument built by Warner and Swasey of Cleveland, Ohio in accordance with the specifications of William Harkness. Although it was modeled closely after the Repsold transit circles in use at West Point, Washburn Observatory in Wisconsin, and Lick Observatory, Harkness tried to incorporate all the experience with previous transit

---

[12] Morgan was in charge of the instrument from 1913 until his retirement on September 30, 1944; PUSNO, **15**, part 5, 117.

instruments. In order to reduce the effects of flexure and temperature changes, for example, the instrument was built entirely of steel, rather than the brass used in the Repsold instruments. The 6-inch diameter objective, made by Brashear, had a focal length of only 72 inches, further reducing the size and weight of the instrument. The right ascension and declination wires for the micrometer were similar to those of the 9-inch instrument. The two steel circles, 26.75 inches in diameter, were divided on silver every two minutes of arc, and read visually using four microscopes that were illuminated by kerosene lamps. Power for the circle and field illumination could be instantly changed from gas to electric.[13]

The 6-inch transit-circle telescope was mounted in December, 1897 in the West transit house, and temporarily used beginning in June, 1899, but found to have a variation in azimuth with temperature, to the extent that, in 1901, the marble piers were finally replaced with brick piers. In 1906 a hand-driven Warner and Swasey traveling-wire micrometer replaced the original one. Looking back from 1950, C. B. Watts wrote that "These and other alterations limited the use of the transit circle appreciably until 1911, when it was placed in regular service. During the ensuing 20 years few changes were made."[14] The names most closely associated with the 6-inch transit circle during its early history are John C. Hammond and Chester B. Watts. The 6-inch transit circle survived its early crises and remained in operation until 1995.

The Pistor and Martins prime-vertical transit (Figure 10.3), with a 4.86-inch objective lens and 77-inch focal length, was one of the instruments purchased by Gilliss in 1844. Aside from a few observations in the early 1880s, the prime vertical had last been used by Asaph Hall in 1863–67 to determine the value of the constant of aberration; Hall believed that its declinations were the most accurate ever obtained with any instrument in the United States. During the move from the old to the new Naval Observatory, the instrument was remodeled by the Washington firm of Fauth and Company, mainly to make the instrument symmetric in form so that it could be more stably mounted. The instrument was mounted inside a transit house about 50 feet north of the central clock house in March, 1893. The astronomer in charge of this instrument, who also made almost all of the observations for this program, was George A. Hill.

Like the 6-inch transit circle, the altazimuth instrument (Figure 10.4) was a new instrument constructed of steel along many of the same principles by Warner and Swasey. The altazimuth, which was not strictly a meridian instrument, was used to observe at small angles on each side of the meridian; in a more restricted mode it could also be used as a zenith telescope. The 5-inch Brashear lens had a focal length of 50 inches. The altazimuth had both vertical and horizontal steel circles, 24 inches in

---

[13] The first description of the 6-inch telescope is in PUSNO, **1**, xii–xiv, along with a more detailed description of the 9-inch telescope. The detailed description of the 6-inch instrument is in its first volume of observations, PUSNO, **3**, part 4.

[14] On the marble and brick piers see PUSNO, **3**, part 4, section D, xxxiii; also Steven J. Dick, "Trials and Tribulations of Operating a New Transit Circle," USNO Star, **2**, no. 2 (Summer, 1990), 3 and 9, USNO Library. For Watts's statement see PUSNO, **16**, part 2, 334.

Figure 10.3. The 4.86-inch Pistor and Martins Prime Vertical Transit. Unlike transit circles, the prime vertical was mounted in an east–west direction. It was used irregularly prior to 1882, underwent extensive repairs in 1892, and observed continuously until 1912. Thereafter it observed sporadically until 1925. The building was torn down in the early 1950s. From PUSNO, volume X, part 1.

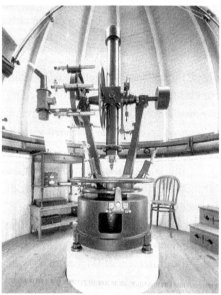

Figure 10.4. The 5-inch Warner and Swasey Altazimuth, one of the instruments purchased for the new observatory. It observed declinations only and could operate a few degrees either side of the meridian. From PUSNO, volume VII.

diameter, and, like on the 6-inch instrument, the two-minutes-of-arc divisions were inscribed on a band of silver. Only vertical-circle observations from this instrument were published in the course of its use at the Observatory; the timing coordinate of right ascension was better obtained with the transit circles. The instrument was mounted in a small wooden house northwest of the clock house in December, 1897.[15] Hill was in charge of the instrument until 1903, when Frank B. Littell took over.

[15] PUSNO, **8** [F. B. Littell, G. A. Hill, H. B. Evans], Introduction, "Vertical Circle Observations made with the Five-Inch Alt–Azimuth Instrument, 1898–1907."

Table 10.2. *Washington Fundamental Catalogs*[a]

| Catalog designation | Dates observed | Published | # Stars | Catalog compilers |
|---|---|---|---|---|
| W25 | 1930–35 | 1949 | 2,383 | Watts and Adams |
| W1(50) | 1936–41 | 1949 | 1,536 | Watts and Adams |
| W2(50) | 1941–49 | 1952 | 5,216 | Watts, F. Scott, and Adams |
| W3(50) | 1949–56 | 1964 | 5,965 | Adams, Bestul, and D. Scott |
| W4(50) | 1956–62 | 1968 | 2,554 | Adams and D. Scott |
| W5(50) | 1963–71 | 1982 | 14,916 | Hughes and D. Scott |
| WL(50) | 1967–73 (south) | 1992 | 23,001 | Hughes, Smith, and Branham |
| W1(J00) | 1977–82 | web | 7,476 | Holdenried and Rafferty |
| W2(J00) | 1985–95 (north) | web | 26,204 | Rafferty and Holdenried |
| | 1986–96 (south) | web | 28,186 | Rafferty and Holdenried |

Note:
[a] The last three entries are not fundamental, but complete the series of 6-inch and 7-inch transit-circle observations.

These four instruments, two modified and two new, carried out the ongoing positional work of the Observatory.[16] The goal remained the most precise determination possible of star positions: the transit circles in both coordinates, the prime vertical and altazimuth in declination only. The transit circles undertook large programs often covering the whole sky; the prime vertical and altazimuth usually concentrated on more specialized goals. Thus the largest program of the 9-inch transit circle would be the observations from 1913–26, a fundamental catalog under the direction of H. R. Morgan. The same is true of the 6-inch transit circle, though for many years instability problems prevented it from doing precise work. During 1911–18 J. C. Hammond directed a program of fundamental observations of the so-called Backlund–Hough stars, which were later used as a basis for Eichelberger's catalog of standard stars. The 6-inch transit circle was also used for determining time during 1918–25. Only in 1925 did it begin its first "fundamental" program under the direction of Watts, the first in a continuous series that would continue to 1995 (Table 10.2).

In contrast to the transit circles, the prime-vertical instrument was restricted largely to observations of the bright star Vega, observations made continuously during 1893–1912 with the view toward improving the aberration constant; from 1900 a list of stars was also observed for precise declinations. The instrument was last used during 1921–25 to observe precise declinations of some 800 stars within a few degrees of the zenith.[17]

So specialized was the work of the altazimuth that even its purchase was somewhat controversial. The Observatory was severely criticized first for purchasing it at all

[16] The 5.3-inch south transit and the photographic zenith tubes were used to determine time rather than position; we discuss them in the next chapter.

[17] PUSNO, **10**, part 1, "Observations made with the Prime Vertical Transit Instrument, 1893–1912," A xi–A xvi; PUSNO, **13**, Appendix II, 5–6. A brief history of the instrument is given in the opening pages of volume 10.

in light of the new 6-inch transit circle, and then for entrusting its construction to an American firm, again Warner and Swasey. After a single year of observations, Astronomical Director Harkness pointedly remarked, "I think we have now had sufficient experience with both instruments to show that the workmanship will compare favorably with anything ever turned out of a European workshop, and that the altazimuth instrument used as a vertical circle will give more accurate declinations than can be obtained with a transit circle."[18] The advantages of an altazimuth over a transit circle were technical in nature, but such technicalities were precisely what positional astronomy was all about. When, in 1908, Littell was called upon to justify the altazimuth program, he wrote that the principal advantage of the vertical circle on the altazimuth instrument was its continuous determination of changes in the reference point of the circle. Whereas in order to determine the zenith reference point a transit circle had to interpolate nadir measurements, the altazimuth determined the zenith point with each observation. "This," he wrote, "is the vital difference between vertical circle and meridian circle work, and is the principle which causes vertical circle work to be so highly valued by the leading astronomers of the world." The altazimuth's last program led to a catalog of declinations of standard stars, as well as those of the Sun, Mercury, and Venus, based on observations made between 1916 and 1933.[19]

With these instruments there was little room for creativity, but much opportunity for meticulous attention to detail. Many of the astronomers who worked on these instruments are therefore better known for the latter. Hammond, who directed the work of the 6-inch transit circle from 1911 until his retirement in 1933, was characterized by Watts as "an accurate and painstaking observer, and his skill in discussing observational material was outstanding."[20] Littell (Figure 10.5), who began his Naval Observatory career as a computer in 1891 and served as a Professor of Mathematics in the Navy from 1901 until retirement in 1933, became a bit better known in astronomy; although he was recognized for his meticulous observations with transit circles, the altazimuth, and the photographic zenith tube (PZT), he was also a delegate to the first meeting of the IAU in Rome in 1922, a sign of some status in the field. Skinner came to the Observatory as an assistant astronomer in 1870, rose to Professor of Mathematics in 1898, and divided his career between meridian astronomy and equatorial telescopes before his retirement in 1907. G. A. Hill (not to be confused with the mathematician G. W. Hill) spent almost 40 years at the Observatory on work so routine that little is known of him today.

By contrast with these now-forgotten astronomers, William S. Eichelberger, Herbert R. Morgan, and Chester B. Watts were also known for their painstaking work,

---

[18] AR (1898), 10.
[19] F. B. Littell, "Altazimuth: Some Advantages and Disadvantages in the Use of the 5-inch Altazimuth Instrument for Fundamental Declinations," USNOA, SF, Altazimuth folder, typescript, p. 2. The instrument's final program is described in PUSNO, **14**, part 3, F. B. Littell, "Vertical Circle Observations made with the Five-Inch Alt-Azimuth Instrument, 1916–1933."
[20] C. B. Watts, "John Churchill Hammond," PA, **48** (August, 1940), 364.

Figure 10.5. F. B. Littell at the eyepiece and W. S. Eichelberger reading the graduated circles of the 9-inch transit circle, about 1900.

but their work on projects of broader significance brought them to the attention of the wider world of astronomy. Like Hammond and Littell, they came to the Naval Observatory within a few years of the move to Observatory Circle, with the exception of Watts, who came in 1911. Eichelberger (Figures 10.5 and 12.5), who received a Ph. D. in astronomy from Johns Hopkins University in 1891, was (like Littell and Skinner) a Professor of Mathematics in the Navy. He is best known not only as the Director of the Nautical Almanac Office during 1910–29, but also as the compiler of a catalog of 1,504 standard stars, which was adopted by the IAU in 1925 as the standard used in the world's national ephemerides until 1940. He was also President of the Commission on Ephemerides of the IAU.[21] H. R. Morgan (Figure 10.6) also made significant contributions to fundamental astronomy beyond the observations. Immediately upon his graduation from the University of Virginia with a Ph. D. in 1901, he came to the Naval Observatory as a "computer," and, aside from a few years teaching astronomy and mathematics at Pritchett College, spent his entire career there. That career is closely connected with the 9-inch transit circle, which he took charge of in 1913 and which was retired upon his retirement in 1944. Morgan not only made

---

[21] H. R. Morgan, "William Snyder Eichelberger," MNRAS, **114** (1955), 289–291.

Figure 10.6. H. R. Morgan at the eyepiece end of the 9-inch Transit Circle, which he ran for much of his career. Note the pendulum clock in the background and (above Morgan's head) the system for illuminating the circles.

many of the meticulous observations himself and published them in well-known fundamental catalogs, but also used these and other observational data to analyze the fundamental constants and determine proper motions of the stars. His status in the scientific community is indicated by the positions he held, as Vice President of the American Astronomical Society (1940–42) and associate editor of its *Astronomical Journal* (1942–48), President of the Commission on Meridian Astronomy of the IAU (1938–48), and recipient of the Watson medal of the National Academy of Sciences in 1952. As early as 1928 he was a delegate to the IAU when it met in Leiden.[22]

C. B. Watts (Figure 10.7), though he had only a bachelor's degree from Indiana University, nevertheless also rose to a prominence equal to that of Morgan; like Morgan he received the Watson Medal of the National Academy (1956) for his lifetime accomplishments. With the exception of four years (1915–19) in the Time Service Division, Watts spent his entire career until his retirement in 1959 in the Six-Inch Transit Circle Division. Watts, one of his fellow astronomers noted, "possessed an unusual combination of talents: those of an astronomer, an experimenter with great imagination, and an expert mechanic," a combination he used to improve the results of the 6-inch transit circle, especially after he was placed in charge of the instrument in 1934. It was Watts who was largely responsible for the series of fundamental catalogs observed with the instrument, which stretched beyond his retirement to the end of the century. Beyond that, perhaps his greatest contribution was the survey of the marginal zone of the Moon, a project we shall place in context in section 10.2.[23]

In summary, at its new site the Naval Observatory was well equipped to carry

[22] Raynor L. Duncombe, "Herbert Rollo Morgan," DSB, **9**, 513–514; F. P. Scott, "H. R. Morgan, Astronomer," *Science*, **126** (1957), 497.
[23] F. P. Scott, "Chester Burleigh Watts," QJRAS, **13** (1972), 110–112; F. P. Scott, "In Memoriam: C. B. Watts," *The Moon*, **6** (1973), 233–234.

Figure 10.7. C. B. Watts at the micrometer end of an experimental photographic transit instrument, about 1926. Watts built the instrument by modifying the old 5.3-inch transit instrument.

on what it did best: determining precise star positions. Despite the inevitable problems associated with a new site and new instruments, a variety of well-known, lesser-known, and obscure astronomers overcame technical problems to do state-of-the-art work in their field; a few participated in the broader community. However, by the very nature of the instrumentation, even a mechanical wizard like Watts could not produce breakthroughs that would greatly increase accuracies. All they could do was maintain the accuracy of the reference frame, improve it slightly (by approximately a factor of five during the century from 1890 to 1990), increase the number of stars whose positions were well known, and produce (in the case of Morgan) proper motions that could in turn be used for a variety of astronomical projects. As Figure 2 in the Prelude shows, the era of revolutionary improvement in positional accuracies lay more than a half century ahead, with the development of new techniques. Moreover, very seldom did Naval Observatory astronomers apply their own work to the practical problems of astronomy during this period – work like the structure of the Galaxy advocated by Boss. They were data producers. The data consumers were for the most part outside astronomers, not bound to a specific mission like their Navy colleagues.

### The large refractors: Standard work and attempts at spectroscopy

By comparison with the routine work of the meridian astronomers, work that consumed the bulk of the Observatory's energies, the goals of the 12-inch and 26-inch refractors were somewhat more glamorous. With the retirement of Asaph Hall on October 15, 1891 the 26-inch equatorial, now almost two decades old, lost its longest and most famous observer. His son, Asaph Hall Jr, was immediately placed in charge of the instrument, observing satellites of Saturn, Uranus, and Neptune. However, when, in July and August, 1892, Mars was closer to Earth than it had been since Hall's

1877 discovery of the Martian moons, the elder Hall was allowed one last observing run, and published his final observations of the satellites, the south-polar cap, and the surface of the planet. (Three years later the publication of Percival Lowell's *Mars* would initiate the furor over Martian canals.)[24] Shortly after the senior Hall's final observations, in December, 1892, the 26-inch refractor was dismounted at the old site. Although a new tube and mounting were awaiting the historic lens, the old tube and mounting were nevertheless taken to the new site, and were to see unexpected use.

Harkness oversaw the remounting of the instrument on the new Warner and Swasey mount, with the objective lens first being attached in June, 1893. By December the telescope was ready for observing. As a piece of machinery, its size and dimensions were still impressive. The cast-iron mounting weighed 18 tons, with the steel polar and declination axes adding another 1.5 tons. The 26-inch lens, with a focal length of 389.66 inches (more than 32 feet), was the original Clark lens composed of a flint-glass element weighing 110 pounds and a crown element of 70 pounds. The total weight of the telescope and mounting was 28 tons, which the clock drive (itself weighing more than half a ton) had to turn at a precise rate. Aside from the tube and mounting, its chief new features were a larger position circle (9.25 inches in diameter divided into half degrees), a more powerful clock drive than before, and both the micrometers and circles illuminated by dim electric lights. The mounting rested on a solid concrete pier extending 12 feet below ground level. The most novel feature was not the telescope itself, but the large elevating floor "to facilitate the use of spectroscopes and other heavy apparatus." An elevating floor – 41 feet in diameter, weighing 15 tons (nearly balanced by counterweights), and rising a maximum of 12 feet – was powered by four hydraulic rams. All of this was placed in a 24 ton dome 45 feet in diameter (Figure 10.8).[25]

A few hundred feet to the east, the dome for the 12-inch refractor sat atop the main administration building (Figure 10.9). The mounting, by George N. Saegmuller of Washington, D. C., had been completed in November, 1892, but, because the 12-inch lens had not yet been completed, the old 9.6-inch refractor was mounted first. On December 19, 1895 the 9.6-inch objective was replaced with the 12-inch lens, with a focal length of 15 feet. One of the novelties of the telescope, which had been designed by Harkness, was its "star dials" for indicating right ascension and declina-

[24] Hall, "Observations of Mars, 1892," *AJ*, 12 (1893), 185–188. Hall remarked that "when the images were good, the usual markings on the planet could be seen, but no duplication of the so-called canals could be made out among the finer markings."

[25] AR (1892), 133–134. For details of how Warner and Swasey procured this contract see Edward Jay Pershey, "Warner and Swasey at the Naval Observatory. A View of the Science–Technology Relationship," in Elizabeth Garber, ed., *Beyond History of Science* (London and Toronto, 1990), pp. 220–230. The most detailed description of the new 26-inch instrument is in PUSNO, 6, A iii ff. The water pressure for the hydraulic floor rams initially came from steam pumps in the basement of the main building, but after 1896 came by direct water pressure from the city mains. The floor is described in detail in PUSNO, 6, A ix, and the change in hydraulic power in AR (1896), 4–5. The floor was not without initial problems; its instabilities caused "anxiety and embarrassment" until it was repaired in 1901 (AR (1901), 10–11). The rising floor is still a favorite of USNO visitors. The dome was initially moved by hand rope until a motor was installed in October, 1907.

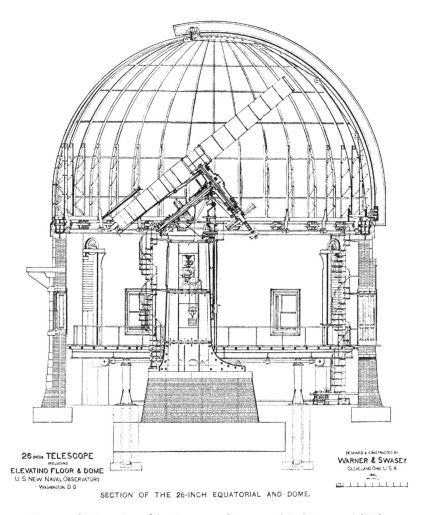

SECTION OF THE 26-INCH EQUATORIAL AND DOME.

Figure 10.8. A section of the Warner and Swasey 26-inch Equatorial Telescope and its dome, 1893.

tion. While the divided circles for this purpose were normally on the polar and declination axes, these were difficult to see at night. Harkness's innovation was to locate additional dials on the pier, where the observer could see them. The 12-inch lens was also provided with a filar micrometer by Saegmuller.[26]

---

[26] The 12-inch instrument is described in PUSNO, **6**, A iii, and in PUSNO, **12**, 53–56. On Saegmuller (1847–1934) see George N. Saegmuller, *The Story of my Life* (privately published, 1929) in the USNO Library. The 9.6 inch refractor had been dismounted at the old site in October, 1892. On the 12-inch star dials designed by Harkness see AR (1893), 6–7, and "Finder Circles for Equatorials," *Nature*, **50** (1894), 64, where the writer stated that they were first placed on the Georgetown University 12-inch telescope. Harkness set the record straight in a letter stating that he designed the first star dials for the USNO 12-inch telescope, on which they were mounted in November, 1892, and that later Saegmuller fitted a second set to the Georgetown instrument that had been constructed several years before. Harkness, "Finder Circles for Equatorials," 12-inch folder, USNOA, SF. The star dials are pictured in an article on the Georgetown instrument in the *Zeitschrift für Instrumentenkunde* (Berlin, 1894), 128–130.

Figure 10.9. The 12-inch telescope, atop the main building. George N. Saegmuller built the mounting in 1892–93 from designs and specifications of William Harkness. The principal novelty of the mounting is a pair of dials that face the observer when he or she is at the quick-motion handwheels, and indicate right ascension and declination. At the side of these dials are the eyepieces of the long microscopes for reading the fine right ascension circle, while the eyepieces of the long microscopes for reading the fine declination circle are at the eye end of the telescope. The rate of the driving clock could be switched from sidereal to solar or lunar.

The general plan of work for the equatorials, as described in the first volume of observations published in 1911, was "to observe (1) the satellites and the diameters of planets, (2) some of the interesting and difficult double stars, (3) the minor planets and comets, and (4) the occultations of stars by the Moon. The 26-inch has been devoted exclusively to observing the fainter and more difficult objects. Continuity in the observations of the satellites of Saturn, Uranus, and Neptune has been considered especially important."[27]

Because of the younger Hall's departure to become Director of the Observatory at the University of Michigan, Stimson J. Brown (Figure 9.15) was placed in charge of the 26-inch equatorial in 1893. He had at his disposal a considerably modified telescope equipped with two kinds of apparatus: a spectroscope and an assortment of visual micrometers for precisely measuring planetary diameters as well as satellite and double-star positions. The Clark I micrometer had been used continuously by Asaph Hall from 1877 to 1892. The Clark II micrometer was placed on the instrument in March, 1900, and used continuously until 1911. A Warner and Swasey micrometer was also delivered with the instrument, and an Alvan Clark double-image micrometer could be used for larger fields of view. Even after the arrival of a new Repsold micrometer in 1914, the Clark II micrometer continued to be used at times; the latter weighed only 18 pounds, compared with 81 pounds for the Repsold. Brown's work on the instrument consisted of observations of Mars, Saturn, and Neptune and their satellites, as well as comets and even the transit of Mercury of November 10, 1894. He made a few drawings of Mars and, just as Lowell's canals-of-Mars controversy was about to erupt, remarked that the 26-inch instrument's observations "are noticeably lacking in the numerous canals which have been sketched with much smaller telescopes."[28]

More unexpected than the astrometric work were the experiments in spectroscopy begun under Brown. Some very minimal spectroscopy had been carried out occasionally at the Naval Observatory, and Harkness had even discovered the important coronal K line. As early as 1867 the Superintendent had written that "The impetus given to scientific inquiry by the recent developments of the spectroscope, in regard to the chemical analysis of the heavenly bodies, renders it very desirable that this institution should be provided with such an instrument," and he followed up by including a spectroscope among the requests for special appropriation. We recall also that a spectroscope, designed by Newcomb, was provided by Clark with the original telescope.[29] Once the 26-inch instrument had been moved to the new site, a direct-vision spectroscope, constructed by John A. Brashear and identical to J. E. Keeler's at Allegheny Observatory, was furnished with the telescope (Figure 10.10).[30]

---

[27] J. C. Hammond, Introduction to volume 6 of PUSNO (1911), A iii.    [28] AR (1895), 110.

[29] PUSNO for 1874 (1877), 38, photo facing p. 33. The 1867 request is in AR (1867), 130.

[30] PUSNO, 6, A xi. The spectroscope could be used with a Rowland grating of 14,438 lines to the inch, but also had a single 60-degree flint-glass prism mounted at minimum deviation, and three 60-degree flint-glass prisms with automatic minimum-deviation apparatus. The spectroscope also included an observing telescope with micrometer, photographic lenses with camera

Figure 10.10. The 16 inch telescope with spectroscope attached.

(footnote 30 cont.)
attachment, and a comparison-spectrum apparatus. A detailed description, with photographs, of the identical Allegheny spectroscope is to be found in James E. Keeler, "The Spectroscope of the Alleghany Observatory," *Astronomy and Astrophysics*, **12** (January, 1893), 40–50. At the end of this article Keeler noted that "A duplicate of the Allegheny spectroscope has been ordered for the new U. S. Naval Observatory, and another will probably be made for the University of Chicago." The spectroscope also had many features similar to those of the one Keeler had designed and used at the Lick Observatory (which was also built by Brashear) as described in James E. Keeler, "The Star Spectroscope of the Lick Observatory," *Astronomy and Astrophysics*, **11** (February, 1892), 140–144. See Don Osterbrock, *James E. Keeler: Pioneer American Astrophysicist* (Cambridge University Press: Cambridge, 1984), pp. 136–148. The 12-inch telescope was also provided with a spectroscope, PUSNO, **6**, A xxvii.

It was with the Brashear spectroscope in its photographic mode that Brown and photographer George Peters began their work. Experimentation began as early as 1896, and Brown reported that spectroscopic work for line-of-sight motion (radial velocities – a work being pioneered at Lick Observatory) was begun in November, 1898. Although 38 measurable plates were secured, many more failed, according to Brown at least in part because the telescope optics were maximized for visual, not photographic, observations. Because of other problems the spectrograph itself was returned to Brashear for "extensive alterations" the following March. Brown also recounted that the spectrum of alpha Aurigae (Capella) showed double lines, which was thought to be a defect. However, "recent discoveries by Professor Campbell at the Lick Observatory indicate that the star is a spectroscopic binary, and the doubling in question was probably due to that fact and not to any defect in the spectrograph."[31] Despite later stories to the contrary, the spectroscopic work seems not to have been abandoned for this reason but, according to Brown, "was never carried beyond the experimental stage, owing to inherent instrumental difficulties, and it was finally abandoned in September, 1900."[32] Peters was more specific about the problem: "It was found that, owing to the strong colors existing in the flint and crown glasses of the equatorial objective, the instrument was almost worthless for spectroscopic investigations by photographic methods in the blue region. The flint lens is of a decided yellow tint, while the crown lens is very green, producing a remarkable degree of absorption in the region of the spectrum employed [the H gamma region]. It was concluded that to overcome this difficulty of the absorption of the blue rays by the large objective it would be necessary to work near the red end of the spectrum, using orthochromatic plates. In view, however, of the decreased dispersion by prisms in this region, it was decided to be impracticable."[33]

Still, it is notable that both Lick and Allegheny pioneered in spectroscopic radial velocities, adapting their telescopes to make the observations, whereas the Naval Observatory never even began such a program. Although a certain W. W. Dinwiddie was specifically transferred from assistant on the equatorial to assistant in spectroscopic work in 1902, the Naval Observatory did not give spectroscopy sufficient priority to purchase new telescopes or equipment, or even modify existing ones.[34] Still reeling from the Board of Visitors' charges that the Observatory received too much money for too little output, the chances of purchasing a new telescope, or

---

[31] AR (1899), 4–6: 6. See also AR (1896), 7, and AR (1897), 4. The latter reports attempts at obtaining Mars spectra.

[32] PUSNO, 6, A xi. The double line story became part of USNO folklore; John Hall cited the story many years later.

[33] PUSNO, 6, A xi; see also George Peters, "The Photographic Telescope . . ." (ref. 50 below), 351.

[34] AR (1902), 4 and 22. Keeler had faced similar problems with the Allegheny and Lick refractors, but had still pioneered in photographic spectroscopy. See Keeler, "On a Lens for Adapting a Visually Corrected Refracting Telescope to Photographic Observations with the Spectroscope," ApJ, 1 (February, 1895), 101–111. For the context, see Osterbrock (ref. 30), pp. 136–148. Keeler, who might have helped the Naval Observatory staff with their spectroscopic problems had they asked, died a young man in 1900.

Figure 10.11. T. J. J. See in his office at Mare Island, 1910.

even a corrector lens, for the purpose were virtually zero. So too were the chances of cooperation with people who might have given advice on the new techniques of astronomy. It would be a quarter of a century before the Observatory would reach out for help in modernization, and more than half a century before it re-entered the field of spectroscopy.[35]

Faced with this situation, the spectroscope was placed in storage and the 26-inch instrument returned to more routine classical functions. On December 15, 1899 T. J. J. See (Figure 10.11) was placed in charge of the 26-inch instrument when Brown succeeded to the position of Astronomical Director on the retirement of Harkness. See, who was only 33 years old at the time he came to the Naval Observatory as a Professor of Mathematics in 1899, had obtained his doctorate from Berlin in 1892, subsequently spent three years at the University of Chicago and two more at Lowell, and was in charge of Lowell's survey of southern double stars. He became involved in a bitter dispute with the celestial mechanician F. R. Moulton, and was banned from publishing in the *Astronomical Journal*. The Naval Observatory's willingness to hire him under such circumstances is something of a mystery, and he did not last long in Washington. Although in 1902 See contemplated expansion of double-star work (he

---

[35] In 1954 a grating spectrograph was constructed for the 40-inch reflector under the supervision of Stewart Sharpless, AR (1954), 364.

Figure 10.12. Asaph Hall Jr at the 26-inch Equatorial Telescope. He is using the Repsold micrometer, which weighed 81 pounds, compared with 18 pounds for the Clark II micrometer. From *PUSNO*, volume XII.

and his collaborators did make about 1,500 measurements at the Observatory), as well as beginning astrophysical and parallax work, his separation on September 20, 1902 prevented this. Perhaps See's most important work during this period was his observations during 1900–01 of Eros as part of the international program to determine solar parallax. See's association with the Observatory was far from over, however; after a year at the Naval Academy, See spent the remainder of his career, spanning the period 1903–39, at the Observatory's Mare Island station. This small station in California was largely devoted to time services, but here See could indulge in his personal work and well-known theories, considered eccentric by most astronomers.[36]

Another factor limiting further work with the Great Equatorial was undoubtedly a shortage in manpower. See wrote that "With the increase of force on the instrument following the assignment of Computer George K. Lawton, June 14 [1901], the hope was entertained of utilizing its powers to their full extent. The lamented death of this talented and accomplished young astronomer, July 25, 1901, after a month's service on the instrument, was not only an irreparable loss to the 26-inch equatorial, but in view of his youth, fine training, high promise, and noble character, altogether one of the greatest losses ever sustained by the Observatory at any period of its history. His abilities were of the highest, his performance of duty beyond all praise."[37]

Following five years of Aaron Skinner and then briefly William Eichelberger, in 1908 Asaph Hall Jr (Figure 10.12) resumed his tenure in charge of the instrument, one that would extend to 1929. Hall, who worked in the shadow of his father's reputation, had obtained his Ph. D. in astronomy from Yale in 1889 and served several stints at the

[36] AR (1901), 11–12. PUSNO, **6**, A iv. The results of See's observations of Eros are in PUSNO, **3**, part 1. The observations were made at the USNO from October 9, 1900 to January 29, 1901, and were reduced by Mr George K. Lawton, who held the position of Computer. See also S. J. Brown, "Feasibility of Obtaining the Solar Parallax from Simultaneous Micrometer Observations of Eros," PA, **8** (August–September, 1900), 353–355. See's life prior to 1913 is described in W. L. Webb, *Brief Biography and Popular Account of the Unparalleled Discoveries of T. J. J. See* (T. P. Nichols & Son: Lynn, Massachusetts, 1913), and his career has recently been analyzed in Thomas J. Sherill, "A Career of Controversy: The Anomaly of T. J. J. See," JHA, **30** (1999), 25–50.

[37] AR (1901), 15. Lawton (1873–1901) had been a student of Asaph Hall Jr at the University of Michigan and of See at the University of Chicago, and died at the age of 27 of typhoid fever. T. J. J. See, "George K. Lawton," PASP, **13** (1901), 182–183.

Naval Observatory during the 1880s and 1890s before settling in as Professor of Astronomy and Director of the Observatory at the University of Michigan (1892–1905). He then returned to the Naval Observatory, was commissioned (like his father) a Professor of Mathematics, U. S. N., and remained until his retirement in 1929.[38]

Although Hall served as the Naval Observatory delegate to the Conference on Organization of International Radio Telegraph Signals in 1912, and the following year played an important role in the Washington–Paris longitude determinations via radio signals, his most important work was with the 26-inch equatorial. It was during his tenure that the Repsold filar micrometer was purchased; it was mounted on June 30, 1914. Most of Hall's work on the instrument was not on double stars, but on micrometric measurements of the positions of the satellites of Mars, Saturn, Uranus, and Neptune. He seems not to have contemplated any spectroscopic work.[39]

The entirety of the work on the 26-inch equatorial up to this point was therefore visual rather than photographic, even though the photographic technique would have been better in many cases had the telescope been adapted to it. Brown reported, for example, that the photographic method was preferred for the Eros work, but could not be used in the United States because "The Naval Observatory has no suitable photographic telescope for this purpose, and with the exception of Harvard College Observatory there is none in the country. Except for this reason there would be no gain in resorting to micrometric observations."[40] The struggle with the application of photography to astronomy at the Naval Observatory was just beginning; to that struggle we turn in the next section.

*Photographic episodes*

Although experiments in astronomical photography had begun as early as the 1840s, it was not until the end of the 1880s that it became clear that photography would play an important role in astronomy.[41] As we saw in chapter 7, the Naval Observatory had experimented with photography for the transits of Venus of 1874 and 1882 with mixed results, and used the photographic technique with the transit of

---

[38] On Hall Jr see Charles P. Olivier, "Asaph Hall," PA, **38** (1930), 531–533. Also Mary Hall Kilpatrick, "Recollections of my Father Forty Years after His Death;" Hall, "Reminiscences of the Naval Observatory;" and "Asaph Hall, Jr.," vita, all in USNOA, BF, Hall, Jr folder.

[39] On Hall's work with the 26-inch instrument see PUSNO, XII, 3–346, on the Repsold micrometer, pp. 31–32, and on the telescope in general Asaph Hall, "Brief Description of 26-inch Equatorial Instrument of the Naval Observatory", PA, **27**, 278–281. The micrometer was ordered from C. L. Berger & Sons of Boston, constructed by A. Respold & Sons of Hamburg, and arrived on August 2, 1913. The delay in mounting it was due to the weight of the new micrometer (81 pounds as opposed to 18 pounds for the Clark II), which necessitated changes to the tailpiece of the telescope. The general Repsold micrometer is described in L. Ambronn, *Handbuch der Astronomischen Instrumentenkunde*, p. 529, and a brief description of the USNO Respold micrometer is also found in J. A. Repsold, AN, 5062.

[40] S. J. Brown, "Feasibility of Obtaining the Solar Parallax from Simultaneous Micrometer Observations of Eros," PA, **8** (1900), 353–355: 354.

[41] John Lankford, "The Impact of Photography on Astronomy," in *Astrophysics and Twentieth-Century Astronomy to 1950*, ed. Owen Gingerich (Cambridge University Press: Cambridge, 1984), pp. 16–39. Lankford says (p. 23) that "The 1880s stand as a watershed in the history of astronomical photography, for as the decade opened it was by no means certain that photography would play more than a peripheral role in astronomy."

Mercury. Photographs were also taken during the many solar-eclipse expeditions sponsored by the Observatory (section 5.4). In 1880 Admiral Rodgers reported that experiments were under way at the Observatory to determine the capabilities of telescopes for various classes of celestial photography. And in justifying a request for a $1,000 appropriation for photographic work in 1885, Superintendent Commodore George Belknap wrote that photographic work "will have to be taken up here in order to keep pace with the requirements of modern astronomical research and observation. It cannot, therefore, be begun too soon . . . Much work is desirable to be done in photographing star clusters, nebulae, and the spectra of sun-spots, stars, &c, and in the production of star maps by photography." For this and eclipse photography, Belknap argued, the Observatory should "have a staff drilled in photographic work, and this might be easily accomplished, as the number need not be great, and it could be made up of officers stationed here." Ensign Albert G. Winterhalter, he reported, had received a course of instruction in photography from Professor Baird, Secretary of the Smithsonian Institution, and was ready to put his knowledge to good use beyond the photographs he had recently taken during an annular eclipse.[42]

Despite these good intentions, three events especially illuminate the Naval Observatory's shaky entry into photographic work: the international astrophotographic project, the routine program of daily photographs of the Sun, and the attempts to acquire a photographic equatorial. Taken together they demonstrate the difficulties encountered by the Naval Observatory in entering the realm of the new astronomy.

One of the first issues to face the Observatory, even as it was preparing to move to the new site, was whether to participate in a new international program for photographic star positions. At the Paris Observatory, where Paul and Prosper Henry had been undertaking stellar photography with increasing success, a grand project to photograph the entire sky was devised, known as the *Carte du Ciel*.[43] Although many countries from around the world participated in this endeavor, the United States did not, despite interest over a period of years from the Naval Observatory. Why this was so demonstrates how, even when the Naval Observatory saw the need for new programs, their implementation depended on securing the necessary funds.

Of the 58 delegates to the International Astrophotographic Congress held in Paris in 1887 (Figure 10.13), three were American: W. L. Elkin (Yale Observatory), C. H. F. Peters (Hamilton College, representing the American Academy of Sciences), and Lt A. G. Winterhalter from the Naval Observatory. Of these three institutions, only the latter seems to have seriously considered undertaking the project. Winterhalter

---

[42] AR (1880), 123–124; AR (1881), 231–234; AR (1885), 9. In 1880 Harkness was undertaking the experiments, assisted by Joseph A. Rogers, chief photographer for the Transit of Venus Commission. Rogers resigned in March, 1883, leaving the photographic experiments incomplete. In 1881 the Superintendent reported that "all our photographic operations are much impeded by the cramped and unsuitable character of our photographic house. It was erected in 1873 as a temporary structure in which to make certain experiments relating to the then approaching transit of Venus, and has been used almost without change ever since."

[43] H. H. Turner, *The Great Star Map* (J. Murray: London, 1912); Lankford (ref. 41), pp. 16–39.

Figure 10.13. Participants in the International Astrophotographic Congress, Paris, 1887. Courtesy of the Royal Astronomical Society Library.

Figure 10.14. Lt Albert G. Winterhalter represented the Naval Observatory at the 1887 International Astrophotographic Congress. Courtesy of United States Naval Academy, William W. Jeffries Memorial Archive.

(Figure 10.14) returned from the Congress, wrote a detailed report, and, on October 9, 1888, strongly urged the Superintendent of the Observatory, Captain R. L. Phythian, to request the necessary funds. In doing so, he argued that the United States, "the home of astronomical photography," would likely be left without a participant in the project if the Naval Observatory failed to obtain funds.[44] Winterhalter specifically recommended a small refractor for solar photography, a refractor for participating in the international program to chart the sky, and a reflector of large diameter and short focal length for photographing faint objects such as comets and nebulae.

Under further pressure from J. Janssen of the French Academy of Sciences and A. A. Common in Great Britain, Phythian replied that "at present we have no facilities for carrying on such work. The new Observatory is now being built, and it is hoped that, before its completion, provision will be made for the necessary buildings, instruments, etc., to enable us to participate in this important branch of science." No promises could be made, he emphasized, until Congress appropriated sufficient funds. On December 17, 1888 Phythian began that process by writing his superior, chief of the Bureau of Navigation, an eloquent plea and the substance of a proposed clause for $50,000 to be inserted into the naval appropriations bill "for the purchase of a photographic telescope and pointer, the construction of buildings and domes and mountings for same,

---

[44] Winterhalter's detailed report is cited in ref. 8 above, pp. 3–71. His immediate recommendations are in Winterhalter to Superintendent, USNO, October 9, 1888, AR ending June 30, 1888, 19–20. Copies of his orders are on pp. 323–332. Winterhalter, pp. 6–7, gives a brief history of celestial photography and notes the photographic work of the Americans Rutherfurd, Draper, Gould, and Pickering, as reviewed in Young, *New Princeton Review* (May, 1887) and Pickering, *Memoirs of the American Academy*, part iv, no. iv. The remainder of the volume is notable for its detailed illustrated descriptions of European observatories.

and material for photographic work."[45] Thirteen instruments were now being constructed for foreign observatories, he noted, and an appropriation would at least allow the Naval Observatory to pledge participation in the project, if not begin it on time. The proposal quickly went through government channels, receiving approval from the Chief on the 19th, the Secretary of the Navy on the 26th, and the Secretary of the Treasury on the 29th, and being referred to the House Committee on Appropriations January 2, 1889. In forwarding the proposal to his superior, Bureau of Navigation Chief J. G. Walker wrote that "It is for such objects as these that the National Observatory has been created and maintained by the Government, and a failure to utilize upon this important occasion its superior appliances and trained staff of observers would be a fatal error."[46]

The proposal, however, reached Congress too late to be acted upon, and, although Phythian suggested that the matter be referred to the Senate Appropriations Committee, where it might be incorporated as an amendment to the sundry civil bill, the Washington *Evening Star* for March 30 reported that Congress had failed to provide the $50,000. "Capt. Phythian, the superintendent and the astronomers at the naval observatory are anxious to take part in this international enterprise," the paper reported, "but by reason of the failure of Congress to provide the $50,000 asked by the Navy department for the work, they have been obliged to delay their preparations until Congress meets again."[47] Approval was expected "early in the coming winter."

That the project was never approved was probably due to several reasons. Lt Winterhalter, undoubtedly a strong lobbying force for the project, was transferred to the Torpedo Station near Newport, Rhode Island in April as part of the Navy's normal tradition of personnel rotation. In July the Observatory itself was transferred from the Bureau of Navigation to the Bureau of Equipment and Recruiting, which may well have been less sympathetic to the project. Furthermore, the construction of the new Naval Observatory a few miles away in Georgetown undoubtedly diverted the energies of the institution to other matters. As late as 1894, however, the Superintendent, Captain F. V. McNair, was still inviting the attention of his superiors to House Executive Document 46, to no avail.[48]

In the end the Naval Observatory never participated in the *Carte du Ciel*, nor did any other U. S. observatory. Some have suggested that this was all for the best, and

[45] Phythian to Janssen and Common, October 23, 1888, RG 78, Phythian to Chief of Bureau of Navigation, December 17, 1888, RG 78, "Letters Sent." The latter was published, along with other relevant documents, in "Charting the Sky," 50th Congress, second session, House Executive Document 46, pp. 3–4, January 2, 1889. As early as March 16, 1887 Phythian had requested from Alvan Clark & Sons an estimate for 12- and 24-inch photographic telescopes; RG 78.  [46] House Executive Document 46 (ref. 45), pp. 2–3.

[47] "To Photograph the Heavens: The Sky Divided Among Observatories of the World," *The Evening Star*, Saturday, March 30, 1889. The subhead read "A Great Co-operative work in which the Naval Observatory is Expected to Take Part."

[48] On Winterhalter's transfer see AR (1889), 414. On the 1894 request, see AR (1894), 167–168. In retrospect, with the increasing use of photography in astronomy by 1890, and especially with Winterhalter's suggestion already in 1889 for the purchase of a photographic telescope, one might well ask why the money for the 12-inch refractor had not been spent on a photographic telescope. The answer seems to be that a 12-inch visual refractor was a standard "workhorse" instrument at any observatory, and the Naval Observatory apparently was not prepared to sacrifice such a standard instrument for a newer technique.

Figure 10.15.  Lucy T. Day at the 40-foot photoheliograph for photographing the Sun, 1947. Erected in 1894 at the Observatory's new site, the instrument was one of eight used for the American transit-of-Venus expeditions in 1874 and 1882. The ornate building was torn down in the early 1950s.

that American observatories in general (in contrast to European ones) may have been able to undertake more creative work unrestrained by the enormous commitment to map the sky photographically. In any case, the Astrographic Catalogue based on the *Carte* was eventually carried through as one of the great photographic projects in astronomical history. As we shall see in section 10.3, the Naval Observatory did play the key role in making the results of the *Carte du Ciel* useful for modern astronomy.

The Observatory fared somewhat better in its solar-photography attempts, undoubtedly because it already had instrumentation on hand that could be used in such a program and also because the move to the new site had been completed. Aside from its solar-eclipse photos, as early as 1879 the Observatory had undertaken an effort to photograph the Sun on every clear day, but it was abandoned seven months later.[49] On October 11, 1898 the Naval Observatory began taking a series of solar photographs that would last for three quarters of a century. The instrument, one of the transit-of-Venus 40-foot photoheliographs (Figure 10.15), had been mounted at the new Naval Observatory in 1894 to observe a transit of Mercury. When, in August, 1898, George H. Peters was hired as the staff "photographer," one of his first duties was to make the photoheliograph operational. Peters (Figure 10.16) would become the single

---

[49]  A series of 108 solar photographs had been taken during the seven-month period, but the work could not be kept up continuously. AR (1881), 233.

Figure 10.16. George H. Peters, seen here at the 12-inch refractor.

most important person at the Observatory in photographic matters for the next three decades.[50]

The photoheliograph that Peters revived produced images of the Sun 4.3 inches in diameter on a 7-inch-square plate.[51] The rationale for this work was not only to study the Sun's surface, but also "since such changes as are observed there are connected very intimately with the changes which take place in the magnetism of the earth, which in turn has a direct influence upon the compasses which guide our ships safely from port to port, the importance of this photographic work to the Navy itself is apparent." This rationale kept the program going until January 1, 1972, when it was discontinued due to budget constraints and the availability of more sophisticated data elsewhere.[52]

Once the solar photography program had become well established, attempts were made to extend the technique to other objects. Photographic methods for

[50] Peters was born in Hartford, Connecticut in 1863, and became interested in astronomy as a youth. He was influenced by articles by Henry Draper and G. W. Ritchey, and began to experiment with photographic work; his first job at the Naval Observatory had been with Brown on the 26-inch spectrographic work. Peters's early experiences are described in Peters "The Photographic Telescope of the U. S. Naval Observatory: On the Construction of a Pair of Triple Objectives: Their Use in Celestial Photography; Together with a few Reminiscences," PA, **27** (1919), 349–358.

[51] The photoheliograph work is described in PUSNO, **12**, 435–526; see also AR (1899), 16. The wet-collodion process was used until December, 1899, when it was replaced by the dry plate. The photoheliograph itself was the one originally used in 1874 at Hobart, Tasmania. Peters further described the work in a paper read before the Astronomical and Astrophysical Society meeting in Washington, January 1, 1903, USNOA, BF, Peters folder.

[52] On the rationale see AR (1904), 11; on the discontinuation AR (1972), 235. The photoheliograph was originally located in its own building south of the dome for the 26-inch telescope. In 1953 it was moved to the South Transit House attached to the main building, and in late 1965 it was

Figure 10.17. The 10-inch photographic-telescope house, with old 6-inch Dallmeyer lenses in place, May 14, 1911 (located just west of the current Time Service building).

observing asteroids were begun at the Observatory in 1902, when a 6-inch Dallmeyer portrait lens with 39-inch focal length was attached to the middle section of the 26-inch equatorial tube. This type of lens was ideal for photographic research; David Gill, astronomer at the Royal Observatory at the Cape of Good Hope, had used a similar 6-inch Dallmeyer $f/9$ doublet portrait lens for his photographic southern extension of Argelander's Bonner Durchmusterung beginning in the 1880s. At the Naval Observatory the plan was to find asteroids on the photographic plate that could then be observed with the 12-inch or 26-inch telescopes; star positions on some of the plates (whose field was $5 \times 6$ degrees) were also measured on a Stackpole measuring engine purchased for the transit-of-Venus work.[53]

Because this was not a very efficient use of the 26-inch telescope, in 1904 W. W. Dinwiddie, the former "spectroscopic assistant," devised another scheme. He took the Clark mount for the 26-inch telescope that had been used at the old site, attached two 6-inch Dallmeyer lenses of 39-inch focal length ($f/6.5$) to the central 10-foot section of the old 26-inch telescope tube, placed the old 9.6-inch telescope inside this tube as a guiding scope, and constructed a shelter around the resulting instrument on the southern part of the Observatory grounds (Figure 10.17). Because the Dallmeyer lenses had been secured for the 1878 total solar eclipse, the entire apparatus was made from available parts, and the Superintendent delighted in reporting that the whole arrangement "will compare favorably with any other that could be duplicated for $10,000 or $15,000." The resulting asteroid observations were published as corrections to existing ephemerides or as approximate positions.[54]

In 1910 Peters, who was in charge of this photographic equatorial, proposed the

moved to the roof of the Simon Newcomb Laboratory. Robert Harrington and Mihran Miranian, the last astronomers associated with the program, summarize it in "Sunspot Areas, 1907–1970," USNO Circular 133 (June 18, 1971).

[53] AR (1902), 15. This arrangement was dismounted from the 26-inch telescope in May, 1906. Gill's work with the Dallmeyer lens is described in Gill, History and Description of the Royal Observatory, Cape of Good Hope (Neill & Co.: London, 1913), p. xlix.

[54] This photographic equatorial telescope was described in AR (1905), 16; AR (1908), 10; and sporadically thereafter. See also PUSNO, **6**; the most connected history is given in PUSNO, **12**, 347–357. The 6-inch photographic telescopic observations were published in AN and AJ; for references see PUSNO, **12**, 355.

Figure 10.18. The 10-inch photographic telescope, devised by Peters. The large tube is the central 10-foot section of the old 26-inch tube, and the smaller tube inside of it is the old 9.6-inch refractor, which served as a guidescope.

construction of a pair of 10-inch triplet photographic lenses, which he would grind, polish, and figure himself at the Observatory. The Superintendent approved, the glass was ordered from Jena, and, after the completion of a new electric driving clock designed by Peters, the resulting "10-inch photographic equatorial" (Figure 10.18) began its long series of observations of asteroids on October 1, 1912. The lenses had 113-inch focal lengths, resulting in a 4-by-5-degree field on 8 × 10-inch plates. To complete the project, the positions of these asteroids were measured on the Stackpole measuring engine. Although it was second-hand, with this equipment Peters not only determined the positions of numerous known asteroids, but also discovered the asteroids 886 Washingtonia in 1918 and 980 Anacostia in 1921. It was used until about 1950.[55]

A final photographic episode in which Peters was involved late in his career was the total solar eclipse of January 24, 1925. Peters had participated in the Observatory's eclipse expeditions in 1901 and 1918, so he knew what was required for success. For the 1925 eclipse he designed and built two special cameras to be used in the dirigible U. S. S. *Los Angeles* (Figure 10.19), one of the Navy's two airships at the time. These cameras, mounted on a polar axis installed in the middle starboard compartment of the passenger cabin, were to record the inner and middle parts of the corona, and prominences. Two motion picture cameras were also installed, one in the cockpit on the top of the ship, and one in the aft engine room to photograph the approaching and receding shadow and the corona. These unusual methods of observation, probably the

[55] PUSNO, **12**, 347–431 describes the instrument and program, and gives observational results from 1912–24. Peters also described the instrument in "The Photographic Telescope" ref. 50 above, and in "The New Electric Driving Clock of the Photographic Telescope of the U. S. Naval Observatory," PA, **30** (February, 1922), 71–79. On the asteroid discoveries see PUSNO, **12**, 429–431, and Lutz D. Schmadel, *Dictionary of Minor Planet Names* (Springer: New York, 1997), pp. 127 and 137.

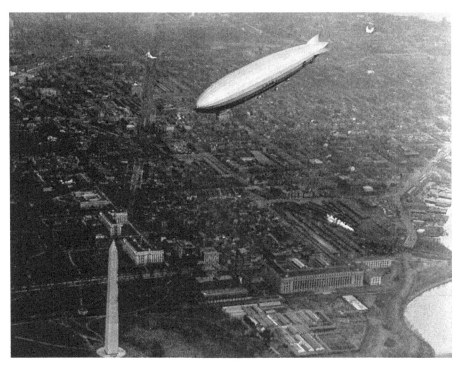

Figure 10.19. Navy dirigible U. S. S. *Los Angeles* over the Washington monument, looking southeast, 1925.

only time an eclipse was ever scientifically observed from a dirigible, were undertaken only after a consensus of opinion of leading astronomers that they should be attempted. Although the observer assigned to the cockpit had his face and fingers frozen while securing motion pictures of the phases, the eclipse was successfully observed from an altitude of 4,500 feet over the Atlantic Ocean east of Long Island. Superintendent Edwin T. Pollock, and a party consisting of Littell, Watts, and C. C. Kiess of the National Bureau of Standards participated (Figure 10.20); Kiess was included as a spectroscopic observer and recorded data from the chromosphere and corona. The successful results, including what must be some of the earliest footage of a solar eclipse with a motion picture camera, were reported in considerable detail.[56] After this colorful and unique event Peters was promoted to full astronomer (1927) and senior astronomer (1928). He was placed in charge of the Astrographic Division in 1930, until his retirement in November, 1931.

In the 1930s the Naval Observatory would acquire a 40-inch photographic telescope of radically new design, followed in 1950 by another new telescope having a 15-inch Cooke triplet. Those instruments were the result of a conscious attempt at

---

[56] *PUSNO*, **13**, Appendix 3, 73–93, including six plates; AR (1925), 2. The dirigible was used with the cooperation of Admiral W. A. Moffett, Chief of the Bureau of Aeronautics; the spectroscopic observer Kiess was assigned to the expedition by Samuel Stratton, Director of the National Bureau of Standards.

Figure 10.20. In the dirigible hangar at Lakehurst, New Jersey. From left to right: C. B. Watts, George Peters, and F. B. Littell were astronomers from the Naval Observatory; Observatory Superintendent Captain Edwin T. Pollock was in charge of the expedition; W. L. Richardson, Bureau of Aeronautics took movie pictures; chief photographer Peterson also took movie pictures from the cockpit and ended up with a frozen face and fingers; and Watson Davis of the Science Service representing the newspapers. A "Richardson camera" is seen behind Captain Pollock.

modernization of the Observatory's instruments during a time of depression and war, a story that we tell in the next section.

### 10.1 Attempts at modernization and origins of the Flagstaff Station, 1927–57

*The modernization program and the 40-inch telescope*

As we have seen in chapter 9, the inspiration for a modernization program came not from the astronomers, but from a new Superintendent, Captain C. S. Freeman. Freeman initiated a comprehensive review of the Observatory upon his arrival in September, 1927, and published its conclusions in the Observatory's *Annual Report* for 1928. The review found the most serious deficiencies in the astronomical equipment of the Observatory. In conjunction with the senior scientific staff, Freeman set forth in the *Report* a new "General Plan," a centerpiece of which was to employ photographic methods "in all cases where astrographic methods may be found to be applicable, and to make such methods primary in securing results" wherever required by astrometric accuracy.

The question of additional equipment was discussed in the Astronomical Council as early as October 25, 1927, when the need to put a younger man in charge of the new Photographic Division was recognized. The new Division, later termed the Astrographic Division, was established on January 1, 1928.[57] Expansion of astrographic work was found to take priority over all other needs of the Observatory, and accordingly members of the Council were instructed to write to experts in photographic astronomy to obtain their opinion as to what type of equipment should be purchased. Asaph Hall Jr, only a year away from retirement, wrote Frank Schlesinger at Yerkes Observatory and H. D. Curtis at the Allegheny Observatory, while Littell wrote Edwin Frost at Yale. Henry Norris Russell was also consulted at Princeton. Further information was solicited at the annual meeting of the American Astronomical Society in New Haven, and, in spring of 1928, Freeman met with Frost, Shapley, and Russell at a meeting of the American Academy of Arts and Sciences.[58]

As a result of this discussion a modernization plan was prepared, showing a total cost of $225,000, most of which was for instruments, with $65,000 remaining for buildings. The major new instrument proposed was a twin photographic telescope with a pair of 24-inch objectives, estimated to cost $76,000. A 6- or 8-inch Ross wide angle lens for a photographic refractor was also included in the plan. Each would be suitable for particular kinds of work. Although it was approved by the Secretary of the Navy, the plan did not make it through the Bureau of the Budget, and Freeman and Robertson found themselves continuing to make a case for funds in 1929 (for the 1930 appropriation), still to no avail.[59]

When, in the summer of 1930, Captain J. F. Hellweg took over as Superintendent, the plan changed considerably. The major items included in what would be the final modernization plan were a 40-inch "Ritchey–Chrétien" telescope (a new type of reflector with a radically different optical design), a 15-inch astrographic telescope and laboratory, a clock vault, and three new precision sidereal clocks, for a total of $160,000. For fiscal year 1932 $50,000 was appropriated and in 1933 another $110,000.[60] With the passing of these appropriation bills, funding for

---

[57] Council Minutes, October 25 and December 13, 1927, USNOA.
[58] The replies of these astronomers are to be found in the Burton papers, USNOA.
[59] "Hearing before Subcommittee of House Committee on Appropriations, consisting of Messrs. Burton L. French (Chairman), Guy U. Hardy, John Taber, William A. Ayres, and William B. Oliver, Navy Department Appropriation Bill for 1930, held Friday, January 11, 1929" (GPO: Washington, 1929), p. 808. For an itemized list and cost breakdown of the $225,000 proposal, see p. 815 of this Hearing. Copy in USNOA, AF, "USNO Modernization" folder.
[60] *Statutes at Large*, volume 46, pp. 556 and 1,453; volume 47, p. 1,445. "Hearing before Subcommittee of House Committee on Appropriations, Navy Department Appropriation Bill for 1933, 72nd Congress" (Washington, 1932), pp. 826–831, copy in USNOA, AF, "USNO Modernization" folder. This document, dated March 10, 1932, states that $50,000 was appropriated in 1932, and that $110,000 was being sought with this hearing. An itemization of the $160,000 total is given on p. 829.

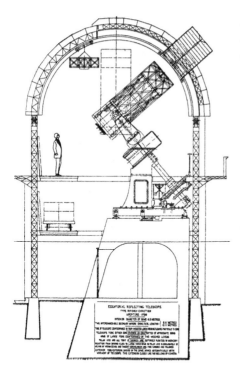

Figure 10.21. A schematic diagram of a 40-inch Ritchey–Chrétien telescope designed for France. From *JRASC*, **22** (1928), 205.

the full modernization program seemed assured. However, other events intervened. As an economy move in the midst of the Great Depression, Hellweg returned $50,000 in 1934, so only the 40-inch reflector and a 15-inch astrograph were completed. These new telescopes nevertheless offered considerable promise in revitalizing the Observatory.

The 40-inch reflector was to be the largest new telescope installed at the Naval Observatory since the 26-inch refractor more than 50 years before. Its origins may be traced to the actions of A. James Robertson, the Assistant Director of the Nautical Almanac Office, soon to become its Director in 1929. Robertson (Figure 12.6) had worked as an assistant to Newcomb, and fancied himself a great astronomer; his publication record, however, shows otherwise. He did, however, have many political connections, including W. L. Jones, an influential Senator from his native state of Washington. It was during Robertson's visit to European observatories following the IAU meeting in Leiden in 1928 that the idea for a Ritchey–Chrétien telescope at the Naval Observatory surfaced. Robertson met with Ritchey at the Paris Observatory, where Ritchey was already designing a 40-inch telescope of this type at the request of the French Duke of Gramont (Figure 10.21). Robertson emerged from his discussions convinced that Ritchey's telescope was what the Naval Observatory needed. Although he did not convince Superintendent Freeman upon his return to the United States, he did convince Freeman's successor, Hellweg.[61]

---

[61] On Robertson see chapter 12. On Ritchey and his telescopes see Donald Osterbrock, *Pauper and Prince: Ritchey, Hale and Big American Telescopes* (University of Arizona Press: Tucson, 1993). Chapter 10 places the Naval Observatory work in the context of Ritchey's work.

The 40-inch Ritchey–Chrétien telescope was a bold gamble on the part of the Observatory managers. Not only was it the Observatory's first reflector, it was also the first operational instrument built of its kind; the only model yet constructed anywhere in the world was a 20-inch used only experimentally by Ritchey in France. The Ritchey–Chrétien telescope was the result of a long attempt to design a telescope with a wide field free of the serious aberration known as "coma." Large telescopes such as the 60-inch and 100-inch Mt Wilson reflectors (also designed by Ritchey) and the 200 inch at Mt Palomar had parabolic mirrors that gave very good images on the optical axis, but serious aberrations a few minutes of arc off-axis. The Ritchey–Chrétien design allowed the telescope to be compact and coma free, eliminating the need for corrector lenses that cause other problems (such as not transmitting well in the ultraviolet). The design gave observers a photographic instrument having a support tube whose weight, length, and cost were approximately half those of older reflectors; the cost of the mounting was also substantially less than those of other designs. Ritchey believed that this design would eventually supersede all other photographic telescopes, except for a photographic refractor with an even wider field designed by Frank Ross. Although its 1.5-degree field was not wide enough for an astrographic survey of the heavens along the lines of the *Carte du Ciel* project, Ritchey wrote in his series of articles on astronomical photography in 1928 that the Ritchey–Chrétien telescope "will enable us to make photographs of fields of stars (fields large or small), of starclusters, of spiral nebulae, etc., which for all purposes of highly accurate measurement of star-positions will surpass the most perfect photographs which can be made with any telescope in use at present." The Naval Observatory had in mind replacing its second-hand 10-inch equatorial, whose main purpose was to photograph asteroids.[62]

Plans forged ahead and, in February, 1931, Ritchey conferred with Superintendent Hellweg and drew up preliminary specifications for the telescope. The telescope designed for the Naval Observatory consisted of a 40-inch concave primary mirror and a 16-inch concave secondary, both composed of low-expansion glass developed by San Gobain Glass Company of Paris. On June 3 Ritchey signed a contract for construction of the 40-inch telescope, to be completed within 27 months at a cost of $76,000. Ritchey was designated the independent contractor, and hired a crew to work with him on the mirrors, tube, operating mechanism, and all other small parts of the telescope at the Observatory. Only the massive parts of the mounting were made at the Baldwin-Southwark Corporation in Philadelphia (Figure 10.22).[63]

---

[62] Ritchey himself places the Ritchey–Chrétien design in historical context in an important series of articles "The Modern Photographic Telescope and the New Astronomical Photography," in JRASC, **22** (1928), 159–177, 207–230, 303–324, and 359–382; and **23** (1929), 15–36 and 167–190, especially Parts II and III on The Ritchey–Chrétien Reflector, JRASC, **22** (1928), 319; the quotation is on p. 230. For a more objective and longer perspective, and particularly why the Ritchey–Chrétien design took so long to be accepted, see John S. Hall, "The Ritchey–Chrétien Reflecting Telescope: Half a Century from Conception to Acceptance," *Astronomical Quarterly*, **5** (1987), 227–251.

[63] On Ritchey see Osterbrock, *Pauper and Prince* (ref. 61). Also Deborah J. Mills, "George Willis Ritchey and the Development of Celestial Photography," *American Scientist*, **54** (1966), 64–93.

Figure 10.22. The 40-inch telescope mounting under construction in Philadelphia, February 26, 1932. G. W. Ritchey stands in front of the mounting.

W. Malcolm Browne, a young staff member who had a background in electrical engineering and none in photography, was the only employee who worked full time with Ritchey on behalf of the Observatory. The South Transit Building adjacent to the main building was used as "an instrument shop, erecting shop for the telescope, wood working and pattern shop, and also would be used to do all of the rough grinding of the discs of glass." In addition a space known as the "Museum Room" on the second floor of the adjacent Administration building was chosen as the office, drafting room, and optical laboratory.[64]

On December 21, 1931, the disk of glass for the mirror arrived at the Observatory. By December 28 the optical machine was running for the first time and the 40-inch disk was on the turntable. By summer 1932 the 40-inch telescope was 60% complete, but it was May, 1934 before Hellweg and Ritchey presided over the

[64] "Construction of 40-Inch Reflector," folder marked "W. M. Browne's Journal of 40-inch Construction," USNOA, SF, 1–4. In addition to two sheets listing workers associated with construction of the 40-inch instrument from May 18, 1931 to December 4, 1932, this item consists of seven pages (both sides) apparently cut from volume 1 of four journals of notes on the progress of the 40-inch instrument. It covers February 6, 1931 to December 28, 1931. Volumes 2–4 of the journals are intact in the USNOA, and cover the period November 5, 1934 to May 24, 1948. Two additional observation journals cover the period March 9, 1935 to fall, 1941. See also John S. Hall and Arthur Hoag, OHI with William M. Browne, "Notes on the construction of the Navy's Forty-inch Ritchey–Chrétien Telescope," December 7, 1982; William M. Browne, OHI, June 24, 1989.

Figure 10.23.  Captain J. F. Hellweg (left) and G. W. Ritchey at the 40-inch telescope in May, 1934, before Hellweg banned Ritchey from the grounds of the Observatory.

installation of the new instrument (Figure 10.23). On June 16 of the same year all members of the Astronomical Council except Morgan and Watts – both experts in instrumentation – signed a statement that the telescope was "complete in regard to mechanical details." When, on October 29, 1934, Ritchey's temporary appointment expired, Browne was put in charge of the 40-inch telescope, under the supervision of Burton, head of the Equatorial Division.

Hellweg had not been slow in boasting about the advantages of the new telescope, even if he was not exactly sure what to do with it. Already in a 1931 article in *The Scientific Monthly* he had quoted Ritchey as boasting "We shall make photographs with it which correspond to 90 percent of the full theoretical photographic resolving power due to the aperture instead of the 10 percent or less represented by the best celestial photographs at present. And we shall do this where, of all places, it should be done: at our National Observatory in Washington."[65] A proposed program of astrometric work was drawn up as early as February 1932.[66] However, a series of events prevented the program from being fully implemented. There were problems with the 7-inch photographic plates, which had to be curved by suction into a concave surface; almost half of the plates broke during this process. On May 4, 1935, while the mirror was being removed to work on the mirror support, the mirror slipped from its clamp and fell about four feet to the steel plates of the observing room. Ritchey had returned to supervise the process; he was not injured but Browne had three fingers on his left hand severely crushed. Though a number of small chips were broken from the outer edges of the mirror, this did not affect

[65] J. F. Hellweg, "The New Reflector Telescope of the U. S. Naval Observatory," *The Scientific Monthly*, **33** (September, 1931), 283–285: 285. Hellweg, USNIP, 1935, pp. 1 ff.
[66] William M. Browne, "Proposed Program of Observing for the 40-Inch Reflector," February 2, 1932, USNOA, SF.

its performance; Hellweg, however, banned Ritchey from the grounds after this incident. Photographs of planetary satellites were successfully undertaken in 1936, but a great deal of tinkering with the instrument seems to have been the main activity.[67]

There is no doubt that the 40-inch telescope did not live up to its reputation prior to the late 1940s, when John Hall used it to discover interstellar polarization. Part of this was due to lack of expertise at the Naval Observatory. Its declared purpose, according to J. E. Willis, who was in charge of it in the late 1930s and early 1940s, was to relate the positions of the brighter zenith stars (used for determining time and variation of latitude) to the fainter stars of about 13th magnitude in order to determine parallaxes and proper motions of the brighter zenith stars and ultimately a value for the constant of precession.[68] This work was hardly designed to elicit the best qualities of the telescope. That the Naval Observatory was still in a defensive mode as of 1939 is evident in Willis's composition of a document "Concerning a statement that no astronomers would think of getting a Ritchey–Chrétien telescope." In a textbook example of how slowly technology may diffuse, only gradually did the Ritchey–Chrétien design become accepted by the wider astronomical community. (The 200-inch Palomar telescope might have been a Ritchey–Chrétien had it not been for the animosity between Hale and Ritchey.) Only in 1956 did Art Hoag publish two spectacular photos taken by the 40-inch telescope, once it had been moved to Arizona (Figure 10.24).

Then, in 1958, with the founding of Kitt Peak National Observatory, astronomer Aden Meinel visited the 40-inch-telescope site. Meinel decided once again to commit government funds to a Ritchey–Chrétien telescope, this time an 84-inch telescope. With the success of this telescope, the subsequent Kitt Peak and Cerro Tololo 4-meter reflectors, the Hubble Space Telescope's 94-inch mirror, and a host of others were of the Ritchey–Chrétien design. As Osterbrock has written, "in the hands of Hall and Hoag it helped convince Meinel to put national resources into a modern Ritchey–Chrétien telescope at an excellent site. From there it swept the world."[69]

---

[67] Browne, Notes (ref. 64), pp. 13–14; Hall and Hoag, Notes, p. 1, and Browne OHI, describe the mirror incident. AR, **32**, 2, and "History of the 40-inch Telescope," manuscript, USNOA, SF. This 15-page history covers the years 1934–48, when John Hall arrived. USNOA, SF, 40-inch history folder.

[68] John E. Willis, "The 40-Inch Ritchey–Chrétien Telescope of the U. S. Naval Observatory," manuscript dated July, 1938, USNOA, SF, p. 6. Astronomers in charge of the 40-inch telescope were as follows: G. W. Ritchey, April 29, 1934–October 28, 1934; William M. Browne, October 29, 1934–January 31, 1936; U. S. Lyons, February 1, 1937–October 31, 1937; J. E. Willis, November 1, 1937–April 30, 1944; and Bevan P. Sharpless, May 1, 1944–?, USNOA, SF, "40-inch" folder. In 1970 the 40-inch instrument was provided with new optics fabricated from ultra-low expansion silica, and the following year the original optics were sent to the Vienna Observatory on extended loan. In 1991 the old 40-inch mirror was still in Vienna, and being contemplated for use on the island of Hvar, Yugoslavia. I have not been able to discover its subsequent disposition. USNOA, SF, "40-inch" folder.

[69] Hall, "The Ritchey–Chrétien Reflecting Telescope "(ref. 62); Willis, "Concerning a Statement that no Astronomers would think of Getting a Ritchey [Ritchey–Chrétien] Telescope," July, 1938, USNOA, SF, "40-inch" folder; Osterbrock, *Pauper and Prince*, pp. 281–282. On the consideration of the Ritchey–Chrétien design for the 200-inch instrument, see Osterbrock (ref. 61), pp. 229–230.

Figure 10.24. The first test photographs from the 40-inch Ritchey–Chrétien telescope in 1956 showed the great promise of the telescope at its new location in Flagstaff. On the left is M51, the Whirlpool galaxy, and on the right M27, the Dumbbell Nebula. The 30-minute exposure of the latter revealed its faint outer portions and intricate structure.

The second major item in the Observatory's 1930s modernization program was a 15-inch telescope, also for photographic use. It was under contract to Warner and Swasey as of 1932; the dome was to be completed March 1, 1933, at the same time as the dome for the 40-inch telescope. The lenses were delivered and mounted in December, 1934, with Robert Lundin (Junior) of Warner and Swasey undertaking the testing. Problems with the lens required its return to Cleveland, where its focal length was changed. By 1940 Hellweg had reported in exasperation that good images had not yet been secured, that the contract was six years overdue, and that "all efforts to have them complete the contract have been unsuccessful." Their procrastination, he continued, "has seriously interfered with the program of this institution for the past 6 years." With the U. S. entry into World War II, Warner and Swasey admitted that they could not finish the contract because of other work. As it turned out, only in 1950 was the 15-inch Cooke triplet (Figure 10.25) mounted to replace the 10-inch photographic equatorial. Although it had considerable problems, and its aperture was reduced in practice to avoid poor images, it was used in the 1950s and 1960s for a program of minor planet observations by J. Gossner and B. Mintz.[70]

In the end, neither the 40-inch nor the 15-inch telescope was used for astrographic work on the large scale, that of surveying the heavens. Upon Peters's retirement as head of the Astrographic Division in 1931, C. B. Watts was placed in charge. The asteroid work begun on Peters's 10-inch photographic equatorial continued until 1950, when it was assigned to the 15-inch telescope.[71] Despite the lack of a substantial plan for its use, however, the 40-inch telescope is the clearest example of Naval Observatory innovation in the era of modernization, even if it is abundantly clear that the right person, one with creativity, training, and insight, is necessary in order to fulfill the promise of any instrument. That right person would finally appear when John Hall arrived in 1948.

[70] AR (1940), 5; AR (1942), 6; AR (1950) in AJ, **55**, 203; and see Hall, OHI. In 1993 the 15-inch telescope was placed on extended loan to John Briggs of the Yerkes Observatory. In 1996 ownership was transferred to Briggs and the Metcalf Astrometric Observatory. USNOA, SF, 15-inch folder.
[71] AR (1931), 16.

Figure 10.25. The 15-inch photographic refractor. Douglas O'Handley is at the controls.

*The drive toward accuracy: Advances in meridian telescopes and related work*

The meridian work of the Observatory played no large role in the modernization program enunciated in 1928. The goals of the effort throughout the era of modernization remained the same: to increase the accuracy of star positions, and to achieve the densification of star catalogs and thus reference stars in the sky. The context of transit-circle work also remained largely unchanged from the nineteenth century, and is still understandable by reference to the hierarchy of star catalogs seen in Table 5.1, now expanded in Table 10.3. At the base of the hierarchy – the observed foundation for the reference system to which all other star catalogs are tied – are the observations for the fundamental catalogs, which we may term "Level 1." Ascending the hierarchy, at "Level 2" are a parallel German and American series of compiled fundamental catalogs, followed by the international reference star catalogs at "Level 3," and finally the photographic catalogs at Level 4. The fundamental catalogs at the bottom of the hierarchy contain relatively few stars with very high accuracy, while the top level has catalogs with larger numbers of stars, but lesser accuracy. The tradeoff of numbers of stars for lesser accuracy was a trademark of astrometry for most of the century. As the century progressed, even the positions of stars at the top level became

Table 10.3. *Hierarchy of nineteenth- and twentieth-century star catalogs*[a]

Level 4: Photographic catalogs

| AC | AGK2 | AGK3 | Yale | CPC | TAC | A1.0/A2.0 | AC2000 | UCAC1 |
|----|------|------|------|-----|-----|-----------|--------|-------|
|    |      | Hamburg | Yale | Cape | USNO | USNO | USNO | USNO/CCD |
|    | 1951–58 | 1975 |      |     | 1996 | 1996/1998 | 1997 | 2000 |
|    | 180,000 N | 183,000 N | S | S | 705,679 | 526 million | 4.6 million | 27 million |

Level 3: International reference star catalogs

| AGK1 | AGK2A | AGK3R | SAOC | SRS | IRS | ACRS | HIPPARCOS | TYCHO-2 |
|------|-------|-------|------|-----|-----|------|-----------|---------|
|      |       | USNO | SAO | USNO/ | USNO | USNO | ESA | USNO/ |
|      |       |      |      | Pulkovo |      |      |      | Copenhagen |
|      | 1943 | 1960s | 1966 | 1988 | 1991 | 1991 | 1997 | 2000 |
|      | 13,747 | 21,499 | 258,997 | 20,448 | 36,027 | 320,240 | 118,218 | 2.5 million |
|      |       | Northern |      | Southern |      |      |      |      |

Level 2: Compiled fundamental catalogs

| BAC | PGC | GC |
|-----|-----|-----|
|     | L. Boss | B. Boss |
| 1845 | 1910 | 1937 |
| 8,377 stars | 6,188 stars | 33,342 stars |

| N1 | N2 | | N30 |
|----|----|----|-----|
| Newcomb | Newcomb | Eichelberger | Morgan |
| 1872 | 1898 | 1925 | 1952 |
| 32 stars | 1,257 stars | 1,504 stars | 5,268 stars |

| FC | NFK | FK3 | FK4 | FK5 | ICRF (radio) |
|----|-----|-----|-----|-----|--------------|
| Auwers | Peters/Auwers | Kopff | Fricke | Fricke *et al.* | |
| 1879 North | 1907 | 1940 | 1964 | 1988 | 1998 |
| 1883 South | | | | | 212 defining |
| 539 N stars | 925 stars | 1,535 stars | 1,535 stars | 4,652 stars | 606 total |
| 83 S stars | | | | | 667 ICRF-Ext. 1 |

Level 1: Absolute observed catalogs
Including Washington, Greenwich, Russian, Cape (South Africa), French, and German observations

*Note:*
[a] Dates are date of publication, not epoch of observation.

known with very high accuracy (one milliarcsecond for the European Space Agency satellite Hipparcos).

Where does the work of the Naval Observatory fit in this four-leveled scheme? During the twentieth century the Observatory increasingly built its reputation for observing the all-important fundamental catalogs situated at the bottom of the hierarchy in Table 10.3; these are the so-called Washington fundamental catalogs, which were observed with the 6-inch transit circle, beginning with the W25 and listed in Table 10.2. At the next level, it continued its compilation of fundamental catalogs that had been begun by Newcomb, and in which Eichelberger and Morgan would play leading roles. Only in the 1950s and later would its work extend to the top levels of the hierarchy of star catalogs, with all the international collaboration that this implied.[72]

The meticulous construction of all these catalogs is a story in itself, but at all levels of the hierarchy of star catalogs it begins with efforts to improve positional accuracies by using improved instrumentation. Beginning in 1932, just at the time the modernization program was getting under way with the equatorial telescopes at the Naval Observatory, C. B. Watts began a series of developments to update the 6-inch transit circle and improve the methods of data acquisition and analysis. The most important were "a motor drive of new design for the traveling wire, a method for the photographic registration of transits and zenith distance measures, the addition of photographic circle microscopes, and a photoelectric device for measuring the circle photographs."[73] Perhaps the most basic of these was motorizing the traveling-wire micrometer, which Repsold had developed in hand-driven form already between 1889 and 1895. The idea was to eliminate personal equation errors, a goal that had been achieved only partially.[74] The original 6-inch transit telescope had only fixed wires, but (as mentioned in section 10.1) by 1906 a hand-driven Warner and Swasey micrometer had been installed. Only in 1933 did a synchronous motor-driven micrometer (Figure 10.26) replace the hand-driven one. Both the hand-driven and the motor-driven micrometer still required the use of the chronograph as a recording device. However, by 1938 Watts had successfully implemented the difficult task of photographic registration of the traveling-wire micrometer. With this innovation the chronograph – instituted a century earlier by John Locke during the Maury era – was discarded for the 6-inch transit circle, and gradually for other instruments as well.

Some of these improvements to the 6-inch transit circle did not necessarily improve accuracy, but introduced uniformity in data acquisition and its reduction, which was also very important. At the same time these improvements relieved the observer of the need to record instrumental readings in a semi-darkened room, and

---

[72] On the development of star catalogs see Hans Eichhorn, *Astronomy of Star Positions* (Frederick Ungar Publishing: New York, 1974). In particular, for the Newcomb, Eichelberger, and Morgan catalogs, see Eichhorn, pp. 189–207, and for the Washington catalogs in the context of other independent catalogs see Eichhorn, pp. 141–163. We recall from chapter 5 that the Naval Observatory had participated in the southern extension of the AGK1 catalog with its transit circle. However, it did not participate in the photographic extension known as AGK2 because it lacked an astrograph of appropriate design. See also H. R. Morgan, "Astronomy of Position," PA, **51** (1943), 527–542.

[73] C. B. Watts, "Description of the Six-Inch Transit Circle: Instrumental Developments: 1932–1948," PUSNO, **16**, part 2 (1950), 333–361. A more general description of the transit circle, and changes to the 6-inch telescope in particular, is given in C. B. Watts, "The Transit Circle," in *Telescopes*, volume 1 of *Stars and Stellar Systems*, ed. G. P. Kuiper (University of Chicago Press: Chicago, 1960), pp. 80–87.

[74] C. B. Watts, "Experiments in the Design of the Traveling-Wire Micrometer," AJ, **50** (29 February, 1944), 179–182. Watts traces the traveling-wire micrometer back to Repsold during the years 1889–95. A comprehensive contemporary summary of the personal equation problem is given in Raynor L. Duncombe, "Personal Equation in Astronomy," PA, **53** (January–March, 1945). Duncombe was just at this time beginning his long tenure at the Naval Observatory.

Figure 10.26. The motorized micrometer for the 6-inch transit telescope. The motor is numbered 1 on the lower left. The bromide paper for registering the micrometer reading is in the cassette at 25. The entire micrometer is five inches square.

Figure 10.27. The Automatic Measuring Engine (AME) for measuring photographic film of transit-telescope circle readings, with astronomer Robert W. Rhynsburger, March, 1958. The measuring engine itself is the device on the table in the foreground, while the box contains associated electronics. AME represented one part of a program for automation of the acquisition and reduction of data.

to spend laborious hours reducing the data. In place of the observer visually reading the graduated circle with "microscopes," Watts introduced photographic registration of the circle in 1941. In 1948 he completed (in his own basement) the Automatic Measuring Engine fondly known as AME (pronounced "Amy" and seen in Figure 10.27) for analyzing the circle photographs. The output of AME was a series of paper tapes containing the readouts from the circle film from each microscope; the readings were then averaged to produce the definitive circle readings, which were combined with the micrometer readings, all necessary calibration readings, and environmental data, to produce the final star position in declination. According to Watts these innovations in both right ascension and declination were first used at the Naval Observatory, and soon spread elsewhere. AME was so successful in automating a tedious process that it was duplicated at other observatories, including Pulkovo in the U. S. S. R.[75]

Further automation proceeded more slowly, however. Despite the introduction of punched cards during the war for the production of almanacs in the Nautical Almanac Office, the transit-circle reductions were still being made by hand at the end of the war. It was only in 1956 that the Automatic Measuring Engine had been furnished with a digitizer, and made capable of operating a card punch. This was not only because of technical problems: "Both Watts and Morgan were a little hesitant to adopt this equipment, because they couldn't see the numbers," one contemporary recalled. "What they were used to was this big sheet, on which every step was written out, and you could see everything that had been applied to the observation and so forth. You couldn't quite do that with the machine computation. It took a little while for them to reconcile themselves to this." This first attempt at digitization involving the transit circle would soon be applied to the acquisition of the data also. Digitization was essential for the other great revolution occurring; by 1955 most of the calculations for reduction of data had been transferred to the electronic computer. As a result of these efforts, Watts's colleague F. P. Scott noted, the "U. S. Naval Observatory transit circles were brought from a 19th century state to a position second to none in the world."[76]

By contrast to the changes to the 6-inch transit circle, the 9-inch transit circle, with H. R. Morgan in charge, generally lagged behind Watts's instrumental innovations. The motor-driven traveling-wire micrometer was utilized on the 9-inch instrument beginning in 1936, and continued in use during the telescope's final program ending in 1945. The difference between the two approaches was in part a management decision not to change both instruments at the same time in case problems occurred. In addition, in contrast to Watts, Morgan's philosophy was to tinker as little as possible with the instrument. In the end the innovations proved themselves. Watts's efforts would result in the construction of a new 7-inch transit circle that incorporated all the new features he had devised over the years. This instrument would play a prominent role following the 1950s.[77]

If Watts was the expert on innovations in instrumentation, Morgan excelled in the use of the positions so meticulously determined. "His philosophy that good observations should be put to use, combined with a thorough knowledge of how to use them," resulted in numerous papers of first order in importance, wrote his under-

---

[75] The details of AME and the reduction process are given in Watts (ref. 73 above). For a description of its operation by a user see Rhynsburger OHI, April 14, 1986, 15–18, and August 20, 1987, 4–10. The replacement of the chronograph on the 6-inch transit circle is described in PUSNO, **16**, part I, 13.

[76] F. P. Scott, "Chester Burleigh Watts," QJRAS, **13** (1972), 110–112; F. P. Scott, "In Memoriam: C. B. Watts," The Moon, **6** (1973), 233–234; A. N. Adams, "Digital Recording System and Motor Drive of the Washington Transit Circle Micrometer," c. 1962, USNOA, BF, Adams folder; Ray Duncombe, OHI, January 11, 1988, 14.

[77] Morgan and Scott, PUSNO, **15**, part 5, 123–124; Watts, AJ, **50** (1944), 179, and Watts, "Experiments" (ref. 74), 179.

study F. P. Scott. Among these were hundreds of thousands of observations resulting in better knowledge of the position of the equator, the motion of the equinox, the constants of aberration and nutation, the motion of the perihelion, and corrections to the orbital elements of the planets. However, his crowning achievement, Scott and others have emphasized, was Morgan's N30 catalog of definitive positions and motions of 5,268 stars (1952). This catalog was used to examine the luminosities of the nearest Cepheids and other research related to the structure of the Galaxy.[78]

All the improvements to the transit circles described above yielded very little improvement in the accuracy of star positions, which still remained at about a tenth of an arcsecond at best. The technology did, however, affect the pace of observations and the pace of innovation. It is also important to emphasize that all of these events occurred in the context of their times. The transit-circle programs, like the Observatory in general, were especially affected by World War II. Morgan and F. P. Scott wrote in the introduction to the last catalog for the 9-inch transit circle that "as the war progressed, the observing staff became so depleted by the release of men to the armed forces or to other activities that shortly after the observing program was closed only one experienced observer [Morgan] remained."[79]

### Watts and the marginal zone of the Moon, 1946–63

Despite the slow pace of innovation with transit circles, there were specific areas in astronomy where accuracies could be improved quite rapidly. One such area was the motion of the Moon, that obsession of Newcomb, whose jagged edge limited its positional accuracy to three seconds of arc. In pursuit of this problem C. B. Watts led a program to define precisely the "marginal zone" of the Moon: those parts of the physical surface that contribute to the limb of the Moon as observed from Earth. Seen through the telescope that limb is far from smooth, its jaggedness caused by mountains, valleys, and craters.

The necessity of defining the lunar edge more accurately is evident in the fact that such charts had been constructed since early in the century. In particular, astronomers were still using the Moon charts of the Austrian astronomer F. Hayn, which had been published in 1914. By 1952 the French astronomer Th. Weimer had already produced refined lunar charts, but that project was much smaller in scope than the effort Watts had in mind. Watts proposed to undertake a survey of the marginal zone of the Moon as early as 1942, but because of the war, it was postponed until 1946. The project was driven by the experiences of Watts and others who had to take the Moon's edge into account for observations such as lunar occultations and transit-circle observations. The results of these observations differed depending on the part of the Moon's

---

[78] The "Catalog of 5,268 Standard Stars, 1950.0 based on the Normal System N30," is in APAE, **13**, part 3 (1952). On the use of Morgan's proper motions for determining the new zero point of the period–luminosity relation, see J. D. Fernie, PASP, **81**, 707, 1969, p. 720; on Cepheids see Blaauw and Morgan, BAN, **12** (1954), 95, and Morgan and Oort, BAN (1951), #431.

[79] PUSNO, **15**, part 5, 122.

edge at which an observation was made; Watts's goal was to define this edge, which changes with time due to libration, so that proper corrections could be made. Supported by the Office of Naval Research, as early as 1954 the Air Force Aeronautical Chart and Information Center also expressed interest in the work because of its own survey work on the Moon. After Watts's retirement in 1959, Yale University also cooperated in seeing the project to its completion, probably as a result of Dirk Brouwer's interest in the theory of the Moon's motion.

Watts based his results on some 700 photographs of the Moon taken between 1927 and 1956 at the Naval Observatory, the Yale–Columbia southern station at Johannesburg, South Africa, and the Lowell Observatory. In yet another case of executing a new project with old equipment, about two thirds of the photographs were taken with one of the 40-foot photoheliographs built by Alvan Clark in 1873 for the transit of Venus, and subsequently set up near the building housing the 6-inch transit circle. The photographs were measured with a Moon automatic measuring engine (MAME; Figure 10.28) designed by Watts specifically for the purpose, and constructed in the Naval Observatory instrument shop by George Steinacker. The photoelectric scanner of this device simultaneously traced the irregular profile of the Moon, a straight base line, and a measure of the density of the photographic plate. These measures were used to construct 1,800 charts, spaced at intervals of 0.2 degrees around the Moon's limb; these charts comprise the bulk of volume 17 of The Astronomical Papers of the American Ephemeris. With the charts the contour of the Moon's limb can be read to an accuracy of within about a tenth of an arcsecond, greatly improving the accuracy of occultation and meridian circle observations of the Moon, not only those made after 1963, but also prior observations. The end result was an improved understanding of the Moon's orbital motion, and everything that depends on knowing it precisely. At the time Watts's work was completed, this included ephemeris time, which was tied to observations of the Moon as determined by the "Markowitz Moon cameras" invented at the Naval Observatory, and discussed in chapter 11. The 951-page tome that resulted from Watts's work in 1963 contained the most detailed and accurate results ever obtained, and has been called "a powerful stimulant to astronomy."[80]

---

[80]  Joseph Ashbrook, "C. B. Watts and the Marginal Zone of the Moon," Sky and Telescope (February 1964), 94–95. The marginal zone survey is published in The Marginal Zone of the Moon, APAE, **17** (Washington, 1963), where references to previous charts of the Moon's limb are given. Almost as early as the date of publication, occultation observers noted that the position angles (orientation) of Watt's published charts were systematically in error by two tenths of a degree. This was confirmed by L. V. Morrison, "On the Orientation of C. B. Watts' Charts of the Marginal Zone of the Moon," MNRAS, **149** (1970), 81–90. An explanation for this offset was given in D. K. Scott, "Position–Angle Discrepancy in Watts' Lunar Limb Charts – an Explanation," AJ, **95** (May, 1988), 1,567–1,568. Scott concluded that a geodetic meridian rather than an astronomical meridian should have been used to orient the Moon plates. The relation between the astronomical and geodetic meridians, however, was not known until the determination of geodetic positions from artificial satellites was perfected. For a personal account by one who worked with Watts on the project see David K. Scott, OHI (26 February, 1988), 17–28.

Figure 10.28. The Moon Automatic Measuring Engine (MAME) used to produce Watt's Moon charts. It was built from the transit-of-Venus measuring engine seen in Figure 7.13, and also has features in common with AME, seen in Figure 10.27. Light from a source at the bottom illuminates the photographic plate of the Moon. As the plate carriage rotates, photocells at the top scan the lunar limb. One of the central pens at the upper right traces the lunar profile; the other pens are for calibration.

*The Equatorial Division and origins of the Flagstaff Station, 1948–58*

The retirement of Asaph Hall Jr as Director of the Equatorial Division in 1929 brought Harry Edward Burton (1878–1949) into that position. Burton (Figure 10.29), who was born just a year after the elder Hall had discovered the two moons of Mars, received his M. S. in mathematics in 1903 from the University of Iowa. He was employed for about a year as a computer at Yerkes Observatory, after which he came to the Naval Observatory as a computer in 1909. By 1929 he had worked his way up to become principal astronomer, and was the logical choice for head of the Equatorial Division, a position he would hold until his retirement in 1948. Although he was the author of some 91 articles on comets, minor planets, occultations, satellites, novae, and double stars, Burton's work nevertheless tended to the routine. In his observations of double stars with the 26-inch telescope, Burton continued the use of the Repsold micrometer, but over the course of his career he is credited with only 229

Figure 10.29. H. E. Burton at the 26-inch equatorial telescope.

measures of double stars (compared with 4,817 for the senior Asaph Hall and seven for Asaph Hall Jr), only one of which was a newly discovered double.[81]

The diverse work of the Division that Burton headed can be found in the Observatory's *Annual Reports* for this period. Aside from satellites and double stars, it continued to include daily solar observations with the photoheliograph, the photographic equatorial, and, of course, the 40-inch telescope, which Burton himself had little to do with. One of the more interesting episodes in the Equatorial Division during this period involved the secular acceleration in the longitudes of the satellites of Mars, a case that involved both visual and photographic techniques, pitting the old 26-inch refractor against the new 40-inch reflector. Although this might seem an esoteric subject, curiously the study resulted in one of the more frequently cited papers from the Naval Observatory, especially after the well-known Russian astrophysicist I. S. Shklovskii used it in 1966 to claim that the satellites of Mars might be artificial.[82] Burton had noted already in 1929 that the orbital longitude of Phobos was not well-behaved, and at the next close opposition of Mars in 1939 Burton and his colleague Bevan Sharpless (Figure 10.30) undertook an observational program to confirm this. Burton made visual observations with the 26-inch refractor, while Sharpless undertook photographic observations with the 40-inch reflector. Although the visual observations were superior to the photographic ones, Sharpless nevertheless had enough data to publish his secular-acceleration results in 1945. The results for Phobos have been confirmed (at about a third the value of Sharpless), but are now considered to be due to gravitational interactions with Mars, known as "solid body tides," rather than to the moons being hollow artificial satellites. Still, the whole episode gave renewed interest to planetary-satellite work.[83]

[81] On the work of Burton and his colleagues on double stars during this period, see *PUSNO*, **15**, part 4; on their satellite work, *PUSNO*, **17**, part 3. On Burton see Burton folder, USNOA.

[82] I. S. Shklovskii and Carl Sagan, "Are the Moons of Mars Artificial Satellites?", in *Intelligent Life in the Universe* (Holden-Day: San Francisco, 1966), pp. 362–376. Shklovskii first proposed the idea in 1959.

[83] Bevan P. Sharpless, "Secular Accelerations in the Longitudes of the Satellites of Mars," *AJ*, **51** (November, 1945), 185–186. Burton's 1929 article is "Elements of the Orbits of the Satellites of Mars," *AJ* (1929), 155–164. Inexplicably, Sharpless does not cite Nautical Almanac Office Director

Figure 10.30. Bevan P. Sharpless, discoverer of the secular acceleration of the moons of Mars, about 1946. Courtesy of Thomas Sharpless.

Despite its role in the moons-of-Mars episode, forefront work in astronomy was not to be expected from an aging 26-inch refractor. Nor had the 40-inch telescope reached its potential despite the years spent on bringing it into operation. Now, as the Washington night skies were becoming even brighter thanks to growing light pollution, the establishment of a dark-sky site for that telescope became an increasing concern in the post-World-War-II era. The man who would be instrumental in the establishment of that dark-sky station was John S. Hall. Hall (1908–91), a graduate of Amherst and Yale, had been influenced by Schlesinger, Brouwer, and Schilt during his graduate studies at Yale, where Schlesinger got him started using an infrared photoelectric cell for his dissertation.[84] His work in photoelectric photometry continued at Amherst beginning in 1938. During the war Hall worked on radar at the MIT Radiation Laboratory, before returning to Amherst.

In late 1947 Hall received a letter from C. B. Watts asking whether he would be interested in taking over from Burton as Director of the Equatorial Division. Hall applied for the position, and in March received a letter from Clemence, advising that he was being offered it and also expressing his personal hope that Hall would accept

Edgar W. Woolard's article of the previous year, "The secular Perturbations of the Satellites of Mars," AJ, **51** (August, 1944), 33–36, nor did Woolard cite the work of his Naval Observatory colleagues Burton and Sharpless. See D. Pascu, "A History of the Discovery and Positional Observation of the Martian Satellites, 1877–1977," *Vistas in Astronomy,* **22** (1978), 141–148. On the modern confirmation and theory see, in the same volume of *Vistas,* A. T. Sinclair, "The Orbits of the Satellites of Mars," 133–140, and J. A. Burns, "The Dynamical Evolution and Origin of the Martian Moons," 193–210. In contrast to Phobos, Sinclair finds the results for an acceleration of Deimos "barely significant."    [84] On Hall see OHIs, April 4, 1983 and June 18, 1989.

it. Clemence wrote "I am certain that this would be in the best interest of the Naval Observatory, and I think that the position offers some attractive possibilities from your own point of view. Although this position is coordinate in rank with that of the heads of the other Observatory divisions, it is in many respects more desirable than any of the others. The load of routine work and the administrative burden are both much less, and the opportunity for original work is therefore greater. There is a good nucleus of instrumental equipment, and you would have virtually a free hand in modernizing it as you think best. You would also have the observatory's resources at your disposal for instrumental development work. I think that you would find your colleagues congenial, the atmosphere not distasteful, and the working conditions good." A letter from the Superintendent sent the same day emphasized that the routine work of the Division, including observation of satellites, asteroids, comets, sunspots, and solar flares, "can be carried out by junior members of the staff, and would not require much of your time, which could be chiefly devoted to research. Administrative duties would probably require less than one-fifth of your time in the long run; but while the Observatory is engaged in moving its instruments to a new location the proportion would be greater." Although Hall expressed reservations about leaving the academic community, after extracting a promise that the Naval Observatory would purchase his unique polarization equipment from Amherst, he accepted.[85]

Hall arrived in August, 1948 in a Division with U. S. Lyons, who had inherited an outmoded tradition of work on double stars; Lucy Day, who performed the routine solar photography; and Bevan Sharpless, who was a creative intellect but whose health greatly impaired his work. Burton died only a month before Hall arrived.[86] Hall wanted to do something new; indeed, it is significant that Clemence and Watts had enticed him to the Observatory just for this purpose. During the early 1950s, Hall hired new staff with backgrounds different from those of most Naval Observatory astronomers, thus launching a small part of the Observatory in new directions. Already in 1950 Arthur Hoag came from Harvard, carrying his still uncompleted dissertation on the photometry of spiral galaxies. In 1953 Stewart Sharpless came from Mt Wilson Observatory, where he had worked with Walter Baade and Edwin Hubble on galaxy photography; Hall wanted him to do spectroscopy. Elizabeth Roemer arrived in 1957, nominally for asteroid work, but she would excel in her photographic observation of comets with the 40-inch reflector.

[85] Clemence to Hall, March 10, 1948, Hall folder, USNO. See also Hall OHI (April 4, 1983), 9–10. The USNO purchased the photometric equipment from Amherst for $5000. Hall was offered the same housing arrangement as Burton; government-owned quarters on the grounds of the Observatory would cost $75 per month, deducted directly from salary.

[86] C. B. Watts to John Hall, July 21, 1948; Hall folder, USNO. Burton died on July 20 after about a six-week illness, and only three weeks after his retirement on July 1. Sharpless (1904–50) had worked as an actuary before coming to the Observatory in 1928. He transferred to the Observatory's Richmond, Florida station in 1949, suffering from the effects of emphysema, and died shortly thereafter. Sharpless folder, USNOA, BF.

Hall was anxious to use the 40-inch telescope for polarization work, but he also recalled a comment from Schlesinger (one of those consulted in 1928 about the Observatory's need for a photographic telescope), who had bemoaned the new telescope's location in Washington, where its inherent power could not be demonstrated because of the encroaching city. Almost immediately upon his arrival in late summer of 1948 Hall proposed to move the telescope. In the meantime he disproved Schlesinger's implication that the 40-inch telescope was useless at its location in the middle of Washington. In spring of 1947 Hall had written to W. A. Hiltner at Yerkes Observatory to see whether they could detect a polarization effect in interstellar space, one that S. Chandrasekhar had proposed in 1946 might be caused by interstellar magnetic fields. Hall took his equipment from Amherst to McDonald Observatory in Texas, where Hiltner had access to the 82-inch reflector operated by the University of Chicago. They found an effect, which Hiltner wished to publish, but Hall believed needed further confirmation. The two then went their separate ways, Hiltner using an ordinary photometer with a rotating Polaroid filter, and Hall taking about nine months to track down and fix a glitch in his Amherst equipment. Hall was, by mid-1948, therefore extremely anxious to get his hands on a reflector, and, upon his arrival in Washington, set to work immediately on the problem of interstellar polarization. Within a few months he had not only shown Schlesinger to be mistaken regarding the use of the 40-inch telescope in Washington, but also made one of the better known discoveries at the Naval Observatory. In the meantime Hiltner had sent a paper announcing his discovery to *Science*; Hall sent one also and they were published back to back in the same issue.[87] The confirmation of the theoretical prediction regarding interstellar polarization awakened astronomers to the idea of galactic magnetic fields, and the discovery led to observations of polarization and magnetic fields in other galaxies as well. Thus, within four months of his arrival, Hall (Figure 10.31) had accomplished more with the 40-inch telescope than anything attempted throughout its previous 15-year history, a testimony to the new telescope design, the new detector Hall attached, and, most importantly, a creative individual. Clemence and Watts must have been elated.

The early success of Hall's work with the 40-inch telescope only increased his desire to see it moved to a dark-sky site, where its performance could only improve. In 1949 Hall was given orders to go to Lick Observatory, Mt Wilson, Inyokern (a Navy base in California), Lowell Observatory, Tucson, and the McDonald Observatory in Texas. In November, 1950, Hall and his colleague in the Equatorial Division, Alfred

---

[87] Hall OHI, 15–16; Hall, "Observations of the Polarized Light from Stars," *Science*, **109** (February 18, 1949), 166–167, preceded by W. A. Hiltner's article "Polarization of Light from Distant Stars by Interstellar Medium," *ibid.*, 165. The discovery involved a priority dispute. Hall had wished to publish a joint paper with Professor Hiltner of Chicago and Yerkes on the discovery, and was surprised to find out from Yerkes Director Otto Struve at an AAS meeting in Cambridge that Hiltner had already submitted a paper. A belated attempt to publish a joint paper failed because Hall was ill and could not work on one. See also OHI with Hall's colleague Alfred Mikesell, 126–129.

Figure 10.91. John Hall at the 40-inch Ritchey–Chrétien reflector in Washington, where he used it to discover interstellar polarization.

Mikesell, drove to Flagstaff, site of Lowell Observatory and location of earlier site testing in connection with the Observatory's politically motivated move described in the previous chapter. Here they consulted the Lowell records, undertook site testing with specially designed scintillation equipment, and finally selected a site some five miles west of Flagstaff, on a hilltop 7,600 feet above sea level. Within a few years Arizona was to become a Mecca for astronomy with the establishment of Kitt Peak National Observatory; in 1953 Hall even suggested that the new National Observatory be centered at the Naval Observatory, but the Superintendent rejected this idea out of hand.[88]

Site selection and acquisition was one thing, obtaining Congressional support for equipping and maintaining a facility was another. Between 1949 and 1954 Clemence, and, on at least two occasions, Hall, testified at the Congressional subcommittee hearings attempting to secure an appropriation for the move. Success came only in 1954, after Hall came to know Roy Elson, the Administrative Assistant to Arizona Senator Carl Hayden, a grand old man of the Senate and one of its most influential members. In 1954 the money was appropriated, and, in late September, 1955, the 40-inch telescope and associated equipment were loaded onto two flatbed trucks and moved to Flagstaff. The site had been prepared in the prior months, and optical alignment of the telescope was begun six days after its arrival. Arthur Hoag became the first Director of the Flagstaff Station, assisted by Joe Egan on the instrumental side. Hall remained in Washington most of the time.[89]

From the beginning, the work of the 40-inch telescope under Hall and Hoag differed in nature from the Naval Observatory's routine astrometric work of the past. It is no coincidence that in, 1956, the Equatorial Division changed its name to the Astrometry and Astrophysics Division. Early work with the 40-inch telescope included additional polarization studies, work using new detectors such as image tubes, and scintillation observations. At the Flagstaff Station photometric, photographic, and spectroscopic techniques quickly became routine, in contrast to the older techniques still being employed in Washington. Hoag undertook a program on the photometry of open star clusters, in cooperation with Harold Johnson at nearby Lowell Observatory. Stewart Sharpless undertook spectroscopy using a spectrograph of his own design, and Elizabeth Roemer began her landmark photographic work on comets. Finally, the 40-inch telescope was living up to its long-asserted reputation, proving once and for all the intrinsic worth of the Ritchey–Chrétien design. In the process, the Naval Observatory gradually became known for more than its traditional

---

[88] Hall OHI, 11; Mikesell OHI, 133 and 137–138. According to Mikesell (p. 139), in order to get the land for the Flagstaff station the Navy arranged for a swap with the Department of Forestry (Interior). It is notable that astronomer John Irwin had published recommendations for Arizona astronomical sites in *Science* prior to 1950. Hall's further site testing made use of the concept of scintillation (twinkling, or variations in the brightness of a star caused by refraction in the Earth's atmosphere), and ideas that Mikesell, Hoag, and Hall developed in a paper that Hall considered one of his best and most cited, Hall OHI (April 4, 1983), 19–20. On the idea of the National Observatory being centered at the Naval Observatory, see Hall OHI (April 4, 1983), 18. In August, 1953 the National Science Foundation sponsored a meeting on photoelectric astronomy in Flagstaff, which eventually led to the concept of a new national observatory. See John B. Irwin, ed., *Proceedings of the National Science Foundation Astronomical Photoelectric Conference*, held at Lowell Observatory, Flagstaff, Arizona, August 31–September 1, 1953, pp. 107–109. For context, see Frank Edmondson, "AURA and KPNO: The Evolution of an Idea, 1952–1958," JHA, **22** (1991), 71, and its expanded version *AURA and Its U. S. National Observatories* (Cambridge University Press: Cambridge, 1997).

[89] Hall OHI (April 4, 1983), 12–13. On the establishment of the new site see John S. Hall and Arthur A. Hoag, "The Flagstaff Station of the U. S. Naval Observatory," *Sky and Telescope* (November, 1956), 4–8.

time, almanac, and routine astrometric functions. At the same time, the routine functions should not be sold short; the Observatory's "traditional" work was becoming increasingly unique and important as other researchers turned to new priorities.[90]

## 10.3 Positional astronomy in the Space Age

There is no doubt that the Naval Observatory benefited from the Space Age, and equally no doubt that the Space Age challenged the Observatory to reach new heights via the need for positional accuracies undreamed of before. Even as the older equatorial telescopes continued their established programs and began new ones, a new 61-inch astrometric telescope, driven by Space-Age concerns and built with its funding, came to dominate the large-telescope functions of the Observatory. Even as meridian telescopes continued their traditional contributions toward a more accurate optical reference frame, by the end of the twentieth century the dynamical reference frame that they produced had been replaced by the International Celestial Reference Frame, an extragalactic frame defined by quasars and radio galaxies rather than by the stars. Finally, in what truly may be said to be the end of an era, meridian instruments were replaced altogether by ground-based interferometers and astrometric satellites, producing a quantum leap in accuracies not seen since the invention of the telescope almost 400 years before.

*Parallax and parsecs: The 61-inch telescope and other equatorial work*

Events at the Naval Observatory must be seen in the broader context of the problems that faced astrometry just prior to the dawn of the Space Age. These problems are encapsulated in several conferences held in the 1950s in which the Naval Observatory played a leading role. The first, "Problems of Astrometry," was held at the instigation of Dirk Brouwer and Gerald Clemence in September, 1953 at Northwestern University in Evanston, Illinois, and was hosted at their request by Kaj Strand (later Scientific Director of the Naval Observatory). The purpose of the meeting, Strand recalled, was to convene the most active astrometrists "to see if we could somehow get astrometry revived. It had been in the doldrums."[91] Many of the leaders in astrometry at this time can be seen in Figure 10.32. Another meeting with a more restricted agenda, "The Cosmic Distance Scale," was held at the University of Virginia in April, 1956; and a broader "Second Astrometric Conference" was held in May, 1959 in Cincinnati. All three meetings were underwritten by the National Science Foundation, which represented a new era in the funding of astronomy, and science in general, in the post-World-War-II era. This funding allowed international participation in the conferences, supplementing the astrometric activities in the International

---

[90] Hall OHI (April 4, 1983), 21–22; Hall OHI (June 18, 1989), 11 ff; Hoag OHI, 11–13; and Roemer OHI. Some of the early work of the Flagstaff station is published in PUSNO, **17**, including Hall's work on interstellar polarization and Hoag's work on galaxy photometry.

[91] Strand OHI, 81. The Conference Proceedings are found in "Problems in Astrometry," A Conference Sponsored by the National Science Foundation and Northwestern University, Evanston, Illinois, September 3–5, 1953, AJ, **59** (1954), 29–104.

Figure 10.32. The 1953 astrometric conference at Northwestern, hosted by Kaj
Strand. First row (sitting): Frank Edmondson, Dan Harris III, F. P. Scott, J. J.
Nassau, A. N. Vyssotsky, R. Glenn Hall, W. W. Morgan, and unidentified. Second
row: Dirk Brouwer, James Baker, R. H. Stoy, C. B. Watts, L. F. Jenkins, Ida Barney,
William Markowitz, Paul Herget, G. van Biesbroeck, Harold Spencer Jones, N. E.
Wagman, Harold Alden, C. D. Shane, Kaj Strand, unidentified, Jan Schildt, S.
Vasilevskis, A. Danjon, W. Krogdahl, W. Eckert, G. M. Clemence, M. Davis,
Hamilton Jeffers, and Douglas Duke. O. Heckmann is partially hidden behind the
goateed van Biesbroeck. Courtesy of Don Osterbrock and the Mary Lea Shane
archives of the Lick Observatory.

Astronomical Union's Commission on Positional Astronomy. Indeed, plans for the
Evanston meeting stemmed from discussions following the IAU General Assembly
meeting in Rome in 1952.

Not only were the needs of meridian astrometry, and the status of work on
proper motions, parallaxes, and double stars discussed at these meetings, but also
resolutions were adopted to advance work in these fields. These resolutions set the
discipline's agenda for a generation or more thereafter, including many activities
undertaken at the Naval Observatory. The most spectacular immediate effect
stemmed from the 1956 Cosmic Distance Scale Conference, at which one of the two
resolutions emphasized "that one of the most urgent needs of astronomy is the determination of the distances of stars fainter than the thirteenth magnitude. Such stars
are too faint to be observed with long-focus refractors; also existing reflectors were
not designed to meet the astrometric requirements. The conference therefore recom-
mends that an engineering study be made with the aim of producing a design for a
reflector that will be suitable for the above-mentioned purpose. A Cassegrain-type

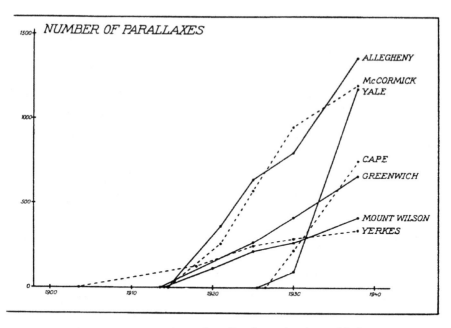

Figure 10.33. The increase in numbers of parallax determinations with time, 1904–37. From Erik Holmberg, "Accidental and Systematic Errors of Modern Trigonometric Parallaxes," Lund *Observatory Publications*, **97** (1938), 10.

reflector is indicated, with the secondary mirror more rigidly mounted than is customary. The aperture should be at least sixty inches so that stars of the eighteenth magnitude can be observed with exposure times not exceeding about twenty minutes. The instrument should be located at a site with a climate that is reasonably uniform throughout the year with regard to percentage of clear sky and quality of seeing. Such an instrument will be usefully employed for many decades."[92]

Two notable points stand out in this statement, the brainchild of Northwestern's Kaj Strand. First, the use of a reflector for parallaxes was an innovative idea. By far most parallaxes in the first four decades of the century (Figure 10.33) had been undertaken by refractors; this was still the case in the 1950s. The only exception is Mt Wilson, which represents the parallax work of Adriaan van Maanen with the 60-inch and 100-inch reflectors. Strand undertook a comparison of those parallaxes with refractors, and found that reflectors could indeed be used for determining parallax, especially if they were designed specifically for that purpose. Secondly, we see already in this statement that the design of the telescope will be a Cassegrain with a flat secondary. No less important for parallax determinations (for which reference

[92] A. Blaauw and A. B. Underhill, eds., "Cosmic Distance Scale," A Conference Sponsored by the National Science Foundation and the University of Virginia, Charlottesville, Virginia, April 5–7, 1956, AJ, **63** (May, 1958), 149–210; the resolution (p. 150) was based on Strand's paper, "Accomplishments and Reliability of Trigonometric Parallaxes," 152–156, especially p. 155. The importance of this meeting is acknowledged in Strand, "The 61-Inch Astrometric Reflector System," PUSNO, **20**, part I, 9.

stars are needed in the field of view), the coma-free field would be a half degree, compared with van Maanen's coma-free field a tenth that size.

The idea of a large telescope for determining parallaxes, driven by the discovery of low-luminosity stars too faint to be seen with existing long-focus refractors, became Strand's personal goal. Returning to Northwestern from that meeting, Strand had formally proposed to the Dean and Vice President that a telescope about 60 inches in diameter be built in Western Illinois, about 120 miles from Northwestern. However, the University did not consider it a high priority. By April of the same year, 1957, Clemence and John Hall had approached Strand about the possibility of assuming Hall's position at the Naval Observatory as Head of the Equatorial Division (which had recently been transformed into the Astrometry and Astrophysics Division), since Hall was moving on to Lowell Observatory. Although Strand had reservations about leaving a university position for a government job, he saw the possibilities for funding for his telescope, especially with the explicit support of Clemence, and, in August, 1958, cast his fortunes with the Naval Observatory.[93]

The work of the Naval Observatory with large telescopes from that moment on therefore becomes closely connected to the administrative actions and scientific work of Kaj Strand. Strand had obtained his doctorate from the University of Copenhagen in 1938, worked under Peter Van de Kamp as a Research Associate at Swarthmore, and, after war work, was associated with Yerkes Observatory. In 1946 he became chairman of the Astronomy Department at Northwestern University, where he remained until coming to the Naval Observatory in 1958. In 1963 he would succeed Clemence as the Observatory's Scientific Director. As the chairman of the 1953 astrometric conference, and a leading participant in the others, Strand was well tuned to the problems of astrometry.

Building on the resolution at the Cosmic Distance Scale Conference, upon his arrival at the Observatory Strand wasted no time in gaining more support for a large telescope. A similar resolution was passed at the Moscow General Assembly of the IAU in August 1958, where Strand served as President of the Commission on stellar distances and motions. In his report upon returning from that meeting, Strand wrote the Superintendent that "I believe this engineering study can best be carried out by the U. S. Naval Observatory with the view of acquiring such a telescope from R and D Funds originating from ONR [Office of Naval Research] or DOD. When the question of R and D funds arises for FY 1961 or a possible reprogramming for FY 1960, this engineering study should be brought up for discussion."[94] All of Strand's subsequent actions were geared toward making the telescope a reality.

In January, 1959, the Astronomical Council of the Naval Observatory supported the construction of the 60-inch instrument, subject to the appropriation of funds. The Superintendent at the time was Captain C. G. Christie, whom Strand described as

---

[93]  Strand OHI, 87 and 92; SAN, Strand folder, USNOA, BF, 1–2.
[94]  "Trip Report, Director, Astrometry and Astrophysics Division [Strand] to Superintendent, undated but following IAU General Assembly August 11–20, 1958," USNOA, BF, Strand folder. Strand OHI, 87–88; SAN, 3.

"knowing his way around the Pentagon, not afraid of raising his voice in presence of superiors, who would give in to his demands just to get him out of their offices."[95] By memorandum of May, 1959, Christie requested the Deputy Chief of Naval Operations for Development to include two million dollars for the fiscal-year 1961 military-construction bill. It was one of his last acts before he retired in June.

In October, the Navy awarded a preliminary architectural and engineering contract to the firm of C. W. Jones in Los Angeles, the same firm that had been involved in the design of the recently constructed 84-inch telescope at Kitt Peak National Observatory, as well as telescopes at Mt Wilson and Palomar. This was followed in March, 1960 by the complete contract. In April Strand discussed telescope construction with Aden Meinel at Kitt Peak and went on to Flagstaff to choose the site for the 60-inch telescope's dome. He also talked with Schwarzschild about the ability of Corning Glass works to fabricate a 60-inch mirror. On July 12, 1960 the military construction bill was passed by Congress with 1.9 million dollars for the 60-inch telescope, but not without considerable suspense. "Before passage there had been a couple of times where the project was in doubt," Strand recalled. "The Navy ran into a problem with the funding of a large radio telescope for communication, bouncing messages via the Moon. This required Vice Admiral Hayward, Deputy Chief of Naval Operations (Development) to refuse using the projected funding of the Observatory telescope project for the instrument. The radio telescope project was later discontinued because of overruns, and technical problems. The other problem arose when there seemed to be little support for the project in the Senate, until Henry Giclas of the Lowell Observatory got the staff of Senator Hayden's office to endorse the project." (Giclas's letter was printed in the Congressional record.)[96] Thus, Senator Hayden was as instrumental in obtaining final funding for the 60-inch telescope as he had been six years before in establishing the Flagstaff station.

In August, 1960 Corning Glass works was awarded a contract to provide the fused silica blanks for the primary and secondary mirrors at a cost of $182,000. Davidson Optronics was awarded a $72,000 contract for figuring the optics. In April, 1961 the glass blanks were shipped to Davidson for rough grinding, and in the end the nominally 60-inch mirror became a 61-inch mirror. On May 25, 1961 groundbreaking took place, and the building was completed in September, 1962 (Figure 10.34). In September, 1963 the assembly of the telescope began (Figure 10.35), and the following March the parallax program was under way.[97]

---

[95] SAN, 1–2.

[96] SAN, 11–12. Public Law 86–500, Military Construction Act of 1960, June 8, 1960, *Statutes at Large*, **74**, 171; Public Law 86–630, Military Construction Appropriation, *Statutes at Large*, **74**, 463. "Military Construction Appropriations, 1961," pp. 467–469 and 939–940. Senator Goldwater also wrote a letter in favor of funding. The 600-foot radio telescope was known as the Sugar Grove project; see "The Navy's Big Dish," *IEEE Spectrum* (October, 1976), 89–91.

[97] SAN, 12–13. More details on specifications and construction timelines are given in Strand, "The 61-Inch Astrometric Reflector System," *PUSNO*, **20**, part I. The telescope is described in detail, with numerous illustrations, in K. A. Strand, "The New 61-inch Astrometric Reflector," *Sky and Telescope*, **27** (April, 1964), 204–209.

Figure 10.34. The Flagstaff station shortly after the construction of the dome for the 61-inch telescope (foreground). The office complex and building for the 40-inch instrument, with its roll-off roof, is behind the dome for the 61-inch telescope. The view is toward the northeast, with the San Francisco peaks in the background.

The dedication of the 61-inch telescope was held June 22–24, 1964, with a conference in honor of Hertzsprung, who was himself present (Figure 10.36). The introduction was given by Martin Schwarzschild, an astrophysicist who spoke of the 61-inch telescope as "a beautiful example of that energy and daring that it takes for scientists to go into new techniques, fraught with possible technical difficulties, but full of promise of magnificent new results." During the conference Strand described the 61-inch telescope itself; Henry Giclas and William Luyten discussed proper motions, and Charles Worley set forth the parallax and proper motion program that the 61-inch telescope would undertake.[98]

The 61-inch instrument represented a triumphant conclusion to at least one of the resolutions from the astrometric conferences of the 1950s. While the work of the long-focus refractors achieved accuracies in the hundredths-of-an-arcsecond regime, the 61-inch telescope began to approach the milliarcsecond region, and by century's end was achieving sub-milliarcsecond accuracies using charge-coupled-device (CCD)

---

[98] *Aspects of Stellar Evolution*, Proceedings of a Conference Held in honor of Ejnar Hertzsprung at Flagstaff, Arizona on June 22–24, 1964, eds. Arthur Beer and K. A. Strand, *Vistas in Astronomy*, **8** (1966). Most of the rest of the conference was taken up with topics in astrophysics; only Walter Fricke and Peter van de Kamp spoke of astrometry, the latter regarding perturbations in stellar proper motions, which was still a hot topic after his announcement the previous year of the discovery of a planet around Barnard's star.

Figure 10.35. The 61-inch astrometric reflector, with Kaj Strand at the controls.

Figure 10.36. Hertzsprung (in white), Clemence, and Strand at the center of the group present for dedication of the 61-inch reflector, June 22–24, 1964. Elizabeth Roemer, Peter van de Kamp, and Bart Bok are standing at the far left; Charles Worley, Allan Sandage, and (in the doorway) Harold Ables are standing in the center; John Hall is seated third from the right. For complete identifications see ref. 98, where this photograph first appeared.

technology for a parallax series on a particular star. Almost four decades after its construction, the instrument continued as a parallax machine, resulting in parallax catalogs every few years with a total of more than 1,300 parallaxes. Even after the Hipparcos satellite produced parallaxes of similar accuracies by the thousands in the 1990s, the 61-inch telescope still proved its usefulness by concentrating on low-mass dwarf stars too faint for Hipparcos to observe.[99]

The future of astrometry seemed brighter because of the 61-inch telescope, but in 1964 Schwarzschild was properly cautious: "If the value of astrometry to astronomy as a whole is this wide and this obvious," he wrote, "is there anything to worry about regarding its future? I feel the answer to this question is still: Yes. The value of a branch of science does not in itself guarantee the continuation of its flourishing. A branch of science, in spite of high intrinsic value, can wither and die if the scientific activities in it appear to consist of nothing but the straight continuation of the techniques and of the aims of the past. However valuable for getting results such a straight continuation of previous activities may be . . . it fails in one decisive respect. It does not provide the explicit challenge necessary to draw into the branch its fair share of bright young scientists." He also spoke of ground-based astronomy "holding its own beside the space effort."[100] Schwarzschild was correct in both respects; the renaissance in astrometry had not quite arrived, but when it did, during the last quarter of the century, it would be linked both to ground-based and to space techniques.

### Double-star work

Even as construction plans for the 61-inch telescope were under way, the transformation of the Equatorial Division into the Astrometry and Astrophysics Division continued in smaller ways using some of the older telescopes. In addition to the 40-inch Ritchey–Chrétien in Flagstaff, the instrumentation consisted of the 15-inch photographic refractor and 26-inch refractor in Washington; the 12-inch telescope had been dismounted to make room for the Markowitz Moon camera, a project of the Time Service Department (section 11.1).[101] In addition to the work of Stewart Sharpless, Arthur Hoag, and Elizabeth Roemer, which was centered on the 40-inch

[99] The parallax catalogs are published in PUSNO beginning with volume 20, part 3, and the astrophysical analyses of the results are published in many journals. The USNO's earliest expertise in CCDs grew out of P. K. Seidelmann's involvement with the Space Telescope, and his use (with Dan Pascu) of the CCD on the 61-inch instrument for planetary-satellite observations; see Westerhout OHI, 69–73, Seidelmann OHI, and Figure 10.38. For parallax work David Monet and Conard Dahn were crucial in importing CCD technology from Kitt Peak.

[100] "New Impetus to Astrometry," in *Aspects of Stellar Evolution* (ref. 98), p. 4; "Summary. Problems and Outlook," ibid., p. 227.

[101] The 12-inch telescope was dismounted from the dome atop the Main Building in the 1950s, and installed in the old 40-inch building after the 40-inch reflector's removal to Flagstaff in September, 1955. It remained there until about 1967, when a 24-inch reflector was put in the dome for the 40-inch instrument. The Markowitz Moon camera was decommissioned in 1974, and the 12-inch telescope was refurbished and remounted in 1980 in its original dome. See T. J. Rafferty, "Refurbishing the U. S. Naval Observatory's 1892 Saegmuller 12″ Refractor," *Telescope Making*, **15** (1982), 24–29.

telescope at Flagstaff, in the early 1950s Gossner was hired to observe minor planets on the 15-inch photographic refractor, Winifred Sawtell Cameron followed by Irving Lindenblad took over from Lucy Day the observation of sunspots, and Otto Franz (who had come with Strand from Northwestern) worked on a variety of astrometric projects. Alfred Mikesell, a mechanically minded astronomer, worked on atmospheric scintillation as applied to seeing conditions.[102]

As we have seen, visual double-star work had formed an intermittent program for the 26-inch instrument over its career, from Asaph Hall's important work in the late nineteenth century to the observations under Burton's supervision during 1929–49, the latter mostly of stars with separations greater than two arcseconds. Upon his arrival Strand initiated a program of photographic double-star observations with the 26-inch telescope, an instrument that had been somewhat neglected under John Hall's tenure. Double-star research was one of Strand's specialties, and when, at the first Astrometric Conference in 1953, he had surveyed the field, he proposed that more emphasis be placed on obtaining photographic observations of wider binaries, while visual observers should be encouraged to observe only the close doubles too difficult for the photographic method. This stemmed from the work of Hertzsprung and others, who had shown that the photographic method was much more accurate than the visual for binaries with separations larger than 2–3 arcseconds. This recommendation Strand carried out upon his arrival at the Naval Observatory. Because the 26-inch telescope had not been modified since 1893, he obtained funds from the Office of Naval Research to update the telescope, with Mikesell in charge and George Steinacker of the Instrument Shop supervising the mechanical changes. Thus after 1958 the telescope underwent its largest change in appearance since its move to the new site in 1893. Not only did a completely new tailpiece replace the original Warner and Swasey one, but also the pier platform and spiral staircase used for access were removed. The telescope controls and console were revamped; already in 1950 the four Warner and Swasey water hydraulic cylinders had been replaced with oil cylinders and a pump that now raised the floor from its lowest position to its full height of 12 feet in 60 seconds.[103]

The photographic double-star program, utilizing the photographic camera Strand brought with him from Northwestern, became one of the long-range programs of the Astrometry and Astrophysics Division. It commenced in 1958 and continued until 1981, with almost 11,000 plates taken. It was prosecuted side by side with the visual double-star program of Charles Worley, which was begun in 1961. A digi-

---

[102] SAN, 2. On Gossner and the 15-inch instrument see Hall OHI (June 18, 1989), 14. Beginning in 1949 the sunspot work was reported in the new USNO circulars authored by Day, Cameron, and Lindenblad; for a listing see http://www.usno.navy.mil/library/index.html.

[103] A brief history of double-star observations at the Naval Observatory is given in in the Introduction to Charles Worley, "Micrometer measures of 1164 Double Stars," PUSNO, **18**, part 6, 18–19, and also at http://ad.usno.navy.mil/ad/wds/ds-history.html. Strand, "The Present Status of Double Star Astronomy," AJ, **59** (1954), 61–66; SAN, 4. The details of the changes to the 26-inch telescope are described in A. H. Mikesell, "Mechanical Improvement of the 26-Inch Refractor," PUSNO, **18**, part IIA (Washington, 1968).

tized micrometer allowed the speed of the visual observations to be increased by a factor of two, so that measures could be carried out at a rate of 2,500 per year. During Worley's tenure at the Observatory, he made over 37,000 measurements of double stars. Under Worley, the Double Star Catalog was transferred from Lick Observatory to the Naval Observatory, was renamed the Washington Double Star Catalog, and more than doubled in size. In addition to the Observatory's own observations, a strong effort was made to add to the Catalog past observations from the literature. Worley also helped maintain the binary-orbit catalog.[104]

In 1990 the visual program was superseded by a more accurate technique known as speckle interferometry, which combines a sequence of short-exposure electronic snapshots of an object, thereby compensating for image irregularities due to the turbulent layers of the atmosphere. The Naval Observatory made use of an improved version of the speckle camera that had successfully been implemented by Harold McAlister and his team at Georgia State. The goal was to provide quality speckle measurements of all double stars within the range of the telescope, thereby providing a bridge between the classical and speckle techniques for these stars. Thus, after a period of neglect in double-star work at the Naval Observatory, during the last third of the century it once again became respectable in the field, and served as the world center for double-star astronomy.[105]

### Planetary satellites and the moon of Pluto

Planetary-satellite work at the Naval Observatory was driven by a need for improved knowledge of the orbits of the Moon and planets, data essential for generating ephemerides in the *Nautical Almanac*. For this, knowledge of planetary masses was required, and, until the Space Age, the most reliable method for determining a planet's mass was from the motions of its satellites. This had been the principal motivation for the construction of the 26-inch refractor and half a century of visual micrometer observations of satellites. This heyday period of satellite studies not only

---

[104] The results of the photographic program were published in PUSNO, **18**, parts 1 and 7, **22**, part 6, and **24**, part 5. The final photographic results were published in F. J. Josties and R. S. Harrington, "Photographic Measures of Double Stars," *ApJ Supplement*, **54** (1984), 103–113. The results of the visual program were published in PUSNO, volumes 18, 22, and 24. William Markowitz, who was influenced by Hertzsprung's work, had observed close visual doubles with the 26-inch instrument from 1949 until 1952. This proved the feasibility of the program under Washington conditions, but the sustained program began with Worley in 1961. The WDS is described in Charles E. Worley and Geoffrey G. Douglass, *Washington Visual Double Star Catalog* 1984.0 (U.S. Naval Observatory: Washington, 1984), and its revision (including 78,100 systems) is described in C. E. Worley and G. G. Douglass, "The Washington Double Star Catalog (WDS, 1996.0)," *Astronomy and Astrophysics Supplement Series*, **125** (1997), 523. The WDS is the successor to the *Index Catalogue of Visual Double Stars*, 1961.0 (Jeffers and van den Bos, 1963), which is usually denoted as the IDS.

[105] G. G. Douglass, Robert B. Hindsley, and C. E. Worley, "Speckle Interferometry at the US Naval Observatory, I," *ApJ Supplement*, **111** (1997), 289–334; Douglass, Mason, and Worley, "Seven Years of Speckle Interferometry at the U. S. Naval Observatory," USNOA, SF, 26-inch speckle folder, abstract in BAAS (1998). The history and accuracy of the speckle technique for double stars is described in H. A. McAlister, "Accuracy of Binary Star Speckle Interferometry," in F. V. Prochazka and R. H. Tucker, eds., *Modern Astrometry* (Institute of Astronomy: Vienna, 1979), pp. 325–337, and some of his results are described in McAlister *et al.*, AJ, **93** (1987), 688.

contributed to determining accurate masses and ephemerides for the planets, but also resulted in the discovery of the moons of Mars, accurate new orbits for planetary satellites, and a valuable new body of knowledge on gravitational theory. Together with Newcomb's work on the motions of the major planets, the satellite studies of this period propelled the Naval Observatory to international stature.

While satellite studies languished worldwide during the inter-war period, they survived at the Naval Observatory, though considerably diminished, culminating with Bevan Sharpless's detection of the secular acceleration of Phobos in the 1940s. The only satellite observations made in the next 20 years were Richard Walker's discovery plates of Epimetheus, an inner satellite of Saturn, which were taken with the 61-inch telescope at Flagstaff during the 1966 ring-plane crossing.

A second-generation observing program was initiated at the Naval Observatory in 1967 in an effort to revise the aging satellite theories by the major almanac producing countries. Amid reports that Phobos was 30 degrees from its predicted ephemeris, due to its secular acceleration, Dan Pascu of the Nautical Almanac Office began photographic observations of the moons of Mars with the 61-inch astrometric reflector, and with the McCormick 26-inch refractor of the University of Virginia. The observations of the Martian satellites were continued with the 26-inch instrument at subsequent apparitions and contributed significantly to the Viking reconnaissance of the Martian system, as well as confirming the secular acceleration of Phobos. A blocking filter was placed in front of the planet's image, thus allowing the Martian moons to be highlighted on the photographic plate (Figure 10.37), and their positions determined to better than 0.1 arcsecond.[106]

NASA's exploration by spacecraft of the outer solar system spurred renewed interest in planetary-satellite work for purposes of space navigation. The Voyager mission, consisting of two spacecraft that visited Jupiter and Saturn in 1979 and 1980, required highly accurate ephemerides of the satellites for the surgically precise close encounters with them. For this purpose an "ad hoc" working group (the Outer Planet Satellite Working Group) was formed and included astrometrists and dynamicists from the United States and abroad. This working group was centered at the Naval Observatory, and co-chaired by Nautical Almanac Office Director P. K. Seidelmann and W. E. Brunk of NASA. Its function was to bring together the JPL scientists who were to deliver the final flight ephemerides to the Voyager project, and the scientists who would be involved in the improvement of those ephemerides. In this effort, the observational load was carried by the Naval and McCormick observatories for the bright satellites of Jupiter and Saturn, and McDonald Observatory for the fainter satellites. When it became clear that the Voyager spacecraft might continue on to Uranus and Neptune, astrometric observations of those systems were begun by Richard

[106] D. Pascu, "An Analysis of Martian Satellite Photographic Observations of 1967," *Icarus*, **25** (1975), 479–483; D. Pascu, "The Naval Observatory Program for the Astrometric Observation of Planetary Satellites," in Paul E. Nacozy and Sylvio Ferraz-Mello, eds., *Natural and Artificial Satellite Motion* (University of Texas Press: Austin, 1979), pp. 17–32.

Figure 10.37. A ground-based image of the Martian satellites, taken with the 26-inch refractor on August 25, 1971. In this 40-second exposure, Phobos, the brighter satellite, is west of the planet, and Deimos is below it. South is up, and the south polar cap is visible at the top. Courtesy of Dan Pascu, U. S. Naval Observatory.

Walker with the 61-inch instrument at Flagstaff. The new orbits derived from these observations were of unprecedented accuracy, and contributed significantly to the successful navigation of the Voyager probes through the outer planetary system.

In 1980 Pascu and Seidelmann began a program of CCD observations of faint, close planetary satellites with the 61-inch telescope, using the Hubble Space Telescope ground-based CCD camera system. This continued for a decade and resulted in 1980 in the discovery of Calypso, a new Lagrangian satellite of Saturn, and ephemerides for the other two Lagrangian satellites in time for the Voyager spacecraft to image them during its passage through the Saturnian system. The CCD program also provided observations of Miranda, a faint inner satellite of Uranus for the Voyager reconnaissance of the Uranian system and observations of Nereid, a faint satellite of Neptune, which were used in the Voyager encounter of the Neptune system. CCD observations of faint inner satellites of the outer planets were continued for another decade by Pascu and J. Rohde with an emphasis on photometry as well as astrometry (Figure 10.38). The Naval Observatory's CCD satellite observations culminated with Hubble Space Telescope observations of the innermost satellite systems of Uranus and Neptune in 1994 and 1997. At the end of the century the demand for astrometric observations of planetary satellites remained high, especially for satellites with poor observational histories.[107]

Meanwhile, routine photography of planetary satellites had led to another famous discovery in the outermost reaches of the solar system. In early 1977, James W. Christy, an astronomer at the Naval Observatory in Washington, requested that photographic plates also be taken of Pluto in order to obtain more precise positions of that

---

[107] D. Pascu, "An Appraisal of the USNO Program for Photographic Astrometry of Bright Planetary Satellites," in *Galactic and Solar System Optical Astrometry*, eds. L. V. Morrison and G. F. Gilmore (Cambridge University Press: Cambridge, 1994), pp. 304–311; D. Pascu, "Astrometric Techniques for the Observation of Planetary Satellites," in Joseph A. Burns, ed., *Planetary Satellites* (University of Arizona Press: Tucson, 1877), pp. 63–86; D. Pascu, "Long-Focus CCD Astrometry of Planetary Satellites," in S. Ferraz-Mello et al., eds., *Dynamics, Ephemerides and Astrometry of the Solar System* (Kluwer: Dordrecht, 1996), pp. 373–388.

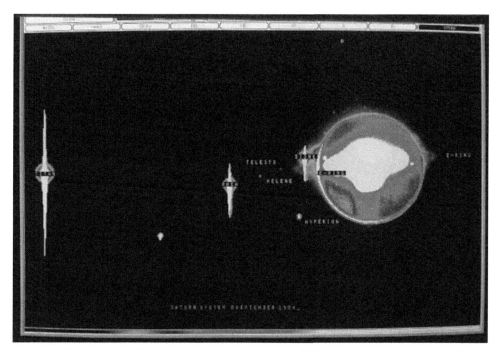

Figure 10.38. A CCD 60-second exposure showing six of Saturn's satellites. The three spiked images are (left to right) Titan, Rhea, and Dione. Saturn and three of its brighter satellites are behind the partially transparent coronagraphic mask. The image was taken with the 61-inch reflector September 8, 1994. Courtesy of Dan Pascu.

distant enigmatic planet. Christy also mentioned that such a study might reveal the perturbation of an unseen satellite.[108] Accordingly, in April and May, 1978, the 61-inch telescope was used on three nights to take six photographic plates, two on each night, of the planet Pluto. These plates were sent back to Washington, where the position of Pluto was to be measured with respect to background reference stars, using the Naval Observatory's measuring machine, known as STARSCAN. It was during this process, on Thursday, June 22, 1978, that Christy noticed that the images of Pluto were elongated, showing a faint southerly extension on April 13 and 20, and a faint northerly extension on May 12. Such elongations are not unusual: astronomical images sometimes trail or are imperfect due to atmospheric conditions. In fact, the plates had been marked as defective before being sent from Flagstaff. The same plates, however, showed an elongated planetary image, and round stellar images, a cause for puzzlement. Examination under the microscope did nothing to eliminate the puzzle. On a busier day Christy might have let the puzzle pass, in fact, might not have even attempted to measure the "defective"

[108] J. W. Christy, "The Discovery of Pluto's Moon, Charon, in 1978," in S. A. Stern and D. J. Tholen, eds., *Pluto and Charon* (University of Arizona Press: Tucson, 1997), pp. xvii–xxii; J. W. Christy OHI, 22–23. The Christy article provides his best first-person account. Owing to Pluto's highly eccentric orbit, between February 8, 1979 and March 14, 1999, Pluto was actually closer to the Sun than was Neptune. However, at the time of the discovery of its moon, Pluto was (just barely!) still the most distant planet.

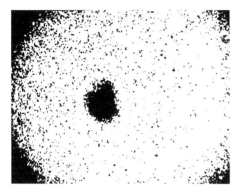

Figure 10.39. The discovery image of Charon, the moon of Pluto, taken with the 61-inch astrometric reflector. The photograph shows both Pluto and Charon, and Charon is seen as a slight bulge on the planet. It would later be resolved with the Hubble Space Telescope's Faint Object Camera.

plates. However, he was about to go on a week's leave, was in a more relaxed mode than usual, and began to consider the possible causes of the elongation. A gigantic eruption on Pluto seemed unlikely to be sustained over a month. The idea of "month" led to the idea of "moon," but at that point Christy "felt a little ridiculous at the thought." He and his colleague F. J. Josties calculated that Pluto's motion was a factor of ten times too small to produce such an elongation over the 1.5-minute exposure. The possibility of a faint background star coincidentally adjacent to Pluto was considered, and an examination of the Palomar Sky Survey actually showed one that was close, but not close enough. At this point the suspicion of a Plutonian satellite became stronger, strong enough that Christy told his colleague and supervisor, Robert S. Harrington.

In order to verify this conjecture, already the following day, June 23, Christy examined plates that had been taken with the 61-inch telescope in 1965 and 1971 for the purpose of measuring the diameter of Pluto, and in 1970 for a project to measure the motion of the photocenter of the planet. Plates taken on two nights in 1965 and five nights in 1970 also showed the elongations, and the latter even permitted an estimate of a period of about six days, approximately the period of Pluto's known light curve of 6.387 days.[109] Equating the putative moon's period with the light curve, Christy now informed Harrington not only that Pluto had a moon, but also that its period was 6.387 days! With this information Harrington calculated an ephemeris for the moon's position, while Christy measured the position angles of the elongation on the old and new plates. Christy recalled what happened next:

> Two hours later, my position angles from two nights in 1965, five nights in 1970 and three nights in 1978 were compared with Bob's ephemeris at a meeting of U. S. Naval Observatory management. The meeting had been called to discuss the work of the previous day; I had just finished my measurements and had handed them over to Bob without seeing his ephemeris. As the meeting started, Bob was bursting at the seams; he interrupted the Director's introduction and started writing numbers on the board. The agreement between measurement and the

[109] According to Christy, the 1965 plates were taken by Otto Franz, who actually noted the elongation on the plate. Conard Dahn had also noted the elongation on the 1978 discovery plates, but believed that it was due to the motion of Pluto. Christy OHI, 23 and 26–27.

Figure 10.40. James Christy (seated) and Robert S. Harrington at the Starscan measuring machine where the Charon image was first identified.

6.387 day period was remarkable. Pluto had a moon, and the moon had an orbit. Twenty-eight hours had elapsed since I had first looked at the Pluto images.[110]

With this information in hand, Conard Dahn used the 61-inch instrument to observe Pluto on every clear night. Elongations were detected on two nights, with the best image on July 2 (Figure 10.39) showing an elongation estimated at 0.8 arcseconds. This elongation corresponds to 17,000 km distance for the satellite. Finally, on July 6, J. A. Graham obtained a confirmation photograph of Pluto with the 4-meter (158-inch) reflector at the Cerro-Tololo Inter-American Observatory in the Andes.[111] On the same day, July 6, 1978, Captain Joseph Smith, Superintendent of the Naval Observatory, announced the discovery to the world. In a letter to the Secretary of the Navy, Smith recalled the Observatory's other famous satellite discovery: "One hundred and one years after the discovery of the moons of Mars by the U. S. Naval Observatory, I have the honor to report that on June 22, 1978, Mr. James W. Christy, an astronomer of the U. S. Naval Observatory, discovered the first satellite of the planet Pluto." The discovery was duly reported in newspapers around the world, and, by July 14, Christy and Harrington (Figure 10.40) had submitted their scientific paper to the Astronomical Journal, where it appeared the next month. Christy named the satellite "Charon," after the ferryman in Greek mythology who conveyed the souls of the dead across the River Styx to the underworld (Hades) ruled by Pluto.[112]

[110] Christy, "The Discovery of Pluto's Moon," p. xx. Present at this meeting, in addition to Christy and Harrington, was Scientific Director Gart Westerhout. A tape of the meeting is in the USNOA.

[111] James W. Christy and Robert S. Harrington, "The Satellite of Pluto," AJ, **83** (August, 1978), 1005–1008. The three plates on which Christy originally noticed the elongation were taken on April 13, 1978 by Tony Hewitt, on April 20, 1978 by Bill Durham, and on May 12, 1978 by Tony Hewitt. Ables to Dick, private communication, July 7, 1994. The two April 20 plates, however, were not of "discovery" quality.

[112] The discovery paper is R. S. Harrington and J. W. Christy, "The Satellite of Pluto," AJ, **83** (August, 1978), 1005–1008. It was followed by Harrington and Christy, "The Satellite of Pluto. II," AJ, **85** (February, 1980), 168–170, and Harrrington and Christy, "The Satellite of Pluto. III," AJ, **86** (March, 1981), 442–443. Christy initially thought of naming the moon Charon, after his wife

The discovery was not without some controversy. Although speckle observations resolved the moon in 1980, "removing any doubt as to the existence of the satellite" in the opinion of Harrington and Christy, some refused to accept the moon for several years. Remarkably, one additional mode of proof was the transit of the moon in front of Pluto, and its occultation by Pluto, events that occur only during an eight-year period every 124 years, but that were predicted to occur in 1985. The actual observation of these events in February 1985 clinched the reality of Charon to the satisfaction of everyone. (The Hubble Space Telescope clearly separated the moon from Pluto in an image taken in the 1990s.)[113]

Christy and Harrington called the moon of Pluto a serendipitous discovery, but one that had many root causes: the excellent performance of the 61-inch telescope, the existence of older Pluto plates taken with the same instrument for other reasons, and the resumption of observations of Pluto with the instrument, again for unrelated reasons.[114] The discovery was of more than passing interest, primarily because for the first time it allowed a determination of the mass of Pluto. The resulting mass was less than that of our own Moon, leading to controversy over Pluto's status as a true planet. The Naval Observatory opposed any downgrading from major-planet status, while others argued that Pluto was the first of a new class of solar-system bodies, termed trans-Neptunian objects or Kuiper-belt objects. The low mass of Pluto also meant that Pluto could not be the cause of certain observed perturbations in the orbits of Uranus and Neptune, leading to another interesting episode: a renewal of interest in the search for a tenth planet in our solar system. Because that search is primarily a matter of celestial mechanics in the tradition of the discovery of Neptune, we will return to it in chapter 12.[115]

*Mapping the sky: The great star catalogs*

As the Space Age began, the attention of positional astronomers was firmly focused on the problems of realizing as accurately as possible the fundamental reference frame of stars, and tying as many stars as possible to this frame, using both transit circles and astrographs. In the service of these goals, the Observatory undertook work

Charlene, whose family called her "Char." However, the tradition is to name planetary satellites after characters in Greek mythology. At the Observatory the name "Persephone," wife of Pluto, was recommended. Soon after Christy learned that Charon was actually a character in Greek mythology! Most astronomers pronounce the name as "Karen" or "Sharon," rather than Charon (rhymes with Lauren), as Christy intended in honor of his wife. Harrington later named one of his daughters Ann Charon.

[113] The first of several speckle observations are reported in D. Bonneau and R. Foy, *Astronomy and Astrophysics*, **92** (1980), L1. CCD images were also obtained in 1983, and are reported in H. J. Reitsema, F. Vilas, and B. Smith, "A Charge-Coupled Device Observation of Charon," *Icarus*, **56** (1983), 75–79. The Harrington and Christy statement is in Harrington and Christy (1980), p. 442. The transit phenomena were reported in R. P Binzel *et al.*, *Science*, **228** (1985), 1,193–1,195, and received media coverage in "Eclipse Proves Existence of Pluto's Moon," *Washington Post* (February 20, 1985), A 8.

[114] Christy and Harrington, "The Discovery and Orbit of Charon," *Icarus*, **44** (1980), 38–40.

[115] For a review of Pluto and its moon, in addition to the Stern volume (ref. 108), see S. A. Stern, "The Pluto–Charon System," *Annual Review of Astronomy and Astrophysics*, **30** (1992), 185–233.

on the several levels seen in Table 10.3: it continued the observed fundamental transit-circle catalogs that would eventually become the backbone of the German FK series, it led the organization and participated in the observations for the much larger International Reference Star catalogs, and for the first time it contributed to the photographic catalogs with its own astrographic telescope. As a punctuation mark to the century, the Observatory produced state-of-the-art catalogs based on the *Astrographic Catalogue* of the *Carte du Ciel* project, more than making amends for its lack of participation a century earlier. Beyond that, it led the way in digitizing techniques applied to existing photographic plates, resulting in a reference catalog of almost half a billion stars.

In the late 1950s, the Observatory was analyzing the just completed observations for its third fundamental catalog (the W350) tied to the equator and equinox of 1950, and had begun observations for the fourth catalog in that series. These catalogs were among the few new ones that fed into the German "Fundamental Star Catalog" series, considered the definitive fundamental celestial coordinate system, and called "the most widely used product of astronomy today."[116] As can be seen seen in Table 10.3, the version still in use at this time was the 1,535 stars of the FK3, but the FK4 was under development. In a review of its uses in 1961, Naval Observatory astronomer F. P. Scott noted that it was a particularly useful reference system "for a great variety of problems arising in geodesy, the accurate determination of time, the study of the motions of star groups, the comparison of star catalogues, and, more recently, in problems arising in space exploration."[117] He foresaw further, more sophisticated uses with the advance of technology.

The fundamental reference frame at this time had the zero points of its coordinates based on the motions of solar system objects, which defined the equinox and equator.[118] The programs therefore involved not only the observation of stars, but also solar-system objects. The W4(50) program (1956–62), for example, included not only FK4 stars, but also observations of the Sun, Moon, all major planets except Pluto, and four minor planets. The W5(50) (1963–71) included again the FK4, as well as special lists of stars, and the solar-system objects necessary to determine the zero points of the coordinates. The W1(J00) (1977–82), so-called because it was tied to the equinox and equator for the year 2000, observed parts of the FK5 catalog and zodiacal stars, and again solar-system objects. The importance of these catalogs may be seen from the fact that, in the preparation of the

---

[116] The German FK series of fundamental catalogs in Table 10.3 had overtaken the American fundamental catalogs such as the N30 in becoming almost universally used, even by the Americans. Although the American series was of high quality (in some ways superior to the FK3 and FK4), it was rather disjointed. The Germans stressed continuity – both of the lists of stars in the catalogs and of the system itself, with one catalog evolving from its predecessor. Toward the end of the century a "Washington Fundamental Catalog" was under way, but it was superseded by the Hipparcos satellite results. T. Corbin to S. Dick, private communication.

[117] F. P. Scott, "Present and Future Programs in Meridian Astronomy," *Proceedings of the International Meeting on Problems of Astrometry and Celestial Mechanics*, La Plata, Argentina (Astronomical Observatory, National University of La Plata: La Plata, 1961), pp. 7–15: 9. Scott also pointed out that, largely to meet the demands of geodesists, a list of 1,987 "FK4 Supplemental Stars," as well as 931 "faint fundamentals" (FKSZ), was prepared for observing by meridian telescopes. Work on the FK4 began in 1953; the catalog was published in 1964.

[118] See G. M. Clemence, "The Equinox and Equator," in *Proceedings* (ref. 117), pp. 17–24.

German FK catalogs, more weight was given to the Naval Observatory's observations than to those of any of other observatory, including Greenwich and Pulkovo in the north, and the Cape Observatory in the south. Even as the W1(J00) catalog was being observed, there was already a hint that some day the "dynamical reference frame," and with it all the intricate problems of determining the equator and equinox, would be superseded by an extragalactic frame, another step on the way to the Holy Grail of an inertial frame. That step would actually occur in 1998, after decades of transition. In the meantime the fundamental reference frame remained tied to zero points defined by solar-system objects.

For these fundamental programs the instrument of observation remained the 6-inch transit-circle telescope, which, as we have seen earlier in this chapter, was mounted in 1897, was placed in regular use in 1911, and had undergone a variety of mechanical improvements in the course of its career. When, in September, 1959, Watts retired after 48 years at the Observatory, it was the end of an era in more ways than one. His successor as Director of the 6-inch Division, A. Norwood Adams, was more attuned to the nascent digital and computer revolutions. The latter techniques were eminently applicable to the massive amounts of data acquired by transit-circle telescopes; until the late 1950s all the data had been reduced by tedious desk-calculating techniques. Whereas the improvements of Watts had been mechanical in nature (or photoelectric in the case of the automatic measuring machine), digital computer technology would now drive the improvements. Already in 1956 Adams had succeeded in digitizing the data from Watts's Automatic Measuring Engine. In 1961, near the end of the W4(50) program (1956–62), the micrometer for the 6-inch instrument, which originally had been constructed by Warner and Swasey in 1906, was fitted with a new motor drive and subsequently a digital data-recording system, the latter again largely as a result of Adams. Adams, assisted by data-reduction chief Sylvan M. Bestul, slowly transformed the reduction procedures from office desk work to those using the more sophisticated punch-card operations. When F. Stephen Gauss joined the transit-circle staff in 1963, his assigned mission was to prepare the 6-inch telescope for a computer-controlled data-acquisition system. Thus, by 1969, near the end of the next fundamental program (1963–71), the data went directly to an online IBM 1800 Data Acquisition and Control System. Again under Gauss, the computerization of data acquisition and reduction were transferred from the IBM 1800 to an HP1000 in 1980. Step by step, a classical transit instrument was being brought into the modern era.[119]

The greatest changes to the 6-inch transit circle, however, occurred prior to its

[119] On the improvements during the W4(50) see A. N. Adams and D. K. Scott, "Results of Observations Made with the Six-Inch Transit Circle, 1956–1962," *PUSNO*, **19**, part 2, 302–303, and A. N. Adams, "Digital Recording System and Motor Drive of the Washington Transit Circle Micrometer," in *Sonderdruck aus Sitzungsberichte der Heidelberger Akademie der Wissenschaften* (Akademie der Wissenschaften: Heidelberg, 1962), pp. 2 and 56–61; on improvements during the W5(50) see J. A. Hughes and D. K. Scott, "Results of Observations made with the Six-Inch Transit Circle, 1963–1971," *PUSNO*, **23**, part 3, 171–172; the computerized acquisition system as of 1970 is described in "High Accuracy Telescope," *Datamation* (February, 1970). The new micrometer motor drive was constructed in the USNO Instrument Shop. The telescope in the midst of these changes is described in Paul D. Hemenway, "The Washington 6-Inch Transit Circle," *Sky and Telescope* (February, 1966), 72–77.

sixth and final fundamental catalog, the W1(J00), which was observed from 1977 to 1982. About 1972 a new temperature-compensated lens, built by Farrand Optical Company, replaced the original Brashear lens. In addition, the graduated circle technology used for determining declinations also underwent a radical evolution. From the mid-1950s until the mid-1960s, the Naval Observatory's instrument shop produced many engraved metal circles, not only for its own instruments, but also for the instruments of other observatories, such as Lund, Ottawa, and Hamburg-Bergedorf. By the end of the 1960s, this capability had been lost at the Naval Observatory due to retirement of key personnel. This, along with rapid developments in the technology of automatic circle scanning systems, caused the Naval Observatory to look to the outside world for a new type of circle. In particular, it was determined that glass circles, rather than metal circles, were distinctly superior when they were used in conjunction with the new circle scanning systems that would now replace the photographic technique instituted by Watts in the 1940s. Such glass circles, usually graduated at 3-minute-of-arc (0.05-degree) intervals, were already in use at Greenwich and Brorfelde, as well as Pulkovo and Caracas. They were procured beginning in 1971, first from Teledyne-Gurley Corporation of Troy, New York, but, beginning in 1976, with circles produced by Heidenhain Corporation in Germany.[120] By 1973 electronic scanners had been developed that would read each of the 7,200 engraved circle divisions to a precision of one part in 5,000. These circles and their scanning system were used throughout the rest of the telescope's history, which ended in 1995.[121]

Throughout this period clock technology, which is necessary for determining right ascensions, was also improving. Early in the W3(50) program (1949–56) a sidereal quartz-crystal clock had replaced the Shortt free-pendulum clock as the standard clock. Some form of sidereal crystal clock remained the standard clock until such clocks were replaced by rubidium frequency standards in 1977, and cesium standards after 1980. These changes will be discussed in more detail in the next chapter.[122]

Despite these efforts to apply new technology to a classical visual instrument, the improvement in accuracy was disappointing. The probable error of a single observation of a star at the equator remained about 0.18 arcseconds in right ascension, and 0.30 arcseconds in declination during the programs observed in the latter half of the century. This could be reduced to under 0.1 arcsecond by repeated observations, but the barriers were formidable, and, after a point, unbreachable. What positional

[120] T. J. Rafferty and B. L. Klock, "Experiences with the U. S. Naval Observatory Glass Circles," *Astronomy and Astrophysics*, 114 (1982), 95–101.

[121] T. J. Rafferty and B. L. Klock, "Circle Scanning Systems of the U. S. Naval Observatory," *Astronomy and Astrophysics*, 164 (1986), 428–432; P. Jones, E. Coyne, J. Davis, and C. Watts Jr, "A Pulsed Photoelectric Scanning System for Angular Measurement in Astrometry," in Brian J. Thompson and John B. DeVelis, eds., *Electro-Optics: Principles and Applications* (Redondo Beach, California, 1973), pp. 105–111. Also B. L. Klock, R. Z. Geller, and M. A. Dachs, "Inductosyn Angular Readout System of the U. S. Naval Observatory Six-Inch Transit Circle," *Proceedings of the Electro-Optical Systems Design Conference* (1969), pp. 633–641. The electronic "Inductosyn" system for measuring the angular position of the telescope varied as a function of temperature, and in the end was used only as an aid to setting the telescope rather than as the primary angle indicator. Instead the primary angle indicator was obtained by reading the circle with the scanners. [122] PUSNO, **19**, part 1, 11; PUSNO, **23**, part 3, 172; W1(J00), PUSNO, in press, 5.

Table 10.4. *Transit-circle calibrations affecting accuracy of absolute star catalogs*

| Quantity | Frequency | Accuracy (arcseconds) |
|---|---|---|
| **Orientation of telescope** | | |
| Azimuth of Marks | Monthly | ±0.08 |
| Marks – one tour | 2–3 hours | 0.30 |
| Collimation | 2–3 hours | 0.25 |
| Level | 2–3 hours | 0.15 |
| Nadir | 2–3 hours | 0.35 |
| **Divided circle** | | |
| Diameter corrections | Four times/program | 0.07 |
| One reading | Each star | 0.06 |
| **Clock correction** | Each tour | 0.10 |
| **Atmosphere** | | |
| Refraction | Published tables | 0.03 |
| Temperature | Each star | 0.10 degrees Centigrade |
| Dew point | Each star | 1.0 |
| Barometer | Each star | 0.1 mm |
| Variation of latitude | BIH and solution | 0.03 |
| Flexure | Program average | 0.02 |
| Screw error in right ascension | One per year | 0.02 |
| Screw error in declination | One per year | 0.02 |
| Pivot errors | Two per year | <0.005 |
| Clamp differences | Reverse monthly | 0.02 |
| Equinox correction | Program average | 0.14 |
| Equator correction | Program average | 0.02 |

*Source:* From Thomas Corbin, "The Determination of Fundamental Proper Motions" (ref. 123).

astronomers were up against is evident in Table 10.4, in which an attempt is made to list the accuracies of the many measurements that go into producing a star position. Some idea of the problems may be given from the fact that a shift of only 10 microns (ten millionths of a meter) in one of the mounting cages of the 6-inch instrument would produce a shift of 2.3 arcseconds in accuracy. Frequent measurements of these shifts were therefore essential in order to apply corrections. Corrections in the graduated divisions of the circle could be measured and applied, but only to an accuracy of 0.07 arcseconds. The refraction correction remained a fundamental barrier, at the level of 0.03 arcseconds. The flexure of the telescope tube under the force of gravity entered in at 0.02 arcseconds. Furthermore, the accuracies of the zero points to the coordinate system also had to be considered.[123] Despite other developments in meridian instrumentation, including the construction of mirror transit circles and the use of photoelectric recording and guiding equipment, it would take entirely new technologies, and a new definition of the fundamental reference frame, to break through these barriers in the 1990s.

[123] T. E. Corbin, "The Determination of Fundamental Proper Motions," in D. S. Hayes *et al.*, eds., *Calibration of Fundamental Stellar Quantities* (Reidel: Dordrecht and Boston, 1985), pp. 53–70.

Finally, no matter how accurately the positions of stars were determined, there was the problem of their proper motions, which needed to be accurately determined so that they could be applied after the observations had been made. Proper motions gave positional astronomers "job security;" no matter how accurately the positions of the stars were determined for use over the short term, their slight motions over decades would change those positions. Proper motions could be measured and extrapolated, but the further removed from the epoch of observation, the greater the error, and any error in fundamental catalogs was sure to propagate to other catalogs based on them. It was this problem of proper motions, as well as the problem of defining a "non-rotating frame" in a dynamical system, that drove astronomers continually toward an inertial reference frame based on extragalactic objects so distant that they had little or no proper motion. Even with an extragalactic frame, although the defining objects are basically free of proper motion, the Hipparcos positions still have proper motions, and it is these brighter stars that must be used for practical applications.[124]

Important as the fundamental catalogs were, they were simply a continuation, with improved techniques, of a program that had been in progress for decades. What was new during this period at the Naval Observatory, however, was its leading role in the International Reference Star Catalogues, a step beyond the fundamental catalogs and a link to the much larger photographic catalogs. While the 6-inch transit circle contributed heavily to the fundamental reference frame, the photographic catalogs were much more useful to the average user from the point of view of numbers of stars. The problem was that the photographic catalogs then in use (the AGK2 in the north and the Cape and Yale catalogs for the south) gave errors on the order of 0.7″ to 3.0″. These errors might be reduced to 0.2″ to 0.8″ if proper motions were applied, but, as errors in proper motions were extrapolated further from the epoch of observation, the situation only worsened. The need for a dense grid of visually observed reference stars to tie the photographic catalogs back to the fundamental system was therefore obvious. Whereas the "Level 3" catalogs in Table 10.3 had been international observing programs with a specific application in mind (reduction of photographic plates), the new program that would become known as the International Reference Stars (IRS) was intended to be an extension for the fundamental frame usable at all epochs. This program would occupy the Naval Observatory during much of the second half of the century. It is of interest not only because of the individuals, strategies, and technologies involved, but also because the program by its very nature required a high level of international cooperation.[125]

[124] F. P. Scott, "The System of Fundamental Proper Motions," in volume 3, *Basic Astronomical Data, Stars and Stellar Systems*, ed. K. A. Strand (University of Chicago Press: Chicago, 1963), pp. 11–29; T. E. Corbin, "The Determination of Fundamental Proper Motions," in *Calibration of Fundamental Stellar Quantities* (ref. 124), pp. 53–70.

[125] F. P. Scott, "Estimates of the Accuracy of Positions taken from Photographic Star Catalogs, Present and Future," in G. Veis, ed., *The Use of Artificial Satellites for Geodesy* (American Elsevier: Amsterdam, 1963), pp. 221–231.

The improvement of photographic catalogs and their reference stars was clearly one of the goals uppermost in the minds of the participants at the 1953 Conference on Problems of Astrometry. The first resolution of that conference emphasized "that it is of the highest importance that a third observation of all AG stars should be undertaken to provide a sufficient number of reliable proper motions of stars in the northern sky." In this so-called AGK3 photographic program, led by Otto Heckmann and Wilhelm Dieckvoss of the Hamburg–Bergedorf Observatory in Germany, the Naval Observatory would play little direct role. However, in its recommended offshoot, "observations of a sufficient number of reference stars for the reduction of the photographic plates . . . by as many meridian circles as possible," the Conference asked the Naval Observatory to play the leading role, not only in observing its share of reference stars for the so-called AGK3R (AGK3 Reference), but also in compiling the entire list of reference stars to be observed by all observatories, reducing from apparent to mean place the positions of all observations as requested by participating observatories, and coordinating the observations and preparation of the catalog.[126] This catalog in turn would be tied to the fundamental reference frame, which was still being put out by the Germans.

The Conference recommendation for an AGK3R project had the blessing of Gerald Clemence, the senior astronomer at the Naval Observatory and one of the organizers. However, the bulk of the organizational work for this endeavor would be carried out by F. P. Scott (Figure 10.41), successor to H. R. Morgan. Scott had come to the Naval Observatory as a computer in 1929, and had worked first under Morgan with the 9-inch, then with the 7-inch transit-circle telescope, before being placed in charge of the latter instrument from 1948 until his retirement in 1970. When he died at the age of 68 in 1974, he was "widely regarded as the Western Hemisphere's leading specialist in the determination of star positions from meridian observations."[127] This despite the fact that his formal training was in mathematics and physics rather than astronomy. Scott had participated in the 1953 conference, and in the 1955 Conférence d'Astrométrie, held in Brussels, where the details of organization for the AGK3R were completed. At the 1955 IAU General Assembly in Dublin, he was appointed chairman of the committee to oversee the AGK3R program, and the program was discussed at the triennial meetings of the IAU, especially in its Commission 8 on Positional Astronomy. When in 1961 Scott delineated "three major problems today which

---

[126] "Problems in Astrometry" (ref. 91), 29–104. The resolutions are on p. 104. The reasons for a third AG catalog are discussed in Otto Heckmann, "The value of a third AG Catalogue," ibid., 31–34. Heckmann, considered the father of the AGK3, was the Director of the Hamburg-Bergedorf Observatory in Germany. It is important to note that, although the AGK3 continued the work of the AGK2, it had no connection with the Astronomische Gesellschaft. The AGK3 was independently organized, and administered by a committee appointed at the IAU in Dublin in 1955, headed by F. P. Scott.

[127] "Cataloguer of Stars," *Sky and Telescope*, **16** (January, 1975), 18; F. P. Scott, "Curriculum Vitae," "Publications and Presentations to Conferences, Symposia and Meetings," F. P. Scott folder, USNOA, BF.

Figure 10.41. F. P. Scott at the 7-inch transit circle, March, 1961.

demand the serious attention of meridian astronomers," the 1953 goals were still uppermost, now stated in more specific terms: "the improvement of the Fundamental Star Catalogue, the establishment of a system of reference stars, and the observation of the star lists recommended to meridian observers at the Moscow meeting of the IAU."[128] The ongoing AGK3R program, he remarked, was addressing those problems for the Northern Hemisphere in establishing an accurate reference frame.

The Southern Hemisphere, however, was another matter entirely, and was in desperate need of attention. In further pursuit of increased accuracy of star positions, densification of the stellar reference frame, and complete sky coverage, the last half of the twentieth century at the Naval Observatory saw an expansion of the work of meridian instruments to include the southern Hemisphere. Southern Hemisphere star positions had always been less well known than their Northern-Hemisphere counterparts, due to there being a smaller population and an accompanying smaller number of observatories. Though Southern-Hemisphere expeditions by other observatories had gone some way toward remedying this, only in the second half of the

[128] F. P. Scott, "Present and Future Programs in Meridian Astronomy," (ref. 117), pp. 7–15. Scott describes the origins of the AGK3 and AGK3R in "The AGK3 – A Co-operative Programme in Astronomy," *ICSU Review*, **3** (1961), 2–7. See also *Transactions of the IAU* (henceforth *Trans. IAU*), volume 10 (1958) (Cambridge University Press, Cambridge, 1960), pp. 124 and 128. The Moscow lists included some 7,000 stars such as doubles and PZT stars.

century was the problem addressed, and once again the Naval Observatory played a leading role. Although a Southern-Hemisphere meridian circle had been mentioned as a desideratum in resolution 4 of the 1953 Conference, it was the Second Astrometric Conference in May, 1959 that initiated the groundwork for a system of Southern Reference Stars (SRS), as the 1953 Conference had for AGK3R in the north. Acknowledging the pioneering work of the Royal Observatory at the Cape of Good Hope, the Conference considered as "urgent the intensification of astrometric observations in the southern hemisphere, especially the absolute determination of fundamental star positions, the differential determination of positions and faint reference stars, and the determination of absolute proper motions by reference to distant galaxies." It recommended a committee, chaired by Yale's Dirk Brouwer and including Scott, "not only to do all that is possible to complete the organization of a reference star program similar to the AGK3R for the southern hemisphere, but also advise on questions concerning programs and methods of observations."[129]

The Naval Observatory carried out its commitment both to the AGK3R and to the SRS not only with the 6-inch transit circle, but also with a new 7-inch transit circle, designed by C. B. Watts and under the direction of F. P. Scott since 1948. The construction of the instrument began in the Naval Observatory Instrument Shop in late 1947, and it was similar in most mechanical details to the 6-inch transit circle.[130] Observations were begun in fall 1955, but problems with the lens delayed the beginning of the AGK3R observations until December, 1957. These observations, however, were only the beginning of the Naval Observatory's role; as Scott had been chairman of the AGK3R committee of the IAU since 1955, Scott's committee organized, supervised, and expedited the program of 21,000 northern reference stars needed for the AGK3 photographic catalog with its 183,000 stars, the latter catalog being the responsibility of Dieckvoss and the Germans. Transit circles at nine observatories in the Northern Hemisphere participated in observing the 21,000 stars of the AGK3R.[131] The Naval Observatory made its observations while observing simultaneously the W4(50) fundamental catalog with the 6-inch transit circle, completing its portion of the observations in 1962. In an early example of the impact of the computer on the field, the full AGK3R catalog of reference stars, encompassing all participating

---

129 Second Astrometric Conference, May 17–21, 1959, AJ, **65** (1960), 167–238: 169. This Conference was spurred on by the Moscow meeting of the IAU in 1958. The sequence of events is described in Scott (ref. 124), pp. 228 ff. As part of the Second Astrometric Conference Fricke reviewed the deficiencies in the Southern Hemisphere, W. Fricke, "The System of Fundamental Stars in the Southern Hemisphere," AJ, **65** (1960), 177–180.

130 "The New Seven-Inch Transit Circle of the U. S. Naval Observatory," in *New Instruments and Methods in the Meridian Astronomy* (Academy of Sciences of the U. S. S. R.: Moscow, 1959), pp. 72–75; also "Reports of Commission 8, X General Assembly, *Trans. IAU*, volume 10 (Cambridge University Press: Cambridge, 1960), 136. The 7-inch transit telescope was made possible by a $5,000 Congressional appropriation for the financial year 1947; AR (1946), 3.

131 Scott, "Report on the AGK3R," AJ, **65** (1960), 175–176. W. Dieckvoss, *AGK3: Star Catalogue of Positions and Proper Motions North of* −2.5 [*Degrees*] (Hamburger Sternwarte: Hamburg-Bergedorf, 1975).

observatories, was published electronically only, one of the first major catalogs to be available on magnetic tape.[132]

In 1962, just as the Observatory was completing its observing for the AGK3R, Scott also became chairman of the SRS committee of the IAU, with similar responsibilities for the Southern Hemisphere. This time 13 transit-circle telescopes were used, six in the Southern Hemisphere (where observations were made from Argentina, Australia, Chile, and South Africa), and seven in the Northern Hemisphere (observing from France, Japan, Rumania, Spain, the United States, and the U. S. S. R.). The overall program ran from 1961 to 1973, producing positions for 20,500 stars.[133]

The Naval Observatory agreed to observe the zone of the Southern Reference Stars from 0 to −20 degrees declination with the 6-inch transit circle, and the entire Southern Hemisphere skies with the 7-inch transit circle; the latter meant a need to establish a station in the Southern Hemisphere, something it had never done before. The location of this station was strongly influenced by the fact that Yale University astronomers, under the direction of Brouwer, had dedicated the Yale–Columbia Southern Observatory earlier in 1965 at El Leoncito, Argentina. To set the wheels in motion for its own program, in November, 1965 Strand and Observatory Superintendent McDowell visited Argentina, conferring with the Naval Attaché and Scientific Attaché of the American Embassy in Buenos Aires, in connection with the installation of the Naval Observatory transit circle on the Yale station's site. The Naval Observatory facility was built at a cost of $51,750, and was to be administered by the Yale–Columbia Observatory through a grant from the Office of Naval Research.[134] The observations, undertaken from December 1966 to August 1973 at El Leoncito (Figure 10.42), included the stars of the German FK4 catalog as well as the SRS stars. Preliminary results of the El Leoncito observations were given in 1978. The Naval Observatory and Pulkovo Observatories compiled the SRS contributions from all telescopes into a single catalog of positions referred to the FK4 system, which was com-

[132] C. Smith OHI, 13. The W4(50), observed simultaneously with the AGK3R at the USNO, was published in PUSNO, **19**, part 2, and the AKG3R observations were turned over to F. P. Scott for compilation with the observations from other observatories. The magnetic tape, F. P. Scott and J. L. Schombert, "Catalog of positions of 21,499 AGK3R Stars," became available in the mid-1960s.

[133] Reviews are given in Scott, "Status of International Reference Star Programs," AJ, **67** (1962), 690; Scott, "The AGK3R, SRS, and Related Projects," in Lubos Perek, ed., *Highlights of Astronomy* (Reidel: Dordrecht: 1968); Scott, "Status of the International Efforts to Improve the Catalogues of Positions and Motions of Stars to the 9th Magnitude," AJ, **72** (1967), 570–571; J. L. Schombert, "Southern Reference Star Program: Progress Report," in W. Gliese, C. A. Murray, and R. H. Tucker, eds., *New Problems in Astrometry* (Kluwer: Dordrecht, 1974), p. 41.

[134] J. A. Hughes, "U. S. Naval Observatory Southern Hemisphere Expedition," AJ, **72** (1967), 566–567; SAN, 38. On the setup of the station see Smith OHI, 17 ff. Yale's goal was to extend the Lick proper motion program to the Southern Hemisphere. The commitments of observatories to zones were made at the La Plata meeting, "Report of the Meeting," p. 5. It should be noted that, for a few years in the first decade of the century, the Naval Observatory had a Southern-Hemisphere station at the Tutuila Naval Station in Samoa, at 15 degrees south, AR (1903), 9. Its purpose was to observe positions of Southern-Hemisphere stars for the *Nautical Almanac*, to determine time, and to make magnetic observations.

Figure 10.42. The Yale–Columbia station at El Leoncito, Argentina, 1969. The Naval Observatory's transit circle is in the small building at far left, and the Yale–Columbia astrograph is in the cylindrical building on the right.

pleted and distributed on tape in 1988. With this accomplished, Thomas Corbin combined it with the AGK3R to produce an all-sky-catalog of positions and proper motions in 1991, known as the International Reference Stars.[135]

By the 1970s, when the international reference stars required reobservation, the Naval Observatory again took a leading role. Under the direction of Astrometry Department Director James Hughes, a pole-to-pole program was planned, in which all 36,000 stars of the International Reference Stars could be observed with the two transit circles of the Naval Observatory, the 6-inch and 7-inch instruments, thus placing under one institution the burden of achieving uniformity in observing and reduction. For this, the first coordinated pole-to-pole program in the history of astrometry, the Naval Observatory established the Black Birch Astrometric Observatory near Blenheim, New Zealand (Figure 10.43). After a ten-year extensive renovation and outfitting with an electronic "image dissector micrometer," the 7-inch telescope arrived in 1984. Extensive testing and development required two years. From

---

[135] Preliminary results of The Naval Observatory's Observations at Leoncito are in C. Smith, "Some Results of Observations with the Scott Seven-Inch Transit Circle at Leoncito," in *Modern Astrometry* (ref. 105), pp. 447–453; final Leoncito results are in J. A. Hughes, C. A. Smith, and R. L. Branham, "Results of Observations made with the Seven-Inch Transit Circle, 1967–1973," PUSNO, **26** (1992), part 2. An overview of the compiled SRS program from all participating observatories is given in C. Smith *et al.*, "The SRS Catalog of 20,488 Star Positions: Culmination of an International Cooperative Effort," in J. H. Lieske and V. K. Abalakin, *Inertial Coordinate System on the Sky* (Kluwer: Dordrecht, 1990), pp. 457–463. The Soviet perspective on the program is given in M. Z. Zverev *et al.*, "On the SRS Catalogue," in H. Eichhorn and R. J. Leacock, eds., *Astrometric Techniques* (Kluwer: Dordrecht, 1986), pp. 691–696. Events at the Leoncito station are described in Smith, OHI, and Corbin, OHI. Corbin's 1991 IRS catalog is described and available at NASA's Astronomical Data Center (NSSDC 91-11, 1991) www.adc.gsfc.nasa.gov, where many other catalogs are also to be found.

Figure 10.43. Black Birch station, near Blenheim, on the South Island of New Zealand. On the left is the dome for the 8-inch double astrograph; the 7-inch transit-circle building is at the center, and the offices, computers and operations center are in the building on the right.

the beginning of stellar observations in 1987 to termination in 1996, in total 400,655 observations of 28,186 stars were made.[136]

Simultaneously with the Southern-Hemisphere observations, the 6-inch transit circle undertook the observations of the northern portion of the IRS, beginning in 1985. Though it was enhanced by electronic circle-reading scanners, unlike the 7-inch instrument it remained a visual instrument with an astronomer actually observing, an increasing rarity in astrometry in the last third of the century. This program, completed in 1995, was to be its last. In the race between man and machine, machine won in terms of numbers of observations: the 6-inch telescope made 317,444 of 26,204 stars, some of them overlapping with the Southern-Hemisphere observations for comparison purposes. However, in the end the results from the 6-inch visual telescope were at least as good as the 7-inch telescope's results that were electronically determined, although the latter was able to observe many more stars during daylight. Moreover, even as the pole-to-pole program was in its final years, the Hipparcos satellite was making observations that were a hundred times more accurate, a success no

[136] The New Zealand program was first described in J. A. Hughes, "SRS Observations: Future U.S. Naval Observatory Meridian Programs," in *Modern Astrometry* (ref. 105), pp. 497–501; also Hughes, *Southern Stars*, 29 (1982), 245–260, and Hughes *et al.*, "The seven-inch transit circle and its New Zealand Program," in H. Eichhorn and R. J. Leacock, eds., *Astrometric Techniques* (Kluwer: Dordrecht, 1986), 483–496. The history of the program is given in Brian Loader, "Black Birch Astrometric Observatory, 1984–1996," *Southern Stars*, 37 (1997), 148–160. The closing of the New Zealand Station is described in F. Stephen Gauss, "Closing the Black Birch Astrometric Observatory," USNO Star, 6, no. 2 (March–April, 1996), 1 and 7; T. J. Rafferty, "Closing the Black Birch Astrometric Observatory: Part Two," USNO Star, 6, no. 3 (October, 1996), 8–10. Under the terms of an agreement with the New Zealand government, in May, 1997 the buildings were demolished and the site was returned to its original appearance.

one could have predicted when the IRS began its program. Had Hipparcos not succeeded, the Naval Observatory's pole-to-pole program would have been by far the most important component in the next fundamental catalog. Whether this would have been an FK6 or a "Washington Fundamental Catalog" is one of the moot points of history. In the end the pole-to-pole program, known as the W2(J00), was a "level-1" observed catalog. In any case, with the IRS, the 6-inch telescope ended an impressive career; its total number of observations in nearly a century of observing was about 930,000, surpassing the 675,000 made by the Airy Transit Circle at Greenwich.[137]

With the end of the IRS programs, the Naval Observatory was out of the transit-circle business, with one interesting exception that also serves as a lesson in the perils of modernization. Even as more accurate methods became available for wide-angle astrometry, an 8-inch transit circle, though not used in its traditional mode, remained in operation for special purposes at the Naval Observatory Flagstaff station. Originally designed and built as a reflector by Farrand Optical Company under a contract awarded in November, 1965 (as plans were being laid for the El Leoncito station), the telescope was a modified Schmidt–Cassegrain design with a 10-inch spherical primary mirror and two corrector lenses. The telescope was moved to Flagstaff station in 1976 but, despite years of testing and modifications, it was never possible to maintain the reflecting optics with the stability required for precision transit-circle work. Only when the mirror was replaced with an 8-inch lens did the telescope become usable. It observed visually from 1983 to 1988, and thereafter was updated with a CCD detector, which could detect stars down to 17th magnitude and tie them to the extragalactic reference frame.[138]

One of the most interesting aspects of the compilation of star catalogs based on transit-circle observations was the use of rapidly developing computer techniques. By 1959 almost all of the transit-circle reductions were being done by computer, using punched cards. For a typical reduction that produced an "observed minus computed" value for a star position, four decks of punched cards were used. One deck contained the circle readings, as output from Watts's automatic measuring engine. A second

[137] Steven J. Dick and Ted Rafferty, "Six-Inch Transit Circle to Complete Last Fundamental Program," USNO Star, **5**, no. 2 (March–April, 1995), 1–2. The results of the pole-to-pole program are described in T. J. Rafferty and E.R. Holdenried, "The U. S. Naval Observatory pole-to-pole Catalog: W2(J00)," Astronomy and Astrophysics Supplement Series, **141** (2000), 423–431. The catalog is available electronically.

[138] H. E. Crull Jr, "A Short History of the 8-Inch Transit Circle," USNOA, SF. Strand, SAN, 39. Plans for the 8-inch instrument were described at the NSF-sponsored Conference on Star Catalogues by B. L. Klock, "Instrumentation Program for Improved Transit Circle at the U. S. Naval Observatory," AJ, **72** (1967), 559–561. See also B. L. Klock and F. S. Gauss, "Instrumental parameters of the U. S. Naval Observatory's Automatic Transit Circle (ATC)," New Problems in Astrometry (ref. 133), pp. 259–257. The CCD telescope and its results are described in R. C. Stone and D. G. Monet, "The USNO (Flagstaff Station) CCD Transit Telescope and Star Positions Measured from Extragalactic Sources," in Inertial Coordinate System on the Sky (ref. 135), pp. 369–370; R. C. Stone, "Recent Advances with the USNO (Flagstaff) Transit Telescope," in Ivan Mueller and B. Kolaczek, Developments in Astrometry and Their Impact on Astrophysics and Geodynamics (Kluwer Academic: Dordrecht and Boston, 1993), pp. 65–70; R. Stone and C. Dahn, "CCD Astrometry," in Astronomical and Astrophysical Objectives (ref. 2), pp. 3–8.

deck contained the zenith-distance micrometer readings ("bisections") and temperature. A third deck contained the readings of the right-ascension micrometer screw (the "ticks") made at 4-second intervals during the observation. A final deck held the star number, date, and a code identifying which telescope had made the observation. The second and third decks were still hand-punched at this point, but Adams and John Schombert developed a synchrosystem by which the micrometer readings were punched directly onto cards.[139]

Entirely different concerns were at the forefront with photographic catalogs, seen at Level 4 in the hierarchy of catalogs in Table 10.3. For reasons rooted in its inability to acquire the required instrumentation through much of the century, the Naval Observatory had never produced such catalogs. Only between 1977 and 1986 did the Naval Observatory observe its first photographic catalog with the twin 8-inch astrograph, utilizing one lens sensitive to blue, the other to yellow. The resulting Twin Astrographic Catalogue (TAC) contains 705,679 star positions to magnitude 11.5.[140] A southern portion was intended to be observed in New Zealand and referenced to the SRS reobserved by the 7-inch transit circle. Although the 8-inch astrograph was moved to New Zealand, that part of the program was abandoned due to financial constraints in the early 1990s. However, a USNO CCD Astrometric Catalogue (UCAC-1) was observed with the same astrograph, now provided with a CCD detector and a red lens, located at Cerro Tololo in Chile between 1998 and 2001. It included positions of millions of Southern Hemisphere stars between magnitudes 7.5 and 16.5, and densified the system based on the Hipparcos observations with an accuracy of about 20 milliarcseconds.

To come full circle from the beginning of the century, the Naval Observatory built on its IRS work to make the best of past photographic catalogs, especially the *Carte du Ciel* program in which it had originally failed to participate. For decades the Astrographic Catalogue had largely been ignored because it was difficult to work with and the plate constants used did not yield the full accuracy inherent in the plates. In 1972 Paul Herget and Thomas Corbin showed that new plate constants could be determined and applied to make use of the full accuracy. In order to carry out this program, all measurements had to be put in machine-readable form; 25% of this had been done by the French in the 1960s, and between the late 1970s and 1996 the remaining data were keyed in under the direction of the Naval Observatory. Meanwhile, between 1989 and 1991 Corbin and Sean Urban combined over 150 meridian and photographic cata-

---

[139] PUSNO, **23**, part 3, 172. F. P. Scott, "The Use of High Speed Computing Machinery in Making Transit Circle Reductions at the U. S. Naval Observatory," in *New Instruments and Methods in the Meridian Astrometry* (Academy of Sciences of the U. S. S. R.: Moscow, 1959), pp. 80–82; "The Reduction of Transit Circle Observations Using High-Speed Computing Machinery," *Trans. IAU*, volume 10 (Cambridge University Press: Cambridge, 1960), p. 136; Scott and Hughes, "Computation of Apparent Places for the Southern Reference Star Program," AJ, **69** (1964), 368. In order to handle the 7-inch data, a second AME was constructed and placed in use in 1967, supplementing the one originally built by Watts in 1948. By this time, however, the readings from both measuring engines had been digitized and automatically punched onto cards.

[140] N. Zacharias *et al.*, "The Twin Astrographic Catalog (TAC) Version 1.0," AJ, **112** (1996), 2,336–2,348.

logs to produce a reference system for the Astrographic Catalogue, tied to the FK5 via the IRS. The Astrographic Catalog Reference Stars (ACRS), completed in 1991, contains 320,000 positions and proper motions of stars to magnitude 10.5, with positional errors of 0.22 arcseconds at epoch 1996. It replaced the Smithsonian Astrophysical Observatory (SAO) catalog. With the accurate reference catalog represented by the ACRS, the entire Astrographic Catalogue could be reduced taking advantage of the intrinsic accuracy of its plates. The result is the AC 2000, a positional catalog of 4.6 million stars down to magnitude 11. The century-old star positions, when they were combined with modern observations, could thus yield high-accuracy proper motions. One of the first applications of an improved version of the AC 2000 was to compare it with the Observatory's Twin Astrographic Catalogue to produce proper motions. The AC 2000 has also been compared with the Tycho catalog from the Hipparcos satellite, yielding positions and proper motions for 988,758 stars. Thus the work of the original *Carte du Ciel* project, whose Astrographic Catalogue of positions had not realized its promise, was proven very useful at the end of the century.[141]

Nor was this the end of exploiting photographic star catalogs. Since the 1950s the Palomar Sky Survey had been the primary source for photographic star positions, and David Monet now carried out the digitization of these photographic plates at the Naval Observatory's Flagstaff station. Beginning in the late 1980s the Observatory built the Precision Measuring Machine (PMM) to measure the Palomar Sky Survey plates with a relative accuracy of 0.15 to 0.25 arcseconds, depending on location on the plate. The result was Monet's USNO A1.0, which is a reference catalog of 488,006,860 reference stars. Its successor, the A2.0, contains more than half a billion stars, and is the largest star catalog ever compiled.[142]

Like all mission-oriented Federal institutions, the dual question of the usage and value of the Naval Observatory's star-catalog work was constantly addressed, even if the answers were not always recorded in print.[143] For example, an Astrometric

[141] Sean E. Urban and Thomas E. Corbin, "The Astrographic Catalogue: A Century of Work Pays Off," *Sky and Telescope*, **95** (1998), 40–44. See also Corbin and Urban, "A Catalogue of Reference Stars for New Reductions of the Astrographic Catalog Plates," in S. Debarbat et al., *Mapping the Sky*, pp. 287–292; T. Corbin and S. Urban, "Proper Motions of the Northern Astrographic Catalog Reference Stars," in A.G. Davis Philip and A. R. Upgren, eds., *Star Catalogues: A Centennial Tribute to A. N. Vyssotsky* (L. Davis: Schenectady, New York, 1989), pp. 59–64; T. Corbin and S. Urban, "Faint Reference Stars," in *Inertial Coordinate System on the Sky* (ref. 135), pp. 433–442. R. L. Duncombe and J. S. Duncombe, "The Star Catalogue Project. A Progress Report," AJ, **72** (1967), 582–583 reports on the conversion of star-catalog data into machine-readable form.
[142] G. Westerhout and D. Monet, "The Flagstaff Measuring Machine," in *Inertial Coordinate System on the Sky* (ref. 135), pp. 491–492. A summary of star catalogs and the Naval Observatory's contributions is Adrian R. Ashford, "Star Catalogs for the 21st Century," *Sky and Telescope*, **102** (2001), 65–67. The same issue contains a description of the UCAC catalog, Roger W. Sinnott, "The Best Star Catalog Ever," 22–23.
[143] "Limitations imposed on Celestial Navigation Due to Inaccuracies of Star Positions," *Navigation*, **11**, (1964), 20; "The System of Fundamental Proper Motions," in K. A. Strand, ed., *Basic Astronomical Data, Stars and Stellar Systems* (Chicago University Press: Chicago, 1963); Scott, "Status of International Efforts to Improve the Position and Motions of Stars to the 9th Magnitude," AJ, **72** (1967), 570; O. Heckmann, "The Value of a Third AG Catalogue," in *Problems in Astrometry* (ref. 91), pp. 31–34.

Conference sponsored by the NSF, held at the University of Maryland in 1966, focused on "The Construction and Use of Star Catalogues." The Conference, organized by Yale and the Naval Observatory (specifically Brouwer, Clemence, Duncombe, Scott, and Strand), sought to "ensure maximum usefulness of the results to non-astronomical users, and to ascertain the actual requirements of those users." However, most of the meeting concentrated on the construction of the catalogs; only one paper discussed star-catalog requirements for satellite geodesy.[144] Nevertheless, toward the end of the century the questions of usage and value were increasingly reevaluated in light of new user requirements. An IAU Symposium in 1992 on "Developments in Astrometry and Their Impact on Astrophysics and Geodynamics" highlighted applications ranging from parallaxes and galactic dynamics to Earth orientation. An accurate reference frame was needed for everything from minor-planet orbits to galactic dynamics; and, for deep-space work, the denser the catalog, the better. (Whereas the early photographic surveys might have had only a few stars per square degree, UCAC would have 1,700.) On the military side, Department of Defense user needs were also emphasized in regular meetings of the Astrometry Forum sponsored by the Naval Observatory in the 1990s. As its mission statement required, the Observatory continued to balance practical needs with the advancement of astronomy.

*Astrometric renaissance: The drive toward accuracy and an extragalactic reference frame*

After only modest gains in positional accuracies for much of the twentieth century, despite literally decades of unceasing efforts, astronomers working in the last decade of the twentieth century truly witnessed a renaissance in positional astronomy. Whereas the best efforts of transit-circle astronomers produced an optical reference frame with accuracies little better than to within a tenth of an arcsecond by the 1980s (for FK5 20–50 milliarcseconds at its epoch), in the 1990s milliarcsecond accuracies became routinely obtainable, first with radio telescopes, and then optically both with ground-based (for some applications) and with space-based instruments. In fact, the 1990s brought an improvement in positional accuracies equivalent to the totality of improvements over the previous two centuries.

In addition, the very nature of the reference frame changed during the decade of the 1990s from the solar-system based reference frame that had been the object of so much work during the century, to an extragalactic frame based on radio sources. As one review put it in 1995, "the traditional reference frame based on the Fundamental Catalog of bright stars is being replaced by the extragalactic reference frame, based on radio sources with accuracies of one milliarcsecond. Thus, astrometry will change from a fundamental reference frame defined in terms of the dynamical reference frame of the solar system with accuracies of 100 milliarcseconds to a space-fixed,

[144] "The Construction and Use of Star Catalogues Conference," 3–5 October, 1966, AJ, **72** (1967), 551–630; F. W. Fallon, "Star Catalogue Requirements for Satellite Geodesy," ibid., 611–616.

extragalactic reference frame with accuracies of one milliarcsecond."[145] These are truly landmark changes, and will surely be seen as such looking back a century from now.

After centuries of hovering at the 0.1-arcsecond level, how was the milliarcsecond revolution brought about? As we have seen, there were limited areas where optical astrometry had reached the milliarcsecond level earlier. The Naval Observatory's own 61-inch telescope led the way in producing parallaxes with that accuracy in the 1960s, and approached sub-milliarcsecond accuracies with CCD technology. However, such small-field astrometry was a long way from producing those accuracies for global astrometry, the inertial reference frame that was the Holy Grail of positional astronomers. In the end the milliarcsecond revolution was begun, not by successive improvements to transit-circle technology, but by an entirely new technology. Indeed, it was in the entirely new-wavelength regime of radio astronomy that these advances were first made.

Observations at radio wavelengths for astrometry have several advantages, the most important of which is the relative unimportance of atmospheric refraction, the nemesis of optical astrometrists. Using the techniques first of connected-element interferometry (CEI) and then of very-long-baseline interferometry (VLBI), steady improvements were made in positional accuracies hand-in-hand with advances in radio-interferometric techniques in the 1970s. The IAU Symposium "New Problems in Astrometry," held in Perth, Australia in 1973, witnessed a crucial turning point: It not only emphasized the importance of Southern-Hemisphere work, but also was the occasion for the first meeting of optical and radio astrometrists, just at the time when radio precisions were overtaking optical precisions. Whereas radio-source positions as of 1950 were uncertain by several minutes of arc, by 1974 uncertainties in the radio positions of about 100 radio sources were quoted at ±0.1″ to ±0.01″. Also at that meeting astronomers expressed hope that precisions of the order ±0.005″ or better might be obtainable through VLBI. Four years later, when another IAU Colloquium on Astrometry was held in Vienna, the astrometry of radio sources was a major topic of discussion. By the time of the meeting on the "Inertial Coordinate System on the Sky" in 1989, not only had milliarcsecond accuracy been achieved for 325 compact extragalactic radio sources, but also an inertial reference frame had been constructed based on these positions.[146]

By the early 1970s, then, radio astrometry was reaching a level that seemed

---

[145] See articles in E. Høg and P. K. Seidelmann, eds., *Astronomical and Astrophysical Objectives of Sub-Milliarcsecond Optical Astrometry* (Kluwer: Dordrecht, 1995), pp. 19 ff.

[146] Bart Bok, "Summary and Conclusions," in *New Problems in Astrometry* (ref. 133), pp. 327–328. Reviews of CEI, VLBI, and radio/optical astrometry were also given at this meeting. Also K. J. Johnston, "Present status of radio astrometry with Baselines ≤35 km," in *Modern Astrometry* (ref. 105), pp. 175–178; C. Ma, "Realization of an Inertial Reference Frame from Mark III VLBI," in *Inertial Coordinate System on the Sky* (ref. 135), pp. 271–280. The use of radio positions for a reference frame had been discussed in the Maryland Star Catalogue Conference in 1966; see B. G. Clark, "Position Determinations for Radio Astronomy," AJ, **72** (1967), 601–603.

certain to impact the long-standing mission of the Naval Observatory, in particular by observing quasars to determine the Earth's rotation. The Navy's expertise in radio astronomy was at the Naval Research Laboratory (NRL), where Kenneth Johnston and others were pioneering in radio astrometry. The NRL was located just a few miles from the Naval Observatory, and, by 1976, a committee was formed linking the two institutions in a study of the application of the new technique to Naval Observatory needs. The conclusions were encouraging, and, upon the retirement of Kaj Strand in 1977, Gart Westerhout was appointed the new Scientific Director with the understanding that he would implement a radio-astronomy program for determining the Earth's rotation. Westerhout had worked under Jan Oort in Leiden during the pioneering days of radio astronomy, and during the previous 15 years had built the astronomy program at the University of Maryland. Within a year and a half of his arrival, Westerhout had arranged for the connected-element interferometer at Green Bank, West Virginia to be used for determination of the Earth's rotation, with a tenfold increase in accuracy. Observations at Green Bank for this purpose began in October, 1978, and would continue there in some form for more than 20 years.[147]

By contrast, with improvements in accuracy in the measurement of the Earth's rotation, the transition from the dynamical reference frame tied to solar-system objects to an extragalactic reference frame composed of distant radio sources did not take place so quickly. The idea of tying star positions and motions to the much more distant (and thus apparently more positionally invariant) galaxies, was discussed in some detail by S. Vasilevskis at the first Astrometric meeting in 1953 in connection with the Lick Proper Motion program.[148] However, the determination of accurate positions in the radio region rather than the optical raised problems. A radio reference frame based on extragalactic sources was not of much use to those who needed star positions in the optical region of the spectrum. What was needed was milliarcsecond accuracy in the optical region, or at least a tie-in of optical to radio. Lacking the technology for such accuracies in the optical region during the 1980s, astronomers were faced with the task of tying the optical reference frame to the more accurate radio frame. As early as 1982 at the Naval Observatory, for example, one project attempted to tie the optical frame to the radio frame by observing the optical counterparts of some of the radio sources that made up the frame. Because of the extreme faintness of the sources, however, this was a boot-strap process, requiring first transit-circle observations of reference stars surrounding the radio sources (part of the W1(Joo) program observed during 1977–82), then photographing these fields with the 8-inch astrograph, and finally using the 61-inch reflector to photograph the faint optical

[147] Westerhout, OHI, 48–60 gives the details of the USNO's entry into radio astronomy, as well as Westerhout's previous work. The NRL–USNO Committee included Johnston from the NRL, and Gernot Winkler and Dennis McCarthy from the USNO Time Service, among others.
[148] S. Vasilevskis, "Some Aspects of the Lick Proper Motion Program," AJ, **59** (1954), 40–43. The objectives of the Lick program had already been expressed by W. H. Wright, Proc. APS, **94** (1950), 1.

counterpart of the radio sources themselves. By 1987 a comprehensive five-year program was begun to establish a radio–optical reference frame; among the participants were the Naval Research Laboratory and the Naval Observatory. Again a two-step process was necessary in order to produce optical positions, using several 4-meter-class telescopes referred to the IRS using plates from wide-field astrographs, including the Naval Observatory's southern station in New Zealand.[149]

Far preferable to such a cascading process of tie-ins, a necessity if the full potential of the radio reference frame were to be realized, was milliarcsecond accuracy by direct observation in the optical region itself. Optical milliarcsecond accuracies in their own right awaited two new technologies, optical interferometry and astrometric satellites, both of which came into being in the 1990s. An astrometric satellite came to fruition first. Although the Naval Observatory had contemplated its own astrometric satellite as early as 1980, it was the European Space Agency's Hipparcos (High Precision Parallax Collecting Satellite) consortium that obtained funding and successful observations with the world's first astrometric satellite from 1989 to 1993. Hipparcos provided positions, parallaxes, and proper motions for about 118,000 stars with an accuracy of about 1 milliarcsecond, and for another million stars at 30 milliarcseconds.[150] Only in 1999 did NASA award the Naval Observatory the first installment of $180 million in funding for its own astrometric satellite, known as the Full-Sky Astrometric Mapping Explorer (FAME). The goal of FAME (Figure 10.44) was to determine the positions of 40 million stars to 50-microarcsecond accuracy, good enough perhaps to discover the effects of planets on the motions of their parent stars. As of 2002, however, cost overruns and technical problems threatened the survival of FAME, an indication of the difficulties of space astrometry.

Meanwhile, although the Naval Observatory had been involved only minimally with Hipparcos, it led the way in applying the radio technique of interferometry to the optical region via its Navy Prototype Optical Interferometer (NPOI), pictured in Figure 10.45. The idea of optical interferometry originated already in the late nineteenth century and was used by Albert Michelson in 1891 to measure the diameters of the Galilean moons of Jupiter. In 1920 Michelson and Francis Pease used an interferometer on the new 100-inch telescope at Mt Wilson to determine the angular diameters of seven red-giant stars. In the 1960s optical interferometry gained momentum with

---

[149] S. Dick and E. Holdenried, "Precise Positions in the FK4 System for 120 Radio Source Reference Stars," AJ, **87** (1982), 1,374; R. Harrington et al., AJ, **88** (1983), 1,376. The project is summarized in T. Corbin, "The Determination of Fundamental Proper Motions," in Calibration of Fundamental Stellar Quantities, D. S. Hayes et al., eds., pp. 53–70: 67–68. J. L. Russell et al., "A Progress Report on the Establishment of the Radio/Optical Reference Frame," in Inertial Coordinate System on the Sky (ref. 135), pp. 281–284.

[150] The Hipparcos and Tycho Catalogues: Astrometric and Photometric Star Catalogues Derived from the ESA Hipparcos Space Astrometry Mission, a collaboration between the European Space Agency and the FAST, NDAC, TDAC, and INCA consortia and the Hipparcos Industrial Consortium led by Matra Marconi Space and Alenia Spazio (ESA Publications Division: Noordwijk, 1997). On Naval Observatory plans for a satellite in the 1980s, see J. A. Hughes, "Absolute Astrometry, Now and in the Future," SOJ, 133–143; G. Westerhout, "An Overview of the U. S. Naval Observatory," ibid., 89–93; and Westerhout OHI, 66–69.

Figure 10.44. The Full-sky Astrometric Explorer (FAME), which was designed to determine star positions to within 50 microarcseconds. The Principal Investigator was Kenneth Johnston, Scientific Director of the Naval Observatory

Hanbury Brown's "intensity interferometer." By the 1980s the Naval Observatory had become involved with Johnston at the Naval Research Laboratory and other collaborators in a series of interferometers built at Mt Wilson, culminating with the so-called "Mark III" in 1986. During the 1960s, 1970s, and 1980s a variety of second-generation interferometers was built. The experiences thus gained were applied to the NPOI, which was dedicated in 1994, and was one of a handful of "third-generation interferometers." At the end of the century, however, optical interferometry was still limited to relatively bright stars in our own galaxy, and milliarcsecond accuracies were not yet routinely being achieved.[151]

Advances in optical astrometry of stars in our galaxy could not overcome the intrinsic problem that had plagued stellar reference frames from the beginning: The stars had proper motions and their positions changed with time. Extragalactic objects suffered from no such problem, at least not at milliarcsecond accuracies, but astronomers did have to worry about the structure of some extended objects. By the end of the century the fundamental reference frame was therefore officially based on radio observations of extragalactic sources, a fundamental shift delineated by the vertical line in Table 10.3. The International Celestial Reference Frame (ICRF), which was formally adopted in 1998, was the result of many years of observations and decisions regarding the best sources. The Naval Observatory played a major role in these deci-

---

[151] J. Thomas Armstrong, Donald J. Hutter, Kenneth J. Johnston, and David Mozurkewich, "Stellar Optical Interferometry in the 1990s," *Physics Today* (May 1995), 42–49. This article also reviews the status of other interferometers under construction at the time. On the USNO's early involvement with the NRL and optical interferometry see Westerhout OHI, 61–66.

Figure 10.45. The Navy Prototype Optical Interferometer (NPOI), a new-generation telescope, located near Flagstaff, Arizona.

sions, and in the maintenance of the new reference frame. Already in the 1980s the IAU Commission 24 Working Group on the Radio/Optical Reference Frame had established a list of 233 candidates.[152] The ICRF as it was implemented in 1998 was based on only 212 radio sources, mostly 16th magnitude and fainter, distributed over the entire sky; thus the importance of the much denser and brighter Hipparcos catalog is evident.[153] That catalog of 118,000 stars defines the optical reference frame to about ten milliarcseconds at the current epoch.

Toward what end were all these efforts expended? Aside from military applications, the benefit to be gained from attaining milliarcsecond accuracies of even selected stars was discussed already by Jan Oort at the 1959 Astrometric Conference. "The ultimate purposes for which this great accuracy is needed," he wrote, "are to get a better distance scale in the universe, a better determination of the constants of galactic rotation, and better determinations of the motions of certain stars of special interest." As we have seen, already by the 1970s the Naval Observatory was using radio astronomy for determining the Earth's rotation, which was important for a variety of

---

[152] A. N. Argue *et al.*, "A Catalog of selected Compact Radio Sources for the Construction of an Extragalactic Radio/Optical Reference Frame," *Astronomy and Astrophysics*, **130** (1984), 191–199.

[153] C. Ma *et al.*, "The International Celestial Reference Frame as Realized by Very Long Baseline Interferometry," *AJ*, **116** (July, 1998), 516–546. The events leading to this reference frame are described in K. J. Johnston *et al.*, "Towards the Definition of a Unified Celestial Optical/Radio Reference Frame," in M. Reid and J. Moran, eds., *The Impact of VLBI on Astrophysics and Geophysics* (Dordrecht, 1988), p. 317; K. J. Johnston *et al.*, "The Extragalactic Radio/Optical Reference Frame," eds. J.A. Hughes, C. A. Smith and G. H. Kaplan, *Proc. IAU Colloq.*, **127** (USNO, 1991), pp. 123–129.

purposes, including Universal Time. Astrometry and time continued to be related at ever-increasing levels of accuracy and sophistication. The Earth's rotation, with all its implications for astronomy and geophysics, is only one aspect of the determination and dissemination of time, a subject that also has a long history, and one that we take up in the next chapter.[154]

[154] Jan Oort, "Very Accurate Positions of Selected Stars," AJ, **65** (1960), 229–233. How the uses for such accuracies had expanded by the 1990s beyond galactic dynamics is detailed in the IAU Symposium, *Astronomical and Astrophysical Objectives of Sub-Milliarcsecond Optical Astrometry* (ref. 2). Already at the 1978 Vienna meeting the implications of a radio reference frame for Earth-rotation parameters were evident. See K. Johnston *et al.*, "Requirements for a radio coordinated reference frame for the determination of Earth rotation parameters," in *Modern Astrometry* (ref. 105), pp. 171–174.

# 11  Time: A service for the world

The appreciation of the value of correct time is a good index to the civilization of a nation, and in this respect the United States is among the very foremost.

Lt Cdr E. E. Hayden, 1906[1]

. . . supported by highly sophisticated clock technology we now have clock time scales in existence with uniformities of 1 nsec [nanosecond] over a day and about one μsec [microsecond] over a year. The fact that time can be measured so precisely, far better than any other physical parameter, represents a technological asset of great importance.

Gernot M. R. Winkler, 1977[2]

And the seasons, they go 'round and 'round
And the painted ponies go up and down
We're captive on the carousel of time . . .

Joni Mitchell, 1966[3]

The twentieth century saw "the acceleration of just about everything" in the words of James Gleick, who used the phrase as the subtitle to his book *Faster*, which detailed just how frenetic modern technological society had become. It is little surprise, then, that time itself became more important during the century. As the era of time balls, telegraphs, and pendulum clocks receded, it gave way to an era of long-distance radio signals, atomic clocks, and satellites. Both time and society were forever changed. Early in the century our concept of time itself also changed, with Einstein's theory of relativity, leading to the "clock paradox," the "twin paradox," and "time dilation." The relation between time and gravitation, the relation between time and the new quantum physics, and the origin and meaning of time became hot topics during the century. Even as time entered new realms in science and philosophy, the practical

---

[1] Edward Everett Hayden, "The Present Status of the Use of Standard Time," PUSNO, **4**, Appendix IV (Washington, 1906), 9.
[2] Gernot M. R. Winkler, "Timekeeping and its Applications," *Advances in Electronics and Electron Physics* (Academic Press: New York, 1977), pp. 33–97: 37.
[3] Joni Mitchell, Lyrics from "The Circle Game," 1966.

aspects of its determination and dissemination remained, and were carried out at national observatories and standards laboratories around the world.[4]

Time service not only remained a central concern of the Naval Observatory during the twentieth century, but also came to dominate the institution in terms of budget, personnel, and perceived importance to the outside world. All three of its major areas of time determination, time maintenance, and time dissemination underwent fundamental changes, even as the primary goal of accuracy remained the same. Transit instruments determining time at the beginning of the century were replaced by the photographic zenith tube in the 1930s, and by radio telescopes observing quasars beginning in the 1980s. Pendulum clocks serving as time standards reached their ultimate development in the form of Riefler and Shortt clocks, only to be replaced by quartz-crystal clocks in the 1940s, and atomic clocks in the 1960s. Time dissemination evolved from telegraph to radio, and finally to a whole host of methods, including ultimately the Transit and Global Positioning System of satellites circling the Earth. During the century, the accuracy of the determination of astronomical time improved from a few hundredths of a second with transit instruments to 20 millionths of a second with radio telescopes; timekeeping accuracy increased a millionfold from a thousandth of a second with the Riefler and Shortt clocks to a billionth of a second with the atomic clock; and the accuracy of time dissemination increased from an average error of a few hundredths of a second with radio signals to ten billionths of a second with GPS.

All of this, while an important part of the story, is only the "nuts and bolts" of time. More fundamentally, during the century new kinds of "time" were defined and determined, driven by the desire for a more uniform time. For centuries the Earth's rotation had served this purpose in an ideal fashion. By early in the century, the motion of the pole on the Earth's surface could be detected, a correction of several hundredths of a second to astronomically observed time. More fundamentally, in the twentieth century man-made clocks caught up with the Earth clock's so-called "Universal Time," and surpassed it. By the 1950s the demonstration that the Earth's rotation itself was variable led to a time scale, known at first as Ephemeris Time and then as Dynamical Time, based on the Earth's orbit around the Sun rather than on the Earth's rotation. Even more fundamentally, by the 1950s the development of the atomic clock had led to a time scale based for the first time on the physics of the atom rather than on an astronomical phenomenon. For the increasingly sophisticated uses of the Space Age both astronomical time and atomic time found their uses, linked together through the mediation of the leap second.

Technological advancements and new forms of time raised deeper issues of

[4] James Gleick, *Faster: The Acceleration of Just about Everything* (Pantheon: New York, 1999). On time in the context of the great scientific and philosophical themes, see Paul Davies, *About Time: Einstein's Unfinished Revolution* (Simon and Schuster: New York, 1995), and for the historical context G. J. Whitrow, *Time in History: Views from Prehistory to the Present Day* (Oxford University Press: Oxford, 1988).

science, policy, and politics: the relation between astronomically and physically based systems of time; the worldwide coordination of time and its dissemination; the relationship between the Naval Observatory and the National Bureau of Standards (later the National Institute of Standards and Technology); and the reasons behind a host of fundamental geophysical phenomena ranging from the variability of the Earth's rotation, to the secular motion of its pole and continental drift.

Administratively, Time Service at the Naval Observatory was dominated by three civilian Directors during the twentieth century: Paul Sollenberger (1928–53), William Markowitz (1953–66), and Gernot Winkler (1966–95). Prior to that time, Navy personnel were in charge of the Time Service (see Appendix 1), usually rotating to another Navy berth after only one or two years, with the exception of Lt Commander E. E. Hayden, who held the post from 1902 to 1910. The appointment of a civilian Director in 1928 indicated the growing importance of the Time Service. Until then, the time-service function had always been combined administratively with the chronometer-rating function, and designated the "Department of Chronometers and Time Service," or some variation thereof. However, while chronometers remained important for navigation (no more so than during World War II), the need for time in the growing United States began to dwarf chronometer usage for ships at sea. Even before the war there was a feeling that chronometers need not be part of an astronomical observatory. The problem was solved in 1950, when the chronometer function was removed to the Norfolk Navy Yard, along with the entire nautical-instrument-repair function. Time Service was left to concentrate on more far-reaching duties, and just in time for the revolution that would occur in the second half of the century, not only in keeping and disseminating time, but even in the very definition of time itself.

## 11.1    Universal Time: Harnessing the Earth clock to 1950

At the beginning of the twentieth century the use of "Standard Time" had been introduced in 36 nations around the world. In his monograph on "The Present Status of the Use of Standard Time" (1906), Lt Cdr Edward E. Hayden, the officer in charge of the Department of Chronometers and Time Service at the Naval Observatory, explained why: "The need of a common and harmonious international system of time becomes greater every year by reason of the rapid extension of railroads, telegraphs, and cables, and the increase of international, diplomatic, and business relations that are conducted by telegraph." The "Universal Time System," as Hayden called it, was based on 24 time zones around the world, each spanning one hour of time, equivalent to 15 degrees of longitude. However, the recommendation of the 1884 Washington Meridian Conference that Greenwich serve as the prime meridian for time had not yet been entirely carried out. Of the 36 countries Hayden listed keeping Standard Time in 1906, 20 were indexed to Greenwich; others were indexed to a variety of meridians, including Paris, Athens, Lisbon, and Bogota. In 1904 Hayden (Figure 11.1) led the

Figure 11.1. Lt-Cdr Edward Everett Hayden, in charge of Time Service, 1902–10, in the time-service room in 1904. His hand is on the time-signal switchboard. Hayden lost a leg while working for the U. S. Geological Survey.

Eighth International Geographic Congress to adopt a resolution favoring "the universal adoption of the meridian of Greenwich as the basis of all systems of standard time." Though this resolution had no international force, the obvious benefits of a system that (with few exceptions) used an integral number of hours indexed to a single prime meridian gradually led to its near-worldwide adoption during the twentieth century.[5]

*Time determination: From transit instrument to PZT*

As before, standard time at the century's beginning was mean solar time based on the rotation of the Earth, as determined by astronomical observation. The second of mean solar time was defined as 1/86,400 of a mean solar day. The problem, pursued especially at national observatories around the world, was how to harness the Earth clock to extract its maximum possible accuracy. With the completion of the move to the Naval Observatory's new site in May, 1893, the telescope employed for the purpose of measuring mean solar time was the 5.3-inch meridian-transit instrument (Figure 11.2). It was located on the west side of the Main Building, along with the observing (sidereal) clock. Every second or third night, depending on the weather, observations

---

[5] Hayden (ref. 1), 8, 12. For a table with a summary of nations using Standard Time see Hayden, 12–13. Much of the rest of the report is a table with more details on national time systems, and on the dividing points between North-American time zones. Hayden also discusses the International Date Line on pp. 5–6. On the gradual adoption of Standard Time indexed to Greenwich see Derek Howse, *Greenwich Time and the Longitude* (Philip Wilson: London, 1997), pp. 145–150; but see also the cautionary note in Bartky, *Selling the True Time* (Stanford University Press: Stanford, 2000), p. 263, note 21. On Hayden (1858–1932), who retired as a rear admiral in 1921, see BDAS, 120–121.

Figure 11.2. The 5.3-inch transit instrument for time determination, which was located in the transit house adjacent to the main building on the west side. This view is from about 1920, after the 6-inch transit circle had superseded the instrument for time determination.

were made of "time stars" (also called "clock stars") so that the corrections and rates of three standard clocks at the Observatory could be determined: a mean-time Seth Thomas, a Parkinson and Frodsham sidereal, and Howard sidereal clock # 404. As the clock star crossed the meridian, the observer pressed a telegraph key, and the exact time of passage was registered on the chronograph; a scale was then used to measure the space between this recorded signal and the clock signal, which was also recorded on the chronograph, in order to determine the correction.[6] This method gave time to within a few hundredths of a second. The time determined by the stars was, of course, sidereal time, but it was mean solar time that was sent out from the Observatory. Mean solar time and sidereal time were rigorously related by a mathematical formula representing the motion of the Sun whose most accurate value dated back to Newcomb, so the conversion between them was straightforward.[7]

Whereas the unit of the mean solar day, the second, was not at all controversial for most of the first half of the century, the definition of the beginning of that day was less obvious. Prior to 1925, as a convenience to astronomers in their nightly work, the mean solar day as used in all astronomy began at noon. Beginning with the national ephemerides for 1925, the International Astronomical Union (IAU) defined the mean solar day to begin at midnight, to bring it into conformance with common civil usage.

---

[6] AR (1896), 19; Day Allen Willey, "The Time of Day," *Scientific American* (October 28, 1905), 336–337. Actually, the transit instrument had several cross-hair wires, and a series of transit signals were registered; subsequent reductions and averaging were performed.

[7] See Newcomb, "Tables of the Motion of the Earth on its Axis and around the Sun," APAE, volume 6, part 1 (U. S. Nautical Almanac Office: Washington, 1898). The formula and procedure as of 1959 are given in Markowitz, "Astronomical and Atomic Times," 9 March, 1959, Markowitz file, USNOA, BF. In practice the calculation did not have to be made every time, and precalculated cards were used; for the method as of the 1930s, see Sollenberger OHI, 69.

Although the term "universal time" had been used before, dating back at least to the 1884 International Meridian Conference and now exemplified by Hayden's "Universal Time System," in fact it was this change in the beginning of the astronomical day that gave rise to its use in official documents. For several years after the 1925 change in the beginning of the astronomical day, the British *Nautical Almanac* still used the term GMT for Greenwich Mean Time, whereas the *American Ephemeris* used the designation Greenwich Civil Time (GCT). In 1928 the IAU adopted the designation "Universal Time" (UT); although this was presumptuous considering possible extraterrestrials, the new terminology was designed to remove all confusion by directing the dropping of both terms, GMT and GCT. Although "GMT" was still widely used informally through the century – especially by the British and navigators – the term is no longer used for precise time purposes.[8]

The accuracies involved in time determination early in the century were central in a controversy that arose in the 1920s involving international comparisons of time. By that time, the 5.3-inch transit instrument had been superseded for the purpose of time determination at the Naval Observatory by the 6-inch transit circle (section 10.1). In 1922 R. A. Sampson, Astronomer Royal for Scotland based at the Royal Observatory Edinburgh, called attention to differences in time determinations on the order of 3–4 tenths of a seconds, based on the reception times at the Greenwich, Edinburgh, and Uccle (Belgium) Observatories of radio time signals transmitted by stations in Washington, Paris, and Berlin. Having eliminated any propagation effects affecting the signal itself, the implication was that the time determination at one or more observatories was in error. Moreover, by comparing the time determination of each observatory with the mean of all of them, it was possible to produce a deviation of each observatory from the mean. Washington, among others, showed unexplained variations from this mean by 0.2 seconds (200 milliseconds). Because Sampson was at this time (and through most of the 1920s) the President of the Commission on Time of the International Astronomical Union, this analysis could not be, and was not, taken lightly. It was, in fact, a beautiful piece of scientific work.[9]

In order to investigate this effect, and naturally sensitive to the possibility of errors in their own time determinations, beginning in 1922 J. C. Hammond (Director of the 6-inch Transit Circle Division) and C. B. Watts also determined time with one of the Prin transits purchased for longitude determinations between Paris and

---

[8] P. Kenneth Seidelmann, ed., *Explanatory Supplement to the Astronomical Almanac* (University Science Books: Mill Valley, California, 1992), p. 76. The attempt to change the astronomical day has a long and contentious history, dating back at least to the Washington Meridian Conference of 1884. See chapter 9, ref. 25; Howse (ref. 5 above), pp. 150–151; and for Newcomb's opinion, *Reminiscences of an Astronomer* (Houghton & Mifflin: Boston and New York, 1903), pp. 227–228.

[9] R. A. Sampson, "On the Determination of Time at Different Observatories," MNRAS, **82** (January, 1922), 215–225. The use of the 6-inch instrument "for the determination of time for the time service" is first explicitly stated in AR (1916), 12, although observations of "time stars" with the instrument are listed beginning in AR of 1915. It was definitely used for time determination from 1916 until 1925.

Figure 11.3. C. B. Watts at the Prin transit used for determining time, 1925–34, and for the Washington–Paris longitude determinations.

Washington (Figure 11.3). The result showed differences between the 6-inch and the Prin instrument amounting to only a few tens of milliseconds, whereas differences between Washington and Paris, as obtained from the *Bulletin Horaire* of the Bureau International de l'Heure (BIH), approached 100 milliseconds. Hammond and Watts concluded that, if the Paris–Washington variations were due to errors in the Naval Observatory determinations, they arose from local conditions that affected both instruments.[10] In a further investigation, however, H. R. Morgan (Director of the 9-inch Transit Division) analyzed the "clock corrections" from the observing programs of the 6-inch and 9-inch transit circles over a period of five years (1913–18), and concluded that some of the discrepancies could be due to instrumental effects in the observations. The interest in these studies is evident in the fact that Morgan read at the 1923 meeting of the American Geophysical Union the results of these and similar studies, "On the Accuracy of Time Determination."[11] As a result of these studies, the Prin instrument was used for time determination until 1934; it appeared somewhat more accurate than the transit-circle determinations, and, after the controversy over time, Hammond was happy to turn the task over to the Prin instrument under the auspices of the Time Service. Wireless telegraphy made possible an international comparison of time that would be the trend in the future of world timekeeping.

Under the Superintendency of J. F. Hellweg (1930–46), the Observatory's time

[10] J. C. Hammond and C. B. Watts, "Comparison of Time Determinations with Different Instruments," AJ, **35** (1923), 106. The use of the Prin is first mentioned in AR (1923), 13.

[11] H. R. Morgan, "On the Accuracy of Time Determination," AJ, **35** (1923), 80–83. The earlier study was Morgan, "A Comparison of Clock Corrections Determined with Large Instruments," AJ, **35** (1923), 4–7. The switch from the 6-inch instrument to the Prin for time is discussed in Sollenberger, OHI, 23.

function was particularly emphasized. Hellweg himself wrote several articles emphasizing improvements in the Time Service, chiefly in the area of clock development and
time broadcasting.[12] However, it was Paul Sollenberger who was in charge of the Time
Service beginning in 1928. Sollenberger had obtained his Bachelor's degree from
Marion Normal College in Indiana in 1913, and came to the Naval Observatory the following year working on the 9-inch transit circle under Morgan. He was largely self-
taught in astronomy, and in 1919 he transferred to the Division of Nautical
Instruments and Time, where he sent out the time signals at noon and 10 pm, assisted
in the testing of chronometers, and later became known for his contributions to the
design of quartz-crystal clocks, chronographs, and the photographic zenith tube. He
remained in charge of Time Service until his retirement in 1953.[13]

When Sollenberger entered Time Service the transit instrument was still the
instrument of choice for time determination. Because all of the transit instruments
worked on the same principle of visually observing the transit times of stars whose
positions were well known, any improvements in precision using that instrumentation were considered minuscule. Beginning in February, 1934, though, the method for
determining Universal Time at the Naval Observatory underwent a fundamental
change when the transit instrument was replaced for this purpose by the
Observatory's Photographic Zenith Tube (PZT). The PZT, which improved on transit-
instrument stability by remaining rigidly fixed pointing to the zenith rather than
moving along the meridian, was originally designed by Frank E. Ross to measure the
"variation in latitude."[14]

Variation in latitude, detected by Friedrich Küstner and Seth Chandler in 1884
and interpreted by Chandler in 1891, was the first variation of star positions due to
motion of the coordinate system found since Bradley's discovery of nutation in the
eighteenth century. It was therefore studied intensely, especially after the
International Association of Geodesy had organized the International Latitude Service
in 1899 and set up several observing stations around the world at latitude 39 degrees
north. In 1911 Ross's PZT was placed into use at the International Latitude Station
(ILS) operated by the U. S. Coast and Geodetic Survey at Gaithersburg, Maryland
(about ten miles north of the Naval Observatory). However, after the ILS had decided
to use a visual nonith telescope instead, the PZT was purchased by the Naval
Observatory and placed into use at the Washington location beginning in October,
1915 for the determination of variation of latitude. Variation in latitude later became
known as "polar motion," a more accurate term since both longitude and azimuth

---

[12] J. Frederick Hellweg, "Time Service of the U. S. Naval Observatory," in American Institute of
Electrical Engineers, *Symposium on Time and Time Services*, January, 1932, pp. 1–3; Hellweg, "United
States Navy Time Service," PASP, **52** (1940), 17–24.
[13] On Sollenberger see S. J. Dick "Paul Sollenberger, 1891–1995," BAAS, **27** (1995), 1,482–1,483;
Sollenberger, OHI.
[14] Frank E. Ross, "Latitude Observations with Photographic Zenith Tube at Gaithersburg, Md.,"
*U. S. Coast and Geodetic Survey Special Publication no. 27* (Government Printing Office: Washington,
1915).

Figure 11.4. J. E. Willis at the "PZT # 1" that he modified for time determination. On the right is the chronograph that recorded times of observations, and on the left a standard pendulum clock. The photo dates from about 1932, two years before the instrument began to be used routinely for time determination.

vary due to motion of the pole. Unlike nutation, with a semiamplitude of 9.2 seconds of arc and a period of 18.6 years, polar motion amounted to only a few tenths of a second of arc, several hundredths of a second of time. Not until the 1950s would polar motion be incorporated as a systematic and rapid correction to Universal Time (although provisional corrections had been published since 1939). Meanwhile the PZT would come to play an essential role in the determination of time.[15]

Although Frank B. Littell was in charge of the PZT, it was John E. Willis (Figure 11.4) who suggested about 1923 (perhaps because of the unsuspected problem found by Sampson) that the instrument could be used for time determination. Willis persuaded Littell to let him make the necessary modifications, and a decade of experiments by Willis and Littell proved the feasibility of this proposal. In addition to the greater stability of the PZT compared with a transit instrument, they pointed to the advantages of the photographic plate as a permanent record, the great reduction of

[15] For the history of polar-motion studies and their applications to time, see Steven J. Dick, Dennis W. McCarthy, and Brian Luzum, eds., *Polar Motion: Historical and Scientific Problems* (Astronomical Society of the Pacific: San Francisco, 2000).

personal equation, and the elimination of the effects of collimation, level, azimuth, pivot-irregularity, and flexure errors, all of which had to be measured and taken into account in transit instruments. Sollenberger made subsequent improvements, including installing a small synchronous motor to drive the plates and to time the reversal between exposures. The PZT output was a small photographic plate whose star positions could be determined from a specialized measuring machine.[16]

The desire to improve time accuracy was the driving force behind the establishment of an alternate time station nearer the equator than Washington. After considering and rejecting Guantanamo Bay, Cuba, in 1949 the Richmond Naval Air Base, near Miami, Florida, then being used by the U. S. Coast Guard, was selected. During World War II the site had included a large hangar, which had since been destroyed by a hurricane. At the latitude of Miami the motion of the stars over the meridian was faster, allowing the increase in precision. For this station Sollenberger and one of his time-service astronomers, William Markowitz, designed a new model PZT (Figure 11.5), which was built in the Observatory's instrument shop. Meanwhile, the French astronomer André Danjon had also designed an astrolabe for the determination of time, and the two types of instruments remained rivals in time determination; although the astrolabe was used at the Naval Observatory for other purposes, the PZT remained its instrument of choice for time determination.[17]

The PZT, though generally recognized as the superior to transit instruments for time determination, did not immediately come into widespread use. Sir Harold Spencer Jones, Astronomer Royal and President of the IAU Commission on Time, remarked in 1955 that small transit instruments, equipped with an impersonal traveling-wire micrometer and reversible during the observation of each star transit to eliminate collimation error, "continue to be the standard instruments for time determination at most observatories." The standard error of a single time observation using ten or 12 stars ranged from 10 to 25 milliseconds, the higher accuracy attained when the traveling wire was motor driven.[18] However, Spencer Jones noted, PZTs were coming into wider use,

16  The best overview of the PZT is William Markowitz, "The Photographic Zenith Tube and the Dual-Rate Moon-Position Camera," in Gerard P. Kuiper and Barbara M. Middlehurst, eds., *Telescopes* (University of Chicago Press: Chicago, 1960), pp. 88–114, volume 1 of *Stars and Stellar Systems*, G. Kuiper, General Editor. The experiments of Littell and Willis are described in F. B. Littell and J. E. Willis, "A New Method of Determining Time," AJ, 40 (1929), p. 9. The interaction between Willis and Littell is described in Sollenberger, OHI, 22–24. The improvements of Sollenberger are found in "Time Determination and Time Keeping," PA, 50 (1942), 74–78; and AJ, 51 (1945), 145. See also Markowitz, "The Photographic Zenith Tube," *La Suisse Horlogère* (April, 1955), Markowitz folder, USNO BF.
17  Markowitz, "Reminiscences of the U. S. Naval Observatory," October 21, 1986, USNO Library, pp. 28–32. The first Richmond instrument was designated PZT # 2; PZT # 3 replaced # 1 in Washington in April, 1954. The PZT that went out of service in 1984 used a 26-inch lens. Markowitz and Glenn Hall instituted the use of the astrolabe at the Naval Observatory in the 1960s for specialized observations; Winkler terminated its use in 1979. Winkler OHI, 56–58.
18  P. Th. Oosterhoof, ed., *Transactions of the International Astronomical Union*, volume 9 (held at Dublin, 1955) (Cambridge University Press: Cambridge, 1957), p. 446; D. H. Sadler, ed., *Trans. IAU*, 10 (held at Moscow, 1958) (1960), p. 488. The sections of these *Transactions* devoted to Commission 31 on Time provide an excellent worldwide overview of issues relating to time at three-year intervals associated with the IAU General Assemblies, beginning in 1922.

Figure 11.5.  The new model of PZT, designated # 2, designed for the Richmond, Florida station in 1949. Numbered components include (1) rotary, (2) lens cell, (3) photographic-plate drive mechanism, (4) photographic-plate access door, (5) level, (6) rotary-reversing-gear cover, (7) rotary-reversing-gear drive shaft, (8), 180-degree stops, (9) setting microscopes, (10) rotary-reversing-gear mechanism, (11) telescoping hood (when operating, hood seats in basin), (12) hood counterweight, (13) mercury basin, (14) operating platform, (15) service steps, (16) rotary reversing motor, (17) rotary-reversing-motor brake, (18) instrument-orienting screws, (19) instrument-leveling screws, and (20) focusing rod.

and their external probable error was only 4 milliseconds. By 1960 nine PZTs were in use around the world. The Naval Observatory itself played some role in their dissemination. Its instrument shop manufactured one in the mid-1960s for La Plata, Argentina and another for the Dominion Observatory in Canada. PZTs would remain in use at the Naval Observatory until 1984 for the determination of Universal Time – exactly 50 years for this type of photographic instrument.[19]

---

[19]  On the history of PZTs at the Naval Observatory see Brent Archinal, "Summary of Photographic Zenith Tube Observations of the U. S. Naval Observatory," unpublished typescript, USNO Library. On the Naval Observatory's role in introducing the instrument in Ottawa, see Malcolm Thomson, *The Beginning of the Long Dash: A History of Timekeeping in Canada* (University of Toronto Press: Toronto, 1978), pp. 88 ff. This volume should be consulted as an interesting study of the cooperation and contrasts between the two neighboring nations.

Table 11.1 *Standard clocks at U. S. Naval Observatory and Greenwich Observatory*

| Type | Accuracy (seconds) | Date invented | Date at USNO[a] | Date at Greenwich |
|---|---|---|---|---|
| Kessels | 0.1 | | 1866 | |
| Parkinson & Frodsham # 611 Howard # 404 | 0.1–0.01 | | 1880 | |
| Riefler | 0.01 to 0.001 | 1889 | 1903 | 1921 |
| Shortt | 0.001 | 1922 | 1931 | 1924 |
| Quartz crystal/ Shortt combined | 0.001 | 1928 | 1946 | 1942 |
| Quartz-crystal Essen ring | 0.0001 | 1938 | 1955 | ? |
| Atomic – cesium | 0.0000001 | 1955 | 1966 | 1966 |
| Master Clock (cesiums and hydrogen masers) | 0.000000001 | | 1990 | Discontinued |

[a] The date at which a clock became the standard may differ from its date of arrival. For example, the Kessels clock arrived in 1842.

### Timekeeping: From pendulum to quartz crystal

When the century began, the primary standard clocks at the Naval Observatory consisted of a mean-time Seth Thomas, a Parkinson and Frodsham sidereal (# 611), and a Howard sidereal clock (# 404), the latter similar to the eight built for the 1874 transit-of-Venus expeditions. For the noon signal the mean of the three was used, with a slight correction for barometric pressure. Beginning in 1903, however, the Riefler clock took over as the primary sidereal standard, and remained so until 1931, when it was replaced by the ultimate development in pendulum timekeepers, the Shortt clock. Despite experimentation with crystal clocks since the 1920s at the Observatory, and the existence of working crystal clocks since the 1930s, only in 1946 were the Shortt clocks completely replaced by quartz oscillators. These in turn were replaced at the Observatory by atomic clocks in 1966. The Observatory's standard clocks are summarized in Table 11.1, which also gives their dates of invention and first use at Greenwich for comparison.[20]

Like time determination, timekeeping was not a clear function of the Time

[20] On the standard clocks, see AR (1902), 35, and for a list of standard clocks at the Naval Observatory see Whitney Treseder, "One Hundred Years of Precise Time: The U. S. Naval Observatory Time Service, 1880–1980," 1998 typescript in USNO Library, Appendix 1. There is some indication that Shortt clocks were still in operation at the Observatory until the mid-1950s, but William Markowitz, the Director of Time Service during this period, stated that the Shortt clocks "remained as primary standards until 1946, when they were replaced by quartz-clocks," in "The Determination of Time by the U. S. Naval Observatory," *Horological Institute of America Journal*, 10, no. 2 (1955), 5–9: 8. As of the year 2000 the Seth Thomas clock was located in the Superintendent's office, and Howard # 404 had just been retrieved from loan at the University of Maryland and was in the USNO instrument collection. A partial compilation of excerpts from the PUSNO regarding clocks is found in James A. DeYoung, "United States Naval Observatory Clocks and other Time-Keeping Apparatus," USNO Library.

Figure 11.6. The new clock vault built in 1932. Riefler clock # 60 is on the left, and # 70 is next to it. Shortt clocks # 40 and 41 are in the center. The periscope allowed visitors to view the vault, which measured 12 feet × ten feet × nine feet. The top of the periscope can be seen in Figure 9.19.

Service Department early in the century. When, in September, 1903, Riefler No. 70 was installed as the standard sidereal clock, the clock was not controlled by the Department of Chronometers and Time Service, but by William S. Eichelberger, head of the Division of Meridian Instruments.[21] This is of interest for two reasons. First, although Lt Cdr Hayden had studied briefly at Harvard College Observatory, he naturally lacked the expertise of Eichelberger, who was a Professor of Mathematics with an astronomy Ph. D. from Johns Hopkins University. In other words, Time Service was not yet a scientific department. Secondly, it indicates that the main reason for obtaining the clock was to improve meridian observations, in accordance with Superintendent C. H. Davis's statement in 1901 that the Department of Meridian Observations was the *raison d'être* of the Observatory. In any case, both for public time and for meridian observations, the care of the clock system was an important consideration. After an experiment during the early history of the Observatory in which a German Kessels standard clock was "nearly sacrificed" after having been placed in an underground vault to control temperature, the standard clocks had been located in chambers above ground. Only in 1901 was a vault dug in the basement of the West room of the clock house at the center of Observatory Circle, where the clocks regulated to mean time (including the Kessels) were placed in airtight containers. By the mid-1930s a second vault was constructed nearby "patterned after the most accurately constructed compartment on a battleship, the ship's magazine," complete with a periscope for visitors (Figure 11.6). In addition to Riefler No. 70, the Observatory operated two other Rieflers, numbers 60 and 151, one taking the place of the other when mechanical difficulties were experienced.[22]

---

[21] AR (1904), 34; AR (1905), 8. Eichelberger places the Riefler clock in historical context in "Clocks – Ancient and Modern," *Proc. AAAS*, **56** and **57** (1907), 359–378; for the arrival of Riefler No. 70 see p. 377. On the history of Riefler clocks see Dieter Riefler, *Riefler-Präzisionspendeluhren: 1890–1965* (Callwey: Munich, 1991), and Klaus Erbrich, *Präzisionspendeluhren: von Graham bis Riefler* (Callwey: Munich, 1978).

[22] Davis, AR (1901), 4–5, and Updegraaf, AR (1920), 19–20. On the periscope for the clock vault see AR (1932), 20. The vault as of the mid-1930s is described in detail in J. F. Hellweg, "Accuracy of Time Broadcasting," *National Jeweler* (December, 1935), 73–75. An early description of the Rieflers in connection with the observations of the 9-inch transit circle is given in PUSNO, **9**, part 1, A 14–15. It is notable that Riefler No. 60 was used at the Observatory's Tutuila, Samoa station from September, 1905 until its closing in July, 1909.

Figure 11.7. One of the Naval Observatory's three Riefler clocks, with free escapement, nickel-steel compensation pendulum, electric self-winding, and electric contact for recording on a chronograph. It is enclosed in an air-tight glass case. From PUSNO, volume 9, part 1 (1920).

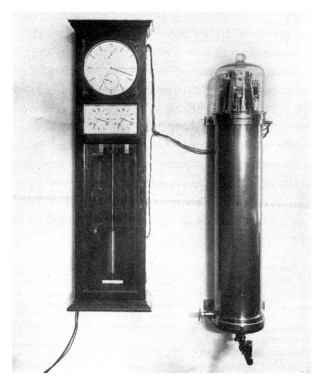

Figure 11.8. A Shortt clock, with the free pendulum on the right and the "slave" clock on the left. The slave clock allows the master pendulum to swing free except for a fraction of a second every 30 seconds, increasing its accuracy from a one-second error in ten days to ten seconds per year.

Aside from quirky intermittent problems (number 151 occasionally dropped whole seconds!), the general problem with the Riefler clocks (Figure 11.7) was that, although they were very accurate over short periods, changes in their rates made them of little use for studies extending over several months. This defect was remedied to a great extent in the Shortt clock (Figure 11.8), which was invented by the British engineer William Shortt working (in collaboration with the Synchronome Company in London) at Edinburgh Observatory at the invitation of Sampson, the Astronomer Royal of Scotland. Like in the Riefler clock, the master pendulum of the Shortt clock was housed in a largely evacuated tube. The master pendulum synchronized the motion of a second "slave" pendulum. The slave pendulum sent electrical signals to the master's drive mechanism and to operate the dialwork on the slave clock. By an elaborate electromechanical feedback the master and slave pendulums kept each other synchronized. Through this mechanism, the Shortt clock could maintain a uniform rate to within one part in ten million for a year or more; over shorter periods the daily rate could be predicted with average errors of 2–3 milliseconds, about three parts in 100 million. The first Shortt (No. 0) was placed in operation in Edinburgh in 1922; Greenwich purchased numbers 3, 11, 16, and 49 in 1924, 1926, 1927, and 1934; and the Naval Observatory lagged somewhat behind with its purchase of numbers 38, 41, and 48 in 1930, 1932, and 1933. By 1935, 54 Shortt clocks (numbered 0 through 53) were in operation, 47 of them under the care of professional astronomers. Shortt clocks could be used both as sidereal and as mean-time clocks; at Greenwich numbers 3 and 11 were sidereal standards, and numbers 16 and 49 were the mean standards.[23]

The Shortt clocks, however, had electrical and mechanical problems that made it difficult to keep them in operation over very long periods (despite the fact that no. 11 at Greenwich operated continuously for nine years). Thus the search for a better timekeeper turned elsewhere, and the answer was found in the quartz-crystal clocks being developed by telecommunications engineers as a reliable standard of frequency. Because of the remarkable electrical property of the quartz crystal known as the piezoelectric effect (electrical charge induced by pressure on the crystal), it could be kept in mechanical oscillation and used in an electronic oscillator, which in turn could be used for time purposes. The frequency of a quartz oscillator, which depended on the size and cut of the crystal, was normally several million cycles per second, but could be divided down by a frequency divider to 1,000 cycles per second. This frequency then drove a synchronous motor, which, with appropriate gearing, moved the hands of a

[23] "The Greenwich Observatory Free Pendulum Clocks," Memorandum published March 1935 on the occasion of stopping No. 11 after a continuous run of nine years untouched," USNOA SF. On the Shortt clocks at the Naval Observatory see Paul Sollenberger, "Time Determination and Time Keeping," PA, **50** (1942), 74–78: 77. In 1984 a comparison of Shortt clock number 41 (which had been kept in the clock vault at the Naval Observatory since 1932!) with an atomic clock showed that it was stable to within 200 microseconds per day over the month-long study, or 2–3 parts per billion, ten times better than the 2 milliseconds (2,000 microseconds) previously reported. Pierre Boucheron, "Just How Good was the Shortt Clock?," manuscript dated November 21, 1984, USNO Library. According to Tony Jones, *Splitting the Second* (ref. 57 below), p. 30, the first Shortt clock was installed at the Royal Observatory in Edinburgh in 1921.

clock display. In 1928 W. A. Marrison and J. W. Horton built the first quartz-crystal clock at the Bell Telephone Labs. As did others around the world, Sollenberger began experiments with quartz clocks, and, as we shall see below, by 1934 he had designed and put into operation a crystal-controlled clock for the automatic transmission of time. Because a transmitting clock was reset just before transmission, it was not as accurate as the standard clock.[24]

Crystal-driven transmitting clocks were one thing, but replacing the standard pendulum clocks was a more drastic step. At the Naval Observatory construction of such clocks began in 1938, in conjunction with the Naval Research Laboratory, and Sollenberger remarked the following year that "It will not be surprising if we find that future developments place the quartz controlled clocks far ahead of all other time-keepers." By 1945 several quartz clocks, rated for mean and sidereal time, were in operation, of which the least accurate was as precise as the poorest of the three Shortt clocks at the Observatory. At that time Sollenberger wrote that, taking into account the combined errors of time determination and timekeeping at the Naval Observatory, it was possible to time an event at about 0.005 seconds, with the error of sending time signals via radio adding a few more thousandths of a second.[25] Still, as of World War II, the time standards both at the Naval Observatory and at Greenwich were based on a combination of pendulum and quartz-crystal devices. Not until 1946 did the quartz crystal become the sole standard clock at the Naval Observatory, and, even then, the quartz clocks used were those of the National Bureau of Standards (NBS). The reason, according to Markowitz, was that the NBS was involved in radio electronics and was able to obtain good experimental quartz crystals from Bell Telephone Labs at Murray Hill. The Naval Observatory, moreover, was an astronomical institution, not a standards laboratory, and thus did not have the same kind of access to developing state-of-the-art equipment as did the NBS. Nor, until Sollenberger's experiments, had the Observatory ever been in the business of building clocks. The situation was similar at Greenwich; its first quartz clock, installed in 1939, had been developed by Dye and Louis B. Essen at the National Physical Laboratory, the British equivalent to the NBS. In 1954 the Naval Observatory obtained from the British Post Office the first of three "Essen-ring" quartz crystals, which also had been developed by Essen at the National Physical Laboratory. Essen-ring crystal clocks became part of the standard clock system at the Naval Observatory in 1955, with an accuracy of one ten-thousandth of a second, ten times better than that of the Shortt clocks.[26]

[24] W. A. Marrison, "The evolution of the quartz crystal clock," *Bell Systems Technical Journal*, **27** (1948), 510–588; Sollenberger OHI, 25 ff; Markowitz, "Time Measurement," EB (1967).

[25] Sollenberger, "Crystal Oscillators as Precision Timekeepers," *Journal of the Horological Institute of America*, **7** (1939), 28–33; Sollenberger (ref. 23 above), 78, and Paul Sollenberger and Alfred H. Mikesell, "Quartz Crystal Astronomical Clocks," AJ, **51** (1945), 123–124. Markowitz describes some of Sollenberger's experiments in his "Reminiscences" (ref. 17 above), pp. 7–9.

[26] Markowitz, OHI, August 18, 1987, 16–22; "The Naval Observatory Time Service," USNO Circular No. 49 (March 8, 1954), pp. 4–5. For parallel events in Canada see Thomson (ref. 19), pp. 11 ff.

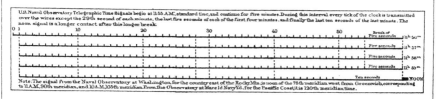

Figure 11.9. The scheme for sending telegraphic time signals as of 1902. From the *Annual Report* for 1902, p. 32.

*Time dissemination: From telegraph to radio*

The determination and maintenance of accurate time is of little use outside an observatory if it cannot be disseminated to users. As we saw in section 5.3, time dissemination at the end of the nineteenth century was dominated by the telegraph, and in fact precipitated a controversy over the control of time between 23 observatories on the one hand, and the Naval Observatory and Western Union on the other. This controversy, however, did nothing to slow the telegraphic dissemination of time. The mean time of the noon (75th meridian) signal continued to be sent to the private telegraph companies daily, except Sundays, according to the scheme in Figure 11.9; in 1902 Hayden reported that the average error of the signal was only 0.14 seconds, although occasional errors of more than a second would creep in as a result of the inability to observe during a prolonged period of bad weather.[27] On New Year's Eve 1903/1904 a telegraphic time signal was even sent around the United States, and expanded the following year to include England, Australia, Alaska, and Argentina, "covering nearly three quarters of the civilized world and traversing nearly 300,000

[27] AR (1902), 35; AR (1903), 29; AR (1904), 52.

miles of wire." This indeed gives an idea of how widespread telegraphy had become, providing an excellent medium for the dissemination of time.

Still, time-dissemination technologies were complementary and overlapping. In 1905 19 time balls were still being dropped by signals in the principal ports of the Atlantic, Pacific, Gulf of Mexico, and Great Lakes coast, using Western Union, the Postal Telegraph Company, and the American Telephone and Telegraph Company. The Washington time ball, initiated in 1845, was dropped until 1936. A few are still ceremonially dropped even to the present day at several locations around the world.[28]

The story of twentieth-century time dissemination, however, is neither the time ball nor the telegraph but the "wireless," better known today as radio, the new medium by which time signals were disseminated at the beginning of the century over very short ranges, and by century's end routinely available worldwide using artificial Earth satellites. Time transmission by wireless was a revolutionary technique, the first new one since the time signals were first transmitted by telegraph more than a half century earlier. For practical reasons, the U. S. Navy pioneered the new technique. Marconi had demonstrated wireless telegraphy in England in 1896, and brought it to the United States in 1899, where the U. S. Navy demonstrated its early interest by having four Naval officers observe Marconi's operation of his ship-to-shore system during the Americas Cup race of that year. Despite the obvious need for communication, not to mention time dissemination for longitude purposes, for three years the U. S. Navy did nothing to adopt the new technology. From 1902 to 1906, however, the Navy made an increasing number of purchases of wireless apparatus, especially after the Roosevelt Board in 1904 determined that the Navy should operate the government's wireless system, rather than the Army Signal Corps or the Weather Bureau, each of which had its own obvious uses.[29]

It was in this context that in the spring of 1904 time signals were first broadcast at low power from the Navy coastal station at Navesink, New Jersey. During these early experiments, the Naval Observatory's noon signal, sent via Western Union lines, was automatically repeated using relays installed in the stations to operate the radio telegraph keys. A special Hydrographic Office Notice to Mariners, dated November 22, 1904, informed potential users of the new arrangement. Although this was not mentioned in the Naval Observatory's Annual Report for that year, while thanking the telegraph companies for their cooperation in disseminating time, Superintendent Colby Chester commented presciently that "Perhaps with the development of wireless telegraphy it is not too much to expect that the day will come when not only every land station but every ship at sea will receive the daily noon signal from our standard clock, in which practically all of the uncertainties of longitude determinations at sea will

---

[28] AR (1905), 14. "Time Ball" files, USNOA, SF.

[29] Joan Mathys, "The Right Place at the Right Time: The United States Navy and the Development of Wireless Time Signaling, 1900–1923," M.A. thesis (George Washington University, 1991), pp. 30 ff; and Susan J. Douglass, "Technological Innovation and Organizational Change: The Navy's Adoption of Radio, 1899–1919," in Merrit Roe Smith, ed., *Military Enterprise and Technological Change* (MIT Press: Cambridge, Massachusetts, 1985).

Figure 11.10. Arlington Towers, located in Arlington, Virginia across the Potomac from Washington, completed in December, 1912 for Navy communications. It played an important role in the transmission of time signals until it was dismantled in 1941.

vanish. When that day comes (and it seems now to be entirely practicable) this clock may be heard in every land and on every ocean of the globe, and be available daily both for standard time and navigation everywhere."[30]

By 1909, 19 Navy coastal radio stations were transmitting to a range of 100 miles.[31] Aware that the French were experimenting with high-power transmissions from the Eiffel Tower beginning in 1908, in 1909 the Navy signed a contract for the manufacture of a 100-kilowatt spark transmitter with a range of 3,000 miles from Washington, as well as two 10-kilowatt transmitters and two receivers for use aboard ship. The primary purpose was of course communications, but the ability to disseminate time at long range was more than a minor piggyback. A site was chosen at Fort Meyer in Arlington, Virginia, just across the Potomac River from Washington. Construction of one 600-foot and two 450-foot towers began in 1911, and was completed in December, 1912, shortly after Joseph L. Jayne, who had been a member of the Roosevelt Board, became Superintendent of the Naval Observatory.[32] With the completion of this powerful transmitting station (Figure 11.10), not only users on land, but also ships were able to receive time signals in large parts of the North Atlantic, the Caribbean, and the Gulf of Mexico. In addition to the noon signal, a night signal was sent from Arlington in order to reach as far as possible. By 1917 the availability of time signals anywhere in the North Atlantic made it possible to reduce the number of chronometers

---

[30] Colby Chester, AR (1904), 14–15. These radio signals were not mentioned in subsequent Annual Reports until 1912, p. 17, where a date of January, 1905 is given for the first radio time signals. The spring, 1904 date is given much later in J. F. Hellweg, "United States Navy Time Service," (ref. 12 above), p. 17. It is likely that experiments began in spring, 1904, and regular service began in January, 1905. The latter date is confirmed in an interesting overview of methods and applications in "Astronomy and Wireless Telegraphy," MNRAS, **74** (1914), 368–369, with references.    [31] Mathys (ref. 29 above), p. 37.

[32] Paul G. Watson, *The Arlington Naval Station: Its Early Years, 1909–1924* (Government Printing Office: Washington, 1925); Mathys (ref. 29 above), pp. 38–42. The Arlington station remained in operation until March, 1941.

aboard naval vessels.[33] The Naval Observatory's Time Station at Mare Island, California (manned by T. J. J. See) began sending a 10-pm signal on July 1, 1913, performing on the West Coast the same function as Arlington did on the East Coast. With the usefulness of radio signals for time proven beyond doubt, time signals began to be sent from transmitting stations in Annapolis, Maryland (NSS) in 1920, in Pearl Harbor, Hawaii (NPM) in 1932, and in Balboa, Canal Zone (NBA) in 1933. At Arlington the number of time broadcasts per day increased from two in 1913 to 24 in 1938.[34]

Even before the Arlington station could be completed, however, the French seized the initiative for international cooperation in matters of time. The International Time Conference, held in Paris in October, 1912, at the invitation of the French government, was precipitated in part by the potential overlap of the radio signals from the new Arlington station with those emanating from the Eiffel Tower. Asaph Hall Jr, one of the delegates along with the Naval Attaché in Paris, H. H. Hough, wrote in his confidential report that, while one reason for the meeting was coordination of the increasing number of strong radio time signals, another (from the French point of view) was to "keep the preeminence of the Eiffel Tower signals," and secure international sanction to "keep Paris as the headquarters of the wireless time signals."[35] The U. S. delegates were instructed to emphasize that, while the Arlington station was designed to reach ships as far at sea as possible, the United States "desires to do this so as not to interfere in any way with the signals from the Eiffel Tower. On the contrary it desires to cooperate with the work of that station so far as it may be mutually advantageous." In a follow-up meeting in 1913 in Paris, representatives of 32 nations agreed to a Convention creating a Commission Internationale de l'Heure (International Commission on Time), to include as part of its structure a Bureau Internationale de l'Heure (BIH). World War I intervened before the Convention could be ratified, but M. Baillaud, as Director of the Bureau and Director of Paris Observatory, carried out many of its recommendations. When the IAU was founded in 1919, the Commission became part of that organization, and the Bureau as its operative organ absorbed fully one half of the IAU budget in its early years.[36]

One means of cooperation with the French, initiated by Jayne and proposed at the 1912 Paris Conference by Cdr Henry-Hughes Hough, was that the Washington–Paris longitude difference be determined via radio signals using the Arlington and

[33] J. L. Jayne, AR (1912), 17; AR (1913), 3; AR (1917), 3.
[34] Mathys (ref. 29 above), p. 50; for a review of time dissemination at the Naval Observatory as of 1913, see J. L. Jayne, "The Naval Observatory Time Service and How Jewelers May Make Use of its Radio Signals," *The American Jeweler*, **33** (1913), 424–432, including many illustrations. For detail on transmitting stations and the number of transmissions per day, see Table 3 in Treseder (ref. 20).
[35] Asaph Hall Jr, manuscript in USNOA, BF, Asaph Hall Jr folder, pp. 7–8. On the Eiffel Tower station see MNRAS (ref. 30).
[36] The quotation and summary of the meetings is found in Jayne, AR (1913), 3–7. The Proceedings of the 1912 meeting are to be found in *Conférence Internationale de l'Heure, Annales du Bureau des Longitudes*, volume 9 (Paris, 1913); the Proceedings for the meeting of October 15, 1913 are in *Conférence Internationale de l'Heure* Paris, October, 1913 (Paris, 1913). A summary of the history of the BIH is given in A. Fowler, ed., *Trans. IAU* (held at Rome, 1922) (London, 1924), 110–111, and more recently in Bernard Guinot, "History of the Bureau International de l'Heure" in Dick *et al.* (ref. 15), pp. 175–184.

Eiffel Tower transmitters. In the 1860s and 1870s the U. S. Coast and Geodetic Survey had made three determinations of longitude differences between Cambridge, Massachusetts and Greenwich, U. K. using the transatlantic telegraph cable, and Canadian and English astronomers had done the same between Montréal and Greenwich. However, the longitude difference between the Naval Observatory and Greenwich or Paris had never been directly determined, and was known only through the Cambridge–Greenwich connections, themselves involving several intermediate stations. Such a direct determination, Jayne emphasized, was especially desirable because of the national observatories located in Paris and Washington. The French had already used radio to make longitude determinations within France to a distance of 960 miles, thus proving the method, but the Paris–Washington distance was 3,840 miles. This was therefore a somewhat daring experiment to determine whether the transmitters were powerful enough, and the propagation effects well enough understood, to obtain a useful scientific result.

When authorities at the Bureau des Longitudes and Paris Observatory approved Jayne's plan, in March, 1913, the French sent five of their leading experts to study the possibilities and make tests (Figure 11.11). On their part, the Americans ordered two 3-inch portable transit instruments from G. Prin in Paris, the same type of instrument as that used by the French in earlier experiments. The observing campaign ran for two periods: from October to December, 1913 two astronomers (including Naval Observatory astronomer G. A. Hill) and two Navy radio experts were stationed in Paris, with a similar team (including Littell) in Washington. From January to March, 1914 the team locations were reversed to reduce personal equation effects. The French had their own observing teams. The result of these elaborate experiments gave a Washington–Paris longitude difference of 5 hours, 17 minutes, 36.653 seconds ±0.0031 seconds, compared with the value used in the *American Ephemeris and Nautical Almanac* of 5 hours, 17 minutes, 36.75 seconds. Whereas cable determinations half a century earlier had differed by as much as 1.5 seconds from results obtained using previous techniques, wireless determinations were now within a few tenths or a few hundredths of a second of the previous cable determinations of Paris–Washington longitude differences, as mediated through the Coast Survey's Cambridge–Greenwich value (Figure 11.12). Even these small errors gave rise to much discussion. Such seemingly minute amounts were important for a variety of scientific and geographic reasons; *Scientific American* even noted that frequent redeterminations were proposed in order to discover possible "creepings of the Earth's crust," a remarkable statement in the days before plate tectonics had been accepted.[37]

---

[37] Jayne, AR (1913), 4–6. The results are published in F. B. Littell and G. A. Hill, "Determination of the Difference of Longitude Between Washington and Paris, 1913–1914," PUSNO, **9**, part 4, Appendix I (Washington, 1916). Jayne's proposal is to be found on p. E7. See also "The American–French Radio Time Signal Experiments," *Scientific American*, **109** (December, 1913), 455; "The Washington–Paris Longitude Campaign," *Scientific American*, **110** (March, 1914), 244; F. B. Littell and G. A. Hill, "Washington–Paris Longitude by Radio Signals: A Valuable Application of Wireless Communications, *Scientific American Supplement*, No. 2,051 (April 24, 1915), 266–267; and Littell and Hill's preliminary results in AJ, **29** (March 15, 1915), 1–10.

Figure 11.11. Members of the French and American wireless parties for determining the Washington–Paris longitude difference, in front of the Naval Observatory, April, 1913. Superintendent Captain Joseph L. Jayne is in the first row second from the right and French Commandant Gustave Ferrie (1868–1932) is to the right of him. Both were wireless pioneers in their respective countries. In the back row on the left is Professor Abraham and F. B. Littell (second left); G. A. Hill is probably also among those in the back row.

So successful was the 1913–14 Washington–Paris operation that even more ambitious longitude operations were planned. Not to be outdone by the Americans, in 1919 General Gustave Ferrié of the Bureau des Longitudes in Paris proposed a World Longitude Operation. This time a number of conferences and extensive tests preceded the actual determinations, as well as approval and recommendations of the newly formed International Astronomical Union (at its 1922 and 1925 meetings) and the International Geophysical Union. The results were two worldwide campaigns carried out in 1926 and 1933, which were meticulously executed and reported. In contrast to the Washington–Paris campaign, the Naval Observatory's role in this worldwide operation was small; its Washington–San Diego observations were only a part of a much larger operation whose results were coordinated by the BIH in France.[38] The opera-

[38] F. B. Littell, J. C. Hammond, C. B. Watts, and P. Sollenberger, "World Longitude Operation of 1926: Results of Observations at San Diego and Washington," PUSNO, **12**, part 4 (Washington, 1929), the recommendations are on p. 531; C. B. Watts, P. Sollenberger, and J. E. Willis, "World Longitude Operation of 1933 at San Diego and Washington," PUSNO, **14**, part 5 (1938).

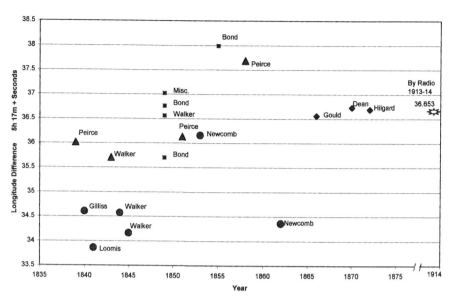

Figure 11.12. Washington–Paris longitude differences, 1839–1914, determined by five different methods; using eclipses and occultations (triangles), Moon culminations (circles), chronometers (squares), cable (diamonds), and radio (oval). Whereas the cable methods in the 1860s and 1870s brought improvements of more than a second in the longitudes, the radio method improved the accuracy of longitude by a few hundreths of a second. Note that the Moon culminations, undertaken by Gilliss and Walker during 1838–43 and by Newcomb as late as 1863, were systematically offset from other methods. The Washington–Greenwich radio determination gave the longitude of Washington at the new Naval Observatory site as 5 h 8 m 15.721 s ± 0.14 s. Data from Littell and Hill (1916, ref. 37).

tion was also an ending. The mission of the Naval Observatory was not longitude determination; that was the role of the U. S. Coast and Geodetic Survey, which, in addition to their participation in the worldwide campaign, used the Annapolis signals for internal longitude determinations.[39]

Time signals via radio also quickly became a popular public service. By World War I "the outbreak of war caused the dismantling of privately owned receiving sets all over the country," the Superintendent, Rear Admiral T. B. Howard, wrote in 1917. "The number of requests for exemption coming in from watch manufacturers, jewel-

[39] George C. Cowie, "Wireless Longitude Determinations of the U. S. Coast and Geodetic Survey," AJ, **35** (1923), 145–147. It is interesting that for the 1926 World Longitude Operation, the Naval Observatory made observations at San Diego, one of three "fundamental" stations, while the Coast and Geodetic Survey observed only at two of the secondary stations near Honolulu and Manila. A comparison of instruments and methods between the two institutions can be gleaned from Clarence H. Swick, *World Longitude Determinations by the United States Coast and Geodetic Survey in 1926* (U. S. Department of Commerce, Coast and Geodetic Survey: Washington, 1931).

Figure 11.13. Early radio-reception equipment, and transmitting clock # 2 in the Naval Observatory Time Room, April, 1927. Astronomer Paul Sollenberger is seated at the radio equipment. The receipt of time signals from the west coast of the United States, or from Paris and other transmitting stations, allowed comparison of time signals, taking into account lag time due to distance.

Figure 11.14. The transmitting clock and time service, April, 1927, showing the transmitting clock on the left, and Paul Sollenberger standing at the switchboard.

Figure 11.15. A closeup of the drum chronograph in the Naval Observatory Time Room, about 1927. The paper on the chronograph drum recorded outgoing and incoming time ticks for measuring the error in the time signal, and for comparisons of clocks.

ers, and scientific laboratories and observatories proved how popular the custom of receiving time signals by wireless has become."[40] Admiral Howard unfortunately gave no data to support his claim, so it is not possible to know just how popular the service actually was by this time.

Radio time dissemination involved a variety of technological adaptations, including the transmitting clocks and associated apparatus used to send the time from the Naval Observatory to the land lines and transmitting stations. Prior to 1934 a pendulum time broadcaster was used, surrounded on one side by the radio-reception equipment (Figure 11.13) and on the other by a "switchboard" and chronograph (Figure 11.14). The two transmitting clocks, which had been purchased from the Self-Winding Clock Company in 1916, were not as accurate as the primary standard clocks, and, after comparison with them, were corrected before each transmission (usually by only a few hundredths of a second) by a magnet acting on the pendulum in the form of current running through a solenoid. The clock transmitted the 75th meridian (Eastern Standard) mean solar time by means of an electric break circuit controlled by the pendulum. The switchboard carried connections to the various clocks, and to the radio sets, for recording the clock ticks and the radio signals on the chronograph (Figure 11.15), and for connecting the transmitting clocks to the broadcasting stations. Private telegraph-company land lines for transmission of time for commercial purposes were also connected to this switchboard. The chronograph also recorded outgoing and incoming time ticks for measuring the error in the time signal. By 1927 time was being sent by radio control circuits not only to the Arlington transmitter, but also to Annapolis, Maryland; Key West, Florida; and San Diego, California.[41]

This method of time transmission was not without problems, especially with the magnetic method of correcting the clock, which took 15 minutes or more. Having convinced Superintendent Hellweg that the old transmitting clocks dating from 1916

[40] AR (1917), 5.
[41] For more details see Sollenberger, OHI, August 17, 1987, 70 ff. Prior to those made by the Self-Winding Clock Company, the transmitting clock from 1879 had been one of the Howard clocks made for the transit of Venus.

Figure 11.16. New automatic broadcasting apparatus, 1934. On the left is the quartz-crystal oscillator. The switchboard is in the center, and the electrical broadcasting "robot" on the right. Paul Sollenberger stands at the printing chronograph, an upgrade from the previous chronograph. The new broadcaster required about five seconds to adjust before transmission, compared with 17 minutes for the old transmitting clock.

had become "a disgrace to the operation," Sollenberger set to work designing quartz-crystal transmitting clocks. By 1934 entirely new automatic transmitting equipment was in operation (Figure 11.16). A quartz-crystal oscillator (on the left in Figure 11.16) stepped down to 1,000 cycles per second drove a synchronous motor-operated clock, which could be quickly adjusted by rotating the field coil of the motor. The automatic broadcaster robot transmitted the time every hour, relaying the signal to the Arlington towers via telegraph line. The average error of all time signals sent at 113 kilocycles via Arlington during 1934–35 was 0.011 seconds, with a maximum error of 0.122 seconds. Sollenberger received the medal of the Franklin Institute for his invention, as did Hellweg for having supported it. Their goal, as war approached, was to run an increasingly reliable and accurate time service.[42]

Despite these improvements, the Naval Observatory faced competition to its time service from the National Bureau of Standards, which had been founded in 1901. With the entry of the United States into the war in December 1941, the Director of Naval Communications sharply cut the daily time broadcasts from the Observatory from 24 to four per day as a security measure. By 1944 Superintendent Hellweg had called the attention of his superiors to "the gradual encroachment on the Navy's time service by the Bureau of Standards." Little by little, he noted, they had developed the standard frequency service (1923), then the standard time inter-vals, standard audio frequencies, and standard musical pitch at 440 cycles per second. In 1944 the NBS sent Hellweg a letter describing their service, intimating that they were going to begin sending time signals. "The fact that the Navy has cur-tailed its time service has given the Bureau of Standards its opportunity to drive an entering wedge," Hellweg wrote his superiors. "It is only a question of time when

[42] Sollenberger describes the work in progress in "Recent Expansion and Developments in the Time-Service of the United States Naval Observatory," *Transactions, American Geophysical Union,* 14th *Meeting* (1933), pp. 52–54. Hellweg, "United States Navy Time Service" (ref. 12 above) states that the first automatic broadcast signal was sent out May 7, 1934, and publicly announced after a six-month trial period. The method of adjusting the clock is described in this article, p. 23, and in Sollenberger OHI, 78–79. On the construction of the clock, see the OHI, 27 ff. For the circumstances of the Franklin medal see Sollenberger, OHI, 34–35.

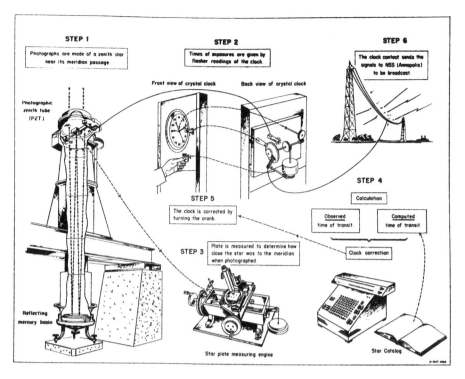

Figure 11.17. Steps in the process of time determination and dissemination, 1955.

the Bureau of Standards will usurp the entire time service and the Navy will lose its control."[43] Indeed, in 1945 the NBS began to transmit time signals, and eventually took over entirely the time transmission function, even as the Observatory built up its Time Service in other areas. It would not be the last skirmish between the Observatory and the NBS over time.

The entire process at mid-century, from time determination via PZT to time dissemination via radio, is summarized in Figure 11.17. The never-ending task of the Time Service at the Naval Observatory was to make improvements at each step in this process, resulting in improved accuracies of the end product for an increasing number of increasingly sophisticated users. By the 1950s the PZTs at the Naval Observatory were determining time to within a few milliseconds, Shortt and quartz-crystal clocks combined were keeping time to within a few milliseconds, and time was being synchronized around the world to within a few tens of milliseconds.[44]

All this activity was in the service of Universal Time, still called by many

[43] AR (1944), 5–6. Hellweg also noted that, in 1932, there was a Congressional investigation of the role of the NBS in time. On the role of the NBS in time see Rexmond C. Cochrane, *Measures for Progress: A History of the National Bureau of Standards* (National Bureau of Standards: Washington, 1966), pp. 291 and 474–475.

[44] For an overview as of 1954 see USNO Circular No. 49 (March 8, 1954), USNO Library.

Figure 11.18. Gerald Clemence at the printing chronograph, 1930s. The electrical broadcasting robot is on the right.

Greenwich Mean Time, or GMT. Its second was 1/86,400 of the mean solar day, based on the rotation of the Earth, even if corrections were made to account for polar motion and variations in the Earth's rotation. This relatively simple situation centered on Universal Time would change during the next decade, which saw not only a new and well-known era for space exploration, but also a new and less appreciated era for the refinement of time.

### 11.2    A variety of times: Turning point at mid-century

Wider issues beyond the mere measurement and dissemination of time came to the fore in the 1950s. The increasingly accurate determination of polar motion (which also continued at the Naval Observatory with the PZT) and the increasingly well-determined variations in the rotation of the Earth, gave rise to new forms of Universal Time. Moreover, as the inherent unpredictability of the Earth's rotation came to the fore, a more uniform time based on something other than the Earth's rotation became increasingly desirable. Beyond these forms of astronomical time, the 1950s saw the development of an atomic time scale, and the need to relate astronomical time to this scale. In all these endeavors, the Naval Observatory played a leading role. It did so largely under the leadership of two scientists: Gerald Clemence, who as Director of the Nautical Almanac Office had a practitioner's interest in the kind of time used for predicting the motions of celestial bodies, and William Markowitz, who in 1953 succeeded Paul Sollenberger upon the latter's retirement after 25 years as Director of Time Service.

*Ephemeris Time*

The move for a uniform time with a more reliable physical basis than the Earth's rotation surfaced first. A key player in this endeavor was Gerald Clemence. Clemence came to the Observatory in 1930 in the Time Service Division (Figure 11.18), worked in the 9-inch Transit Circle Division from 1937, was hired into the Nautical Almanac Office by Wallace J. Eckert in 1940, and was Director of the Office during

1945–58; we have already mentioned his service as Scientific Director from 1958 to 1963, and we shall see his broader role in chapter 12. It was during Clemence's *Nautical Almanac* period that the issue of a more uniform time came to a head. Clemence had an interest in time from his days in the Time Service; one of his earliest papers was written with Paul Sollenberger on lunar effects on clock corrections. In an important article in 1948 Clemence raised the question of a variety of inconsistencies in the system of astronomical constants. Foremost among them was a discrepancy in time: the national ephemerides made use of Newtonian time, which satisfied the equations of celestial mechanics, whereas the time astronomers used in practice depended on the variable rotation of the Earth. The use of two different times for theory and observation naturally gave rise to discrepancies between theory and observation. This inconsistency, Clemence declared, "cannot be allowed to persist; it defeats the principal purpose for which tables of the sun, moon, and planets are constructed, which is to predict their positions."[45]

The discrepancy in the use of time came to a head at a meeting on fundamental constants in Paris in 1950. At this meeting the Director of the Paris Observatory, André Danjon, had suggested that the fundamental unit of time known as the second should be defined on the basis of the motion of the Earth around the Sun. The name "ephemeris time" was suggested by Yale astronomer Dirk Brouwer at the same meeting, because it was the kind of time (based on the laws of Newton) used to calculate the motions of celestial bodies as given in the national ephemerides. As Clemence reiterated, "the solar and planetary theories give the coordinates of the sun and planets as functions of Newtonian time, whereas the observed positions are functions of variable time; hence the theoretical coordinates differ from the observed."[46] At the same meeting Astronomer Royal Sir Harold Spencer Jones broached the subject of a clock based on the measurement of a selected atomic frequency, an early allusion to the "atomic clocks" then under development.

Following up on these deliberations, in 1952 the International Astronomical Union recommended that the second of Ephemeris Time (ET), based on the orbital motion of the Earth around the Sun, be used as the basis for uniform astronomical time, replacing the second based on the rotation of the Earth, defined as 1/86,400 of the mean solar day. In 1956 the International Committee of Weights and Measures (Comité International des Poids et Mesures, or CIPM) redefined the second in terms

---

[45] G. M. Clemence, "On the System of Astronomical Constants," AJ, **53** (1948), 169–179: 171. The article with Sollenberger is G. Clemence and P. Sollenberger, "Lunar Effects on Clock Corrections," AJ, **48** (1939), 78–80.

[46] Clemence's fascinating account of the history of Ephemeris Time is given in "The Concept of Ephemeris Time: A Case of Inadvertent Plagiarism," JHA, **2** (1971), 73–79; the quote is from p. 75. Clemence also shows here that Danjon had the general idea already in 1929. The 1950 meeting was Colloques Internationaux du Centre National de la Recherche Scientifique, xxv: *Constants fondamentales de l'astronomie* (Paris, 1950). On Ephemeris Time from the perspective of the head of the British Nautical Almanac Office, see D. H. Sadler, "Ephemeris Time," *Occasional Notes of the RAS*, **3**, no. 17 (1954), 103–113.

of Ephemeris Time, so that "the second is the fraction of 1/31,556,925.9747 of the tropical year for 12 h E.T. of January 0, 1900." In 1958 the IAU added a definition for the precise instant for the beginning of Ephemeris Time. These definitions were approved by the General Conference on Weights and Measures (Conférence Général des Poids et Mesures, or CGPM), and went into effect in 1960, when the second based on Ephemeris Time became the conventional standard of time. For the first time in history the second was defined as part of the International System of Units (Système International d'Unités, or SI), just as length and mass had been.[47]

Although Ephemeris Time was defined by the orbital motion of the Earth about the Sun, in practice it was obtained from the Moon's motion around the Earth. The crucial figure in making this possible was William Markowitz. Markowitz had obtained his Ph. D. in astronomy from the University of Chicago in 1931 (where he had worked with Frank Ross), joined the Naval Observatory in 1936, and became Director of the Time Service in 1953 after Sollenberger's retirement. Markowitz had practical experience at the Naval Observatory with telescopes and instrumentation; it is therefore not surprising that he devised a practical way to measure Ephemeris Time. As Markowitz recalled, after a session of the American Astronomical Society meeting held at the National Bureau of Standards in June, 1951, in which it was claimed that accurate atomic clocks might eliminate the necessity for astronomical time, Clemence called him into his office to discuss the matter. Clemence believed that there was still a need for astronomical time in the sense of the Ephemeris Time then being proposed; "what he was afraid of was that the physicists on their own might adopt some value for the frequency of an atomic clock that did not correspond to the Ephemeris Time. That is to say, the two kinds of seconds would differ in value, and this would cause considerable confusion and other problems. What he wanted was a method of obtaining Ephemeris Time rapidly." At the same time, this would ensure that astronomers maintained a role in time.[48]

Because the Sun itself could not be accurately observed, and because the planets moved too slowly in their orbits for accurate time determination, Markowitz proposed to use the Moon, and for this purpose developed the dual-rate Moon camera (Figure 11.19). As with the navigational method of lunar distances, the Moon was to serve as the hands on the face of a clock. However, these much more accurate observations were to be made from the Observatory, not from shipboard. A tilting dark-glass filter effectively held the Moon fixed relative to the background stars during a simultaneous exposure of 20 seconds. The first such camera, which was built in the Observatory instrument shop by George Steinacker, was placed in operation on the 12-inch refractor at the Naval Observatory in June, 1952, and 20 cameras of a second

[47] W. Markowitz, "The Second of Ephemeris Time," *Berichtsbuch des VI. Internationalen Kongress für Chronometrie* (1959), volume 1, pp. 91–96, copy in USNOA, BF, Markowitz file.

[48] Markowitz, "Reminiscences" (ref. 17 above), pp. 10–12. On Markowitz, aside from his "Reminiscences," see Markowitz OHIs, August 18, 1987, and August 9, 1988; also Steven J. Dick and Dennis W. McCarthy, "William Markowitz, 1907–1998," BAAS, **31** (1999), 1,605.

Figure 11.19.  Markowitz at the Moon camera he developed, March, 1961.

model were used during the International Geophysical Year around the world, having been deployed from the Observatory.[49] An observed position of the Moon gave the Ephemeris Time of the epoch of observation by interpolation in the *Improved Lunar Ephemeris*, which was used after 1960 in the national ephemerides, and based on the lunar theory of E. W. Brown. The Moon-camera program had mixed success. The analysis of the international observations proved difficult, and the lasting contribution of the instrument was its determination of Ephemeris Time for the cesium atomic-clock frequency as described below. The use of the Moon camera was terminated in the mid-1970s, when international cooperation flagged.[50]

Although the introduction of Ephemeris Time inevitably caused some confusion, Clemence reminded everyone that, in accordance with its origins, highly precise predictions must be referred to Ephemeris Time, whereas highly precise observations had to use Universal Time.[51] Put another way, any function requiring knowledge of the rotational position of the Earth on its axis had to use Universal Time, whereas Ephemeris Time was required primarily for studying the motions of celestial bodies. The reduction from UT to ET gave rise to the term "delta T," the difference between the two times determined by using the Earth's rotation and its orbital motion, a difference that increased with time.

[49] Markowitz, in Kuiper and Middlehurst (ref. 16 above), pp. 107–114, where references are given to previous methods used by Henry Norris Russell among others; Markowitz, "Photographic Determination of the Moon's Position, and Applications to the Measure of Time, Rotation of the Earth, and Geodesy," *AJ*, **59** (1954), 69–73; Markowitz, "Reminiscences," pp. 10–15. R. Glenn Hall supervised the reduction of the Moon observations for this program, for which advance copies of Watts's Moon charts were also used. Although occultations could have been used for Ephemeris Time over years or decades, the Moon camera was much quicker.

[50] Winkler OHI, 64. The last observations are reported in *BAAS*, **7** (1975), 212. The Moon cameras were funded by the Office of Naval Research and built in the Naval Observatory instrument shop. At the Naval Observatory they were always used in the dome for the 12-inch instrument, though the 12-inch telescope itself was dismounted and a different tube and lens used for many years for the Moon camera. The *Improved Lunar Ephemeris, 1952–1959* (Government Printing Office: Washington, 1954) was produced by the Nautical Almanac Office.

[51] G. M. Clemence, "The Practical Use of Ephemeris Time," *Sky and Telescope* (January, 1960), 148–149.

*Refinements of Universal Time*

Because celestial navigation, precise surveying, satellite tracking, and daily life in general are regulated by day and night based on the Earth's rotation, Universal Time could not be abandoned in favor of ephemeris time, atomic time, or any other variety of time; to the contrary attempts to improve it were accelerated as knowledge of polar motion and variation in the Earth's rotation increased.

By the 1950s knowledge of these motions was considerable. By that time polar motion had been monitored continuously for 50 years by the stations of the International Latitude Service. It had been well established that the effects of polar motion amounted to as much as several hundredths of a second, equivalent to 0.3 arcseconds. Already the 12-month period and the 14-month "Chandler wobble" were well documented, and there was also evidence for a long-term secular motion of the pole.[52] Earth's variable rotation had not been so easily pinned down. Although already in 1900 Newcomb suspected that discrepancies observed in the Moon's motion compared with predicted values were due to irregularities in the Earth's rotation, and although in the 1920s E. W. Brown and W. de Sitter believed that discrepancies in the motions of the Moon and the planets were due to the same effect, this was only definitively shown by Harold Spencer Jones in 1939.[53] Quartz-crystal clocks finally demonstrated the seasonal variation in the Earth's rotation, and the Markowitz moon camera also played its role.[54]

There were problems, however, in incorporating these refinements into the daily dissemination of time. Polar motion could be determined only months after the fact, and for many years the BIH applied polar-motion corrections only in its annual analysis of time signals. The Royal Greenwich Observatory first applied polar-motion corrections to time as a daily practice in 1947, and the Naval Observatory began applying provisional corrections in 1953.[55] With the IAU's establishment of a Rapid Latitude Service in 1955, however, and its provision for the BIH to extrapolate Earth-rotation corrections a year in advance, the timing community was poised to incorporate polar motion and seasonal variations in the Earth's rotation into the disseminated

[52] On the history of polar motion studies see Dick, McCarthy, and Luzum (ref. 15 above), and, for a succinct history, William Markowitz, "Polar Motion: History and Recent Results," *Sky and Telescope*, **52** (August, 1976), 99–108. Values of polar motion from 1970 on are given in The *Astronomical Almanac*, section K. Current values are published by the IERO.

[53] E. W. Brown, *Transactions of the Yale University Obervatory*, **3**, part 6 (1926); W. de Sitter, "On the Secular Accelerations and the Fluctuations of the Longitudes of the Moon, the Sun, Mercury and Venus," BAN, **iv** (1927), 21–38; Harold Spencer Jones, "The Rotation of the Earth and the Secular Accelerations of the Sun, Moon and Planets," MNRAS, **99** (1939), 541.

[54] L. Essen, J. V. L. Parry, W. Markowitz, and R. G. Hall, "Variation in the Speed of Rotation of the Earth since January, 1955," *Nature*, **181** (April 12, 1958), 1,054; Markowitz, "Variations in Rotation of the Earth: Results Obtained with the Dual-Rate Moon Camera and Photographic Zenith Tubes," AJ, **64** (April, 1959), 106–113. According to Markowitz (OHI, 21) Stoyko came up with the first good seasonal variation about 1937 at about 60 milliseconds, which was confirmed at the 25 ms level by Markowitz and others in 1958.

[55] *Explanatory Supplement to the Astronomical Ephemeris and The American Ephemeris and Nautical Almanac* (Her Majesty's Stationery Office: London, 1961), pp. 86 and 445, and on the Naval Observatory time signals, USNO Circular No. 49, March 8, 1954, p. 9.

time in a coordinated and systematic way. At the meeting of the IAU in Dublin in 1955, Markowitz, the incoming President of Commission 31 on Time and by now the Director of Time Service at the Naval Observatory, wrote that "The introduction of corrections for polar variation and for annual fluctuation allows a quasi-uniform time to be determined. The corrections applied by various observatories are not on the same system, however. It is proposed that the corrections be applied in a uniform manner, so that the improvements in time-keeping may be obtained without having a multiplicity of time systems." Markowitz then introduced four resolutions, all of which were adopted by the IAU. These resolutions instructed the BIH to compute, for those observatories cooperating in the international time service, the longitude corrections due to polar motion, as determined by the ILS; to publish in advance corrections for the annual fluctuations in the Earth's rotation, to be used by all observatories in the determination of UT; and that the bulletins of those observatories publish the corrections to be added to the reception of radio time signals.

Although Markowitz believed that it might be some years before this plan could be implemented, in fact it occurred in only a few months. After correspondence between A. Danjon and N. Stoyko of the BIH, and Harold Spencer Jones and Markowitz (outgoing and incoming Presidents of the IAU Commission on Time, respectively), beginning on January 1, 1956 Universal Time (UT) as observed by the visual and photographic instruments of the time was designated UT0, while UT1 was corrected for polar motion and UT2 was further corrected for seasonal variations in the Earth's rotation. In effect, each step from UT0 to UT2 produced a more uniform time scale. Time adjusted to UT2 was broadcast until 1972, when a leap second system (described in section 11.3) went into effect.[56]

### Atomic time

As the astronomers were refining Universal Time and inventing Ephemeris Time during the 1950s, physicists were developing a fundamentally different way of measuring time. Atoms and molecules could be harnessed for a clock, making use of microwave technology developed for radar during World War II. By counting the resonance frequency of a particular atom or molecule in the microwave range, a clock of unprecedented accuracy could be produced, since the resonance frequency of a particular atom or molecule never varied. The development of such clocks was not in the realm of expertise of astronomers at the Naval Observatory, or anywhere else; but it was very much within the technical capabilities of creative physicists. In January, 1949, a team led by Harold Lyons at the NBS demonstrated a clock using the ammonia

---

[56] P. Th. Oosterhoof, ed., *Trans. IAU*, **9** (held at Dublin, 1955) (Cambridge, 1957), 452 and 456–459; D. H. Sadler, ed., *Trans. IAU*, **10** (held at Moscow, 1958) (Cambridge, 1960), 489–490. Time Service Notice # 4 (January 5, 1956) notes that the BIH adopted this notation. According to this notice, beginning January 1, 1956, times of reception were published on the basis of UT2. Variation-of-latitude determinations at USNO were reported regularly in the AJ, e.g. **50** (1942), 7–9.

Figure 11.20. A quartz-crystal clock designated "B 12" with VLF receivers and recorders at the Naval Observatory, March, 1961.

molecule. Further research by I. I. Rabi and others showed that the so-called hyperfine transition of cesium-133, resulting from the "spin flip" of cesium's outer electron, was an ideal choice for atomic-clock technology. Naval Observatory astronomers played no role in these developments, and could only watch from the sidelines. Only in 1955, when the National Company of Boston introduced (under contract to the Army Signal Corps) the first commercial cesium atomic clock, with the trade name of Atomichron, could the Naval Observatory have acted on the new developments in clock technology. In the five years 1956–60 some 50 Atomichrons were sold, 90% of them to military agencies.[57]

Among those agencies were the Naval Research Laboratory which received its first Atomichron in September, 1956, and furnished data from its Atomichron to the Naval Observatory from the beginning. Even as the latest quartz clocks were still being used as standards (Figure 11.20), these early atomic clocks were used to calibrate quartz-crystal clocks whose frequencies "were matched daily to that of the one megacycle per second output of the cesium standard, which is not operated continuously." Because the Naval Observatory did not have its own electronics capability, cesium atomic resonators (Figure 11.21) were not placed in operation at the Naval Observatory in Washington until June 29, 1960, and at the Richmond (Florida) station on the following day.[58]

[57] Paul Forman, "Atomichron: the Atomic Clock from Concept to Commercial Product," *Proc. IEEE*, **73** (1985), 1,181–1,204. For a recent history of atomic time see Tony Jones, *Splitting the Second: The Story of Atomic Time* (Institute of Physics Publishing: Bristol and Philadelphia, 2000).
[58] USNO Time Service Notice No. 6, January 1, 1959, p. 2; AR in AJ, **62** (1957), 313. Winkler OHI, 28.

Figure 11.21. William Markowitz with the 2.5-megacycle quartz oscillator (left) and the Atomichron cesium atomic resonator (right) at the Naval Observatory, November 13, 1962. Note that the term "Master Clock" is already in use.

Ephemeris Time and Universal Time coexisted side by side. As a uniform measure of time Ephemeris Time was a great improvement over Universal Time. It was, however, not very accessible; Ephemeris Time based on Moon-camera observations was generally available only a month after the fact. One problem that remained was the calibration of an atomic clock in terms of the ephemeris second. By 1954, Markowitz recalled, it seemed evident that someone was going to construct an operational atomic clock of high precision, using either the ammonia molecule or the cesium-beam technology as its basis. Accordingly, Markowitz approached physicists working on such clocks at MIT and Columbia, but could not interest them in a calibration project tied to an astronomical institution. Ironically, the pioneering NBS had fallen behind in atomic clock technology. However, at the 1955 meeting of the IAU in Dublin, Louis Essen announced that an atomic clock was in operation at the National Physical Laboratory in Teddington, U. K., the British equivalent to the NBS; its clock was operating with a precision exceeding one part in $10^9$. Ephemeris time could therefore be defined in terms of the frequency of a cesium atomic clock, a definition that would make the second of ET immediately available. This was exactly the task undertaken by the joint team of Markowitz and Glenn Hall at the Naval Observatory, and Essen and J. V. L. Parry at the National Physical Laboratory beginning in 1955.[59]

---

[59] Louis Essen, "The Measurement of Time," *Vistas in Astronomy*, **11** (1969), 45–67: 50–51; Markowitz, "Reminiscences" (ref. 17 above), pp. 15–18; Markowitz–Essen correspondence, 1956–1959, USNOA, SF. On the stalling of atomic-clock development at the NBS see Jones (ref. 57 above), pp. 45–46. On the context at the National Physical Laboratory see Jones (ref. 57), pp. 53–67, and Edward Pyatt, *The National Physical Laboratory: A History* (Adam Hilger: Bristol, 1983), especially pp. 103–105, 160–161, and 174–175.

The period between 1955 and 1958, Markowitz recalled, was "a bit of a hectic time. There was pressure on the Observatory to produce the Ephemeris second. If not there was a danger [which Clemence had expressed already in 1951] that the physicists would on their own adopt a frequency not related to the ephemeris second, and we would have two seconds fairly close to each other that would differ enough to cause problems. I was asked several times by Dr. Essen when we expected to have a value of the ephemeris second. He stated that if one was not forthcoming soon enough, it was probable the physicists would act." Both Markowitz and Essen were members of the Consultative Committee for the Definition of the Second (CCDS), which met at the International Bureau of Weights and Measures (Bureau International des Poids et Mesures, or BIPM) in Sèvres, France, near Paris, at which this was a hot topic. Finally, in summer 1958, they announced that the number was 9,192,631,770 cycles per second (plus or minus 20). This number provided the link between astronomical time and atomic time, and thus the basis for our modern standard of time.[60]

As of January 1, 1958 the Naval Observatory introduced a system of atomic time, denoted A.1, based on this frequency and obtained from the cesium standards of the Naval Research Laboratory. At that time Ephemeris Time and Universal Time agreed; both were 0 h, 0 min, 0 s at the same instant. Soon other atomic-time laboratories joined this time scale. However, having an atomic time scale was not the same as using that time scale as the primary standard, now termed the Master Clock. Only in 1966, with the retirement of Markowitz and the arrival of Gernot Winkler as Director of the Time Service Department, did the atomic clock become the Master Clock of the United States.

Despite the proliferation of varieties of time during this period (Table 11.2), only one time could be designated the conventional standard. Ephemeris Time remained the conventional standard of time for seven years. For nine years after the calibration of atomic clocks based on Ephemeris Time, there was much discussion about whether atomic time should be the basis for the definitive SI second. Ephemeris Time was finally superseded in 1967, when the second was redefined by the Thirteenth General Conference of Weights and Measures as "the duration of 9,192,631,770 periods of the radiation corresponding to the transition between the two hyperfine

---

[60] W. Markowitz, R. Glenn Hall, L. Essen, and J. V. L. Parry, "Frequency of Cesium in Terms of Ephemeris Time," *Physical Review Letters*, 1 (August 1, 1958), 204L, 1–2, and 105–106. 9,192,631,770 cps was provisionally accepted as the standard in 1960, and officially accepted in 1967. See also L. Essen. J. Parry, W. Markowitz, and R. Hall, *Nature*, **181**, 1,054 and 1,958, and Markowitz, "The Second of Ephemeris Time," (ref. 47 above). The calibration of the atomic clock in terms of the ephemeris second was first suggested at the ninth General Assembly of the IAU in Dublin (1955). See also Markowitz, "Reminiscences," pp. 15–20. In October, 1956, when Winkler visited Markowitz, the latter thought that the last three digits of the frequency would be 840, or 70 hertz higher. As Winkler remarked, 70 hertz at 9 gigahertz is about one part in $10^8$ difference. "That was the uncertainty due to the difficulty of measuring Ephemeris Time accurately." Winkler, OHI, March 30, 1989, 20–21. The Markowitz–Essen correspondence during the period 1956–59 is in USNOA, SF.

Table 11.2. Varieties of time

| Class | Type | Introduced |
|---|---|---|
| Astronomical Time (Earth's rotation) | Mean Time (Mean Solar Time) | Seventeenth century (Flamsteed) |
| | Greenwich Mean Time (GMT) | 1880 (Great Britain) |
| | | 1916 (U. K.) |
| | | 1918 (United States officially adopts Greenwich-linked meridians) |
| | Universal Time (UT) | 1928 (IAU) |
| | UT0, UT1, UT2 | 1956 (IAU) |
| | Coordinated Universal Time (UTC) (within 0.1 seconds of UT2) | 1960, January 1 |
| | Coordinated Universal Time (UTC) redefined with leap second; (within 0.9 seconds of UT1) | 1972 |
| Astronomical Time (Earth's orbital motion) | Ephemeris Time | 1960 (CGPM) |
| | Terrestrial Dynamical Time (TDT) | 1976 (IAU) |
| | Barycentric Dynamical Time (TDB) | 1976 (IAU) |
| | Geocentric Coordinate Time (TCG) | 1991 (IAU) |
| | Barycentric Coordinate Time (TCB) | 1991 (IAU) |
| | Terrestrial Time (TT) | 1991 (IAU) |
| Atomic Time | AT (NPL) | 1955 (July) |
| | AM (NPL, NRL, NBS, CNET, etc.) | 1956– (BIH) |
| | A.1 (USNO) | 1958 |
| | Temps Atomique International (TAI) | 1972 (CGPM) (available since 1955) |

Notes:
IAU, International Astronomical Union; CGPM, Conférence Général des Poids et Mesures; NRL, Naval Research Laboratory (United States); NPL, National Physical Laboratory (U. K.); NBS, National Bureau of Standards (United States); CNET, Centre National d'Etudes des Télécommunications (France).

levels of the ground state of the undisturbed cesium-133 atom."[61] This definition remains in use today.

In the space of a decade, three official definitions had been given for time (Table 11.3). Now, for the first time, the second was based not on astronomy, but on physics. Yet, in a sense astronomy still held the upper hand, because astronomy was the basis for the redefinition: in the end the length of the atomic second had to be calibrated to the real world. That real world was represented by the Earth's motion around the Sun, the Ephemeris Time invented by Danjon and Clemence among others, and realized by the measurements of the Markowitz Moon camera developed at the Naval Observatory. Moreover, the real world remained interested in astronomical

---

[61] Markowitz, "The Atomic Time Scale," IRE Transactions on Instrumentation, 1 (December, 1962); Markowitz (1959), ref. 7 above, and Clemence (1971), p. 76. The redefinition of the second took place October 13, 1967.

Table 11.3. *Definitions of the second*

| Basis | Defined | Effective date | Definition |
|---|---|---|---|
| Rotation of Earth | | To 1960 | 1/86,400 of Mean Solar Day (Greenwich Mean Time and Universal Time) |
| Orbital motion | 1956 (CIPM) 1958 (IAU) 1960 (CGPM) | 1960–67 | 1/31,556,925.9747 of tropical year at 12 h ET, 1900, January 0 d 12 h (Ephemeris Time) |
| Atomic vibrations | 1967 (CGPM) | 1967–present | 9,192,631,770 vibrations of radiation in cesium, resulting from the hyperfine transition due to electron-spin flip (Atomic Time) |

*Notes:*
IAU, International Astronomical Union; CGPM, Conférence Général des Poids et Mesures (General Conference on Weights and Measures); CIPM, Comité International des Poids et Mesures (International Committee of Weights and Measures).

time, as determined by the rotation of the Earth, with all of its irregularities, now being monitored well by atomic clocks. This symbiotic relationship between astronomy and physics defined the world of time for the rest of the century.

## 11.3   Time service in the Space Age

As the Space Age began, Universal Time and polar motion were still determined with the PZT, and the definition of the second was still based on the rotation of the Earth. A number of Markowitz Moon cameras had been deployed around the world to determine the more uniform time known as Ephemeris Time, which would become the basis for the definition of the second and the conventional basis of time in 1960. Functioning atomic clocks had been in use for several years, but had not yet superseded quartz-crystal clocks as the standard clocks at the Naval Observatory – or anywhere else. Ground-based radio signals were still the primary means of time dissemination.

All of this would change after 1957. PZT observation of stars for determining the Earth's rotation and Universal Time would be gradually phased out in the last quarter of the century in favor of extragalactic quasar observations using radio astronomy – a field that barely existed in the first half of the century. For more technical purposes, Ephemeris Time evolved into "Dynamical Time" based on the orbital motion of the Earth, Moon, and planets; now relativistic effects, unknown at the beginning of the century, were included. Not only would atomic clocks become the standard clocks for the Naval Observatory and all other countries as well, but also a highly coordinated International Atomic Time (TAI) based on these clocks around the world would provide the basis for the realization of the second. The second was not only the most precisely reproducible unit in the SI, but also became the basis for the meter, the unit

Figure 11.22. Gernot Winkler, Director of Time Service, 1966–95.

of distance. Methods for time dissemination also changed dramatically; the innovation of radio time dissemination at the beginning of the century was transformed at its end by the ability to transfer time via satellites circling the Earth. Time accurate to within a few tens of a billionth of a second became available anywhere on the Earth via the Global Positioning System of satellites, whose system time was kept on Naval Observatory time. The Naval Observatory both affected, and was affected by, all of these developments.

*Atomic time: From atomic clocks to International Atomic Time*

Although the Naval Observatory had obtained atomic clocks under Markowitz, it was not until the arrival of Gernot Winkler as Director of the Time Service in 1966 that the atomic clock was made the standard Master Clock. Winkler (Figure 11.22), who was born in Austria in 1922, came to the United States in 1956 and worked with F. Reder in the microwave resonance branch at the U. S. Army Signal Corps, which was supporting the work of the National Company on the Atomichron. In 1956 Winkler received the first prototypes at the Signal Corps, and when the Signal Corps purchased 12 more, he helped establish a worldwide application and testing program. As part of this effort, he worked in Essen's laboratory in Teddington to test how well a commercially produced Atomichron compared to the laboratory cesium clock.[62]

Winkler's action on his arrival is therefore not surprising. As he recalled, on his arrival "we had a master clock, a Western Electric 2.5 megahertz [quartz-crystal] clock, which once a day was calibrated against the cesium clock, and I changed it immediately. The next day we put all our faith into that cesium clock." The clock accuracy at the time was about 100 nanoseconds over one day.[63] During Winkler's tenure, one of the primary goals was building a Master-Clock system second to none. This was done not only by incorporating the latest atomic clocks as they were developed, but also by increasing the number of clocks and applying sophisticated statistical techniques to

[62] On Winkler see Winkler, OHI, March 20, 1989; his work in Essen's lab is described on pp. 16–17.
[63] Winkler, OHI, 65–67. Winkler emphasized, however, that every atomic clock still has a quartz crystal oscillator, which is used as a kind of "flywheel."

produce the time scale. For the remainder of the century a combination of cesium-beam and hydrogen-maser atomic clocks served as the Master Clock. Overall, Winkler's goal was to improve timekeeping and the control of the Navy Time Service.

As more and more time services produced atomic time scales, the international coordination of these time scales became increasingly desirable. Individual national time scales were produced from the first atomic clock, beginning in July, 1955 with the atomic clock at the National Physical Laboratory, which produced a scale called "AT." When other laboratories joined, including the Naval Research Laboratory (1956), the Observatoire de Neuchatel in Switzerland (1957), the National Bureau of Standards (1958), and the Centre National d'Etudes des Télécommunications in France (1958), the scale was denoted AM. The Naval Observatory's A.1 time scale, which had been introduced in January, 1958, was not immediately included in the BIH international time scale.[64]

The production of a true international time scale, however, was hampered by the lack of accurate means for time comparisons over large distances. The means for such comparison appeared only in 1968 with the synchronization of the Atlantic LORAN-C chains for radio navigation. The reception of LORAN-C signals, along with portable cesium atomic clocks, made clock comparison within one microsecond possible between North America and Europe.[65] As a result of these new synchronization means, the BIH scale was redefined as of January 1, 1969 as the mean of three national atomic time scales, those of the German Physikalische-Technische Bundesanstalt (PTB), the French Commission Nationale de l'Heure, and the Naval Observatory. Later in the same year, they were joined by the British, Swiss, Canadians, and the National Bureau of Standards. Although atomic time scales had been available since 1955 with Essen's clock, only in 1967 did the CIPM and the IAU recommend the new time scale. Approval followed from the International Union of Radio Science (URSI) in 1969, the Consultative Committee for International Radio (CCIR) of the International Telecommunications Union (ITU) in Geneva in 1970, and finally the General Conference of Weights and Measures in France in 1971. It was the latter that gave the official name of International Atomic Time (Temps Atomique International, TAI) to the resulting time scale, when it was officially introduced in January, 1972. Time had become a truly international commodity, requiring not only consultation and cooperation, but also diplomacy. By the end of the century about 50 laboratories were contributing to International Atomic Time, kept at the International Bureau of Weights and Measures (BIPM) in Sèvres. These laboratories represented some 250 clocks, with the Naval Observatory making the largest single contribution to this time scale, amounting to about 40%.[66]

---

[64] Bernard Guinot (ref. 36 above), pp. 180–181.   [65] Ibid., pp. 180–181.

[66] On the role of these international agencies in time, see Winkler, OHI, 79–83 and R. H. Nelson et al., "The Leap Second: Its History and Possible Future," Metrologia, **38**, no. 6. In March, 1985 the administration and time section of the BIH was moved to the Bureau International des Poids et Mesures (BIPM) in Sèvres, France, and in January, 1988 the BIPM took over responsibility for TAI and other time activities.

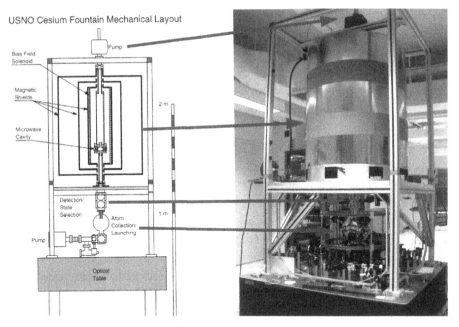

USNO Cesium Fountain Mechanical Layout

Figure 11.23. The cesium fountain atomic clock under development at the Naval Observatory. It will increase the precision of the Master Clock by a factor of ten.

Even as this operational system provides time to the world, new primary frequency standards are undergoing research and development. Among the foremost is a new form of atomic clock known as the cesium fountain. The first cesium fountain went into operation at the Paris Observatory in 1995; a second fountain at the National Institute of Standards and Technology (NIST, formerly the National Bureau of Standards) in the United States began contributing to TAI in 1999. Other laboratories in Germany, the U. K., and the United States are in various stages of developing these new frequency standards, including the Naval Observatory (Figure 11.23).[67] The cesium fountain can determine the length of the second with an accuracy of about two parts in $10^{15}$. Because the Riefler clock could keep time to about one hundredth of a second per day, or one part in $10^7$, over the last century time accuracy has improved by eight orders of magnitude, or 100 million times.

An accurate international time scale had many uses, but, with the unprecedented stability of atomic clocks, interesting experiments could be undertaken even with a few individual clocks. In 1972, the same year that TAI was introduced, the results were published of a famous experiment involving atomic clocks at the Naval Observatory, in particular the measurement of the relativistic effects of moving atomic clocks. Among the predictions of Einstein's Special Theory of Relativity was that a moving clock would record less time than a clock at rest, giving rise to the "clock

[67] Demetrios Matsakis, "USNO and GPS: It's About Time," *GPS World* (February, 2000), 32–40. Matsakis became Director of Time Service upon Winkler's retirement in 1995.

paradox" and the time-dilation effect. In order to test this the physicists J. C. Haefele of Washington University in St Louis and Richard Keating of the Naval Observatory undertook just such an experiment using cesium beam atomic clocks flown on regularly scheduled commercial flights. One clock flew eastward around the world, and one westward, while two other clocks remained at the Naval Observatory in Washington. They determined that the moving clocks lost time (ran slower) during the eastward trip by $59 \pm 10$ nanoseconds, and gained time (ran faster) by $273 \pm 7$ nanoseconds during the westward trip, verifying special relativity within the experimental uncertainties.[68]

The developments in atomic time brought an unprecedented need for international cooperation, which is not always an easy task. Ironically, in the United States national cooperation in the area of time became an issue; one of Winkler's goals on his arrival in 1966 was to improve relations with the NBS, which shared responsibilities for time in the United States. The Time and Frequency Division of the NBS had moved to Boulder, Colorado in the mid-1950s, substantially interrupting their progress on atomic clocks. Gradually, however, they built up their atomic-frequency-standard capability again with a large group of prominent scientists. More to the point, the NBS's WWV time signal begun in 1945 had become more popular, and there was growing confusion among the public as to the official source of time in the United States. That the NBS aspired to be the official source of time is clear from an inquiry in 1961 from the Legal Department of the NBS regarding the statutory authority for the Naval Observatory to publicize time. The opinion from the Navy's Judge Advocate General stated that the authority "does not derive from any explicit statutory enactment; rather it is based on a combination of historical practice, Congressional recognition and approval, and practical utilization of information." During Markowitz's tenure the interagency dispute had erupted openly in a meeting of the CCIR in Oslo in the mid-1960s, causing the representatives both of the Naval Observatory and of the NBS to be recalled by the Department of State. After the NBS had placed Winkler on its Time and Frequency Division Advisory Committee, more professional interactions were developed. Winkler and his counterparts at the National Bureau of Standards finally settled on functional statements that it was the responsibility of the NBS to provide a standard of frequency, and of the Naval Observatory to provide a standard of time. Yet, the NBS continued to provide both to the public, and, in 1994, near the end of his tenure as Time Service Director, Winkler was still trying to clear up the confusion even among professionals. Responding to a statement in a publication of Hewlett-Packard (the primary manufacturer of cesium-beam atomic clocks), that "The NIST provides the official basis for standard time in the United States," Winkler

[68] J. C. Haefele and Richard E. Keating, "Around the World Atomic Clocks: Predicted Relativistic Time Gains," *Science*, **177** (1972), 166–168, and its companion paper, Haefele and Keating, "Around-the-World Atomic Clocks: Observed Relativistic Time Gains," *Science*, **177** (1972), 168–170. The predicted values were $40 \pm 23$ nanoseconds lost going eastward, and $275 \pm 21$ nanoseconds gained going westward. The rotation of the Earth and other effects also had to be taken into account.

called this misleading. "The legal arrangements under the Uniform Time Act [of 1966]," he wrote, "are under the administrative responsibility of the Department of Transportation (Office of the General Counsel). The standard of frequency and the unit of time are the responsibility of the NIST (as successor of the National Bureau of Standards (NBS)). The official Master Clock of the United States is the Master Clock of the USNO. The USNO also is the authority that determines Universal Time (of which the UTC is a real time realization to within 0.9 s)."[69] In testimony on the Green Bank interferometer in the 1990s, Congress also implicitly recognized the Naval Observatory as the location of the official Master Clock.

*Astronomical time (UTC) and its coordination with atomic time*

As we saw in section 11.2, astronomical time based on the Earth's rotation had been distinguished into three kinds of Universal Time (UT) during the 1950s: UT0, UT1, and UT2. UT2, which included both polar motion and the Earth's rotation, was the time broadcast, but without any international coordination.[70] Following the cooperative arrangement between the British and American Nautical Almanac Offices in 1958, it was also proposed that the two countries' time signals should be coordinated so as to be sent out at nearly the same instant. In an informal meeting at Essen's house with Markowitz, Donald Sadler (Superintendent of the British Nautical Almanac Office) and Humphrey Smith of the Royal Greenwich Observatory, it was decided that the Naval Observatory would send its PZT results to Humphrey Smith, who would take them into account along with the Greenwich PZT results, and decide whether or not the clocks should be changed on the basis of the combined American–British observations. A cablegram to the British Embassy allowed the United States and U. K. to act in concert in sending out the same time.[71] An expanded arrangement became formalized beginning January 1, 1960, when the Naval Observatory, NBS, Naval Research Laboratory, Royal Greenwich Observatory, and National Physical Laboratory began to coordinate their time signals. Beginning with this effort, standard time on the prime meridian was designated UTC (Coordinated Universal Time), and UTC was the time broadcast by time services, rather than UT2.[72]

Also at issue was how often to make changes in the times and rates of the

[69] "Memorandum of Law concerning the Authority of the Naval Observatory to Publicize Time," Department of the Navy, Office of the Judge Advocate General, May 4, 1961, USNOA, SF, "Statutory Authority, Time Service" folder. Winkler to Dr Leonard S. Cutler, March 24, 1994, USNOA SF, "NBS" folder. See also Winkler, OHI, 29–30 and 40 ff. At least as early as June 10, 1965 SecNav Instruction 4120.15, citing DoD Directive 5160.51 of February 1, 1965, ordered that "the epoch and the interval of time (frequency) as determined by the U. S. Naval Observatory shall be utilized as standards within the Department of the Navy." A series of DoD instructions since that time has expanded this directive to the entire Department of Defense.
[70] *Explanatory Supplement* . . . (ref. 8), p. 87.
[71] Markowitz, "Reminiscences," pp. 29–31. The time steps implemented by the USNO Master Clock from 1956 to 1971 are found in *Explanatory Supplement* . . . (ref. 8), pp. 86–87.
[72] D. H. Sadler, ed., *Trans. IAU*, XIA – Reports (held in Berkeley) (London and New York, 1962), 362–364; *Explanatory Supplement* . . . (ref. 8), pp. 85–86. UTC differed from UT2 by less than 0.1 seconds (Markowitz, "Time Measurement," in *Encyclopedia Britannica*, (1967), p. 1,161. Today, the International Bureau of Weights and Measures generates UTC from TAI after inserting leap seconds. Guinot (ref. 36), p. 182; see also *Explanatory Supplement* . . . (ref. 8), p. 40.

clocks. As late as 1953, Sollenberger changed the transmitting clocks at Annapolis immediately after every time sight. When Markowitz took over he had changes made "very infrequently," because he felt that the accidental errors of the daily observations did not justify such accuracy. At the same meeting in August, 1959, at which Markowitz, Essen, Humphrey Smith and Donald Sadler had developed a method to coordinate British and American time signals, they also decided that the changes made to clocks should be in steps of a tenth of a second (100 milliseconds) because "for the purposes of navigation it wasn't necessary to use a closer one."[73]

Also involved with the 100 millisecond change was a change in the offset frequency of the clocks, so as to steer them to the Earth's current rotation rate. The method of synchronizing the clocks to the Earth's rotation has varied. While some stepped or steered the clocks to adjust for the Earth's rotation, different institutions used different approaches to keeping the atomic clock synchronized to the Earth clock. By 1966 the matter was becoming more complicated. Ten or 15 countries were involved in coordination of time signals, and the matter of coordination was turned over to IAU Commission 31 on Time. One of the issues was how much difference could be tolerated between astronomical and atomic time. Some, representing navigators, argued for great accuracy, even though the sextant could not use that accuracy.

The controversy over when and how to account for the differences between atomic time and astronomical time came to a head during the late 1960s. Both Winkler and Essen proposed the idea of a "leap second" at a meeting of the Consultative Committee for the Definition of the Second (CCDS) in 1968. Finally, after numerous discussions between the navigators (who wanted no change) and the physicists (who wanted no offsets), and with the official approval of the International Astronomical Union and the Consultative Committee for International Radio (CCIR) of the International Telecommunications Union, in 1972 the "leap second" was implemented. According to this system, only an integral second would be added or subtracted, and only once or twice per year in June or December. Furthermore, the frequency would not be offset, but strictly that of TAI, so giving the official SI second continually. This compromise between the needs of celestial navigation and physicists lasted through the end of the century, when there were once again proposals for change.[74]

After 1971 all time transmissions were also coordinated in reference to the BIH. Since the introduction of the leap second in 1972, TAI has been a continuous time scale that differs from UTC by an integral number of seconds, increasing by about 1 second every 18 months. In short, UTC takes the Earth's rotation into account; TAI

---

[73] Markowitz, "Reminiscences," pp. 29–32; Winkler OHI, 18–19, describes these clock adjustments he observed while visiting the Naval Observatory in 1956.

[74] Markowitz, "Reminiscences," pp. 30–31; Winkler OHI, 51–53, Nelson *et al.* (ref. 66). On new proposals for the future of leap seconds see Dennis D. McCarthy, "The Definition of UTC," in Kenneth Johnston *et al.*, eds., *Towards Models and Constants for Sub-Microarcsecond Astrometry,* Proceedings of IAU Colloquium 180 (U. S. Naval Observatory: Washington, 2000).

Table 11.4. *Transmitted time*

| Date | Type |
| --- | --- |
| To 1956 | UT (uncorrected for Polar motion and Earth rotation) |
| 1956–60 | UT2 |
| 1960 | UTC within 0.1 seconds of UT2 |
| 1972 | UTC within 0.9 seconds of UT1 (leap-second system) |

does not.[75] Whereas prior to 1972 the broadcast UTC was within 0.1 seconds of UT2, after 1972 the leap-second system kept UTC within 0.9 seconds of UT1. These developments are summarized in Table 11.4.

In the midst of all the new definitions of time and the coordination among the varieties of time, the matter of determining astronomical time still remained essential. UTC, a clock time approximating UT1 and UT2, required knowledge both of polar motion and of the variations in the Earth's rotation, which together came to be known as earth-rotation parameters. Visual and photographic zenith tubes and astrolabes determined polar motion at a sustained level of accuracy of a few milliseconds, and continued to be the sole source of these parameters until the 1960s. In fact, on his arrival in 1966 as Director of the Time Service, Winkler began a large PZT program, which resulted in a 26-inch (65-cm) instrument known as PZT 7. The idea was to be able to observe fainter stars, and to have a wider field of view.[76] Beginning in the mid-1960s Doppler satellite tracking by the Navy Satellite Tracking System was used to determine polar motion at an accuracy equal to or better than that of the optical techniques. Data obtained using optical techniques and Doppler data were combined by the BIH for a number of years. With the introduction of satellite laser ranging (SLR) in the late 1970s, the use of Doppler data for this purpose was phased out (Figure 11.24).

Beginning in the late 1970s radio telescopes employing the technique known as connected-element interferometry (CEI) came into use for measuring the Earth's rotation. As we have seen in chapter 10 in the context of astrometry, in conjunction with Kenneth Johnston at the Naval Research Laboratory, at the Naval Observatory Scientific Director Gart Westerhout was decisive in implementing this technique, which combined the simultaneous observations of several nearby "connected" radio telescopes. Westerhout, who had come in 1976, was a radio astronomer, in contrast

---

[75] *Explanatory Supplement . . .* (ref. 8), pp. 6 and 724. A table of differences between UTC and TAI is to be found in the *Astronomical Almanac*, Section K9.

[76] The 65-cm PZT is described in W. J. Klepczynski, "The 65-cm PZT at the U. S. Naval Observatory," in IAU Colloquium No. 26, *Reference Coordinate Systems for Earth Dynamics*, B. Kolaczek and G. Weiffenbach, eds. (The Union: Warsaw), pp. 309–313. Modern and historical techniques for determining the Earth's rotation are described in *Explanatory Supplement . . .* (ref. 8), pp. 251–265. On PZT 7 see Winkler, OHI, 39 and 48–50. PZT 7 never officially made a contribution to Universal Time, and was superseded by other techniques.

## CONTRIBUTION OF THE TECHNIQUES TO THE IERS COMBINED SOLUTIONS

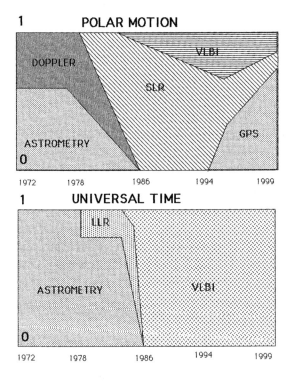

Figure 11.24. The evolution of polar-motion and Earth-rotation techniques, as reflected in data used by the International Earth Rotation Service (IERS). From Ivan I. Mueller, "The First Decade of the IERS," in Dick, McCarthy and Luzum, *Polar Motion* (ref. 15), 201–212. SLR is satellite laser ranging, LLR is lunar laser ranging, VLBI is very-long-baseline interferometry, and GPS is Global Positioning System.

to his predecessor, Kaj Strand, who had experience only in optical astronomy. From 1978 to 1987 the Naval Observatory was the sole user of the connected-element interferometer at Green Bank for earth-rotation parameters, resulting in polar-motion accuracies of 0.01 arcseconds and data for the Earth's rotation accurate to within 1 millisecond. At that time very-long-baseline interferometry (VLBI), using radio telescopes scattered around the world rather than physically connected, superseded CEI with a tenfold improvement in accuracy, to one milliarcsecond for polar motion and a tenth of a millisecond for the Earth's rotation (UT0 minus UTC). During the 1990s GPS made an increasingly significant contribution to polar-motion observations. Just as in astrometric work, in which the use of transit instruments, astrometric satellites, and optical interferometers overlapped for a period of time, so in the area of the Earth's orientation and polar motion did the technologies of PZT and Doppler, and then SLR, VLBI, and GPS overlap (Figure 11.24), and with varying degrees of accuracy (Table 11.5).[77] The resulting knowledge of polar motion was important not only for

[77] D. M. Matsakis *et al.*, "The Green Bank Interferometer as a Tool for the Measurement of Earth Orientation Parameters," AJ, **91** (1986), 1,463–1,473. On Westerhout's role implementing radio-astronomy techniques at the Naval Observatory, see Westerhout OHI, 48–60, and Winkler OHI, 59–60; on the relative roles of all methods worldwide over time, see I. Mueller, in Dick *et al.* (ref. 15 above).

Table 11.5. *Accuracy of techniques for measuring Earth's rotation*

| Technique | Polar motion (arcseconds) | Universal time (milliseconds) | Averaging time | Measurement conditions |
|---|---|---|---|---|
| Astrolabe | 0.06 | 4 | 2 hours | One instrument |
| PZT | 0.04 | 4 | 2 hours | One instrument |
| Optical method | 0.01 | 0.8 | 5 days | Network of 80 instruments |
| Doppler | 0.01 | | 2 days | One satellite, 20 stations |
| CEI | 0.01 | 1 | 3 days | One baseline |
| VLBI | 0.001 | 0.02 | 24 hours | Network of 5–6 stations |
| | | 0.1 | 1 hour | Two stations |
| Satellite laser ranging (LAGEOS) | 0.001 | 0.1 | 3 days | Network, 20 stations |
| Lunar laser ranging | | 0.1 | 1 day | One instrument |

*Notes:*
PZT, photographic zenith tube; CEI, connected-element interferometry; and VLBI, very-long-baseline interferometry.
*Source:* Adapted from P. K. Seidelmann, ed., *Explanatory Supplement to the Astronomical Almanac* (1992), p. 62, with permission.

accurate time, but also for a variety of geophysical problems including continental drift.[78]

Just as with international atomic timekeeping, the coordination of these observational techniques required an extraordinary level of international cooperation, and it is not surprising that, with changing observational techniques, data analysis and combination, the administration of earth-rotation parameters evolved. The main part of an international program known as MERIT (Monitor Earth Rotation and Intercompare Techniques of Observation and Analysis) began in 1983, with the cooperation of many institutions, including the Naval Observatory. The results of this campaign led to the need for a new institution, and, on January 1, 1988, the International Earth Rotation Service (IERS) began operation. It replaced the BIH and the International Polar Motion Service (IPMS) as the organization for determining and predicting the orientation of the Earth.[79] The Central Bureau of the IERS was located at the Paris Observatory until January, 2001, when it was transferred to the Bundesamt für Kartographie und Geodäsie in Frankfurt am Main, Germany. The IERS has a complex international structure with a number of product centers, including the Naval Observatory, which serves as the subbureau for rapid prediction. Although it performs many functions, the IERS is best known as the institution that determines when the leap second will be added.

The other type of astronomical time, Ephemeris Time, has had a checkered

[78] W. Markowitz, "Latitude and Longitude and the Secular Motion of the Pole," in S. K. Runcorn, ed., *Methods and Techniques in Geophysics* (London, 1960); Dennis McCarthy, "Polar Motion and Earth Rotation," *Review of Geophysics and Space Physics,* **17** (1979), 1,397–1,403; Markowitz, "Redeterminations of Latitude and Longitude," *Trans. AGU,* **26** (October, 1945), 197; Markowitz OHI, 35.
[79] G. A. Wilkins, "Project MERIT and the Formation of the International Earth Rotation Service," in Dick *et al.* (ref. 15 above), pp. 187–200, and Ivan I. Mueller, "The First Decade of the IERS," ibid., pp. 201–213. The details of the IERS structure are given at http://www.iers.org/iers/.

history. There were, in fact, problems with ET as a uniform time. The tropical year on which ET was based depended not only on Newcomb's theory of the Sun, but also on his astronomical constants, whose determinations were now being improved. Moreover, since the Sun could not be observed accurately, Markowitz had used the Moon. Furthermore, ET did not take into account relativistic effects. Eventually, improvements in the lunar theory resulted in different varieties of ET, just as refined measurements introduced new varieties of UT.[80] With the introduction of a new system of astronomical constants in 1984, Ephemeris Time was replaced by Dynamical Time (TDT and TDB), which resolved many of these issues, especially the failure to take relativity into consideration. "Delta T" was redefined as the difference between UT and Dynamical Time, specifically the difference between UT1 and Terrestrial Dynamical Time.[81] At the IAU General Assembly in Grenoble, France in 1976, recommendations for the definition of "dynamical time" were given. It was still based on orbital motion, representing "the independent variable of the equations of motion of the bodies of the solar system," just as Clemence had stated, but now taking relativity into account and a variety of reference frames, including geocentric and barycentric. In 1991 the time scales (TCB, TCG, and TT) based on dynamical time were defined. The relationship between TDT and TAI provides continuity with Ephemeris Time. From an observational point of view, once again the rapid motion of the Moon was important in extrapolating prior to 1955 when atomic time became available. Table 11.2 shows how the varieties of time had expanded by the 1990s.[82]

### Time for the world

During the second half of the twentieth century, new methods combined with the old to make time available worldwide with unprecedented accuracy. Radio time signals continued to proliferate, but in new forms. By the 1960s impressive transmitters such as the Arlington towers had given way to time transfer by ground-based radio navigation systems, and by the 1980s space-based radio-navigation systems were taking over as the primary means for time transfer. In addition the era of portable atomic clocks allowed time to be literally carried from one place to another, with high precision. All of these methods constituted a revolution in the dissemination of time, no less important than the revolution in the timekeepers themselves. The Naval Observatory played roles of varying significance in each of them.[83]

[80] Explanatory Supplement . . . (ref. 8), p. 82; G. M. R. Winkler and T. C. van Flandern, "Ephemeris Time, Relativity and the Problem of Uniform Time in Astronomy," AJ, **82** (1977), 33–97.

[81] On considerations leading to Dynamical Time see G. M. R. Winkler and T. C. Van Flandern, "Ephemeris Time, Relativity, and the Problem of Uniform Time in Astronomy," AJ, **82** (1977), 84–92, and S. Aoki et al., "The New Definition of Universal Time," Astronomy and Astrophysics, **105** (1982), 359–361. For a review of the varieties of Astronomical Time developed as of 1991 see Dennis D. McCarthy, "Astronomical Time," Proc. IEEE, **79** (1991), 915–920.

[82] Explanatory Supplement . . . (ref. 8), pp. 41–48 and 63.

[83] Reviews are given in W. J. Klepczynski, "Time Transfer Techniques: Historical Overview, Current Practices and Future Capabilities," Proc. 16th Annual Precise Time and Time Interval (PTTI) Meeting, Washington (Washington, 1984), pp. 385–402; P. K. Seidelmann, B. Guinot, and L. E. Doggett, "Methods of Time Transfer" section in "Time" chapter of Explanatory Supplement . . . (ref. 8), pp. 64– 69.

Figure 11.25. A portable clock awaits loading onto a plane for a "clock trip", June, 1964.

As new methods of time transmission were developed, the Master Clock itself was improved; by the 1990s it was based on a system of some 60 independently operating cesium-beam atomic clocks and ten hydrogen masers. The hydrogen masers are extremely stable clocks over short periods of less than a week; thus the combination of cesium beam and hydrogen maser clocks is analogous to the combined use of Riefler and Shortt clocks earlier in the century. To ensure their stability, these clocks were distributed over 20 environmentally controlled clock vaults, spread over the Naval Observatory grounds in Washington. These clocks were automatically intercompared every hour to compute a reliable and stable time scale with a day-to-day resolution of 100 picoseconds.

For a variety of reasons, time itself could not be disseminated to this accuracy. Perhaps the simplest concept for transferring accurate time, though only to very specialized customers, was use of the portable atomic clock, which did not involve the propagation delays and other uncertainties that had to be taken into account with radio signals. F. Reder and Winkler first experimented with Atomichrons as portable atomic clocks in 1959 and 1960. Although these 7-foot-high, several-hundred-pound instruments strained the concept of "portable," they could be transported while still operating. By the early 1960s the cesium-beam atomic clocks had so reduced in size that they could be handled by two persons.[84] In 1965 the USNO established a portable clock service, whereby atomic clocks transported in commercial airplanes could synchronize time to 1 microsecond. The clock often occupied a first-class seat aboard the aircraft (Figure 11.25), accompanied by two Naval Observatory employees. Although it was still characterized in the early 1990s as "the most reliable and accurate method" for the comparison of time scales, the last portable clock trip was taken in 1992, the method having been superseded by GPS.[85]

---

[84] F. H. Reder and G. M. R. Winkler, "Preliminary Flight Tests of an Atomic Clock in Preparation of Long-Range Clock Synchronization Experiments," *Nature*, **186** (1960), 592–593; Reder and Winkler, "World Wide Clock Synchronization," *IRE Trans. Military Electronics*, **4**, no. 2/3 (1960), 366–376; A. S. Bagley and L. S. Cutler, "A New Performance of the Flying Clock Experiment," *Hewlett Packard Journal*, **15** (1964), 11.

[85] *Explanatory Supplement . . .* (ref. 8), pp. 65–66. The last portable-clock trip at the Naval Observatory was undertaken on October 30, 1992 by George Luther; G. Luther, private communication.

Of use to a far greater number of users were radio signals, which were now being sent out in many forms. Some Arlington-type ground stations did remain. Beginning in 1956, time signals from naval radio stations at Annapolis, San Francisco and Balboa (Canal Zone) were broadcast from clocks placed at the transmitters, thus eliminating variations in land-line lag. In 1959 station NBA in the Canal Zone began continuous transmission of very-low-frequency (VLF) time signals (18 kc s$^{-1}$), but, due to propagation errors and errors in time-of-arrival measurement at the Naval Observatory, it was some time before they improved in accuracy over the high-frequency (HF) signals. By 1964 the Naval Observatory controlled five Navy VLF stations, with a time-transfer accuracy of half a millisecond, while time synchronization could be achieved with HF radio signals to an accuracy of 1 millisecond at a distance of 10,000 km. It is important to note that, unlike the earlier days, the Naval Observatory did not itself send out these radio signals, but monitored time systems such as VLF, obtained a correction to the time of emission, and published a correction for the users of that particular signal. By contrast, the National Bureau of Standards had gradually increased its role in this area; as we have already mentioned, in 1923 it had begun to send out frequency signals from WWV (then in Beltsville, Maryland) to enable radio stations to hold their assigned frequencies. Starting in fall, 1945, it sent shortwave time signals from its WWV station at Ft Collins, Colorado; beginning in 1956 it sent out a low-frequency time signal from WWVB, and later operated WWVH from Hawaii.[86]

However, it was the navigational systems that were the wave of the future for radio time transfer. LORAN (Long-Range Navigation) was such a system, which was developed during World War II and coordinated by the U. S. Coast Guard; an improved LORAN-C came into use in 1958. It consists of chains of radio transmitters, each with a master transmitter and two to four secondary transmitters. Each station has a cesium standard for deriving time signals, and each chain is overseen by a monitor station that measures and controls the emission delays. In 1961 the Naval Observatory began to control the timing pulses from the master station of the East Coast LORAN-C chain at Cape Fear, N. C. By 1969 the LORAN-C navigation system had become the routine time-transfer standard for time and frequency coordination, synchronized to UTC (USNO). By the early 1990s the LORAN-C system had expanded to 16 master transmitter stations in the Northern Hemisphere. Although their main purpose was for navigation, they also served the timing community, and during this time the synchronization with the Naval Observatory Master Clock was achieved with increasingly

---

[86] AR (1956), 347; Time Service Notice No. 5; Markowitz, "Time Determination and Distribution: Current Developments and Problems," manuscript, USNOA, BF, Markowitz folder (1964), p. 9. On the history of the involvement of the National Bureau of Standards with time, the use of different frequencies for time dissemination, and numerous other time-related issues, see James Jesperson and Jane Fitz-Randolph, *From Sundials to Atomic Clocks* (second edition, National Bureau of Standards: Washington, 1999), pp. 117–137. For a brief summary see http://www.boulder.nist.gov/timefreq/museum/847history.htm.

sophisticated means. These techniques brought time transfer into the microsecond regime.[87]

Advances early in the Space Age brought the possibility of disseminating time to accuracies even beyond the microsecond regime. Satellite time-transfer experiments had already begun in 1962, when the Naval Observatory used the Telstar communications satellite to synchronize clocks in the United States and U. K. to one microsecond. By 1965 the Observatory had used the Relay 2 satellite to synchronize time between the U. S. and Japan to 0.1 microseconds. The Navy Transit navigational satellites, developed by the Applied Physics Laboratory of the Johns Hopkins University, carried clocks that could synchronize separate ground stations to about 30 microseconds. Regular use of what came to be known as "two-way satellite time transfer" (TWSTT), used largely for time transfer between laboratories and military and civilian remote users, came only in the 1990s with the use of communications satellites and low-cost terminals. In its simplest form, two time standards send each other a signal over a communications circuit. Because the transmission delays cancel out in both directions, the offsets between the two time standards can be measured very precisely; by the 1990s the precision was down to 0.2 nanoseconds and the accuracy (in an absolute sense) to one nanosecond.[88] Methods of time transfer and their improvement in accuracy over the 20 years from 1964 to 1984 are summarized in Table 11.6.

In the 1980s and 1990s the Global Positioning System became the primary method for time transfer. Each GPS satellite has four atomic clocks, which the Naval Observatory monitors; the Air Force then uses the data to model the clocks. By the end of the century GPS consistently provided time transfer to ten nanoseconds (ten billionths of a second), and one nanosecond with carrier phase, the most accurate of a whole range of time-dissemination methods. Beginning about 1985 the BIPM used common-view GPS to transfer time around the world, and in 1999 began phasing in TWSTT for laboratory synchronization.[89] For those lacking GPS receivers, during the 1990s the Naval Observatory provided time via the internet, but with a million times less accuracy. Thanks to the Network Time Protocol developed at the University of Toronto and the Defense Advanced Research Projects Agency (DARPA), 71 million online computers at the end of the century could receive time accurate to one millisecond on a local-area network, or a few tens of milliseconds on wide-area networks.

[87] Markowitz, "Time Measurement Techniques in the Microsecond Region," *The Engineer's Digest*, no. 135 (July–August, 1962), 9–18; *Explanatory Supplement . . .* (ref. 8), pp. 66–67; Laura G. Charron and Carl F. Lukac, "Evolution of LORAN-C Timing Techniques," *Proceedings of the Twelfth Annual Wild Goose Association Technical Symposium*, October, 1983; Laura G. Charron, "U. S. Naval Observatory and LORAN," *Naval Oceanography Command News* (September, 1992), 25–26. The Omega navigation system was also used.

[88] The early experiments are described in Markowitz (ref. 86), p. 11; Markowitz, *Encyclopedia Britannica* (1967), p. 1,163. Two-way time transfer is described in James A. DeYoung, "Two-Way Time Transfer Using Earth Orbiting Geostationary Communications Satellites," *Naval Meteorology and Oceanography Command News* (June, 1994), 26–27 and 30.

[89] Gernot M. R. Winkler, "Changes at USNO in Global Timekeeping," *Proc. IEEE*, **74** (1986), 151–155.

Table 11.6. *Accuracy of time-transfer methods,*
*1964–84*

| | Accuracy | | |
|---|---|---|---|
| | 1964 | 1974 | 1984 |
| Radio techniques | | | |
| VLF | 500 μs | 10 μs | 10 μs |
| LF | 50 μs | 40 μs | 20 μs |
| HF | 1 ms | 0.2 ms | 0.2 ms |
| TV | – | 100 ns | 10 ns |
| Navigation systems | | | |
| LORAN-C | 5–10 μs | 0.5–5 μs | 40–700 ns |
| OMEGA | – | 5 μs | 5 μs |
| TRANSIT | 500 μs | 25 μs | 3–20 μs |
| GPS | – | 100 ns | 5–40 ns |
| Satellite systems | | | |
| GOES | – | 50 μs | 30 μs |
| DSCS | – | 100 ns | 50 ns |
| Commercial | 1 μs | 10 ns | 200–600 ps |
| Special systems | | | |
| Portable clocks | 1–5 μs | 30–1000 ns | 5–500 ns |
| VLBI | – | 150 ps | 60 ps |
| Laser ranging | 0 | – | 200–600 ps |

*Notes:*
VLF, very low frequency; LF, low frequency; HF, high frequency;
GPS, Global Positioning System; DSCS, Defense Satellite
Communication System; VLBI, very-long-baseline
interferometry; ms, milliseconds; μs, microseconds; ns,
nanoseconds; and ps, picoseconds.
*Source:* Adapted from Klepczynski (1984) (ref. 83).

GPS, and thus the Naval Observatory, is the source of time for most of the 91 "stratum
1" NTP servers around the globe.[90] In 2000 about 18 billion time requests were
received, representing several million direct users. During the last decades of the
century, time became a large and growing concern. Improvements in timekeeping and
time dissemination, as well as the increasingly sophisticated uses of time, were a
major focus of the Annual Precise Time and Time Interval (PTTI) meetings, begun by
the Naval Observatory in 1969. The proceedings of these meetings provide one
window on how important time had become for the world during the last quarter of
the twentieth century.[91]

[90] Richard Schmidt, "GPS Times the Internet," GPS World (February, 2000), 42–47.
[91] For a review of the PTTI meetings see Gernot M. R. Winkler, "Twenty-Five Years of PTTI," in 25th
Annual Precise Time and Time Interval (PTTI) Applications and Planning Meeting, NASA Conference
Publication 3,267 (Goddard Spaceflight Center: Greenbelt, Maryland, 1994), pp. 1–5. Another
important series of meetings was the annual Frequency Control Symposium, which
concentrated on clock engineering rather than users.

By the end of the century a user of time could obtain it in a variety of ways depending on the needs. If no more than a few seconds' accuracy was required, radio or television broadcasts gave time good enough, and this is undoubtedly where most of the public obtained its time. If one-second accuracy was required, the voice announcer on the Naval Observatory Master Clock sufficed, and the WWV shortwave signal from the NBS (now the National Institute of Standards and Technology) could give time to a few milliseconds. At the other extreme of accuracy, a $100 GPS receiver gave any user time to ten billionths of a second.[92]

What had begun in 1845 as a time ball dropping at noon, accurate to about a second and visible to the few who could see the falling ball atop the Naval Observatory dome, ended 150 years later with an Earth-circling system of satellites, whose time was provided by the Naval Observatory and could reach the user with an accuracy of ten billionths of a second anywhere in the world. Moreover, the Naval Observatory was by the end of the century by far the largest single contributor to the official world time, kept in Sèvres. Nor was any of this an academic exercise of scientists tinkering in the laboratory. Geodesy, space tracking, communications systems, and various forms of metrology all required precise time at various levels. However, as we shall see in chapter 12, although time had many uses to many users, its most demanding requirement was for navigation.

---

[92] For a review of the uses of time, see Winkler (ref. 2), and Winkler OHI, 70–79.

# 12   Navigation: From stars to satellites

Fortunately for the navigator, astronomers have been deeply interested in [the navigator's] questions on their own account, and the results of centuries of observations and of an enormous mass of mathematical analysis and numerical calculations are boiled down into the Nautical Almanacs.

Henry Norris Russell, 1943[1]

A chart of the solar system is as necessary for interplanetary navigation as one of the surface of the earth is for earthbound navigation. [The chart] consists of a timetable or *ephemeris* . . . The construction of such tables has been one of the principal duties of the Naval Observatory for more than a century.

Gerald Clemence, 1960[2]

It is not much of an exaggeration to say that today navigation is virtually synonymous with GPS [the Global Positioning System]. This is a development of the present decade, which has seen the completion of the GPS satellite constellation, the shutdown of other electronic means of navigation, and a drastic reduction in the prices of GPS receivers. For Department of Defense vehicles, GPS is the principal means of navigation.

George H. Kaplan, 1999[3]

We end our history as we began – with navigation, the practical application that has been the province of the U. S. Naval Observatory over 170 years. During that time navigation has advanced dramatically, and the role of the Naval Observatory in that capacity has changed accordingly. The rating and care of chronometers and other instruments, the original purpose of the Depot of Charts and Instruments, remained a function of the Observatory until 1951, when the Nautical Instruments Division was disbanded, and the chronometer function transferred to the Norfolk Navy Yard. The Observatory's role in navigation, however, remained rooted in the Nautical Almanac Office, whose primary function was still the production of the nautical and astronom-

---

[1] Henry Norris Russell, "In Flight, Troubles Begin," *Scientific American* (April, 1943), 164–165. Russell, the dean of American astronomers at this time, taught navigation during the war and in this article described the problems of aerial navigation, which he had also worked on during World War I. For the latter context see David H. DeVorkin, *Henry Norris Russell, Dean of American Astronomers* (Princeton University Press: Princeton, 2000), pp. 157–161.

[2] Gerald Clemence, "Interplanetary Navigation," *Proc. IRE*, **48** (April, 1960), 497–499.

[3] George H. Kaplan, "New Technology for Celestial Navigation," in NAOSS, 239–254.

ical almanacs. During most of the twentieth century that office remained intact, until emerging needs late in the century reduced its scope to a small office of the new Astronomical Applications Department. Toward the end of the century, celestial navigation took an increasingly secondary role as navigation evolved, paralleling the evolution of time dissemination from celestial to radio to satellite technologies. The Global Positioning System of Earth-encircling satellites completely transformed the landscape of navigation.

During the century the balance between the Almanac Office's dual functions of almanac production and research in celestial mechanics changed with changing times and new Directors. Though the calculations were routine, research stemming from navigational considerations included determination of the astronomical constants, improvements to the theories of planetary motion, refined empirically determined masses of the planets and their satellites, all of which are among the most important basic data to astronomy. Moreover, this research led Observatory staff to more specific investigations, including possible variations in the gravitational constant, the question of the shrinking Sun, and the controversial search for a tenth planet in the solar system. The history of the Nautical Almanac Office during the twentieth century is therefore also an important part of the history of astronomy, especially celestial mechanics, ephemerides, and related fields. However, there can be no doubt that, no matter what the more purely scientific research, Navy patronage depended on practical results.

## 12.1  Chronometers and nautical instruments

Although the modern Naval Observatory is not usually associated with navigational instrumentation, for a large part of its history (until 1951) this was one of its integral functions. The *Nautical Almanac*, after all, was only one element needed for the process of celestial navigation; it was useless without the chronometer and sextant, and these two crucial pieces of technology, along with navigational instruments of increasing variety and complexity, were still very much subject to naval scrutiny and quality control.

We have seen in section 5.3 how chronometer rating and repair were still important at the Observatory at the end of the nineteenth century. With the outbreak of the Spanish–American War, for example, the demand for chronometers for the "mosquito fleet" quickly exhausted the supply. Although 26 repaired chronometers and 13 new ones were on hand at the beginning of 1898, some 25 more had to be ordered from London.[4] In the first half of the twentieth century the repair and rating of chronometers remained important functions that were largely routine except on occasion. In 1902 E. E. Hayden reported that there were 691 box chronometers in use in the U. S. Navy, of which 134 were "hacks" of low accuracy.[5] As we have seen in

---

[4] AR (1898), 16.
[5] AR (1904), 48–49. Hayden also discussed the possible use of torpedo-boat watches to replace box chronometers.

Figure 12.1. Chronometer repair was a major function of the Observatory until 1951. Here an instrument maker identified only as "Whitten" is working on Negus 630, used on the U. S. S. *Jeanette*, which had been lost in the Arctic on an expedition to reach the North Pole.

chapter 9, one of the great administrative challenges at the Observatory during World War I and World War II was the scarcity of chronometers and the search for a method of mass production. Along with the increase in usage during times of war came an increase in staff for chronometer repair. This meticulous work (Figure 12.1) would eventually become almost a lost art, first as disposable quartz chronometers replaced the classical box chronometer, and then as satellite navigation inevitably decreased the number of chronometers in general use.[6]

However, the instrument duties at the Naval Observatory involved much more than chronometers. The Observatory, which had its origins in the Depot of Charts and Instruments in 1830, had become deeply involved in a variety of navigational instrumentation by the first half of the twentieth century. A century after the Depot's founding, as the Observatory was contemplating its modernization program in 1929, Superintendent Freeman summarized its work in nautical instruments: "The observatory provides for every gyro and magnetic compass, every sextant and chronometer, and every nautical instrument employed in the navigation of the ships of the Navy. A certain representative portion of this immense stock of material is handled directly at the observatory in order that there may be available first-hand information on the quality and capabilities of various instruments, especially those of the highest precision. For instance, last year more than 60,000 instruments and parts were received at the observatory and about 40,000 instruments and parts were shipped. In its particular line of work the observatory repair plant is one of the most effective in the country. In it is done work on precision instruments for other Government activities, including the Bureau of Standards and the Bellevue Research Laboratory."[7] Contemplating

---

[6] Despite radio and satellite navigation, the Navy still requires chronometers on board its ships in case of failure of electronic navigation. The current chronometer is a battery-powered quartz-crystal movement in a water- and impact-resistant case. The normal number of chronometers on each ship is three, although some small ships have only one. At the end of the century celestial navigation remained an active issue as part of the Navy/Marine Corps Positioning, Navigation and Timing Policy.

[7] Hearing before Subcommittee of House Committee on Appropriations, consisting of Messrs. Burton L. French (Chairman), Guy U. Hardy, John Taber, William A. Ayres and William B. Oliver, Navy Department Appropriation Bill for 1930" (Government Printing Office: Washington, 1929), p. 813, USNOA, AF.

Figure 12.2. Naval Observatory Instrument Shop personnel, 1951, just after the Navy's chronometer function had been moved to Norfolk. Among those identified in the photo are the head instrument maker, George Steinacker (center back row with mustache), Sam Bellintende and Al Gates (back row left), and Joe Egan (far right). In the front row are Grayson Bell (second right) and Carl Engle (first right).

the amount of labor implied by that statement, one is impressed with the massive instrumentation effort undertaken at the Naval Observatory at quite a different level than the astronomical observations with which it is usually associated.

Nautical-instrument work reached its peak during World War II, when Hamilton Watch Company mass-produced chronometers, a project extremely important to the war effort. As with World War I, another important part of the war effort was the collection, repair, and distribution of binoculars, which were gathered from the public. One measure of how the instrument effort had increased "beyond all reasonable expectation" is the fact that, whereas the value of the navigational instrument stock on hand at the Observatory was $424,344 in 1930, it was almost $24 million in 1945. Furthermore, whereas the instrument shop employed 34 people in 1930, in 1945 it employed 294.[8]

Important though this work was, the Observatory was open to the criticism, especially from astronomers, that navigational instruments should be no part of the work of an astronomical observatory. This was undoubtedly part of the reason that the chronometer function was transferred in 1951 to the Norfolk Navy Yard, leaving only a small group of instrument makers who repaired and constructed the Observatory's telescopes and other astronomical instrumentation, rather than navigational instruments (Figure 12.2).[9] The departure of this function left the Observatory free to concentrate on its more astronomical duties. For navigation, these duties were largely invested in the Nautical Almanac Office, which was approaching its half-century mark as the Observatory moved to its new site.

## 12.2 The Nautical Almanac Office, 1893–1958

The move of the Nautical Almanac Office to Observatory Circle in 1893 saw Simon Newcomb, still four years from retirement, with much unfinished business.

---

[8] AR (1945), 2–3.
[9] For more on the instrumentation aspect at the Naval Observatory during World War II, see Marvin Whitney, OHI. On the beginnings of the machine shop, see AR (1903). See also History of the Material Department folder, USNOA, AF.

His astronomical constants had not yet been adopted internationally, his research on the motion of the Moon remained unfinished, and other loose ends dangled at the end of a busy career. Not surprisingly, even after his retirement, Newcomb's legacy dominated the Nautical Almanac Office, especially until his death in 1909. Beginning in 1910 two figures directed the Office prior to World War II, William S. Eichelberger and A. James Robertson. Though their contributions were very different (Eichelberger's scientific and Robertson's political), their tenure saw no radical changes in the Office or its work. By contrast, World War II set in motion large and irrevocable changes both in production and in research. Prior to the Space Age Wallace J. Eckert and Gerald Clemence oversaw these changes, which were driven by advances in automation and the beginnings of the computer revolution.

### Early evolution of the Office at Observatory Circle

The Nautical Almanac Office, according to conventional wisdom, became a part of the Naval Observatory when the former moved from the northwest corner of 19th and Pennsylvania Avenue to Observatory Circle in 1893. Both politics and personalities, however, made the actual case far from straightforward. The Office did indeed move to Observatory Circle on October 20 of that year, but only a year later, on September 20, 1894, did the Secretary of the Navy issue a regulation making the Nautical Almanac Office a "branch" of the Naval Observatory. Even then the Office was absorbed into the Observatory only over a period of years. According to Superintendent C. H. Davis II at the turn of the century, "In 1894 the Nautical Almanac Office, on account of the crowded state of the Navy Department building, was accommodated at the new Observatory, which was first occupied in 1893; but the Almanac has remained a distinct organization, having its own director and independent appropriations. It has never been merged with the Observatory and should not be. This point should be distinctly noted."[10]

One needs to remember here that the son of the first Superintendent of the Nautical Almanac Office, as well as the Superintendent of the Naval Observatory, is speaking. Indeed, one finds the Nautical Almanac Office incorporated into the Observatory's *Annual Report* beginning September, 1894, with Newcomb's title transformed from Superintendent to Director of the Nautical Almanac Office. However,

---

[10] [C. H. Davis II], "Memorandum," USNOA, AF, Davis to Colby Chester folder, p. 2. From internal evidence the document is certainly by C. H. Davis II, and the date around November 1, 1902, when Chester assumed the Superintendency. On the location of the Almanac Office prior to its removal to Observatory Circle see Weber, 27. On its move to Washington in 1866 it shared quarters with the Hydrographic Office, in the building known as the "Old Octagon House," on the northeast corner of New York Avenue and 18th Street, NW, near the Navy Department. In 1877 Newcomb secured rented quarters at the Corcoran building. The office moved to the east wing of the new State, War and Navy Building in 1883 (where the Hydrographic Office had moved in 1879). It moved to the Navy Yard in December, 1889, and to rented quarters at 1901 Pennsylvania Avenue in 1890 (although Newcomb remained in State, War and Navy, room 566; see ref. 25 below). Weber gives the date of the Almanac Office move to Observatory Circle as October 20, 1893 (p. 27). See also C. B. Watts, "C. B. Watts Recording his Recollections of the Naval Observatory," USNOA, BF, Watts folder.

ambiguity remained as to whether the Office was a Department of the Observatory. One can well imagine that Newcomb, who did not retire until 1897, chafed at becoming a part of the Naval Observatory, which he had enthusiastically departed 20 years earlier. At issue was not only the natural sentiment that the Superintendent of an independent institution did not wish to become subsumed under another institution. There was also the personal matter that the Astronomical Director at the Naval Observatory was William Harkness, long ago Newcomb's best man at his wedding, but now a bitter enemy thanks to the transit of Venus and other controversies. It is thus no surprise that, as Astronomical Director, Harkness was in charge of the "Department of Astronomical Observations," as distinct from Newcomb's "Department of the Nautical Almanac."

Even after the Nautical Almanac had been incorporated into the Observatory's *Annual Reports* as a separate Department, and though Harkness doubled as Director of the Nautical Almanac from 1897 to 1899, the status of the Office with regard to the Observatory's organizational structure remained in question. In a Navy Department decision rendered January 19, 1905, the Nautical Almanac Office was held not to be a separate shore station, and this ruling seems to have settled the matter. Writing in 1928, Superintendent C. S. Freeman stated that "In 1904, the Nautical Almanac Office, which for 10 years had been located in the observatory grounds under general observatory supervision, was definitely incorporated as an integral part of the observatory organization and has functioned as a department of the organization ever since." In recognition of this fact Freeman, with approval of the Astronomical Council and the Bureau of Navigation, gave the Almanac Office the administrative designation "Ephemeris Department."[11]

After Newcomb's retirement in 1897, the position of Director was held by a succession of four Professors of Mathematics in four years (see Appendix 1), including Harkness and the soon-to-be-infamous S. J. Brown; these Professors held the post in addition to their regular duties at the Observatory. By law dating back to 1857 the post had to be filled by a Professor of Mathematics or other naval officer. A full-time Director, Professor Walter S. Harshman, was appointed in 1901, followed by Professor Milton Updegraff in 1907. During these transition years, little in the way of new research was undertaken; though Harshman had Masters and Ph. D. degrees from George Washington University, much of the activity was based on inertia remaining from Newcomb's whirlwind tenure. Newcomb himself still supervised the completion of the new tables of the planets Mars, Uranus, and Neptune, and continued with his research on the motion of the Moon until his death in 1909. Indeed, until 1905, he remained (with the exception of G. W. Hill) virtually the sole contributor to the *Astronomical Papers Prepared for the Use of the American Ephemeris*, which he had founded in 1882. Harshman "systematized and regulated the office and brought the work up to

---

[11] AR (1928), 22; Weber, 28.

date," Naval Observatory Superintendent C. H. Davis II noted. "The office is now in a high state of efficiency, and general harmony and enthusiasm prevail in its staff," the son of the Almanac Office founder proudly wrote in 1901.[12]

We have seen in chapter 8 the importance of the 1896 Paris Conference, where Newcomb's astronomical constants were adopted, with much subsequent controversy, especially in America. Ironically, Harkness was left with the task of incorporating Newcomb's constants in the Ephemeris for 1900. He was also left with the ensuing controversy; the new constants, he wrote, "met so much opposition among prominent American astronomers that it has been thought best to give in the Ephemeris for 1901 sufficient data to enable either the constants of Struve and Peters or those of the Paris conference to be used with equal facility, and thus each astronomer is left free to choose for himself which he will employ." This was hardly in the spirit of the intended standardization, and eventually Newcomb's constants won out; beginning with the volume for 1912, only Newcomb's constants were used in the body of the book.[13]

Before his retirement, and in the midst of his other work as Astronomical Director, Harkness did manage to make progress on the determination of new elements for the satellites of the outer planets. This work, as well as work on a new zodiacal catalog, was continued through Harshman's tenure. The *Catalogue of Zodiacal Stars* (1905), carried out by Henry B. Hedrick, was particularly important because it provided the reference stars against which the motions of the Moon and planets could be measured. This also explains why it was a product of the Nautical Almanac Office rather than of the Astronomical Department of the Observatory; it was a compilation of other catalogs rather than based on new observations. An update of Newcomb's *Catalogue of 1098 Standard and Zodiacal Stars* (1882), it was the only one of its kind to be adopted by all national almanacs, and served as the standard for 35 years until it was updated by the Nautical Almanac Office in 1940. It also holds the distinction of being the first research in many years published in the *Astronomical Papers Prepared for the Use of the American Ephemeris* not authored by Newcomb or G. W. Hill.[14]

It was also during this transition era that women came to play an increasingly important role on the staff. Hanna Mace Hedrick, a Vassar graduate who married her fellow Nautical Almanac Office colleague Henry Hedrick in 1896, had begun her career as a computer in the Office in 1894, and would continue there sporadically until 1940. Milton Updegraff hired the first woman assistant astronomer in the Nautical Almanac Office (and in the Naval Observatory) in 1908. Isabel Lewis (Figure 12.3), who

---

[12] AR (1901), 7; Weber, 80.

[13] AR (1898), 13. Other changes to the volumes during these years are enumerated in AR, 1899, 19–20. On the 1912 date, see AR (1908), 14.

[14] Henry Benjamin Hedrick, "Catalogue of Stars for the Epochs 1900 and 1920 Reduced to an Absolute System," APAE, **8**, part 3 (U. S. Nautical Almanac Office: Washington, 1905). Newcomb's Zodiacal Catalogue had been published in APAE, **1**, part 4. On Hedrick see H. R. Morgan, "Henry Benjamin Hedrick," *Science*, **84** (November 27, 1936), 477. Hedrick is also well known as the co-author of E. W. Brown's Tables of the Motion of the Moon; see Dorrit Hoffleit, *Astronomy at Yale*, 1701–1968 (Yale University Press: New Haven, Connecticut, 1992), pp. 97–104.

Figure 12.3. Isabel Lewis in front of the Sikorsky Amphibian Transport PS2 from the U. S. Naval Air Station in San Diego, at the total solar eclipse of April 28, 1930 at Honey Lake, California. A Land Plane 02U1 took a motion picture of the approach of the Moon's shadow. The expedition is described in PA, **38** (1930), 455–460.

had an A. B. (1903) and A. M. (1905) from Cornell, had first been hired by Simon Newcomb as a computer in 1905, and learned eclipse-prediction work under his guidance. At the Almanac Office she specialized in eclipse work, and became a prolific writer of popular astronomy. At about the same time Eleanor A. Lamson (Figure 12.4) was working her way up the ladder elsewhere at the Observatory. Lamson had obtained her B. S. (1897) and M. S. (1899) from George Washington University, and joined the Observatory as a Miscellaneous Computer in 1900. By 1903 she was a full-time computer and by 1907 an assistant in the Computing Division. In 1921 she became the first female supervisor at the Naval Observatory as the head of the Computing Bureau, a position in which she served until her death in 1932 at the age of 57. As the century progressed, the percentage of women on the staff of the Nautical Almanac Office increased, reaching a peak in the 1970s.[15]

---

[15] On Hedrick, Lewis, Lamson, and others, see Merri Sue Carter, "The Contributions of Women to the United States Naval Observatory: The First 150 Years," in NAOSS, 165–177. See also http://maia.usno.navy.mil/women_history/lewis.html.

Figure 12.4. Eleanor Lamson (far left) and Etta Eaton, human computers, seen here in January 1905 in the "computer room."

*War and depression: The Eichelberger and Robertson era, 1910–39*

The appointment of William S. Eichelberger as Director in 1910 brought stability back to the Nautical Almanac Office; during a tenure of almost 20 years, Eichelberger earned the respect not only of his colleagues but also of the wider astronomical community, extending to his activities in the nascent International Astronomical Union (where he was President of Commission 4 on Ephemerides in 1925). Eichelberger (Figure 12.5) had obtained his Ph. D. in astronomy from Johns Hopkins University in 1891, and came to the Naval Observatory in 1896. In 1900 he passed the competitive exam to become a Professor of Mathematics (taking the place of Harkness) and advanced to the rank of captain in 1920. As we have seen in chapter 10, Eichelberger is well known for his contributions to fundamental meridian astronomy, and especially for his catalog of 1,504 standard stars (1925), which was adopted as the standard by the IAU in 1925 and used by the national ephemerides until 1940. Before coming to the Almanac Office, he had played a crucial role in reorganizing the meridian work of the Observatory, and in obtaining modern clocks. His colleague H. R. Morgan characterized him as "an honest and talented scientist, a very rapid and accurate computer, and a hard worker."[16]

Already at the beginning of Eichelberger's tenure, the issue of international

---

[16] H. R. Morgan, "William Snyder Eichelberger," MNRAS, **114** (1955), 289–291. Eichelberger's catalogue is "Positions and Proper Motions of 1504 Standard Stars for the Equinox 1925.0," APAE, **10**, part 1 (1925).

Figure 12.5.  William S. Eichelberger, Director of the Nautical Almanac Office. 1910–29, in 1915.

cooperation came to the fore. A program of exchange of data had been recommended at the International Congress on Ephemerides held in Paris in 1911, a meeting that was in many ways a follow-up to the 1896 Paris conference at which Newcomb's system of constants had been adopted. The following year the naval-appropriations bill approved by Congress authorized the Secretary of the Navy "to arrange for the exchange of data with such foreign almanac offices as he may from time to time deem desirable, with a view to reducing the amount of duplication of work in preparing the different national nautical and astronomical almanacs and increasing the total data which may be of use to navigators and astronomers available for publication in the American Ephemeris and Nautical Almanac." The United States did have some reservations, however, as is evident in a clause stating that the agreement could be terminated on one year's notice. One of the reservations was the use of the Greenwich meridian, which had been used from the beginning for nautical purposes. The Navy wished to reserve the right to use the meridian of Washington for certain ephemerides. On the positive side, however, Eichelberger noted that data exchanges should allow more time to devote to original research. In fact, beginning with the volumes for 1916, the computations were shared by the nautical almanac offices of France, Great Britain, Germany, and the United States.[17]

Changes made to the *Almanac* during Eichelberger's years were mostly technical or stylistic, but interesting landmarks nonetheless. One of the most noticeable was the discontinuation of the lunar-distance tables beginning in the *Nautical Almanac* for 1912. Inquiries made in 1907 by the Chief of the Bureau of Equipment showed that "these tables are practically no longer used by the navigators either of the naval service or of the merchant marine."[18] Thus, the chronometer method, which had become the

---

[17]  AR (1912), 12–14. Woolard, "The Centennial of the NAO," *Sky and Telescope* (December 1951), 28. The Proceedings of the 1911 Paris Congress are found in *Procès-Verbaux, Congress International des Ephémérides Astronomiques* (Bureau des Longitudes: Paris, 1912).

[18]  *American Ephemeris*, Preface by Updegraaf, November 1909. See also Alan Fiala, "Evolution of the Products of the Nautical Almanac Office," NAOSS, 202–225.

primary method of navigation already by the late nineteenth century, completely superseded the use of lunar distances, ending the battle between the two methods begun by John Harrison and Neville Maskelyne in the eighteenth century.

In 1916 Eichelberger initiated another change, tailoring the Nautical Almanac to the use of the navigator. The American Ephemeris from its beginning had been divided into two distinct parts: the ephemeris for navigators (also reprinted as the Nautical Almanac) and the ephemerides for the Sun, Moon, and planets for meridian transit at Washington, together with data for eclipses, occultations, and other phenomena. After 1881 the "Phenomena" were grouped into a third part. Since 1916 the Nautical Almanac was prepared separately from the Ephemeris and therefore designed especially for navigators. The precision required for astronomers was replaced by the lesser precision needed for navigation, and the form and arrangement of the tables were changed.[19] Perhaps the biggest change in content was in the Almanac beginning in 1925, where the civil day beginning at midnight was introduced rather than the day beginning at noon.[20] We have already seen this issue in connection with the designation "Universal Time" in chapter 11.

Under Eichelberger work also continued on improving the elements of planetary satellites. In the late 1920s, as the Eichelberger period ended, Superintendent Freeman emphasized the "direct and mandatory connection" between the Ephemeris Department and the Observation Department, "especially because of the utilitarian nature of the work turned out by the ephemeris department."[21] He also emphasized that data were still being exchanged directly with foreign nautical almanac offices, and that the Department was engaged in research and investigatory work, largely published in Astronomical Papers of the American Ephemeris. Among that research was Eichelberger's work on the orbit of Neptune's satellite, which was published in 1926. In 1927, Eichelberger began work on a new Zodiacal Catalog to contain over 3,000 stars, including all those in Hedrick's 1905 Zodiacal Catalog and in Gill's Catalogue of 2,798 Zodiacal Stars. It was not to be finished before his retirement; Eichelberger's 1925 Catalogue of Standard Stars remained his most important contribution.[22]

With Eichelberger's departure in 1929 a considerable controversy erupted over his successor. That successor would turn out to be A. James Robertson (Figure 12.6), whom we have already seen in connection with his political allies and the 40-inch Ritchey–Chrétien telescope in chapter 10. Because Robertson was the Assistant Director of the Office, his succession to the Directorship would seem natural. In fact, Robertson had already removed the chief stumbling block: the law stating that the Director must be a Professor of Mathematics or other naval officer. According to Robertson, already in 1912 a law was passed on his behalf by which he might act as, or

---

[19] Woolard (ref. 17), 27–28. Woolard also noted that the Greenwich hour angle was added to the ephemeris of the Moon in 1932, and to the ephemerides of the Sun, planets, and stars in 1934. No further changes were instituted until 1950.  [20] AR (1923), 12.

[21] AR (1927), 23; AR (1928), 21.  [22] AR (1927), 10; AR (1928), 23–24.

Figure 12.6. A. James Robertson, Director of the Nautical Almanac Office, 1929–39.

be appointed, Director. Meanwhile, the Professor-of-Mathematics rank had been abolished by the Navy; Eichelberger and Littell were the only two with that rank left at the Observatory. In 1926, as Eichelberger's tenure neared an end, the position of Assistant Director was created for Robertson.[23]

With Eichelberger's retirement imminent, in early 1929 Frank Schlesinger of Yale wrote Lick Observatory Director W. W. Campbell (now also President of the University of California) explaining the situation and asking for Campbell's help in preventing Robertson's appointment. Through his influence with Senator W. L. Jones of Washington, Schlesinger wrote, Robertson had himself appointed Assistant Director without the knowledge of the Director of the Office, or of the Astronomical Council, and perhaps even the Superintendent. "In my opinion," he wrote further, "and in that of every one else who is conversant with the facts, it would be in the nature of a scientific calamity if he should direct the Nautical Almanac . . . Mr. Robertson is a good computer in the sense of being able to apply a formula that is presented to him. So far as I know, and so far as one can gather from the very meager list of papers that he has published, he would be totally unable to set up a formula in any new case that may arise, or in any new phase of a known application." Robertson's attendance at Leiden and Heidelberg meetings the previous summer, Schlesinger continued, had

---

[23] The July 12, 1927 Astronomical Council minutes record that Robertson "was introduced as an additional member of the Council," USNOA. On Robertson's title as "Assistant Director" of the Almanac Office see Council Minutes of December 13, 1927, p. 3. See also Robertson, "Memorandum for Doctor L. M. Lucas," December 30, 1936, USNOA, BF, Robertson folder.

been a source of embarrassment to his American colleagues, because of his lack of technical expertise. Schlesinger suggested that, if Campbell would have a word with the Secretary of the Navy Wilbur, the matter might be settled very quickly. In fact it proved not so easy. Though the well-known and well-connected astronomers Walter S. Adams, E. W. Brown, George E. Hale, A. Leuschner, and Harlow Shapley also wrote letters to the Secretary and the National Academy of Sciences echoing Schlesinger, their appeals were to no avail. On September 19, 1929 Robertson was appointed Director of the Nautical Almanac Office.[24]

For the decade before World War II James Robertson therefore served as Director of the Nautical Almanac Office. Robertson, the son of one of the first settlers of the state of Washington, had received his B. S. from the University of Michigan in 1891. He became an assistant in the Nautical Almanac Office in 1893, working under Simon Newcomb, and gradually rose through the ranks to astronomer (1924–27) and senior astronomer (1927–29) before becoming Director of the Office.[25] Perhaps his greatest claim to fame was his work on the fifth satellite of Jupiter. Shortly after entering the Almanac Office, Newcomb gave him E. E. Barnard's observations of this satellite, which Barnard had made at Lick Observatory. Robertson derived the elements of its orbit "by the use of formulae he derived for that purpose;" those elements were used in the *Nautical Almanac* until 1959, and in the *American Ephemeris* for several years, before being replaced by those of Tisserand and then Cohn. Leuschner later praised Robertson's work, a statement he came to regret because Robertson used it in high places to further his career. Robertson also computed eclipses and occultations, and in 1933 was awarded an honorary doctorate by Georgetown University.[26]

As Schlesinger suggested might be the case, however, Robertson seems to have contributed little original to the Office. He did see to it that the almanacs were produced on time and with accuracy, but he did no research. The *Zodiacal Catalogue*, which had been initiated by Eichelberger as an update to Hedrick's 1905 catalog, was pursued further by Robertson and his staff; not quite finished at the end of his tenure, it carried Robertson's name when it was published in 1940. However, not everyone

[24] Schlesinger to Campbell, January 15, 1929, USNOA, BF, Robertson folder. The letters from other astronomers are also in this folder. See also John S. Hall, "Talk about A. James Robertson presented before the HAD of the AAS," January 8, 1987, John Hall folder, USNOA, BF. Hall also reports that Robertson was responsible for the departure of Frank Ross, who had worked part or full time at the Almanac Office from 1902 to 1915. Citing a letter Ross wrote to Schlesinger, Hall reports that, when Ross was about to ascend to the Assistant Director position, Robertson had it abolished, and Senator Jones on the floor of the Senate denounced Ross as a grafter.

[25] Robertson recalled his first meeting with Newcomb, who was located in Room 566 of the State, War and Navy Building, while his staff labored two blocks away at 19th and Pennsylvania Avenue. James Robertson, "Highlights in the Career of Simon Newcomb," *PA*, **44** (November, 1936), 471–475; "Recollections of Simon Newcomb," *JRASC*, **30** (December, 1936), 419–421.

[26] Robertson enumerated his accomplishments, including the Jupiter work, in the "Memorandum to Lucas," December 30, 1936, USNO BF, Robertson folder. On Leuschner's letter see Campbell to Schlesinger, USNO BF, Robertson folder. A synopsis of the Jupiter work is AJ, **35**, no. 840 (August 28, 1934). On the use of Robertson's elements for the fifth satellite of Jupiter, see P. Kenneth Seidelmann, *Explanatory Supplement to The Astronomical Almanac* (Universtiy Science Books: Mill Valley, California, 1992), p. 644.

had a low opinion of Robertson; Hellweg wrote in the *Annual Report* for 1939 that "it is with the deepest reluctance and regret that the Naval Observatory has lost the unusual and valuable knowledge and services of Doctor Robertson, gained through forty seven years of continuous effort."[27] As the Observatory's Superintendent, Hellweg no doubt appreciated Robertson's political contacts, which were very useful in budget fights. It has even been suggested that Robertson's ten-year incumbency "was not quite the calamity predicted by the astronomical leaders in 1929," and that, in the midst of the Great Depression, "a politically-minded director of the Almanac could better serve the Observatory than a scientist." The scientific community, however, remained skeptical to the end; at his death in 1960 at the age of 92, the man who had boasted of his work with Newcomb, worked at the Nautical Almanac Office for 46 years, and served as its Director for a decade, earned no obituary in any scientific journal.[28]

*World War II and its aftermath: Eckert and Clemence, 1940–58*

The retirement of Robertson on May 31, 1939 left a gap in leadership at a crucial time as war was stirring in Europe. The Directorship was offered to Yale astronomer Dirk Brouwer, who declined because the research possibilities at Yale were better. However, there was no lack of promising applicants. Fred Whipple at Harvard wrote Schlesinger at Yale asking to be recommended for the Directorship. "If by any chance the application should be successful," he wrote, "my aim would be to attempt to bring more life to an organization that was once of prime importance in American Astronomy." Schlesinger endorsed the application and commented, with obvious reference to Robertson, that "It is a matter of deep concern to us all that the institution that was once so active and fruitful should have degenerated into a mere computing bureau. What the situation needs, is, of course, a greater freedom from political constraints."[29]

There was, however, a specific need at the Almanac Office that drove the selection process. The methods of the Almanac Office at this time were increasingly antiquated, a later Director of the Office recalled: "slide rules, desk calculators, logarithms, Crelle's multiplication tables, things of that sort were being used in order to produce the *American Ephemeris* and the *Nautical Almanac*" (Figure 12.7).

---

[27] AR (1939), 2. Hellweg was no doubt grateful for Robertson's political connections in securing funding; see Hall's typescript on Robertson (ref. 24), 3 and 6; Robertson, "Catalog of 3539 Zodiacal Stars for the Equinox 1950.0," APAE, **10**, part 2 (1940); a second edition was published in 1967.

[28] John S. Hall (ref. 24), 7. Biographical sources for Robertson are very slim, undoubtedly a reflection of his reputation in the astronomical community. Aside from short notices, probably written by Robertson, in *Who was Who in American History – Science and Technology*, and *American Men of Science*, **9**, 1,620, the chief source of biographical detail is his obituary in *The Washington Post*, January 21, 1960, B2.

[29] Dorrit Hoffleit, "Yale and USNO Cooperation Especially in the Brouwer and Clemence Era," *Comments on Astrophysics*, Part C, **16** (1992), 17–30: 21. Hoffleit cites Whipple to Schlesinger, October 9, 1939, Yale Department of Astronomy Records, YRG 14-E. Brouwer had been recommended by a California Congressman, and, although the position was offered to him, there were worries about his temperament, "Notes from Administrative Committee Meetings," Eckert folder April 28, 1939, April 8/May, 1939; June 5, 1939.

Figure 12.7. Nautical Almanac Office personnel, *circa* 1915, in the days of hand calculators and multiplication tables. The setting is Room W of the Observatory's Main Building. From left to right: Joseph Arnaud, Arthur Snow, Frank Langelotti, Louis Lindsey, James Robertson, and Clifford Lewis. Robertson was at this time the Assistant Director of the Office, and was in the room only for the photograph.

"Eichelberger was of the old school . . . in the sense that he used multiplication tables instead of multiplication machines, which were not then available." When Ralph Haupt, later an Assistant Director of the Nautical Almanac Office, arrived at the Observatory in 1928, he recalled that "there were very few desk calculators. In fact, some people worked their whole life there with never touching a desk calculator. And some of them, many years after I joined the Observatory, never used a desk calculator." Although the Observatory had a few "Millionaire" machines in 1928, it was slow to add new machines such as the Mercedes-Euclid, the Marchant, and the Burroughs.[30]

The decade of Robertson's leadership had done little to change the situation. The immediate driving force for change was the burgeoning Army Air Corps (transformed into the U. S. Air Force in 1947), which required a means of navigation as aircraft range became longer and longer. An "Air Almanac" was needed, indeed had already been experimented with, but with the current methods it would require a large increase in staff. The differences between ocean and air navigation were enormous. Distances were traversed so rapidly that frequent position plotting was necessary; with lengthy calculations out of the question the process had to be more oriented

---

[30] Ray Duncombe, OHI, January 11, 1988, 5; Ralph Haupt, OHI, May 18, 1987, 3–4. Haupt came from the National Bureau of Standards in 1928, where, he noted, "everyone had a desk calculator on his desk" at the time. See also Morgan, "Eichelberger," (ref. 16), 290. "Crelle's Tables" refers to A. L. (August Leopold) Crelle (1780–1855), *Dr. A. L. Crelle's calculating tables, giving the products of every two numbers from one to one thousand, and their application to the multiplication and division of all numbers above one thousand* (G. Reimer: Berlin, 1908). The Naval Observatory Library contains numerous editions, beginning with the German edition of 1869 and the English edition of 1908. 1930 is the last edition in both languages in the Observatory Library. The Naval Observatory purchased its first "electrically driven 8 × 8 Millionaire Calculating Machine with stand" from W. A. Morschauser in New York City in 1912. The price was $550. Another was purchased in 1916. See "Telescopes, Clocks, etc. Belonging to Naval Observatory: Cost and Date of Purchase," USNOA, SF, "Instruments" file.

Figure 12.8. Wallace J. Eckert, Superintendent of the Almanac Office during World War II, introduced punched-card techniques.

toward table look-up. Already in the late 1920s in the United States there had been interest in producing such a volume tailored to the needs of aircraft navigators. This interest resulted in the production of the *Aeronautical Supplement* for September, 1929 to December, 1930; the *Supplement* for 1931; the revised *Nautical Almanac* for 1932; the *Air Almanac* for 1933; and the revised *Nautical Almanac* for 1934 and 1936. However, despite the publication of German (1935), French (1936), and British (1937), air almanacs, the Americans had not yet succeeded in producing a regular annual volume.[31]

With the German invasion of Poland in September, 1939, these needs became increasingly urgent. The solution was to hire, on February 1, 1940, Wallace J. Eckert (1902–71) to head the Office. Eckert (Figure 12.8), who obtained his Ph. D. in astronomy from Yale in 1931 under E. W. Brown, was one of the pioneers of automatic equipment for astronomical computations.[32] While he was a Professor of Celestial Mechanics at Columbia, he had become familiar with the punched-card work of Leslie J. Comrie (1893–1950), the leader of punched-card methods in astronomy. Comrie, an employee of the British Nautical Almanac since 1925, and its head since 1930, revolutionized its methods of production. Like the American Nautical Almanac Office, its British counterpart had used highly trained freelance computers for the laborious calculations. Comrie's innovation was to use clerical workers to run standard calculating machines used in business practice; the product was scientific rather than commercial. Furthermore, Comrie brought in punched-card machines; although he had to transfer Brown's *Tables of the Motion of the Moon* onto half a million punched cards, in seven months he could turn out the tables for the next 20 years, at a fraction of the

---

[31] On the differences between air and ocean navigation see especially J. E. D. Williams, *From Sails to Satellites: The Origin and Development of Navigational Science* (Oxford, 1992), pp. 119–120.
[32] On Eckert see Henry S. Tropp, "Eckert", DSB, 15, 128–130; Martin C. Gutzwiller, "Wallace Eckert, Computers, and the Nautical Almanac Office," NAOSS, pp. 147–163.

cost. Brown himself visited Comrie, and took back to the United States the idea of using punched cards for scientific computation.[33]

Working under Brown, Eckert applied Comrie's adaptation of commercial business machines to astronomical calculations, using the new punched-card equipment at Columbia's Statistical Bureau beginning in 1929. In 1933 he established the Thomas J. Watson Astronomical Computing Bureau, which was operated jointly by Columbia, the American Astronomical Society, and IBM. Here he used an IBM 601 multiplying punch, a credit-balancing accounting machine, and a "summary punch controlled by a plugable relay box" from a statistical tabulator, giving him the capability for mechanical reading, writing, and arithmetic. Astronomical problems were addressed from the beginning, especially the numerical solution of the equations of planetary motion.[34]

Even before the entry of the United States into World War II with the Japanese bombing of Pearl Harbor on December 7, 1941, Eckert had initiated substantial changes at the Nautical Almanac Office. With war looming it was necessary to be able to carry on independently of all foreign sources.[35] This meant revolutionizing the production methods for the American *Nautical Almanac* just as Comrie had a decade earlier for the British. This is exactly what Eckert did with the introduction of punched-card machines, including an IBM tabulator, 601 summary punch, and sorter for the production of the almanacs (Figure 12.9).[36] The contrast with the computational methods could hardly be greater.

The *American Air Almanac* was the first "guinea pig" for the punched-card method. Despite the attempts mentioned earlier, only under Eckert in 1941 did the *American Air Almanac* become a regular publication.[37] The importance of accuracy in this task is obvious from the fact that erroneous figures could quickly result in the loss of a plane and its crew, with no time for cross-checks or recovery as in ocean navigation. With three volumes of the *American Air Almanac* per year, each with hundreds of pages and thousands of figures, the need for reliable production methods was not a luxury, but a necessity. Some editions ran to almost 200,000 copies. As Eckert and his colleague Ralph Haupt later wrote, "The assurance of perfection in millions of copies

---

[33] On Comrie and concurrent developments in the British Nautical Almanac Office see George Wilkins, "The History of H. M. Nautical Almanac Office," NAOSS, 55–81: 58–60. On Comrie see also H. S. W. Massey, "Leslie John Comrie, 1893–1950," *Obituary Notices of the Royal Society*, **8** (November, 1951), 97–105. For further context, see Mary G. Croarken, *Early Scientific Computing in Britain* (Oxford University Press: Oxford, 1990).

[34] Henry S. Tropp, "Eckert", DSB, XV, 128–130; Eckert, "Computing in Astronomy," in Preston C. Hammer, ed., *The Computing Laboratory in the University* (Madison, 1957), pp. 43–50. For the context of the history of computing, see Martin Campbell-Kelly and William Aspray, *Computer: A History of the Information Machine* (New York, 1996), pp. 66–69. Also, Eckert, "Calculating Machines," *Encyclopedia Americana* (1958).   [35] AR (1940), 1–2.

[36] Eckert, *Punched Card Methods in Scientific Computation* (Thomas J. Watson Astronomical Computing Bureau, Columbia University: New York, 1940).

[37] AR (1940), 9. W. J. Eckert, "Air Almanacs," *Sky and Telescope*, **4** (1944), 4–8. It was renamed *The Air Almanac* in 1953, when the British and American air almanacs were merged to become the standard air almanac for the Western world.

Figure 12.9. The punched-card machine room, 1941. Helen Smith and Rubye Barnes are running the machines, which include a punched-card sorter (left). The tabulator added, subtracted, accumulated, and differenced punched-card data and made a printed record. The reproducer transmitted punches from one set of cards to another. When it was used with the tabulator, it could punch answers given by the tabulator. The high accuracy of the *Air Almanac* and *Nautical Almanac* was attained by using these electric punched-card machines to prepare and check the numbers.

with billions of figures requires great care, especially under wartime conditions."[38] With the methods developed by Eckert in the Almanac Office, the *American Air Almanac* was produced entirely by the automatic punched-card machines, not only to perform the calculations but also to print the results in such a form that they could be reproduced by photo-offset.

The success of automated methods with the *Air Almanac* led to an expansion of these methods to other areas. The job of automating the preparation of the *American Ephemeris and Nautical Almanac* was led by Paul Herget (Figure 12.10), an assistant

---

[38] Wallace Eckert and Ralph Haupt, "The Printing of Mathematical Tables," *Mathematical Tables and Aids to Computation*, **2** (January, 1947), pp. 187–202. This paper traces the evolution of the production of the *Air Almanac* during the war years. After the war the British and American Air Almanacs were unified; see G. M. Clemence and D. H. Sadler, "Unification of the Air Almanac and the American Air Almanac," *Journal of the Institute of Navigation*, **3** (1950), 9–11 and **4** (1951), 386–388.

Figure 12.10. Paul Herget automated the preparation of the *American Ephemeris and Nautical Almanac* during World War II, and subsequently worked on many projects in celestial mechanics with Eckert, Clemence, and Brouwer.

professor of astronomy at the University of Cincinnati who took emergency leave from 1942 until 1946 in order to help out the Office. As a student at Cincinnati, Herget, like Eckert, had been much affected by Comrie's work on punched-card machines. For his dissertation on the computation of orbits, Herget became expert at using mechanical desk calculators, and eager to employ any improvements. The needs of the Almanac Office, under wartime conditions, gave him his chance. "We had each started out in the era of logarithm tables and lead pencils, and then used desk calculators," he wrote, speaking of his collaboration with Eckert and Clemence during the war years. "It was an interesting time for the three of us, because the use of standard punch card machines was something new in scientific computing and there was a wealth of fresh problems to occupy our best efforts."[39]

One of the problems was related to heavy Allied ship losses during the war. By 1943 thirty percent of Allied convoys were being lost to the "wolf-pack" tactics of German submarines. Owing to fuel shortages, these submarines did not return home immediately after firing their torpedoes, but lay in wait in shipping lanes observing Allied convoys and then radioing to German headquarters the positions of Allied ships. In order to counter this threat, the Allies established more than a hundred listening posts around the world, each keeping constant surveillance for incoming radio messages on a wide spectrum of frequencies. With the solutions of about a quarter of a million spherical triangles, these observations could locate German submarines to within five miles. Because the Nautical Almanac Office had one of the few scientific computation laboratories in the Washington area, in August, 1943, Naval Communications officers visited Eckert and Herget to explain the problem and the possible solution. Herget was assigned the task, assisted only by two Waves from

---

[39] Paul Herget, "The Keeper of Mars," *Sky and Telescope* (April, 1975), 215–216; Donald Osterbrock, "Paul Herget, 1908–1981," BMNAS, **57**, 59–86, including bibliography. See also David DeVorkin, OHI with Herget, April 19–20, 1977, USNO Library.

Naval Communications, and the punched-card machinery. They carried out the work 12 hours a day over three months, working at night so that the equipment could be used during the day for production of the *Air Almanac*. By November the "submarine book" was finished and by December the Allied casualty rate was down to 6%. The computations for the "submarine book," Herget stated, "gave him the greatest satisfaction of his lifetime."[40]

Also during the war years, Herget and his Almanac Office colleagues tested a system for computing LORAN (Long-Range Navigation) tables for the U. S. Coast Guard, using punched cards and introducing the concept of optimum interval tables. Even after the war, Herget continued to work with the Naval Observatory, publishing three important studies on the coordinates of the asteroids, the Sun, and Venus over long time periods.[41]

Not everyone favored the new methods, and a generational divide separated the old from the new: Long-time employees like Isabel Lewis, Glen Draper, and Charlotte Krampe "represented the old style of doing things as they'd been done traditionally. They were somewhat upset at this change being wrought by Eckert, and at least Mrs. Lewis and Draper never took it up. They showed no interest in trying to do computations that way; although some of the computations that they were responsible for were transferred over to punch cards, they would check them in the traditional way as their responsibility." Nevertheless, others in the office had to learn the new techniques, among them the young Gerald Clemence, with important consequences for the Observatory at large.[42]

During the war years Eckert had revolutionized Almanac Office production methods, but as the war neared its end he decided to return to the Watson Laboratory at IBM, now called the Watson Scientific Computing Laboratory. Well before Eckert's departure in February, 1945, Brouwer was once again offered the position. He wrote the Yale Provost his reasons for turning down the position a second time: "The temptation of the appointment is that in the organization of the Naval Observatory I would be able to cooperate with several capable men in my own field, whereas at Yale Observatory I stand rather alone. In this respect the conditions in Washington are considerably more favorable than in 1939. In addition, the equipment and resources of the Naval Observatory are far beyond those available here. On the other hand, the Director of the *Nautical Almanac* is burdened with administrative duties of considerable extent, and finds his freedom of action in the organization of his office restricted by

---

[40] Osterbrock (ref. 39), BMNAS, 66. On the submarine work, see Herget, "The Submarine Book" June, 1977, Herget papers, USNOA. The "Final Coordinates Tracking Charts HO-5405 . . . Series," are in the USNOA, Herget papers.

[41] Herget, "The Keeper of Mars," 216. Punched-card methods also spread to other parts of the Observatory. They were used in conjunction with measuring engines in the transit-circle divisions, where the measures were recorded on punch cards (see chapter 10). W. J. Eckert and Rebecca Jones, "Measuring Engines," in W. A. Hiltner, *Astronomical Techniques*, volume II of *Stars and Stellar Systems* (University of Chicago Press: Chicago, 1962), pp. 424–439: 437–438.

[42] Duncombe OHI, June 18, 1983, 7–9, and Duncombe OHI, January 11, 1988, 39. On Burroughs machines before Eckert, and other interaction with Eckert, see Jack Belzer OHI. For Brown, Hedrick, and earlier use of punched-cards in astronomy, see Hoffleit (ref. 14), pp. 99–100.

Figure 12.11. Gerald Clemence, Director of the Nautical Almanac Office, 1945–58. Clemence also served as the first modern Scientific Director of the U. S. Naval Observatory, 1958–63.

Civil Service regulations."[43] Brouwer also noted that an important consideration was the opening of the greatly expanded IBM Watson Laboratory at Columbia, an event that would allow opportunities for research at Yale.

In any case, Brouwer's declining the offer (and also the decision of Assistant Director Paul Herget to return to his academic position at Cincinnati) paved the way for Gerald Clemence to take over as Director of the Office in 1945.[44] Clemence (Figure 12.11) had obtained his undergraduate degree in mathematics from Brown University in 1930, and came to the Observatory in the same year. He began as a junior astronomer in the Time Service Division through 1937, then became an assistant astronomer in the 9-inch transit-circle Division until 1940, working under Morgan. In 1940 Eckert offered Clemence a position in the Almanac Office just below the Assistant Director; here Clemence worked with Eckert and Herget in introducing the new "punch-card" machines.[45]

Though Clemence was now in charge of the almanacs, his interests went far beyond the routine tasks of their production, tasks that had dominated the office since Newcomb, and that the war had imposed on Eckert. In 1948 Clemence enumerated three distinct purposes for the national ephemerides: to serve as the basis for the navigational almanacs; to predict astronomical phenomena such as eclipses, occultations, and the rising and setting of the Moon; and finally to allow comparison of the theory on which they are based with observations, permitting improvement of the

[43] Brouwer to Provost E. S. Furniss, November 8, 1944, Yale University Astronomy Department Records, YRG 14–E, cited in Hoffleit, *Astronomy at Yale* (ref. 14), p. 147. On Eckert's subsequent work see Gutzwiller (ref. 32), NAOSS, 150 ff.

[44] AR (1944–45), in AJ, **51** (1946), 214 noted that, on Eckert's resignation in March (actually February 28) 1945, he was "succeeded by G. M. Clemence, Paul Herget becoming assistant director." In March, 1946 Herget "terminated his leave of absence from the University of Cincinnati . . . when he resigned as Assistant Director of the Nautical Almanac Office to assume active direction of the Cincinnati Observatory." Edgar Woolard became Assistant Director, AR (1945–46) in AJ, **52** (1947), 142. On the transition to Clemence see Duncombe, OHI (1988), 15–16.

[45] On Clemence see Raynor L. Duncombe, "Gerald Maurice Clemence, 1908–1974," BMNAS, **79** (2001), 2–13; Paul Herget, "The Keeper of Mars," *Sky and Telescope* (April, 1975), 215–216; Clemence, "How to Get Medals," USNOA, BF, Clemence folder. The latter constitutes a brief but interesting autobiography.

Table 12.1.  *Classical steps in construction of ephemerides*

**Theory**
A mathematical expression giving coordinates as a function of time, sometimes containing hundreds of terms. Incorporates observations, including mean orbital elements and astronomical constants.
Example: G. W. Hill, "A New Theory of Jupiter and Saturn," *Astronomical Papers of the American Ephemeris*, volume 4 (1890).

**Tables**
Constructed from the theory. Tabulates groups of terms as functions of arguments increasing linearly with time. High-speed electronic computers obviate the need for tables, since ephemerides can be generated directly from the theory.
Example: G. W. Hill, "Tables of Jupiter, Constructed in Accordance with the Methods of Hansen," *Astronomical Papers of the American Ephemeris*, volume 7 (1898).

**Ephemerides**
Positions of celestial bodies, listed for certain dates, obtained by interpolating the tables.
Example: *The American Ephemeris and Nautical Almanac*

astronomical constants or removal of the inadequacies in the theory. While the first two were of passing interest to the astronomer, Clemence remarked, the last was of first-class importance.[46]

The hallmark of the Clemence era was thus a return to research on the theories of planetary motion and the associated refinement of the astronomical constants. We recall that there were three classical steps in obtaining an ephemeris for any given body: theory, tables, and the printed ephemeris itself (Table 12.1). The theory for a particular celestial object was a mathematical expression giving the coordinates of a planet as a function of time. It was based on the orbital elements produced from observations fit to an ellipse; it incorporated the astronomical constants, and could contain hundreds of terms, each of which had to be evaluated and summed with the others to obtain the coordinates at a single date. Because this process, for each date, was extremely laborious, tables were constructed from the theory giving the values of groups of terms as functions of arguments increasing linearly with time. (With the advent of high-speed computers such tables are no longer necessary.) The ephemeris was generated by interpolating the tables for each date required and summing the results. Clemence remarked that using tables reduced the labor of calculating a one-year heliocentric ephemeris of Saturn from a year to a few days. Hill's theory of Jupiter and Saturn occupied 576 pages, and the tables derived therefrom 285 pages. Even so, the construction of the ephemerides of the eight planets and the Moon kept four skilled computers continuously occupied.[47]

It is not too much to say that Clemence picked up where Newcomb and Hill left off, employing not only a half-century's worth of new observations, but also the vastly improved reduction methods, involving first punched cards and then computers.

---

[46]  Clemence, "On the System of Astronomical Constants," AJ, **53** (May, 1948), 169–179.
[47]  Clemence, "The System of Astronomical Constants," *Annual Review of Astronomy and Astrophysics*, **3** (1965), 93–112.

Mercury, an object of special interest because the advance of its perihelion by some 40 arcseconds per century had been the best proof of Einstein's general theory of relativity, held Clemence's early attention. Already by 1943 Clemence had compared thousands of observations of Mercury from 1765 to 1937 with Newcomb's orbit in order to derive new orbital elements; this research was published in the same *Astronomical Papers* series where Newcomb's research had appeared.[48] He then tackled the motion of Mars, Newcomb's last and most incomplete planetary project. Finding that the task needed a complete overhaul because Newcomb had not adequately taken into account second-order terms caused by Jupiter's strong perturbing effect, Clemence started from scratch to construct a new theory. By 1949 he had published a first-order theory, noting that the calculations were entirely undertaken with punched cards; the final theory was not published until 1961. Upon its publication Brouwer commented that Clemence's work on Mars represented "a new standard of accuracy for a general planetary theory that surpasses previous efforts by a considerable margin." In 1975, after extensive comparison with observations, Herget characterized the Mars theory as "the most accurate of the general theories for any of the principal planets." Using the U. S. Naval Ordnance Research Calculator, Herget found that the differences of theory from observation in latitude never exceeded 0.008 arcseconds, and that those in longitude never exceeded 0.04 arcseconds.[49] In order to compare theory with observation Clemence had to grapple with the problems of time introduced by the variable rotation of the Earth; we have already seen in chapter 11 in this connection that the concept of Ephemeris Time became an issue in which he took the lead.

Though much of Clemence's work was undertaken alone, he also had the benefit of a strong collaboration with Dirk Brouwer of Yale and Eckert at the Watson Scientific Computing Laboratory. This collaboration was greatly strengthened in 1947 when the Office of Naval Research (ONR) awarded a long-term contract to Yale, the Naval Observatory, and the IBM Watson Laboratory to undertake work on a variety of solar-system problems. The rationale behind the work was that more accurate theories and tables could be produced in light of the new computing machinery. In particular, the first step in the construction of ephemerides via a general theory using analytical techniques could potentially be replaced, or supplemented, by numerical methods. Whereas general theory made use of terms in a Fourier series, into which time and date were input resulting in a position for that specific time, numerical integration made use of equations of motion, which were numerically integrated forward for future positions.

---

[48]  Clemence, "The Motion of Mercury, 1765–1937," APAE, 11, part 1 (1943); AJ, 50 (1943), 126–127; "The Motion of Mercury," *Sky and Telescope*, 7 (December, 1947), 31–33. Duncombe, OHI (1988), 21–22.

[49]  Clemence, "First-Order Theory of Mars," APAE, 11, part 2 (1949); "Theory of Mars (Completion)," APAE, 16, part 2 (1961); Paul Herget, "The Keeper of Mars," (ref. 45), 215–216. Clemence employed the method of Hansen, as Hill did for Jupiter and Saturn. Brouwer's comment is in his "Review of Celestial Mechanics," *Annual Review of Astronomy and Astrophysics*, 1 (1961), 219–234: 221.

The ONR contract, which set the research agenda of the Office for more than a decade, centered on a revision of the motions of the principal planets, including Mars. More specifically, the program consisted of six parts: measurement of photographic plates of Saturn's satellites in order to re-evaluate the mass of the system; improvement of the theory of Jupiter's Galilean satellites; work on the secular perturbations of Pluto; research on the theory of motion for Jupiter and Saturn to see whether the theories of motion of the principal planets could be developed with the same degree of accuracy as the lunar theory; accurate orbits of the first four asteroids; and a new theory of the motion of Mars by Hansen's method. Between 1949 and 1970, some 22 papers were published in the *Astronomical Papers* as a result of this significant collaboration. One of the first products of the collaboration was *Coordinates of the Five Outer Planets, 1653–2060* (1951), which quickly became the standard source for all orbital research involving the planets from Jupiter to Pluto, and the source for the *American Ephemeris* of the ephemerides for the outer planets from 1960 until 1983. The numerical integration for this work was performed on the famous IBM Selective Sequence Electronic Calculator (SSEC), which Wallace Eckert pioneered at IBM after leaving the Observatory. The outer planet calculations were its first major test.[50]

An important aspect to the improvement of theories of planetary motion was the determination of a self-consistent and accurate set of astronomical constants, since the accuracy of all reduction computations for celestial positions depended on the accuracy of values of the astronomical constants used. Thus the Holy Grail of comparing gravitational theory with observation at ever-increasing levels of accuracy would be dashed without accurate constants. The determination and application of these astronomical constants, a prominent astronomer once remarked, is "an extremely exciting basic problem" camouflaged "by a superbly boresome name."[51]

The introduction of new constants, was, however, a delicate task, as Newcomb had discovered 50 years earlier. Clemence emphasized that a self-consistent system was more important than the highest accuracy. Speaking as a man who had to produce almanacs, he noted that it was necessary only that the square of the errors be negligible. The desire of astronomers that the value be the best possible, he noted, "has occasionally led to changes in the official system of constants that were productive of more harm than good. The introduction of Newcomb's value of the precession, for example, has not lightened the labor of any astronomer, but on the contrary has

[50] W. J. Eckert, D. Brouwer, and G. M. Clemence, "Coordinates of the Five Outer Planets, 1653–2060," APAE, **12** (1951); Hoffleit, *Comments on Astrophysics* (ref. 29), pp. 23–24. On the close personal relation of the three authors, see Herget, "The Keeper of Mars," (ref. 45), 216. The first report on the ONR work is in AR, **53** (1948), 151. Progress in ONR work is reported in the ARs at least through 1958. For the context of the SSEC see Charles Bashe *et al.*, *IBM's Early Computers* (MIT Press: Cambridge, Massachusetts, 1986), pp. 47–59; and on the work, Duncombe OHI, 23–24; Seidelmann OHI, 20.
[51] Martin Schwartzschild, in *Aspects of Stellar Evolution*, Proceedings of a Conference Held in honor of Ejnar Hertzsprung at Flagstaff, Arizona on June 22–24, 1964, eds. Arthur Beer and K. A. Strand, *Vistas in Astronomy*, **8** (1966), 3.

caused much useless labor and difficulty for everyone who has tried to compare observations made before the change with those made afterward. I think that no one who has experienced these difficulties has felt himself recompensed by the knowledge that Newcomb's value was a little 'better' than the one it superseded."[52] Although some saw the current system as not completely satisfactory either from the point of view of accuracy or in terms of consistency, the practical problem was keeping the amount of recalculation in ephemerides, and in comparison of theory with observation, to a minimum.

However, in 1950 the time was ripe, for many astronomers were now ready to begin a discussion on adopting new constants. In this endeavor Brouwer and Clemence played a leading role, along with their British, French, and Russian counterparts. The problems and potentials of new constants were argued at a seminal meeting in Paris in the spring of 1950 (Figure 12.12) – the third international meeting on astronomical constants (after the 1896 and 1911 Paris meetings), and the same meeting in which Danjon had raised the question of Ephemeris Time. The observations of the Sun and planets employed by Newcomb, the participants pointed out, extended over a 140-year period to 1890, the first 60 years of which were of low accuracy with both systematic and random "accidental" errors introduced at a time when Repsold's traveling-wire micrometer had not yet been introduced and observations by eye-and-ear method caused large personal equation errors to enter. In addition, the non-uniformity in the Earth's rotation needed to be taken into account.[53] Moreover, Newcomb's constants had been developed before Einstein's theory of relativity. Any changes, Clemence, Spencer Jones, and others at this conference emphasized, should not be made lightly. The Conference recommended to the IAU no immediate change in the value of any constant, that the ephemeris of Mars be based on work at the Naval Observatory, that the motions of the five outer planets be based on numerical integrations then under way in the United States, and that Ephemeris Time be used. So controversial was the recommendation, and so cautious the participants, that only well into the Space Age would new astronomical constants be introduced.[54]

The Naval Observatory's role in astronomical constants was not limited to its leading participation in international meetings. It also had a long tradition of determining better values for the constants and the theories associated with them; the transit-of-Venus expeditions, we recall, were mounted with the express purpose of

---

[52] Clemence, "On the System of Astronomical Constants," AJ, **53** (May, 1948), 170.

[53] See more in Spencer-Jones (ref. 54), p. 88, including the solution to Newcomb's fluctuations in the motion of the Moon. Spencer Jones called this anomaly in the Moon's motion "The most enigmatic phenomenon presented by the celestial motions."

[54] *Constantes Fondamentales de L'Astronomie, 27 March–1 April, 1950* (CNRS: Paris, 1950). In this volume see especially Sir Harold Spencer Jones, "The System of Astronomical Constants," pp. 85–101, and the recommendations on pp. 128 ff. Spencer Jones recommended (p. 100) that the revision of constants be undertaken at the Naval Observatory, but this was not among the general recommendations of the conference. Clemence gives the background to the introduction of new constants in "The System of Astronomical Constants," *Annual Reviews of Astronomy and Astrophysics*, **3** (1965), 93–112.

| 1 M. STOYKO. | 7 M. SPENCER JONES. | 13 M. CHAZY. |
|---|---|---|
| 2 M. AMBARTSUMIAN. | 8 M. CLEMENCE. | 14 M. DANJON. |
| 3 M. SADLER. | 9 M. NEMIBO. | 15 M. BROUWER. |
| 4 M. OORT. | 10 M. MORGAN. | 16 M. FAYET. |
| 5 M. KULIKOV. | 11 M. KOPFF. | 17 M. JEFFREYS. |
| 6 M. ZVEREY. | 12 M. BATROUCHEVITCH. | 18 M. MINEUR. |

CONSTANTES FONDAMENTALES DE L'ASTRONOMIE.

OBSERVATOIRE DE PARIS. 27 Mars–1st Avril 1950.

Figure 12.12. The 1950 Paris Conference on Astronomical Constants. The frontispiece from *Constantes Fondamentales de l'Astronomie* (Paris, 1950).

determining one of those constants – the solar parallax. Another example – cautionary in this case – was observational work related to the constant of nutation. In order to make this determination, a star needed to be observed for 18.6 years in order to obtain the full period of nutation. George A. Hill had done this by making observations of Vega (alpha Lyrae) with the prime-vertical instrument from 1893 to 1912. However, Hill found that declinations obtained during the night were systematically larger than daytime observations, and so discarded the daytime observations. In doing so he produced a value for nutation of 9.2498 ± 0.0143 arcseconds; because the current value was 9.21 arcseconds, his value – based on almost two decades of effort – was considered much too large.[55] Despite the difficulties, one of the themes running through Naval Observatory history concerned efforts to determine better values for precession, aberration, and other constants.

Not all Clemence's problems, of course, were as far-reaching as the astronomical constants. He was still responsible for *Almanac* production, played an important role in producing a nearly universal set of tables of sunrise and sunset, and introduced the idea of supplementing observations of the Sun with observations of minor planets to determine the position of the ecliptic and equator. His appointment in 1958 as Scientific Director of the Naval Observatory, the position being filled after a 55-year hiatus, was the culmination of a successful career during which he became a leader in American astronomy and a member of the National Academy of Sciences, one of the few Naval Observatory members of the Academy in modern times. Clemence's books, with Dirk Brouwer on *Methods of Celestial Mechanics* (1961), and with Edgar Woolard on *Spherical Astronomy* (1966), were widely regarded as the standards for their subjects. His work continued after his departure for Yale in 1963, where he actively continued research until Brouwer's untimely death in 1966. The subsequent dissolution of the Brouwer–Clemence–Eckert–Herget collaboration constituted the end of an era in celestial mechanics.

## 12.3    From stars to satellites: Into the Space Age

The beginnings of the Space Age brought the immediate realization that techniques that astronomers had long applied to celestial bodies would now be applied to artificial satellites. The launch of Sputnik in 1957 brought an urgency to the computation of orbits that had never before existed. On recalling the upsurge of interest in dynamical astronomy, Clemence noted that, whereas in previous decades the number of active workers in the field in the United States was about six (and only slightly more around the world), "now within a single year scores of students were seized with the desire to study dynamical astronomy, and the demand for teachers of the subject far exceeded the supply." The Yale Department of Astronomy, with Dirk Brouwer at its

---

[55]   K. A. Kulivov, *Fundamental Constants of Astronomy* (translation, NASA: Washington, 1964), p. 126. On the determination of one of the astronomical constants see, for example, H. R. Morgan and J. Oort "A New Determination of the Precession," BAN, 11, no. 431.

head, had more graduate students apply than it could accommodate. With National Science Foundation sponsorship Brouwer created a summer institute in dynamical astronomy, offering a six-week course of lectures open to college teachers and employees of industry and government.[56]

Aside from the new-found interest in the computation of orbits, as Clemence implied in the passage at the beginning of this chapter, the Naval Observatory was well situated to extend its realm to space navigation, once probes left Earth orbit. The use of satellites in Earth orbit as an aid to navigation on Earth took longer, but once a full system was in place, navigation changed radically. Meanwhile, solar system dynamics still offered a variety of perplexing problems, some of which were addressed following the long-standing philosophy of the American Nautical Almanac Office that research should go hand-in-hand with production.

The rise of electronic computers brought with it the potential to employ more extensively techniques of celestial mechanics that heretofore had been too laborious. Such was the method of numerical integration, as opposed to the general "analytical" techniques that had been used by Newcomb, Hill, and their successors at the Naval Observatory. As Brouwer and Clemence wrote in their classic text *Methods of Celestial Mechanics* (1961), numerical integration was the most powerful technique for calculating the motion of any solar system body for a few revolutions around its primary. To determine an orbit for many revolutions, analytical techniques were more likely to be effective, except for highly eccentric orbits. Thus numerical methods had been used for most comets and minor planets, while analytical techniques were mostly used for other solar-system objects. "Whether this condition will long persist cannot be foretold," they wrote in 1961. They saw punched cards and electronic computers as making both numerical and analytical techniques more efficient, but "it cannot be known as yet whether either method will gain at the expense of the other." The practical celestial mechanician, they believed, would always profit from a judicious combination of numerical and analytical methods.[57] Although this has proven true, the efficacy of the numerical integration technique for planetary positions needed for spacecraft navigation would play an essential role in shifting ephemeris production from the Naval Observatory to agencies more directly involved in spacecraft navigation. In the end, this even resulted in the ephemerides for the *Astronomical Almanac* originating outside the Naval Observatory.

### Early impact

The first impact of the Space Age on the Naval Observatory was in the computation of orbits, which had long been the purview of the nautical almanac offices of the world, and now became a matter of urgent national concern. The Vanguard launch

---

[56] Gerald M. Clemence, "Dirk Brouwer, 1902–1966," BMNAS, **41** (1970), 69–87: 73.
[57] Dirk Brouwer and Gerald M. Clemence, *Methods of Celestial Mechanics* (Academic Press: New York and London, 1961), p. 167.

Figure 12.13. Raynor Duncombe and Ralph Haupt in 1969, discussing star charts for Apollo 12. The charts, developed at the Observatory, show positions of the stars as seen from the Command Module and were used in each Apollo flight since Apollo 8. They were used in navigation and visual orientation.

vehicle and satellite was a project of the Naval Research Laboratory – located only a few miles from the Observatory. It was therefore natural for Clemence and his Almanac Office colleague, Raynor Duncombe, to serve as consultants to that project, for which Herget was already the principal consultant for orbital computations. By the time the Sputniks were launched, Duncombe, who had come to the Observatory in 1942 and entered the Almanac Office in 1945, was spending almost 100% of his time at the Vanguard project.[58] (Later, NASA wanted to hire Duncombe, but he remained at the Observatory.) More generally, the Almanac Office "met increasing demands for astronomical data and ephemerides arising from space age requirements of other government agencies and industry."[59] This eventually included support to the Apollo program, for which the Almanac Office (at the request of NASA) developed a series of star charts for navigation and visual orientation. Duncombe and Ralph Haupt (Figure 12.13) were primarily responsible for this activity, which began already with Apollo 8.

In meeting the demands for the Space Age, however, the Almanac Office was joined by new players; highly accurate ephemerides of the planets and satellites, which are critical for space missions, were supplied largely by the Jet Propulsion Laboratory (JPL) in California. Because the standard ephemerides of the early 1960s were not accurate enough for the navigation of spacecraft, the JPL initiated its own work on ephemeris development, using the methods of numerical integration and the best observations available. In this endeavor they were aided by the Ephemeris Working Group, consisting of personnel from the Naval Observatory, Naval Weapons Laboratory, NASA, JPL, and MIT, as well as Paul Herget from the Cincinnati Observatory. The working group grew out of informal discussions by users and producers of optical and radar observations of the Moon and planets. Its first meeting was held in February, 1967 at the Naval Weapons Laboratory, and subsequent meetings were held every few months at the participating agencies. During the course of these meetings the decision was made that the JPL would develop ephemerides; at least one

---

[58] AR, **62** (1957), 312; Duncombe OHI (1988), 18 and 32–34. Constance Green and Milton Lomask, *Vanguard: A History* (Smithsonian Institution Press: Washington, 1971), especially chapter 9, "The Tracking Systems," pp. 145–182: 159–161.    [59] AR, **65** (1960), 3.

Naval Observatory employee, Douglas O'Handley (who received his Ph. D. under Brouwer and Clemence at Yale), had already transferred to the JPL in 1966.[60]

Part of the reason for the shift in expertise to the JPL was that the Naval Observatory had been slow to adopt new precise observing techniques applicable to ephemerides – radar ranging, very-long-baseline interferometry (VLBI), lunar-laser ranging, spacecraft ranging and Doppler techniques – or to apply results from such techniques to ephemerides. Perhaps more to the point, the Observatory did not have the same access to spacecraft data as did the JPL, the NASA-funded agency responsible for many of the missions and their navigation. This lack of data was crucial; whereas modern transit-circle observations determined the position of Mars to within about 100 km, radar ranging to the planets determined distances to a precision of a few hundred meters. Analysis of signals from the Viking spacecraft on Mars gave distances to within less than ten meters, and lunar-laser ranging yielded a precision of a few centimeters. Traditional planetary theories – the analytical techniques traditionally employed at the Naval Observatory – did not meet these levels of precision. (The Eckert, Brouwer and Clemence numerical integrations of the outer planets in the late 1940s were spearheaded by Eckert, and the Naval Observatory could no longer use the software of that early IBM computer.) Under these circumstances the JPL was only one of several centers that stepped in; others were MIT and Goddard Spaceflight Center. Even within the Navy, the Naval Surface Weapons Center eventually became the Navy agency for orbit computation, rather than the Naval Observatory, because it used numerical integration for its ballistic-weapons tests. In the JPL ephemerides the Observatory did, however, remain the crucial source for optical data in the form of transit-circle observations taken with the 6-inch and 9-inch transit circles (Table 12.2). In a series of planetary "Development Ephemeris" products stretching back to 1968, the JPL improved its ephemerides, until its Development Ephemeris 200 (DE 200) became the basis for planetary and lunar positions in most national almanacs, including the Naval Observatory's *American Ephemeris and Nautical Almanac*, which was renamed, with the 1981 edition, *The Astronomical Almanac*.[61]

Part of the Observatory's reluctance to jump into the Space Age with full force was deliberate. When he was asked to become closely involved with NASA, Clemence, as the Scientific Director from 1958 until 1963, made the crucial decision to limit the Naval Observatory's role. As Duncombe recalled, "It was not our charter or our

---

[60]  NASA JPL Technical Report 32–1296, "Card Format for Optical and Radar Planetary Data," May 1, 1968, p. vii, USNOA, BF, O'Handley folder; D. O'Handley, H. Fliegel, and M. Standish OHI; Seidelmann OHI. The Naval Observatory members of the Working Group were Duncombe, R. Glenn Hall, Alan Fiala, Thomas Van Flandern, Paul Janiczek, William Klepczynski, and P. Kenneth Seidelmann.

[61]  *Explanatory Supplement* . . . (ref. 26), p. 290. E. Myles Standish, "The JPL Planetary Ephemerides," *Celestial Mechanics*, **26** (1982), 181–186, and "The Observational Basis for JPL's DE 200, the Planetary Ephemerides of the Astronomical Almanac," *Astronomy and Astrophysics*, **233** (1990), 252–271. LeRoy Doggett, "Celestial Mechanics," in John Lankford, ed., *History of Astronomy: An Encyclopedia* (Garland: New York, 1997), pp. 131–140: 138; Seidelmann OHI, 19–20.

Table 12.2. *Transit-circle observations from the U. S. Naval Observatory used in JPL ephemerides, DE 118*

| Catalog | Time span | Telescope | Volume/part | Number of observations used |
|---|---|---|---|---|
| W(10) | 1911–18 | 6-inch | XI | 2,436 |
| W(20) | 1913–25 | 9-inch | XIII | 3,381 |
| W(25) | 1925–35 | 6-inch | XVI/1 | 6,911 |
| W(40) | 1935–44 | 9-inch | XV/5 | 4,547 |
| W(50) | 1935–41 | 6-inch | XVI/1 | 3,777 |
| W2(50) | 1941–49 | 6-inch | XVI/3 | 3,444 |
| W3(50) | 1949–56 | 6-inch | XIX/1 | 3,678 |
| W4(50) | 1956–62 | 6-inch | XIX/2 | 4,051 |
| W5(50) | 1963–77 | 6-inch | XXIII/3 | 5,811 |
| Circle | 1975–77 | 6-inch | | 1,543 |
| Total | | | | 39,579 |

Note:
The volume/part refers to the second series of USNO Publications.
*Source:* From P. K. Seidelmann, ed., *Explanatory Supplement to the Astronomical Almanac* (1992), p. 291, with permission.

mission at the Naval Observatory, and he [Clemence] didn't see any purpose in trying to do this. Here was NASA, newly created to handle all this. Why should a little place like the Naval Observatory try to spread out its funds to compete with something like that? It made a lot of sense not to do that."[62] It was a decision that affected the Observatory for decades.

An important trend at the Observatory during this period – and one that affected the techniques it could use for ephemerides – was the increasing use of electronic computers. For the second time in two decades a major change was in store for the methods of computation. The automatic tabulating equipment introduced by Eckert in the early 1940s was now to be in turn largely superseded. Because of its need for computing power, and the innovative tradition that Eckert had set as head of the Office, the Almanac Office was responsible for computers at the Observatory. However, despite its pioneering role in the introduction of punched-card machines for scientific computation, and despite the Navy's early interest in electronic computers, the Observatory was not one of the first government agencies to obtain an electronic computer.[63] In the early 1950s IBM was working on a small reliable machine with a "magnetic, drum" memory that could store a program and work in a punched card environment. The company delivered its first "650" computer in December,

[62] Duncombe OHI (1988), 37–38; Seidelmann OHI, 15–16.
[63] An early example of Navy–IBM cooperation in electronic computers was the Naval Ordnance Research Computer. John von Neumann calculated that the ENIAC in the 1940s represented an improvement in speed by a factor of 100–1,000 over previous state-of-the-art machines; the IBM 701 (about 1952) another factor of 30 over that, and the NORC (1954) another factor of five; see John von Neumann, "The NORC and Problems in High Speed Computing," Address by Dr John von Neumann on the occasion of the first public showing of the IBM Naval Ordnance Research Calculator, December 2, 1954.

Figure 12.14. Sol Elvove with IBM 650, March, 1961.

1954, by which time there were more than 450 orders. Clemence reported in July, 1955 that arrangements had been made for Naval Observatory personnel to use the IBM 650 at the Naval Ordnance Laboratory for eight hours per month. Some 300 IBM 650 machines had been installed by mid-1956, when the Chief of Naval Operations approved the purchase of one for the Naval Observatory. It was delivered in July, 1957 and became fully operational in August, shortly before the launch of Sputnik (Figure 12.14).[64]

Clemence played a leading role in transitioning the staff to the new methods. In the early months of 1956, even before the Observatory owned its own computer, he taught a daily class for 15 staff members on the use of the 650. Applications within the Observatory were widespread; the *Annual Report* for 1957 reported that, preparatory to the arrival of the 650 at the Observatory, "programs have been prepared for a number of computations, including apparent place reduction, reduction of meridian observations, photographic lunar observations, photometric lunar observations, photometric measurements, and solar and lunar eclipses." The utility of the new machine was obvious; Clemence pointed out that one particular problem, that had taken one to three hours of computation before, could now be performed in two seconds. Given this impetus, most of the calculations undertaken in the Nautical Almanac Office had been programmed for the 650 by 1958, and other parts of the Observatory were soon to follow.[65]

As with other scientific institutions, the Observatory was then caught up in the never-ending process to keep up with the latest in computing power. The last IBM 650 was manufactured in 1962, the same year that the Observatory moved on to the next model, the IBM 1410. By 1966 it had acquired an IBM 360 (model 40), and in March 1980 a 4341 replaced the 360. By 1990 the Observatory was engaged in moving all applications from its two central computers (an IBM 4381 and a Dec VAX 8530) onto Unix work stations within each Department. By 1994 the computer-support functions

[64] Charles Bashe *et al.*, *IBM's Early Computers* (MIT Press: Cambridge, Massachusetts, 1986), pp. 165–172.
[65] AR (1957), 313; AR (1958), 379; Astronomical Council minutes, 1956–57, USNOA.

Figure 12.15. Edgar Woolard, Director of the Nautical Almanac Office, 1958–63.

were assumed by a new Information Technology Department at the Observatory. These events – and the associated need to transfer programs from older to newer technology, and to develop new software – increasingly consumed the time of Observatory astronomers, and represented an important change in their daily work.[66]

### Research efforts

Improvements of knowledge of planetary orbits and astronomical constants remained important themes under the Directorships of Edgar Woolard (1958–63), Raynor Duncombe (1963–75), and his successor, P. Kenneth Seidelmann (1976–90), the latter of whom had obtained his Ph. D. in 1968 under Herget at Cincinnati. Woolard (Figure 12.15), who had obtained a Ph. D. in meteorology from George Washington University in 1929, worked for the Weather Bureau prior to coming to the Naval Observatory in 1945. He served as Assistant Director of the Nautical Almanac Office under Clemence from 1945 until 1958, and during those years constructed a new theory of the motion of the Earth relative to its center of mass, which was adopted as the basis for computing nutation for the national ephemerides.[67] Woolard once remarked that, if he had known the difficulties of the nutation theory, he never would have started on the problem. When he became Director of the Office upon Clemence's

---

[66] AR (1993–94), 661.
[67] Edgar W. Woolard, "Theory of the Rotation of the Earth Around its Center of Mass," APAE, **15**, part 1 (1953); "A Redevelopment of the Theory of Nutation," AJ, **58** (1953), 1–3. On Woolard see Woolard folder, USNOA, BF, and Duncombe OHI; the quote is on p. 42.

promotion to Scientific Director, Woolard placed *Almanac* production under the direct charge of Assistant Director Ralph Haupt, and continued to publish not only scientific research, but also historical research, in which he had a deep interest. "Woolard was a fantastic mathematician, a very conservative man," Duncombe recalled. "He's the closest to a scholar of any man I've known, absolutely devoted to his work."[68]

Difficult as it was, Woolard's work on nutation was only a small part of the problem of astronomical constants. As both Woolard and Clemence neared retirement, in 1963 the Fourth Paris Conference on astronomical constants was held. The results of space technology, in Clemence's words, "provided the immediate incentive for the meeting." These applications included the refined astronomical unit based on Earth–Venus radar bounces, a much more accurate Earth–Moon mass ratio, and a very precise value of the Sun–Venus mass ratio from Mariner II. Since the 1920s the International Astronomical Union's Commission 4 on Ephemerides had played an increasingly important role in the field of astronomical constants, as well as in international coordination of efforts in the nautical-almanac offices of the world. In 1964 the IAU adopted what was known as the "1968 IAU System of Astronomical Constants." These constants, which did not include a new value for precession, went into effect in 1968. Recalling the problems in introducing Newcomb's constants at the turn of the century, Clemence noted that it was one thing to adopt new constants, and quite another to bring them into actual service. "It cannot be done in an instant," he wrote, "for some years the old system and the new one will necessarily exist side by side, since many astronomical data for future years have already been compiled and published."[69]

The problem of new constants was a major part of the work of the Almanac Office under Duncombe (seen in Figure 12.13), who had taken over from Woolard in 1963. Duncombe had his Master's degree in English, and obtained his Ph. D. in 1956 under Brouwer at Yale some 14 years after coming to the Observatory. At Clemence's urging, Duncombe published his corrections to Newcomb's Venus theory in 1958, analogous to Clemence's own work on Mercury – both using analytical rather than numerical techniques. When he assumed the Directorship in 1963, Duncombe encouraged research on methods for improving knowledge of the mass of Jupiter by studying the motions of the minor planets, as well as Jupiter's satellites. These problems were studied by several members in the Office; the staff as of 1966 is seen in Figure 12.16.[70]

Astronomical theory and practice were advancing so fast that by 1970 it was recognized that the ephemerides in national almanacs required improvements not only in constants, but also in the fundamental star catalog and the definition of time, and even required the replacement of the B1950.0 epoch for the celestial reference

---

[68] Duncombe OHI, 42. On Haupt see LeRoy Doggett, "Ralph Haupt, 1906–1990," BAAS, **23** (1991), 1,490, and OHIs.     [69] Clemence (ref. 46), pp. 95 and 111.
[70] Duncombe, "Motion of Venus," APAE, **16**, part 1 (1958); Duncombe, OHI (1988), 45–46.

Figure 12.16. Nautical Almanac Office staff and Observatory administrators, summer, 1966. Front row (from left to right): Ralph Haupt, Cdr Stanfill (Deputy Superintendent), Captain McDowell (Superintendent), Raynor Duncombe (Director of the Office), Kaj Strand (Scientific Director), and Ruth Meyers. Second row: Jean Hampton, Doug O'Handley, Julie Duncombe, Barbara McMorris, Armstrong Thomas, and Berenice Morrison. Third row: Alan Fiala, Sol Elvove, Gertrude Johnson, Vivian Holland, Louise Weston, Louise Long, and Victoria Meiller. Back row: Ken Seidelmann, Gerald Larson, Dan Pascu, Harry Heckathorn III, Lawrence Buc, William Klepczynski, Peter Schultz, George Brown, Diana Simmons, Judy Wise, and Joan Bixby. Seidelmann and Fiala would become Directors of the Office after Duncombe.

system. In 1970, IAU Colloquium No. 9 on the IAU System of Astronomical Constants was held in Heidelberg, and recommended the establishment of three working groups: on planetary ephemerides, precession, and units and time scales; these working groups were formally established at the 1970 IAU General Assembly in Brighton, U. K. The issues were further discussed at meetings in Washington in 1974, as well as at numerous meetings on specific subjects; they were the issues that would dominate Seidelmann's tenure as Director, which began in 1976 (Figure 12.17). Resolutions were formally adopted at IAU meetings in 1976 and 1979; the outcome of these meetings was the culmination of deliberations on astronomical constants that

Figure 12.17. P. Kenneth Seidelmann, Director of the Nautical Almanac Office, 1976–90, seen here about 1980. A portrait of Simon Newcomb hangs in the background.

stretched back almost a century (Table 12.3). The levels of accuracy achieved are indicated by the differences in the values for prior to 1968, from 1968–84, and after 1984 (Table 12.4).[71] Even then, vigorous discussions continued in the IAU regarding the constants and their relation to time and reference systems.

As part of the new arrangements, as a result of a cooperative agreement worked out between the British and American Offices under Seidelmann and George Wilkins, in 1981 *The American Ephemeris and Nautical Almanac* and *The Astronomical Ephemeris* were replaced by a single title, *The Astronomical Almanac*. The early volumes under this title included changes to the layout of the book and the format of the tables. By international agreement, however, not until the 1984 editions were all these changes, including a new "1976 IAU System of Astronomical Constants," introduced at one time into the national almanacs. In the end Newcomb's constants, and his theories and tables for the Sun and the inner planets, were not completely superseded until 1984. Table 12.5 shows the effect in 1984, when the JPL's DE 200 (Development Ephemeris 200) replaced the theories and tables previously based on the work of Newcomb and his colleagues for the Sun, Moon, and inner planets, and replaced as well those based on the work of Eckert, Brouwer, and Clemence for the outer planets.[72]

Just as agreement on new constants required international cooperation, so also international cooperation in the preparation of the *Almanacs* increased; this increase in cooperation constitutes a final hallmark of the Clemence, Woolard,

---

[71] *Explanatory Supplement* . . . (ref. 26), pp. 617–621; Seidelmann OHI, 29–32.
[72] *IAU Transactions,* volumes XVI (1977), XVII (1980), and XVIII (1983) provide details. On international cooperation, and the difficulties in renaming the *Almanac,* see P. K. Seidelmann, "International Cooperation," in NAOSS, 297–303, especially 299–300; *Explanatory Supplement* . . ., pp. 317 ff.; Seidelmann OHI, 29–32.

Table 12.3. *Conferences on astronomical constants and their adoption*

| | | |
|---|---|---|
| Paris | 1896 | Newcomb's system adopted |
| Paris | 1911 | International cooperation agreement |
| Paris | 1950 | No immediate change in constants |
| Paris (IAU Symposium) | 1963 | Resolutions passed for new system of constants |
| Hamburg (IAU General Assembly) | 1964 | 1968 IAU System of Constants adopted |
| Heidelberg (IAU Colloquium) | 1970 | Working groups on planetary ephemerides, precession, and units and time scales formed |
| Washington (Naval Observatory) | 1974 | Working groups draft report |
| Grenoble (IAU General Assembly) | 1976 | Resolutions adopted for 1976 System of Astronomical Constants with changes to be introduced in 1984 resulting in an order-of-magnitude improvement to ephemerides |
| Montréal (IAU General Assembly) | 1979 | More resolutions adopted for 1976 System of Astronomical Constants including theory of nutation, effective January 1, 1984 |

Note:
After the adoption of the new system of astronomical constants, discussions of new standards and best estimates of constants, how often they should be changed, and the related subjects of reference frames and time systems, became a regular feature of the triennial meetings of the IAU, especially for the working groups on astronomical constants and reference frames.

Table 12.4. *Values of selected official astronomical constants*[a]

| | Velocity of Light (meters/second) | Equatorial radius of Earth (meters) | Constant of aberration (arcseconds) |
|---|---|---|---|
| Prior to 1968 | 299,860,000 | 6,378,388 | 20.47 |
| 1968–1983 | 299,792,500 | 6,378,160 | 20.496 |
| 1984– | 299,792,458 | 6,378,140 | 20.49552 |

| | Constant of nutation (arcseconds) | General precession in longitude (arcseconds) | Solar parallax (arcseconds) |
|---|---|---|---|
| Prior to 1968 | 9.21 | | 8.80 |
| 1968–1983 | 9.210 | 5,025.64 | 8.794 |
| 1984– | 9.2025 | 5,029.0966 | 8.794148 |

Notes:
[a] The full IAU (1976) System of Astronomical constants is given on pages K6 and K7 of *The Astronomical Almanac*. Historical information is found in the *Explanatory Supplement*, pp. 656–657. The current value of precession is expressed in arcseconds per Julian century at standard epoch J2000.

Table 12.5. *Authority (theory/tables) for ephemerides in* The American Ephemeris[a]

**Sun**

| | |
|---|---|
| 1855–57 | Carlini (1810), with Bessel's revisions (1828) |
| 1858–99 | Hansen and Olufsen (1853) |
| 1900 | Newcomb |
| 1901–80 | Newcomb, tables (1895) |
| 1981–83 | Newcomb, theories (1895) |
| 1984– | DE 200/LE200 (JPL) |

**Moon**

| | |
|---|---|
| 1855–82 | Peirce (1853) |
| 1883–1900 | Hansen (1857) |
| 1901–22 | Hansen (1857) with Newcomb's corrections to right ascension (1878) |
| 1923–59 | Brown (1919) |
| 1959–1983 | Brown's theory as given in the *Improved Lunar Ephemeris* (1954) |
| 1984– | DE 200/LE200 (JPL) |

**Mercury**

| | |
|---|---|
| 1855–99 | Winlock (1864), based on theory of Leverrier (1845) |
| 1900–80 | Newcomb, tables (1895) |
| 1981–83 | Newcomb, theories (1895) |
| 1984– | DE 200/LE200 (JPL) |

**Venus**

| | |
|---|---|
| 1855–75 | Lindenau (1810) |
| 1876–99 | Hill (1872) |
| 1900–80 | Newcomb tables (1895) |
| 1981–83 | Newcomb theory (1895) |
| 1984– | DE 200/LE200 (JPL) |

**Mars**

| | |
|---|---|
| 1855–99 | Lindenau (1811) |
| 1900 | Newcomb (1895) |
| 1901–02 | LeVerrier (1861) |
| 1903–21 | Newcomb (1898) |
| 1922–80 | Newcomb, tables (1898) with Rosa's corrections (1917) |
| 1981–83 | Newcomb, theory (1898) with Rosa's corrections (1917) |
| 1984– | DE 200/LE200 (JPL) |

**Jupiter**

| | |
|---|---|
| 1855–97 | Bouvard (1821) |
| 1898–1959 | Hill (1895) |
| 1960–83 | Eckert, Brouwer, and Clemence (1951) |
| 1984– | DE 200/LE200 (JPL) |

**Saturn**

| | |
|---|---|
| 1855–82 | Bouvard (1821) |
| 1883–99 | Hill (1890) |
| 1901–59 | Hill (1895) |
| 1960–83 | Eckert, Brouwer, and Clemence (1951) |
| 1984– | DE 200/LE200 (JPL) |

**Uranus**

| | |
|---|---|
| 1855–75 | Bouvard (1821) and others |
| 1876 | Newcomb |
| 1877–1900 | Newcomb (1873) |
| 1901–03 | Leverrier (1877) |
| 1904–59 | Newcomb (1898) |
| 1960–83 | Eckert, Brouwer, and Clemence (1951) |
| 1984– | DE 200/LE200 (JPL) |

Table 12.5. (cont.)

| | |
|---|---|
| Neptune | |
| 1855–69 | Peirce's theory (1848) and Walker's elements (1848) |
| 1870–1900 | Newcomb (1865) |
| 1891–04 | Leverrier (1877) |
| 1904–59 | Newcomb (1898) |
| 1960–83 | Eckert, Brouwer and Clemence (1951) |
| 1984– | DE 200/LE200 (JPL) |
| Pluto | |
| 1950–59 | Bower (1931) |
| 1960–83 | Eckert, Brouwer, and Clemence (1951) |
| 1984– | DE200/LE200 (JPL) |

*Notes:*

[a] Full references are given in P. K. Seidelmann, ed., *Explanatory Supplement to the Astronomical Almanac* (1992), pp. 631–642 and 657–664.

Duncombe, and Seidelmann years. For years Clemence had collaborated with his British counterpart, Donald Sadler, to unify the preparation of the British and American nautical almanacs. This task was made more difficult by the need to persuade the military forces of both countries to change their practices in order to arrive at a common content and format. Only in 1960, after much consultation, were the contents of the *American Ephemeris* and of the British *Nautical Almanac* unified, in accordance with resolutions of the IAU. In 1961 an *Explanatory Supplement to the Astronomical Ephemeris and The American Ephemeris and Nautical Almanac* was also produced; Seidelmann edited a new and completely rewritten *Explanatory Supplement to the Astronomical Almanac* (1992).[73]

The upshot of all this activity was increasingly accurate ephemerides, which raised increasingly refined research problems, in particular the problems of solar-system dynamics based on the current gravitational model. Although research by necessity often took a back seat to production at the Naval Observatory, following in the Clemence tradition the Nautical Almanac Office tried to remain involved in research efforts beyond the needs of navigation. One area of research, pursued by Thomas C. van Flandern, was the question of the possible variation in the gravitational constant "G." This possibility had been proposed by P. A. M. Dirac; van Flandern tested the idea using observations of the Moon's orbital motion and came up with a positive result that is still open to interpretation.[74] Solar eclipses were another area of

[73] On unification of the almanacs see Clemence and Sadler, "The Conformity of the American Ephemeris and the (British) Nautical Almanac," *Observatory*, **75** (1955), 176; "Unification of the Abridged Nautical Almanac and the American Nautical Almanac," *J. Institute of Navigation*, **9** (1956), 171–176; "Unification of Astronomical and Navigational Ephemerides," *ICSU Review*, **2** (1960), 1–5, and USNO Reprint # 9 (1955); AR, **62** (1957), 313. Also Wilkins (ref. 33), especially pp. 60–66. On earlier changes see Cdr Edwin Beito, "The New Type 1950 Nautical Almanac", USNIP **75**, pp. 1,394–1,401.

[74] T. Van Flandern, "A Determination of the Changing Rate of G," MNRAS, **170** (1975), 333–342; Van Flandern, "Is Gravity Getting Weaker?", *Scientific American*, **234** (1976), 44–52.

Figure 12.18. Alan Fiala, Chief of the Nautical Almanac Office, 1996–2000, seen here at his desk in 1982.

research. The prediction of eclipse events had continued unbroken in the American Nautical Almanac Office since the founding of the Office, when eclipses were used as a test of the *Almanac*'s accuracy. In the twentieth century such predictions were the responsibility in turn of Isabel Lewis, Simone Gossner, Julie Duncombe, and Alan Fiala (Figure 12.18), a student of Clemence who took over as Chief of the Almanac Office in 1996 after a long career in the Office. The Office continued to sponsor eclipse expeditions through much of the century, sometimes to test the accuracy of predictions, but also to perform more specialized research. Fiala, for example, set up a program to observe eclipses at the outer limits of the shadow path in an attempt to determine whether the Sun was shrinking.[75] Although John Bangert and Robert Miller continued eclipse predictions after Fiala, in the 1990s solar-eclipse research and circulars were discontinued due to shifting priorities, and were taken over by NASA. Research into the motions of the minor planets has been a continuing theme, involving Duncombe, Seidelmann, Paul Janiczek, van Flandern, William Klepczynski, and, more recently, James Hilton and Marc Murison.

---

[75] A. D. Fiala, D. W. Dunham, and S. Sofia, "Variation of the Solar Diameter from Solar Eclipse Observations, 1715–1991," *Solar Physics*, **152** (1994), 97–104.

Perhaps the most highly visible research effort was the search for a tenth planet in the solar system, appropriately labeled "Planet X." The effort was led by Robert S. Harrington, an astronomer in the Astronomy and Astrophysics Division with both observational and dynamical interests; his work was in collaboration with Nautical Almanac Office personnel, especially Seidelmann and van Flandern. Although considerations on the stability of the solar system had led Harrington, van Flandern, and others to speculate about a planet beyond Pluto, interest in such a planet was heightened by the discovery at the Naval Observatory of the moon of Pluto in 1978 (see chapter 10). This discovery yielded an extremely small mass for Pluto, and Harrington realized almost immediately that it could not be causing the observed perturbations in the orbits of Uranus and Neptune. His computerized numerical integrations indicated the existence of a planet some three to five times the mass of the Earth, more than five billion miles beyond Pluto, with a period of about 1,000 years. Such an object, which was predicted to be found in the region of Scorpius, would be extremely difficult to detect, but Harrington and others made targeted observations anyway, most notably during the 1990s from the Observatory's southern station in New Zealand. Despite "blinking" more than 300 photographic plates, nothing was found. Although many small bodies have now been found in the "Edgeworth–Kuiper belt," it is possible that the perceived perturbations in the orbits of Uranus and Neptune were actually due to an incorrect mass of Neptune, which was subsequently refined with Voyager 2 data sent in 1989.[76]

*Navigation changes course*

Even as changes were being made in the production of almanacs and the application of celestial navigation, much broader changes were occurring in navigational technology. Through the opening decades of the Space Age, the newly unified British and American *Almanac* retained much of its original importance; in Woolard's words in 1951 on the occasion of the centennial of the founding of the American Office, "It is the principal source of the astronomical data required for surveying, accurate time determination, and other practical purposes; it contains the fundamental ephemerides that are essential to astronomers for current use in planning, making, and reducing astronomical observations, and for permanent future reference in conducting astronomical investigations; and it provides a permanent annual record of astronomical phenomena for general reference."[77]

The use of satellites in Earth orbit as an aid to navigation on Earth changed all

[76] For a review of the subject, see P. K. Seidelmann and R. S. Harrington, "Planet X – the Current Status," *Celestial Mechanics*, **43** (1988), 55–68, and Ken Croswell, *Planet Quest* (The Free Press: New York, 1997), chapter 4, "Planet X." Also Harrington OHI, 26–28; Harrington, "The Location of Planet X," AJ, **96** (1988), 1,476–1,478. For the case against planet X based on the new Neptune data see E. Myles Standish Jr, "Planet X: No Dynamical Evidence in the Optical Observations," AJ, **105** (1993), 2,000. Van Flandern's involvement is described in his book *Dark Matter, Missing Planets & New Comets* (North Atlantic Books: Berkeley, 1993), chapter 18.

[77] Woolard (ref. 17 above), 28.

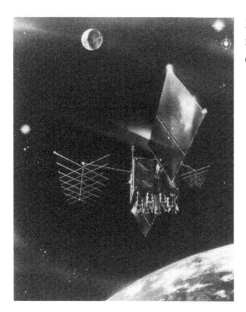

Figure 12.19. GPS satellites revolutionized navigation and time dissemination.

of this radically, and with it the Nautical Almanac Office. A taste of things to come was the Navy-sponsored Transit satellite system (the Navy Navigation Satellite System), developed at the Johns Hopkins Applied Physics Laboratory. The Transit satellites became fully operational in 1964, were available worldwide by 1967, and ceased operations in 1996. In order to develop a more efficient system, including inexpensive and compact user receiver equipment, the Global Positioning System (GPS) of satellites was conceived already in the 1960s. Whereas Transit had been based on positioning by the Doppler shift of a satellite of known orbit, GPS (Figure 12.19) used the concept of "ranging" – the measurement of distances to several satellites. Each satellite transmits a radio signal at specific known times, and the ranging is determined by receivers that measure the exact times the pulses arrive from the satellites. In 1974 a joint-service Department of Defense program was established to implement the GPS concept, with the Air Force as program manager. Satellites were launched beginning in the 1980s, and the system of 24 satellites reached full operational capability on July 17, 1995. They orbit in six circular orbital planes 20,200 kilometers above the Earth, at an inclination of 55 degrees with a nearly 12-hour period. The satellites are spaced in orbit so that at any time a minimum of six satellites will be in view to users anywhere in the world. This satellite navigation system provides position to within a few meters.

With the widespread success and adoption of the Global Positioning System (GPS) of satellites in the 1990s, celestial navigation as supported by the Naval Observatory for almost 150 years became a secondary system. Satellite navigation, however, depended on time in a critical way. The Naval Observatory's Time Service, which had originated because of the need to rate chronometers for celestial navigation, now became the heartbeat of the new primary navigation system. The

Observatory's Master Clock, its time known as UTC (USNO), was the sole source for the time on the atomic clocks onboard each satellite; these satellites were constantly monitored, and the data used by the Air Force to model the clocks. The GPS ground station clocks were similarly sychronized to the satellite clocks. Moreover, the Naval Observatory's newly formed Earth Orientation Department used data from 70 GPS ground stations to determine GPS orbits, and relayed these determinations to the International GPS Service for the final orbital solution. Finally, as mentioned in chapter 11, the Naval Observatory used GPS to determine UT1 and polar motion. The GPS system thus greatly affected not only the Nautical Almanac Office, but also the Time Service and Earth Orientation functions of the Observatory.

With the development of satellite navigation, the value of the once indispensable *Nautical Almanac* came into question. However, continued usage showed that claims of its demise were premature. Upon Newcomb's retirement at the end of the nineteenth century, some 1,571 annual copies of the *American Ephemeris* and 2,266 copies of the American *Nautical Almanac* were in circulation, either sold or distributed each year.[78] By the end of the twentieth century nearly 9,000 copies of the *Astronomical Almanac* and almost 10,000 copies of the *Nautical Almanac* were still in print each year, despite the advent of GPS. Moreover, as computers became more widespread, the Almanac Office undertook the most revolutionary change in its history – the transition of almanac products from print to software. This change was foreshadowed already with the *Almanac for Computers*, under the direction of LeRoy Doggett (Figure 12.20), from 1977 to 1991. This was a paper product designed to facilitate the use of the computer or calculator for astronomical applications. The *Floppy Almanac*, a product produced by George Kaplan and Tim Carroll for 1986 to 1999, was the first software distributed for general use by the Naval Observatory. The *Floppy Almanac* was in turn superseded by the *Multiyear Interactive Computer Almanac* (MICA), a 15-year computer-based almanac.

Thus, even after satellite systems had become the primary method of navigation in the 1990s, celestial navigation still played a significant role. "There is one principle, neither geometric nor physical, as old as navigation beyond familiar landmarks and still applicable to modern navigation," wrote Paul Janiczek of the Almanac Office. "The principle itself is simple, but cardinal; A navigator should use every available means to determine his position."[79] There was an additional factor: The Observatory and the Navy realized that "GPS has operational characteristics and vulnerabilities (including jamming) that may render it unusable or unreliable under certain conditions." Much work was therefore devoted to developing strategies for coping with GPS outages, in particular, updated methods for celestial navigation.[80] Even before GPS vulnerability had become a hot topic, however, at the end of the century the Naval

[78] AR (1897), 10. (Also p. 1,292 of the *Pacific Coaster's Almanac* and p. 515 of the AE.)
[79] P. M. Janiczek, "A Brief Survey of Modern Navigation," NAOSS, 179–200.
[80] George H. Kaplan, "New Technology for Celestial Navigation," in NAOSS, 239–254.

Figure 12.20. LeRoy Doggett, Chief of the Nautical Almanac Office, 1990–96.

Observatory developed a computerized celestial navigation package, known as STELLA (System to Evaluate Latitude and Longitude Astronomically), for DoD clients. STELLA was developed because Navy celestial-navigation instructors were spending too much of their classroom time correcting math blunders. In addition, fleet navigators, many of whom had been using the *Floppy Almanac*, were interested in software that provided additional capabilities. This led to a formal request for STELLA in 1993. To this day, Navy regulations continue to require proficiency in celestial navigation and the use of the sextant. Celestial navigation, though no longer in its prime after centuries of use, was far from dead.[81]

The radical changes in navigational methods and the rise of the computer resulted in organizational changes in the Almanac Office as it attempted to adjust to the new world around it. In September, 1990 the Nautical Almanac Office underwent a major change "to respond to emerging, specialized needs of the Department of Defense (DoD), the civilian departments of the U. S. government, and the astronomical community for astronomical data." The result was the formation of the Astronomical Applications Department (of which the Nautical Almanac Office was a Division supervised by LeRoy Doggett), and the Orbital Mechanics Department. The Astronomical Applications Department retained the *Almanac*-production duties and

---

[81] John Bangert, "The Astronomical Applications Department," in NAOSS, 227–238; Bangert, "Set your Sights on STELLA: New Celestial Navigation Software from USNO," CHIPS (April, 1996), 5–7. See also ref. 6 on the continuing use of the chronometer in the Navy.

the Product Development Division was created at an equal level to design new software products. The Orbital Mechanics Department continued the research function "to develop accurate planetary, lunar and satellite ephemerides and theories, to provide expertise in celestial mechanics and solar system astrometry." By 1994 the Orbital Mechanics Department had been disbanded, and its functions were again subsumed under a research branch of the Astronomical Applications Department.[82] At the end of the twentieth century the Astronomical Applications Department, with the Nautical Almanac Office as a small component under Sethanne Howard, still oversaw the production of the *Astronomical*, *Air*, and *Nautical Almanacs*. This still required a research effort of formidable magnitude.

Looking back over the century, one can see that the American Nautical Almanac Office, like other such offices around the world, was involved in much more than the production of almanacs for navigation and a variety of other uses. In the spirit of its founder C. H. Davis, research on planetary theory and the associated astronomical constants was also a primary theme of the work of the Office. Much labor was involved in the annual production of the navigational almanacs, but underlying it all was the theory of the motions of the planets, requiring also knowledge of the reference frames against which the positions of those bodies were measured, as well as the careful use of new time scales, judicious application of new astronomical constants, and international diplomacy as well as scientific acumen. As the twentieth century closed 150 years after Maury and Davis had founded the American Nautical Almanac Office, its work was being undertaken in the radically new environment of the Space Age and electronic computers. Yet the goals of the Office remained the same, and over time had become even more closely intertwined with the positional astronomy and timekeeping functions of the Observatory. As with those functions, new technologies brought unprecedented accuracy in the service both of the Navy's practical needs and of more general scientific purposes, a process that continues today.

---

[82] John Bangert, "The Astronomical Applications Department Today," in NAOSS, 227–238; AR (1990–91), 589–590.

# Summary

The history of the U. S. Naval Observatory is a tiny slice of the history of science, limited to the contributions of a small government agency to one discipline, but rooted deeply in time as American scientific institutions go. One must therefore be cautious about generalizations, either to other government agencies or to American science, especially with the rapid growth of scientific institutions in the twentieth century. Nevertheless, several important themes have emerged in this history, ranging from issues of government and military patronage to the relations among science, technology, and the incessant drive for accuracy in science. Over 170 years six generations of Naval Observatory astronomers have observed, calculated, and analyzed their data, and disseminated their results to a broad array of users, a sustained effort unparalleled in science in the United States. In this summary we offer some final reflections on these themes and accomplishments.

### Government and military patronage for astronomy

The Observatory's history is a small but significant example of sustained government and military patronage for science, in particular astronomy and navigation. A wider investigation of this theme, even within the Navy, would encompass the histories of the Naval Research Laboratory, the Office of Naval Research, and the Oceanographer of the Navy, among others, and disciplines as diverse as oceanography, mapping, and ballistics. Beyond the Navy the scope would broaden enormously, in proportion to the growth of military and government support for science in the United States. By the end of the twentieth century, Federal funding for astronomy in the United States came from the National Science Foundation, NASA, the Department of Energy, the Smithsonian Institution, and the National Institute for Standards and Technology, as well as the Department of Defense. Although a few other DoD agencies (the Naval Research Laboratory and Air Force Phillips Laboratory) undertook research in astronomy, the Naval Observatory remained by far the Defense Department's major commitment to astronomy.[1]

In this context, the Naval Observatory and its predecessor Depot of Charts and

---

[1] Harvey M. Sapolsky, *Science and the Navy: The History of the Office of Naval Research* (Princeton University Press: Princeton, New Jersey: 1990); Gary Weir, *An Ocean in Common: American Naval Officers, Scientists, and the Ocean Environment* (Texas A&M University: Houston, Texas, 2001); David K. Allison, *New Eye for the Navy: The Origin of Radar at the Naval Research Laboratory* (Naval Research Laboratory: Washington, 1981). Current Federal astronomy funding is detailed in *Federal Funding of Astronomical Research* (National Research Council: Washington, 2000), pp. 66–75.

Instruments represent the earliest sustained effort of the U. S. government to contribute to astronomy and harness its practical applications. The tension between pure and applied research – a perennial theme in science – is evident already in Matthew Fontaine Maury's work, and in the very fact that he transformed the Congressionally mandated "Depot" not only into a "naval observatory" but also into a "national observatory." Both James Melville Gilliss and Maury undertook research programs beyond the immediate needs of the Navy, and C. H. Davis encouraged such programs in his role as the first Superintendent of the Nautical Almanac Office. That the Navy was interested in more than astronomy as applied to navigation was evident when, in 1873, Superintendent Sands and Simon Newcomb secured the purchase of the 26-inch refractor, the largest telescope in the United States; when Superintendent Hellweg and James Robertson secured the installation of the 40-inch Ritchey–Chrétien reflector in Washington even in the midst of the Great Depression; when Gerald Clemence and C. B. Watts brought John S. Hall to the Observatory in 1948 to institute a program of astrophysics; and when Superintendent Christie and Kaj Strand won the Congressional appropriation for the 61-inch reflector at the beginning of the Space Age.

The broader interests of the nation are similarly evident in the expansion of the Observatory's time service beyond the rating of chronometers to providing time to cities and railroads as the nation grew westward, and in the fact that the Naval Observatory became an important player in disseminating time for the world in the twentieth century. It was always clear that the Observatory, like all government agencies, was driven by a mission that first and foremost had to be accomplished, in this case, to keep the fleet navigating. Contributing to the advancement of astronomy, however strongly it was desired by military or civilian personnel, was of necessity a secondary goal. The allocation of resources to one or the other was sometimes a synergistic process, but more often an administrative choice that had to favor the Observatory's mission. Most of the Observatory's more widely known accomplishments – the discovery of the moons of Mars and Pluto, and of interstellar polarization, and its work on double stars, comets, and extragalactic astronomy – were byproducts of mission-related work rather than of monies specifically allocated to research projects aimed at new discoveries. This is commonplace for government institutions, but stands in sharp contrast to the two American national observatories founded in the 1950s.

Government patronage brought both advantages and disadvantages. Astronomers at universities and private observatories such as Lick, Yerkes, and Harvard College Observatory could only look with envy at the relatively large facilities and appropriations of the Naval Observatory in the nineteenth century. They could bitterly complain (as some did with regard to time service), when the government's services impinged on their own – and on their ability to generate funds. They could not hope to match the government's largess in supporting the far-flung transit-of-Venus expeditions, elaborate solar-eclipse expeditions, and sustained decades-long programs of astrometric observations. The government's expenditure of hundreds of thousands of dollars for the transit-of-Venus expeditions in 1874 and 1882, with the narrow goal of determining one of the astronomical constants known as the solar parallax, is a remarkable testimony both to the persuasive power of high-ranking naval officers and to the nation's sometimes wavering resolve to catch up to European science as the nation grew in the nineteenth century. (One should remember that $100,000 is equivalent to more than $1.5 million today.) The government's failure to appropriate the much smaller funding necessary to publish the results of the transit-of-Venus expeditions is an example of how quickly that same patronage could be withdrawn, subject to the political winds. Similarly, the expenditure of hundreds of thousands of dollars in the 1880s for the purchase of a site and construction of the new Naval Observatory is a remarkable example of government support for science – even as it wrestled with the question of the best administrative structure for the Observatory.

On the other side of the coin, however, non-government institutions were free to pursue their own research interests unfettered by mission. The Observatory was often denied swift appropriations for new technologies, including photography and spectroscopy, precisely because they were perceived as having little to do with the practical applications that were the Navy's primary interest. Once it had been convinced that a program was strongly related to its mission – as in the case of Newcomb's complex work on astronomical constants requiring a sustained effort over decades – the Navy more often than not followed through with personnel and funding in a way that private institutions could not. This tension – and occasional synergy – between pure and applied science is one of the hallmarks of the history of the Naval Observatory, and an issue that has been played out in a much more general sense throughout the Federal government and, as the government-grant process grew after World War II, throughout the nation's science establishment.

### The Naval Observatory as a national observatory

As discussed in the Introduction and as seen in Table P.1, the Naval Observatory falls into that class of institutions known as national observatories, and was the first and only national observatory in the United States for more than 100 years, the founding of the quasi-governmental Smithsonian Astrophysical Observatory in the 1890s notwithstanding. Membership in this exclusive club had several implications. First, by virtue of this status the Naval Observatory was in a unique position in the United States with respect to other observatories. It was the nation's premier observatory in the nineteenth century, a standing that it self-consciously fostered and savored. At the same time, this status brought intrinsic problems: Astronomers around the country believed that it was their duty to ensure that their national observatory was well administered, giving rise to a long and bitter controversy over civilian versus military control. More than that, with the rise of astrophysics at the end of the nineteenth century, the Observatory was faced with a momentous decision: whether to embrace the new astronomy or to stick more closely to its mission centered on time, positional astronomy, and navigation. Although it dabbled in astrophysics, the failure to embrace fully the "new astronomy" meant that the Observatory was left behind in the important advances at the opening of the twentieth century. As the importance of astrophysics grew, the Observatory did not – could not – maintain the same status in astronomy it had attained in the nineteenth century. Although it floundered in the eyes of many astronomers in the first half of the century in some aspects of its work, the Observatory fulfilled its mission during two world wars of navigating the fleet, with important work on navigational technology, mass production of chronometers, and punched-card machine techniques, accomplishments that deserve to be more appreciated today. It emerged in the second half of the century as the nation's premier Observatory in its mission-mandated niche of time, navigation, and positional astronomy, even as the scope of astronomy exploded around it. Even though it declined to be absorbed by NASA, the Observatory was reinvigorated by the Space Age, and at the end of the century was fully engaged in state-of-the-art astronomical observing technology both on the ground and in space.

Secondly, national-observatory status implied a large degree of international cooperation in each of its areas of expertise. This is especially true of the Nautical Almanac Office, which reached the peak of such cooperation when the British and American almanacs were unified at mid-century. However, it was also true in the area of time, which in the second half of the century was governed by at least five international agencies, with which the Naval Observatory, as all national time services, had to engage. It was equally true with positional astronomy, as the Observatory contributed to international projects to catalog stars, and at the end of the century played an important role in moving toward an international celestial reference frame. Indeed, it is no exaggeration to say that, in terms of pure science, during the twentieth century the Naval Observatory was often more appreciated internationally than by scientists in its own country, as measured by participation and interaction with colleagues in the International Astronomical Union as compared with the American Astronomical Society. To some extent this was inevitable, since the Naval Observatory's mission-imposed niche in astronomy isolated it from other institutions specializing in astrophysics. Toward the end of the century, however, astrometry and astrophysics were increasingly synergistic, as precise star positions were required in order to reveal the motions of the stars and galactic dynamics, the presence of dark matter, and the existence of planetary companions.

As a national observatory, the Naval Observatory invites comparison with the two national observatories in the United States founded in the 1950s, the Kitt Peak National Observatory (now known as the National Optical Astronomy Observatory, or NOAO) and National Radio Astronomy Observatory (NRAO). From the beginning both of these observatories had research as their prime objective, with little in the way of applied science. To be sure, keeping up with pure astronomical research was another avenue of advancing the national interest, but very different from the Naval Observatory's goals.

Similarly, the Naval Observatory suggests comparison internationally with Greenwich, Paris, and Pulkovo Observatories, among others, all of which were founded to meet the practical needs of national governments. Greenwich was the quintessential example of a national observatory founded to advance navigation; although its storied history is filled with some of the most famous names in the history of astronomy, it was administered until 1965 not by astronomers, but by the British Admiralty. In that year control passed from the Ministry of Defence (Navy) to the Science Research Council, and in 1994 to the Particle Physics and Astronomy Research Council (PPARC). The demise of the Royal Greenwich Observatory as an active scientific institution in 1998 was due in part to the fact that, having lost the patronage of the Admiralty, the Observatory had to compete for funding along with other research institutions in the U. K. This "survival-of-the-fittest" strategy, combined with political conditions, proved too much even for one of the world's oldest scientific institutions.[2] Paris Observatory (which combined pure and practical research) has fared better, and Pulkovo Observatory at the end of the twentieth century still survived, if diminished by the tumultuous events of the 1990s in Russia. Such comparisons, general as they are, remind us that the first American national observatory need not have developed as it did under naval patronage; indeed, if John Quincy Adams had had his way, astronomical history – and the history of the Smithsonian Institution – would have unfolded very differently. Although the task of a full-blown comparison of national observatories is beyond this volume, it is one that would repay close attention as the histories of these astronomical institutions become better known.

### Military versus civilian control

The problem of administrative control of the Observatory, which was so important in the nineteenth and early twentieth centuries, gradually lost its force. Despite the post-World-War-II founding of the National Science Foundation (NSF) – which funded the two new American national observatories in the 1950s – the Naval Observatory was not a candidate for such patronage precisely because of the practical nature of its work. The primary goal of the NSF was to fund pure research, which, to be sure, might later be applied.[3] However, the NSF was quite happy to let the Navy pay

---

[2] On the changing status of the Royal Greenwich Observatory (which had been renamed from Royal Observatory Greenwich after its move to Herstmonceux in 1948), see A. J. Meadows, *The Royal Observatory at Greenwich and Herstmonceux, 1675–1975*, volume 2, *Recent History (1836–1975)* (Taylor and Francis: London, 1975), chapter 5, "The Status of the Royal Observatory," pp. 107–129. On its closing see "Royal Greenwich Observatory to Close," *Sky and Telescope*, **94** (1997), 21.

[3] J. Merton England, *A Patron for Pure Science: The National Science Foundation's Formative Years, 1945–1957* (NSF: New York, 1982).

for its own navigational needs, either through annual appropriations or through grants from the Office of Naval Research. Short of using the Naval Observatory as the core of a new kind of national observatory – an idea that John Hall broached but the Observatory rejected in the 1950s – the long-term nature of the Naval Observatory's programs rendered funding from anywhere outside the government unlikely.

The continued ties of both positional astronomy and time to navigation made the U. S. Navy the sensible, if not the only possible, patron within the U. S. government. Earlier in the century it had been suggested that the Smithsonian administer the Observatory, and, although that idea would have had a certain historical symmetry in terms of the earlier efforts of John Quincy Adams, time had passed that idea by after World War II. Because time and frequency had come under the rubric of "standards," the National Bureau of Standards (NBS) might have absorbed the timekeeping functions of the Observatory. However, the Navy was proud of its longstanding tradition of time, which is so closely related to navigation, and in any case the NBS could not reasonably have overtaken the Observatory's expertise in positional astronomy and almanac production. The Coast and Geodetic Survey (later the National Geodetic Survey) overlapped in expertise with the latter functions, but – busy enough with its own mission – made no attempt to secure them. NASA might have absorbed the Observatory at the beginning of the Space Age, but Gerald Clemence, in his role as Scientific Director, chose not to take that course.

Two administrative innovations served to provide a balance between military administration and a decision-making role for civilian astronomers. The formation of the Astronomical Council in 1908 gave the highest-ranking astronomers more input into decisions that affected the staff. Also – after an abortive attempt early in the century – the appointment of a civilian Scientific Director since 1958 not only provided a more effective means to work with the military Superintendent, but also became a practical necessity with the increasing sophistication of astronomical technologies.

By the end of the century the Naval Observatory remained firmly in the hands of the Navy, and was not only its oldest scientific institution, but also the oldest continuously operating scientific institution in the U. S. government. The wisdom of remaining in the Department of Defense was underlined with the Observatory's crucial role in providing time to the Global Positioning System satellites, a DoD system whose atomic-clock heartbeat, the government felt, best originated from within the DoD. No less important were the star positions and Earth-orientation parameters used for classified DoD purposes.

### Technology, science, and the drive for accuracy

Instrumentation, the drive for accuracy, and their relation to scientific advance have been pervasive themes throughout this history. From the Observatory's beginning as a Depot of Charts and Instruments, improvements in instrumentation have played an essential role: Navigators needed the best chronometers, sextants, and compasses; astronomers needed bigger and more efficient telescopes, whether refractors, reflectors, or zenith and meridian instruments; and time-service personnel lobbied incessantly for more accurate standard clocks, from the early pendulum clocks to the Riefler, Shortt, quartz-crystal, and atomic varieties. Increasingly accurate techniques for finding and disseminating time, polar motion, and other Earth-orientation parameters were no less crucial.

Because of the Naval Observatory's concentration on positional astronomy, time, and navigation, the history of the Observatory has been the history of this drive for accuracy to a far greater extent than for other observatories. Accordingly, the improvement of instrumentation to achieve these goals has been a critical activity, whether based in developments within the Observatory or procured from the outside. The transit circle, the workhorse instrument of the Observatory for most of its history, is the prime example in the field of positional astronomy. During the Maury era transit instruments were used for the determination of right ascension, and the mural circle for declination, but by 1866 the two functions had been combined in a single instrument at the Observatory, the 8.5-inch Pistor and Martins transit circle. This instrument was modified to the 9-inch transit circle in the late nineteenth century, joined by the 6-inch transit circle in 1897, and superseded by the 7-inch transit circle after World War II. Together the 6- and 7-inch transit circles charted the northern and southern skies, culminating in an unprecedented coordinated pole-to-pole program of star observations, which ended in 1995.

In the course of their history these transit circles embodied a whole host of innovations designed to improve the accuracy of star positions. The engraved circles for declination were constantly improved, even as the methods of reading them went from visual to photographic to photoelectric, and the methods for determining and applying their errors were refined. Similarly, the micrometer in the focal plane of the eyepiece was improved, as were methods for reading it, culminating in electronic techniques late in the century. The more accurate determination of the right-ascension coordinate depended not only on improvements in clocks, but also on methods of registering and permanently recording time. These methods, too, advanced from the eye-and-ear method to the chronograph then to photographic registration in the 1930s, and finally to digital readouts directly input to computers, which also analyzed the data. After 150 years of such incremental innovations, the end result of all this was the improvement in accuracy of star positions by a factor of ten, from a few tenths to a few hundredths of an arcsecond. As can be graphically seen in Figure P.1, the improvement over that time was excruciatingly slow.

Similar improvements were seen in the detectors of the equatorial telescopes. The 26-inch telescope, which Asaph Hall used visually to discover the moons of Mars in 1877, eventually progressed to photographic techniques for its double star program, and in the 1990s inaugurated electronic techniques in the form of the speckle interferometer. The 61-inch reflector in Flagstaff, which began with photographic techniques for stellar parallaxes in the early 1960s, by the end of the century produced sub-milliarcsecond accuracies with the new charge-coupled device (CCD) technology. By the end of the century the astrographic telescope was using not only CCD detectors, but also new lenses sensitive to regions of the spectrum that yielded better data.

Progress was faster in the arena of precision timekeeping than it was in precision astrometry. As Figure P.2 shows, the curve of improvement was much steeper. We have seen the reasons for this in the progression from pendulum clocks, to the precision pendulum Riefler and Shortt clocks, to the quartz-crystal and atomic clocks. The fundamental reason was that these clocks represented breakthrough technologies in timekeeping, as contrasted with the transit circle, which advanced only through the incremental changes described above.

Only in the last two decades did breakthrough technologies occur that would supersede the transit circle in the way that pendulum clocks were eventually superseded. The Hipparcos satellite, which orbited above the Earth's atmosphere and its refraction effects, achieved in a few years the same tenfold increase in accuracy that had taken 150 years of innovations on transit circles. The Naval Observatory played only a minimal role in this European project, but at the end of the century was working on its own astrometric satellite. Radio astronomy brought milliarcsecond accuracies from the ground, and optical interferometry promised the same at the end of the century. The effect is seen in the dramatic downturn in the curve in Figure P.1. The resulting shift from the dynamical reference frame based on the solar system, to the international celestial reference frame based on extragalactic objects, was truly a landmark event.

Navigation, the ultimate goal of this work, benefited tremendously from these improvements in accuracy, both to time and to positional accuracy. Whereas Lt Goldsborough complained of ships being miles off course due to poorly rated chronometers, at the end of the century the Global Positioning System could provide positions on land, sea, or air to within a few meters. Figure P.3 shows the dramatic improvement. GPS satellites radically altered the status of and need for celestial navigation, rendering secondary a process that had been the origin of the Observatory and central for most of its history. At the end of the century the same satellite system provided the most reliable and accurate method for disseminating time around the world, rendering mostly obsolete previous radio methods.

New observation and measurement technologies in themselves were hardly enough for scientific advancement, especially in the twentieth century. The bulk of the Observatory's work was calculation-intensive; faster and more efficient methods of calculation were essential both to timeliness and to accuracy. Part of our story has therefore been to detail the transition from hand calculations using logarithms and Crelle's tables, to electric machines such as the Millionaire and Marchant, to the punched-card accounting machines introduced by Eckert in World War II, and finally to the electronic computer. The development of measurement technology to record and measure telescope data, from the chronograph for time, to transit-circle readings for angles, and photographic-plate-measuring machines for images, has also been essential. Finally the setting and data acquisition of the telescopes themselves were computerized, as was the analysis of the data. Human "computers" in the nineteenth century were replaced by electronic computers in the late twentieth, with landmark consequences that are still being played out into the twenty-first century.

### Six generations, and three levels of history

Because of its long history, the Naval Observatory provides a unique opportunity to examine the activities, interactions, and career paths of six generations of scientists spanning 170 years (Appendix 4). This is another of the tasks only begun in this history, which has discussed many astronomers whose work has largely been forgotten, yet who did important work in the field. To undertake a full discussion would require another book that would examine such facets of social history as intellectual and educational background, the evolution of careers, status in the scientific community, external rewards, and similar questions pioneered in the broad work of Marc Rothenberg and John Lankford.[4] Although such a study focused on a single astronomical institution with such longevity as the Naval Observatory would undoubtedly pay substantial rewards, we can only point to the general questions in this summary.

It is important to note that the practical work of the Naval Observatory was appealing only to a certain kind of scientist. In his work on the astronomer Henry Norris Russell, David DeVorkin has distinguished astronomical "foxes" from "hedgehogs." The fox seeks out hot topics in the field, making effective use of the data produced by the hedgehogs. In other words, a hedgehog is a producer of data and a fox is a user.[5] If Russell is the quintessential example of an astronomical fox, Naval Observatory astronomers, and the Naval Observatory as an institution, largely play the role of hedgehogs, producers of data, whether in positional astronomy, time, and Earth-orientation parameters, or almanac production. Nor is such a designation pejorative; many observatories have specialized in data production, in whole or in part. The production of spectra at Pickering's Harvard, radial velocities at Campbell's Lick Observatory, or star positions at national observatories such as the Royal Greenwich Observatory are prime examples.[6] In a broader sense, many government

---

[4] John Lankford *American Astronomy: Community, Careers and Power, 1859–1940* (University of Chicago Press: Chicago, 1997); Marc Rothenberg, "The Education and Intellectual Background of American Astronomers," Ph. D. dissertation (Bryn Mawr College, 1974).

[5] D. H. DeVorkin, "A Fox Raiding the Hedgehogs: How Henry Norris Russell Got to Mt. Wilson," in *The Earth, the Heavens and the Carnegie Institution of Washington*, ed. Gregory A. Good (American Geophysical Union: Washington, 1994), pp. 103–111.

[6] On Campbell's work on radial velocities, see Donald E. Osterbrock, John R. Gustafson, and W. J. Shiloh Unruh, *Eye on the Sky: Lick Observatory's First Century* (University of California Press: Berkeley, 1988), chapter 8, "The Creative Scientist Who Became a Factory Manager, 1900–1923," pp. 130–148. Robert Smith, "A National Observatory Transformed: Greenwich in the Nineteenth Century," JHA, **22** (1991).

institutions, quintessentially the National Institute for Standards and Technology, produce routine, but important data for other users.

The vast majority of positional astronomers at the Naval Observatory specialized in observing and producing catalogs of the positions of celestial objects. A few went a step further and produced stellar proper motions. However, proper motions were not an end in themselves; yet rare was the Naval Observatory astronomer who went beyond such data to actual scientific or practical applications, unless it was to determine astronomical constants, planetary masses, or other basic data useful for navigation. This is not surprising; after all, the primary mission of Naval Observatory astronomers was to produce data, not to use it themselves for basic research. Whether for navigation or astrophysics, the chief users were the Navy, other astronomers, and the public. The same was true for time, with rare exceptions, such as Naval Observatory involvement in experiments to verify the time-dilation effect in relativity in the early 1970s. An examination of the types of astronomers who sought employment at the Naval Observatory, how they changed over time, and how they compared with astronomers at other institutions would undoubtedly prove illuminating.

The internal movement of Naval Observatory astronomers once they became members of the institution is cause for further study. It is clear that, prior to the Space Age, astronomers tended to move among different Divisions more than they did later, undoubtedly due to increasing specialization. Thus, Clemence began as a junior astronomer in the time service, and worked with H. R. Morgan in the 9-inch Transit Circle Division before entering the Nautical Almanac Office, becoming its Director, and eventually the Observatory's Scientific Director. Such broad experience with the Naval Observatory's work clearly helped Clemence in his career. Similar flexibility marks the careers of Newcomb and Harkness in the nineteenth century, and C. B. Watts in the twentieth century, all of whom attained high positions within the observatory, and high stature in the astronomical community. The decline of such mobility with increasing specialization, and the subsequent effect on careers, is one of the themes awaiting study of an astronomical community where the institutional variable is held constant. The question of how such career phenomena bear on the findings of Rothenberg and Lankford has yet to be analyzed.

Aside from the work of astronomers, their interactions with each other are also notable. Since the influx of astronomers hired by Gilliss around the Civil War, whole generations of Naval Observatory astronomers tended to play out their careers together, retiring at about the same time, and passing on the torch to a new generation (Appendix 4). There are, of course, exceptions to this rule, but in general careers at the Observatory tended to be long and the astronomers close friends or protracted enemies. In the second generation Newcomb and Harkness, and Hall and Eastman are prime examples. In the third and fourth generations G. H. Peters, Asaph Hall Jr, H. R. Morgan, F. B. Littell, J. C. Hammond, C. B. Watts, A. N. Adams, and F. P. Scott stand out among others. Some astronomers stayed much shorter than their career generation, passing through, as did H. S. Pritchett on his way to the Presidency of MIT, E. S. Holden before becoming Director of Lick Observatory, and John Hall on his way to becoming the Director of Lowell Observatory. Some came only to make special contributions in time of national emergency, such as Wallace Eckert and Paul Herget, before returning to their home institutions. Occasionally, staff members spanned significant parts of two generations, as did Thomas Harrison, who, in his role as the Clerk of the Observatory (1848–1921), served almost 75 years, and H. R. Morgan, who had a foot equally in the Observatory's third and fourth generations.

The study of the interactions of astronomers, and the subsequent benefit or detriment to science, is no easy task. Nor, in fact, are many of the conclusions of institutional history easy to draw, especially in the sociological and political realms. In this respect, and in respect to all history, three levels may be distinguished in terms of what the historian may know at some distance from actual events. The first level is laid bare in the printed record, the primary trace the author leaves in the way that he or she intended to convey it to the scientific community or the general public. Quite aside from the all-important data, historians have long recognized that this public face is a sanitized version of events that may be far from the full story both in terms of motivations and regarding how a discovery was actually made. The second level of history is sometimes revealed in the archival record, where correspondence between astronomers, administrators and others may illuminate the motivations of the players and often help explain the content of the public record. However, there remains a third level, which is often hinted at in oral histories but in many cases forever inaccessible, of "what actually happened in history." There is little hint in the printed record of the animosity between Newcomb and Harkness – the latter once Newcomb's best man at his wedding. This animosity becomes abundantly clear in the archival record, where Harkness bitterly objects to Newcomb's administrative manipulations and his scientific work. Nonetheless, the source of this animosity – the day-to-day interaction that at some point first split scientific friends and over time converted them into enemies – remains forever hidden, unless it is discoverable in some archive. The same may be said for the origins of the Observatory in 1842, its reincarnation in the 1880s, and virtually all of its landmark events (not to mention those of lesser importance). The testimony of the written record notwithstanding, we will never know entirely the personal and political alliances, the bureaucratic tangle of influences, and the informal wrangling on the Congressional floor that form the stuff of history and yet go unrecorded in any manner. Despite the importance of archives and oral histories, these sources in turn are also open to interpretation. In this history of the Naval Observatory we have most often achieved the first level, sometimes the second level, but perhaps very rarely (to a degree we will never know) the third level where the "truth" of the matter resides. History most often asymptotically approaches the truth, but rarely, if ever, reaches it.

**Sky and ocean joined**

This history has documented the relation between astronomy and navigation, along with a delicate balance of pure astronomical research, as played out in one scientific institution over a span of 170 years. Over that extended period the Naval Observatory reflects the increasing specialization of science, as it lost first its hydrographic function (now the purview of the Oceanographer of the Navy), then its mapping function (now incorporated into the National Imagery and Mapping Agency, or NIMA), and finally its chronometer and navigational instrument function (now largely superseded by GPS satellites). What remained – the determination of time and Earth orientation parameters, positional astronomy, and a variety of astronomical data services – was still essential to navigators on land, sea, and air, and in space. Even as traditional methods of celestial navigation were radically altered by the Space Age, sky and ocean remained inextricably joined in their historic relationship, and seemed certain to maintain that relation into the foreseeable future.

# Select bibliographical essay

This essay deals only with published sources; for an overview of sources see Appendix 1. Aspects of the history of the Naval Observatory have appeared in J. E. Nourse, *Memoir of the Founding and Progress of the United States Naval Observatory* (Government Printing Office: Washington, 1873); Gustavus A. Weber, *The Naval Observatory: Its History, Activities and Organization* (Johns Hopkins Press: Baltimore, 1926); Jan K. Herman, *A Hilltop in Foggy Bottom: Home of the Old Naval Observatory and the Navy Medical Department* (reprinted from *U. S. Navy Medicine*, 1984); Steven J. Dick and LeRoy Doggett, eds., *Sky with Ocean Joined: Proceedings of the Sesquicentennial Symposia, U. S. Naval Observatory* (U. S. Naval Observatory: Washington, 1983); Gail S. Cleere, *The House on Observatory Hill* (Government Printing Office: Washington, 1989); and Alan Fiala and Steven J. Dick, eds., *Proceedings Nautical Almanac Office Sesquicentennial* (U. S. Naval Observatory: Washington, 1999). For a compilation of information about clocks and other matters related to time at the Naval Observatory see James A. DeYoung, "United States Naval Observatory Clocks and other Time-Keeping Apparatus found in the Washington Observations and the Publications of the USNO, 2nd series," 35-page typescript (c. 1983), Naval Observatory Library, and the same author's "United States Naval Observatory Clocks Mentioned in the Annual Reports, 1842–1930" (1982), 18-page typescript, Naval Observatory Library.

Many books provide insight into the broad context of Naval Observatory history. They are too numerous to mention here, and many have been cited in the references. However, I have found particularly useful Robert Bruce, *The Launching of Modern American Science, 1846–1876* (Knopf: Ithaca, 1987); A. Hunter Dupree, *Science in the Federal Government* (Harvard University Press: Cambridge, Massachusetts, 1957); Daniel Kevles, *The Physicists: The History of a Scientific Community in Modern America* (Knopf: New York, 1971); Nathan Reingold, *Science in Nineteenth Century America: A Documentary History* (Hill and Wang: New York, 1964), and Nathan Reingold, ed., *The Sciences in the American Context: New Perspectives* (Smithsonian Institution Press: Washington, 1979). *The Papers of Joseph Henry*, edited by Nathan Reingold and Marc Rothenberg (Smithsonian Institution Press: Washington, 1972– ), is an invaluable resource on American science during the nineteenth century. Thus far eight volumes have been published, covering the years 1797–1853. On the American scientific community I have benefited from Sally Gregory Kohlstedt, *The Formation of the American Scientific Community: The American Association for the Advancement of Science, 1848–1860* (University of Illinois Press: Urbana, Illinois, 1976), and Sally Gregory Kohlstedt, Michael M. Sokal, and Bruce V. Lewenstein, *The Establishment of Science in America: 150 years of the American Association for the Advancement of Science* (Rutgers University Press: New Brunswick, New Jersey, 1999). On the American astronomical community essential references are John Lankford, *American Astronomy: Community, Careers and Power, 1859–1940*

(University of Chicago Press: Chicago, 1997), and Marc Rothenberg, "The Education and Intellectual Background of American Astronomers," Ph.D. dissertation (Bryn Mawr College, 1974). The formation and rise of the American Astronomical Society provides context for our story at the turn of the twentieth century; see David DeVorkin, ed., *The American Astronomical Society's First Century* (American Institute of Physics: Washington, 1999).

The history of American astronomical institutions is finally receiving attention. Among recent examples are Donald Osterbrock, John Gustafson, and W. J. Shiloh Unruh, *Eye on the Sky: Lick Observatory's First Century* (University of California Press: Berkeley, California, 1988); Donald E. Osterbrock, *Yerkes Observatory, 1892–1950: The Birth, Near Death, and Resurrection of a Scientific Research Institution* (University of Chicago Press: Chicago, 1997), and Frank K. Edmondson, *AURA and Its U. S. National Observatories* (Cambridge University Press: Cambridge, 1997). The Smithsonian Astrophysical Observatory centennial issue of *Journal for the History of Astronomy,* **21** (February, 1990) also contains important papers. Substantive histories exist for only a few national observatories; see for example the Greenwich Tercentenary volumes *Greenwich Observatory: The Royal Observatory at Greenwich and Herstmonceux, 1675–1975*; volume 1, Eric G. Forbes, *Origins and Early History (1675–1835)* (Taylor and Francis: London, 1975); volume 2, A. J. Meadows, *Recent History (1836–1975)* (Taylor and Francis: London, 1975); and volume 3, Derek Howse, *The Buildings and Instruments* (Taylor and Francis: London, 1975).

Among the many useful works on general naval history are Kenneth J. Hagan, *This People's Navy: The Making of American Sea Power* (Free Press: New York, 1991); Stephen Howarth, *To Shining Sea: A History of the United States Navy, 1775–1991* (New York, 1991); and Dudley W. Knox, *A History of the United States Navy* (New York, 1948). On the history of navigation and navigational instruments see J. E. D. Williams, *From Sails to Satellites: The Origin and Development of Navigational Science* (Oxford University Press: New York, 1992); W. J. H. Andrewes, *The Quest for Longitude* (Collection of Historical Scientific Instruments, Harvard University: Cambridge, Massachusetts, 1996); Dava Sobel and William J. H. Andrewes, *The Illustrated Longitude* (Walker and Company: New York, 1998); Peter Ifland, *Taking the Stars: Celestial Navigation from Argonauts to Astronauts* (Krieger Publishing Co.: Malabar, Florida, 1998); Allan Chapman, *Dividing the Circle: Development and Critical Measurement of Celestial Angles, 1500–1850* (E. Horwood; New York, 1990).

For Part I of this history, by far the most comprehensive source on Louis Goldsborough is Jean Ponton, *Rear Admiral Louis M. Goldsborough: The Formation of a Nineteenth Century Naval Officer,*" Ph.D. Dissertation (Catholic University: Washington, 1996). Goldsborough's full paper arguing for the establishment of a Depot appears in Steven J. Dick, "Louis M. Goldsborough's Proposal to Establish a Depot of Charts and Instruments in the U. S. Navy: Text and Commentary," *Rittenhouse: Journal of the American Scientific Instrument Enterprise,* **4** (May, 1990), 79–86. The best biographical source for Charles Wilkes is his own autobiography, published as *Autobiography of Rear Admiral Charles Wilkes, U. S. Navy, 1798–1877*, eds. William James Morgan *et al.* (Department of the Navy: Washington, 1978). See also Daniel M. Henderson, *Hidden Coasts: A Biography of Admiral Charles Wilkes* (New York, 1953).

The best primary source for the origins of the Depot of Charts and Instruments and the Naval Observatory during the crucial period 1842–1844 is James Melville Gilliss's own account in *Report of the Secretary of the Navy, Communicating A Report of the Plan and Construction of the Depot of Charts and Instruments, with a Description of the Instruments*, 28th Congress, second session, February 18, 1845, Senate Executive Document 114; reprinted in I. B. Cohen, *Aspects of Astronomy in America in the Nineteenth Century* (Arno Press: New York, 1980). The most comprehensive, if somewhat hagiographical account of Matthew Maury is Frances Leigh Williams, *Matthew Fontaine Maury, Scientist of the Sea* (Rutgers University Press: New Brunswick, 1963), which makes heavy use of archival sources. The problem of bias is also evident in the accounts of C. H. Davis by his son, C. H. Davis (Jr), "Memoir of Charles Henry Davis," BMNAS, **4** (1902), 25–55, and C. H. Davis (Jr), *Life of Charles Henry Davis, Rear Admiral* (Houghton, Mifflin and Co.: Boston and New York, 1899), p. 86. The best source of Gilliss's life is Gould's account in "Biographical Notice of James Melville Gilliss," BMNAS, **1** (1867). It too must be used with care, both for its glowing account of Gould's friend, Gilliss, and in its obvious bias against Gould's enemy, Maury, in the wake of the Civil War. The best source of Gilliss's genealogy and family is Frances Howard Ford Greenidge, *Ancestors of Raymond Oakley Ford and Frances Howard Ford* (privately published, 1994), volume 1, Naval Observatory Library. Gilliss's most substantial writing is his six-volume *The U. S. Naval Astronomical Expedition to the Southern Hemisphere during the Years 1849–50–51–52* (Government Printing Office: Washington, 1855–56).

For Part II of this volume, the post-Civil-War era, Asaph Hall's, *Biographical Memoir of John Rodgers*, BMNAS (Judd and Detweiler: Washington, 1906) is essential for understanding the crucial figure in transitioning to the new Naval Observatory. A modern treatment is Robert E. Johnson, *Rear Admiral John Rodgers, 1812–1882* (United States Naval Institute: Annapolis, 1967). My treatment of time at the Naval Observatory during the nineteenth century has been influenced greatly by Ian Bartky, *Selling the True Time: Nineteenth-Century Timekeeping in America* (Stanford University Press: Palo Alto, California, 2000). Also relevant is Michael O'Malley's *Keeping Watch: A History of American Time* (Viking: New York, 1990). Asaph Hall's discovery of the moons of Mars is detailed in Hall's *Observations and Orbits of the Satellites of Mars, with Data for Ephemerides in 1879* (Nunn and Company: Washington, 1878). Modern accounts of the discovery of the moons of Mars are given in the centennial volume *The Satellites of Mars, Vistas in Astronomy*, **22** (1978). The story has even been novelized in Thomas Mallon, *Two Moons* (Pantheon Books: New York, 2000). Among the most useful obituaries of Hall is G. W. Hill's *Biographical Memoir of Asaph Hall, 1829–1907* (Washington, 1908). Also interesting for different reasons are the accounts of his sons, Percival Hall, *Asaph Hall: Astronomer, A Biographical Sketch of Asaph Hall*, 3d (1945), privately printed and distributed by its author, and Angelo Hall, *An Astronomer's Wife: The Biography of Angeline Hall* (Baltimore, 1908). The best account of Alvan Clark and his instrumentation is Deborah J. Warner and Robert B. Ariail, *Alvan Clark & Sons: Artists in Optics* (second edition, Wilman-Bell: Richmond, Virginia, 1995). The history of the 26-inch refractor on the occasion of its centennial is given in Robert W. Rhynsburger, "A Historic Refractor's 100th Anniversary," *Sky and Telescope*, **46** (October, 1973), 208.

The nineteenth-century transits of Venus are of increasing interest because, for the first time since the events described in chapter 7, the phenomenon will occur again in 2004 and 2012. The eighteenth-century expeditions are described in Harry Woolf, *The Transits of*

*Venus: A Study of Eighteenth-Century Science* (Princeton University Press: Princeton, New Jersey, 1959). A contemporary account of the nineteenth-century transits of Venus is R. A. Proctor, *Transits of Venus: A Popular Account of Past and Coming Transits* (fourth edition, Longmans, Green & Co.: London, 1882). Modern historians have tackled the subject; see Albert van Helden, "Measuring Solar Parallax: The Venus Transits of 1761 and 1769 and their Nineteenth-Century Sequels," in R. Taton and C. Wilson, eds., *Planetary Astronomy from the Renaissance to the Rise of Astrophysics*, Part B (Cambridge University Press: Cambridge, 1995), *General History of Astronomy*, volume 2. Part of the interest of the transits of Venus derives from the early use of photography for astronomy. See John Lankford, "Photography and the 19th-Century Transits of Venus," *Technology and Culture* (1987), 648–657. On the rise of photography in astronomy, see Lankford, "The Impact of Photography on Astronomy," in Owen Gingerich, ed., *Astrophysics and Twentieth-Century Astronomy to 1950* (Cambridge University Press: Cambridge, 1984), *General History of Astronomy*, volume 4A, pp. 16–39. The upcoming transits of Venus, with a brief history of those in the past, is given in Eli Maor, *June 8, 2004: Venus in Transit* (Princeton University Press: Princeton, New Jersey, 2000).

The primary sources for the American transit-of-Venus expeditions are as follows. In preparation for the 1874 Transits the U. S. Commission on the Transit of Venus published *Papers Relating to the Transit of Venus in 1874, Prepared under the Direction of the Commission Authorized by Congress and Published by Authority of the Hon. Secretary of the Navy*, Part I (Government Printing Office: Washington, 1872). Part I comprised 25 pages of letters relating to preparations in 1872, and Newcomb's paper on photographic techniques. Part II (Government Printing Office: Washington, 1872) comprised 48 pages of G. W. Hill's charts and tables, dated June 29, 1872, giving his predictions for times and places of visibility. Part III, comprising 51 pages, was an *Investigation of Corrections to Hansen's Tables of the Moon; with Tables for their Application*, by Simon Newcomb (Government Printing Office: Washington, 1876), to be used in analyzing the occultation observations for determining the longitudes of the various stations. The Commission published separately 28 pages of *Instructions for Observing the Transit of Venus, December 8–9, 1874* (Government Printing Office: Washington, 1874). After the 1874 transit the Commission published *Observations of the Transit of Venus, December 8–9, 1874, Made and Reduced under the Direction of the Commission Created by Congress*, edited by Simon Newcomb, Secretary of the Commission. Part I, "General Discussion of Results" (Government Printing Office: Washington, 1880), consists of 157 pages plus illustrations. Part II consists of two volumes of page proof only, deposited in the Naval Observatory Library. Parts III and IV do not exist.

For the 1882 transits see the 50 pages of *Instructions for Observing the Transit of Venus, December 6, 1882* (Government Printing Office: Washington, 1882); four pages of *Instructions Respecting Time Signals to be used in Connection with the Transit of Venus, December 6, 1882* (Government Printing Office: Washington, 1882); and an 18-page *List of Articles furnished to the U. S. Transit of Venus Parties in December, 1882*. Newcomb's *The Elements of the Four Inner Planets and the Fundamental Constants of Astronomy* (Government Printing Office: Washington, 1895) places the transit results in context in the sense that it shows the extent to which they entered into the actual set of astronomical constants, compared with other methods for determining the solar parallax.

The broadest study of Simon Newcomb is Albert E. Moyer, *A Scientist's Voice in*

American Culture: Simon Newcomb and the Rhetoric of Scientific Method (University of California Press: Berkeley, California, 1992). More detailed analysis of Newcomb's scientific work is found in Arthur L. Norberg, "Simon Newcomb and Nineteenth-Century Positional Astronomy," Ph.D. dissertation (University of Wisconsin-Madison, 1974); Norberg, "Simon Newcomb's Early Astronomical Career," Isis, **69** (1978), 209–225; and Norberg, "Simon Newcomb's Role in the Astronomical Revolution of the Early Nineteen Hundreds," in Dick and Doggett, eds., Sky with Ocean Joined (U. S. Naval Observatory: Washington, 1983), pp. 74–88. A fascinating account of his many activities is given in Newcomb's own Reminiscences of an Astronomer (Houghton & Mifflin: Boston and New York, 1903).

For Part III of this volume, two essential primary documents related to the location of the Observatory's new site are John Rodgers, Reports on the Removal of the United States Naval Observatory (Government Printing Office: Washington, 1877), also in Report of the Secretary of the Navy, 1877, 45th Congress, second session, House Executive Documents, Volume VII, no. 1, part 3, pp. 307–324, and Daniel Ammen et al., Report of the Commission on Site for the Naval Observatory (Government Printing Office: Washington, 1879), also in Report of the Secretary of the Navy, 1878, 45th Congress, third session, House Executive Documents, Volume VIII, no. 1, part 3, pp. 292–307. An important but neglected primary source on astronomical institutions in the late nineteenth century is Lt Albert G. Winterhalter, The International Astrophotographic Congress and A Visit to Certain European Observatories and Other Institutions (Government Printing Office: Washington, 1889). The military–civilian controversy at the Naval Observatory was first detailed in Howard Plotkin, "Astronomers vs. the Navy: The Revolt of American Astronomers over the Management of the United States Naval Observatory, 1877–1902," Proceedings of the American Philosophical Society, **122** (December 18, 1978), 385–399. Two important primary sources for this tumultuous period in the Observatory's history are Report of the Board of Visitors to the United States Naval Observatory (Government Printing Office: Washington, 1899), Report of the Board of Visitors to the Naval Observatory for the Year 1901 (Government Printing Office: Washington, 1901).

The history of star catalogs has been neglected, but a good starting point by a working astronomer is Hans Eichhorn, Astronomy of Star Positions (Ungar: New York, 1974). No comprehensive history of astrometry has been written, but the outlines may be glimpsed in general histories of astronomy such as A. Pannekoek's A History of Astronomy (Interscience Publishers: London and New York, 1961). A brief history of astrometry is given in Steven J. Dick, "Astrometry", in John Lankford, ed., History of Astronomy: An Encyclopedia (Garland: New York and London, 1997), pp. 47–60; the article includes further references and a detailed table on the improvement in precision of astrometric measurements. The same volume contains many articles on related subjects. For context on the 40-inch reflector, G. W. Ritchey and his telescopes are discussed in Donald Osterbrock, Pauper and Prince: Ritchey, Hale and Big American Telescopes (University of Arizona Press: Tucson, 1993).

An overview of the concepts of time is James Jesperson and Jane Fitz-Randolph, From Sundials to Atomic Clocks (second edition, 1999). The interaction of time and navigation is emphasized in Derek Howse, Greenwich Time and the Longitude (Philip Wilson: London, 1997). Two definitive works on the history of chronometers have been written by an instrument maker employed at the Naval Observatory from 1940 until 1950: Marvin E. Whitney, *Military Timepieces* (American Watchmakers Institute Press: Cincinatti, 1992), Whitney, *The Ship's Chronometer* (American Watchmakers Institute Press: Cincinatti, 1985). One of the few histories of timekeeping in a national context is Malcolm M. Thomson, *The Beginning of the Long Dash: A History of Timekeeping in Canada* (University of Toronto Press: Toronto, 1978), written by one who was a participant for 40 years. For the history of polar-motion studies and their applications to time, see Steven J. Dick, Dennis W. McCarthy, and Brian Luzum, eds., *Polar Motion: Historical and Scientific Problems* (Astronomical Society of the Pacific: San Francisco, 2000). Essential works on the history of atomic time and atomic clock technology are Tony Jones, *Splitting the Second: The Story of Atomic Time* (Institute of Physics Publishing: Bristol and Philadelphia, 2000), and Paul Forman, "Atomichron: the Atomic Clock from Concept to Commercial Product," *Proc. IEEE*, **73** (1985), 1,181–1,204.

Both past and present aspects of the Nautical Almanac Office are treated in A. Fiala and S. Dick, eds., *Proceedings Nautical Almanac Office Sesquicentennial* (U. S. Naval Observatory: Washington, 1999). Much technical data and some history are provided in the standard reference, P. Kenneth Seidelmann, ed., *Explanatory Supplement to the Astronomical Almanac* (University Science Books: Mill Valley, California, 1992). No comprehensive history of celestial mechanics has been written, but very useful are the articles in R. Taton and C. Wilson, eds., *Planetary Astronomy from the Renaissance to the Rise of Astrophysics, Part B: The Eighteenth and Nineteenth Centuries* (Cambridge University Press: Cambridge, 1995), volume 2 of *The General History of Astronomy*, M. Hoskin, ed. See especially Bruno Morando, "The Golden Age of Celestial Mechanics," pp. 211–239, and Morando, "Three Centuries of Lunar and Planetary Ephemerides and Tables," pp. 251–259. Also very useful as a general overview is LeRoy Doggett, "Celestial Mechanics," in John Lankford, ed., *History of Astronomy: An Encyclopedia* (Garland: New York and London, 1997), pp. 131–140, including further references. Many works on celestial mechanics provide brief historical perspective, including Dirk Brouwer and Gerald M. Clemence, *Methods of Celestial Mechanics* (Academic Press: New York and London, 1961). Such works also provide a snapshot of where celestial mechanics stood at the time of publication.

# Appendix 1: Sources

1. **Published sources**

The primary published sources for U. S. Naval Observatory history are its *Annual Reports*, its published volumes of observations and research, and its *Circulars*. In addition, Congressional documents are essential to tracking Congressional debate and legislation.

A. *Observatory publications*

The *Annual Reports* of the U. S. Naval Observatory, and its predecessor, the Depot of Charts and Instruments, have gone under several names and appeared in several formats. Beginning in 1842, and through most of the nineteenth century, they were published in the *Report of the Secretary of the Navy*. Through most of the first half of the twentieth century they were printed separately as *Report of the Superintendent of the U. S. Naval Observatory*. From 1945 until 1967 the reports were published in the *Astronomical Journal*, volumes 51–72, and from 1968 to the present they are published in the *Bulletin of the American Astronmical Society* as part of its "Observatory Reports."

Many published observations are to be found in *Astronomical Observations* (also referred to as *Washington Observations*), 37 volumes published from 1845 until 1899, and the *Publications of the U. S. Naval Observatory*, 25 volumes published from 1900 to 1992. Other observations have been reported in a variety of journals. The *Astronomical Papers Prepared for the use of the American Ephemeris*, 23 volumes published from 1882 to 1986, include much of the research work of the Nautical Almanac Office, beginning with Newcomb. The USNO Circular series includes 173 items published from 1949 to 1990. A listing of the tables of contents of these Observatory publications is available at http://www.usno.navy.mil/library/index.html. A content analysis of the volumes of *Washington Observations* and *Publications of the U. S. Naval Observatory* may be accessed at the same site.

Earlier listings of USNO publications are found in M. Yarnall, J. E. Nourse, J. R. Eastman, and E. S. Holden, "Instruments and Publications of the United States Naval Observatory, 1845–1876," Appendix I to *Washington Observations for 1874* (Government Printing Office: Washington, 1877), pp. 8–9; E. S. Holden, "A Subject-Index to the Publications of the United States Naval Observatory, 1845–1875," Appendix I to *Washington Observations for 1876* (Government Printing Office: Washington, 1879); and William D. Horigan, *List of Publications Issued by the United States Naval Observatory, 1845–1908* (Government Printing Office: Washington, 1911).

B. *Congressional documents*

The *United States Statutes at Large*, commonly referred to as the *Statutes at Large*, is the official source for the laws and resolutions passed by Congress. Publication began retroactively in 1845, and legislation is published in order of the date of passage. At the time of writing, 17 volumes had been placed online, covering the laws of the first 42 Congresses, 1789–1873. They are accessible at the Library of Congress web site http://memory.loc.gov/ammem/amlaw/lwsl.html.

The *Congressional Globe*, commonly referred to as the *Globe*, contains the debates of Congress from the 23rd Congress, first session through the 42nd Congress (1833–73). There are 46 volumes in the series, printed as 110 books. The *Globe* was preceded by the *Annals of Congress* and the *Register of Debates*, and is followed by the *Congressional Record*. The *Globe* is also accessible at the Library of Congress web site.

## 2.   Oral-history interviews (cited as OHI)

Transcripts and tapes of oral histories are deposited in the Naval Observatory Library, Washington. Most have detailed tables of contents. They range in length from 30 minutes to ten hours, with most averaging 2–3 hours. All were conducted by the author except as indicated below. Interviews were conducted from 1983 to 2001, with most in the late 1980s.

| *Interviewee* | *Date* | |
|---|---|---|
| Harold Ables | June 19, 1989 | |
| Jack Belzer | June 9, 1988 | (LeRoy Doggett and Dick) |
| W. M. Browne | June 24, 1989 | |
| James W. Christy | June 16, 1989 | |
| Thomas Corbin | June 11 and 15, 1998 | |
| | December 6 and 13, 2001 | |
| Harry Crull | January 25, 1991 | |
| Julena S. Duncombe | June 18, 1983 | (LeRoy Doggett) |
| Raynor Duncombe | January 11, 1988 | |
| Joseph Egan | June 20, 1989 | |
| Otto Franz | June 21, 1989 | |
| John Hall | April 4, 1983 | |
| | June 18, 1989 | |
| Robert S. Harrington | November 4, 1992 | |
| Ralph Haupt | September 23, 1986 | (LeRoy Doggett) |
| | May 10, 1987 | |
| Arthur Hoag | June 15, 1989 | |
| Edward S. Jackson | January 23, 1991 | |
| William Markowitz | October 21, 1986 | |
| | August 18, 1987 | |
| | August 9, 1988 | |
| Alfred Mikesell | August 3, 4, 10, and 15, 1988 | |
| | September 14, 1988 | |
| Doug O'Handley, | December 4, 1999 | |
| Myles Standish, | | |
| and Henry Fliegel | | |

| Interviewee | Date | |
|---|---|---|
| Ida Ray | August 21, 1987 | |
| Dennis Robinson | September 10, 1996 | |
| Robert W. Rhynsburger | April 14, 1986 | |
| | August 20, 1987 | |
| Elizabeth Roemer | June 16, 1989 | |
| David Scott | February 26, 1988 | |
| P. K. Seidelmann | July 20 and 28, 2000 | |
| Stewart Sharpless | October 5, 1988 | |
| Clayton Smith | May 23 and 24, 1991 | |
| Paul Sollenberger | August 17, 1987 | |
| Kaj Strand | December 8, 1983 | (David DeVorkin and Dick) |
| | January 3, 1984 | (David DeVorkin and Dick) |
| John Watkins | January 29, 1989 | |
| Gart Westerhout | February 19, 1993 | |
| | March 5, 1993 | |
| | July 19, 21, 27, and 29, 1993 | |
| Marvin Whitney | May 2, 1988 | |
| Gernot M. R. Winkler | March 30 and 31, 1989 | |
| | May 18, 1989 | |
| Margaret Woodward | March 4, 1988 | |
| Charles Worley | December 12, 1987 | (David DeVorkin and Dick) |
| | August 2, 1995 | (Leroy Doggett and Dick) |

## 3. National Archives of the United States, Washington (cited as NA)

As of 2001, most records related to the Naval Observatory, including Record Groups 45 and 78, are located at the downtown Washington location of the National Archives (Archives I).

### A. Record Group 45 (Archives I)

Records related to the Depot of Charts and Instruments, 1830–42, are to be found in Record Group 45 – Naval Records Collection of the Office of Naval Records and Library. The inventory of this record group is found at http://www.nara.gov/guide/rg045.html. It is based on the "Preliminary Checklist of the Naval Records Collection of the Office of Naval Records and Library, 1775–1910," compiled by James R. Masterton, December, 1945.

Part II of the checklist contains an inventory of the Board of Navy Commissioners, 1815–42. Most directly related here is Entry 228 "Letters from the Depot of Charts and Instruments," seven volumes as follows.

Volume 1  June, 1831–December, 1836
Volume 2  January, 1837–November, 1837
Volume 3  January, 1838–September, 1838
Volume 4  October, 1838–May, 1839
Volume 5  January, 1840–January, 1841

Volume 6   February, 1841–December, 1841
Volume 7   January, 1842–August 27, 1842

B.   *Record Group 78 (Archives I)*
The chief archival source for U. S. Naval Observatory history is Record Group 78, comprising some 660 linear feet of records, including records of the Nautical Almanac Office, 1849–1911. The inventory of RG 78 may be found in http://www.nara.gov/guide/rg078.html, which is based on the *Guide to Federal Records in the National Archives of the United States* compiled by Robert B. Matchette et al. (National Archives and Records Administration: Washington, 1995), three volumes, 2,428 pages. Related record groups are cited here also.

C.   *Record Group 78 (Archives II, College Park, Maryland)*
Some 201 still images and one motion-picture reel dating back to 1930, as well as a number of miscellaneous records, are found at the College Park Archives II.

4.   **Library of Congress (cited as LC)**
The Library of Congress Law Library houses all the Congressional documents described in category 1B. The Naval Historical Foundation collection of correspondence of the Naval Depot of Charts and Instruments, 1833–54, of the Naval Observatory, 1866–95, and of the Nautical Almanac Office is in the Manuscript Division, Library of Congress, Washington. This material is supplementary to much more voluminous records in the National Archives. Among the personal papers in the Manuscripts Division are those of Louis Goldsborough, Charles Wilkes, James Melville Gilliss, Cleveland Abbe, Asaph Hall, and Simon Newcomb.

5.   **Navy Historical Center (cited as NHC)**
The biographical "ZB" file contains much information on naval officers. The holdings of the Center's Navy Department Library include a massive collection of the *Annual Register of the Commissioned, Warrant and Volunteer Officers of the Navy of the United States.* These volumes are essential for determining personnel, ranks, and much else about the Navy for any particular year. A list of Secretaries of the Navy, and much other information, is available at the Center's web site http://www.history.navy.mil.

6.   **U. S. Naval Observatory Archives On-Site (cited as USNOA)**
Records management in Federal institutions, especially management of historical records, is a perennial problem, and the U. S. Naval Observatory is no exception. While the Observatory's records prior to World War II are safely deposited in the National Archives in Washington, an important part of my task in preparing this history has been to collect, inventory, and preserve the records scattered around the Observatory since that time. They are now gathered in the Naval Observatory archives at its present location, and are referred to in this volume as USNOA.

A. *Written records*

USNOA citations in this volume are further divided into administrative files (USNOA, AF), arranged chronologically; biographical files (USNO, BF), arranged alphabetically; and subject files (USNO, SF), arranged by subject. In addition, on-site archives include Nautical Almanac Office Outgoing Correspondence, 1897–1982; records of the Time Service dating from the 1950s onward; records of the Transit Circle, Astrometry, and Equatorial Divisions; records and inventories related to instruments; Administrative Division records, including those relating to Navy versus civilian control of the Naval Observatory and relocation of the USNO to Charlottesville, Virginia, 1947–50, and personal papers of Naval Observatory astronomers. The detailed inventory of these records is at http://www.usno.navy.mil/library/index.html.

B. *Photographic records*

The Naval Observatory Library holds an extensive collection of photographs related to the Naval Observatory. Almost all of the photographs in this volume are from the collection, which was organized by Mabel Sterns in the 1980s.

7. **Current information**

The Naval Observatory web site is http://www.usno.navy.mil.

# Appendix 2: Superintendents, Scientific Directors, and Department Directors of the U. S. Naval Observatory

**Officers-in-Charge of the Depot of Charts and Instruments**

| | |
|---|---|
| Lt L. M. Goldsborough | December 6, 1830–February 11, 1833 |
| Lt Charles Wilkes | March 12, 1833–June 14, 1837 |
| Lt James Melville Gilliss | June 14, 1837–July 11, 1842 |
| Lt Matthew F. Maury | July 11, 1842–October 1, 1844 |

**Superintendents of the Naval Observatory**
(Ranks are the highest rank achieved during tenure as Superintendent)

| | |
|---|---|
| Cdr Matthew F. Maury | October 1, 1844–April 20, 1861 |
| Capt James M. Gilliss | April 23, 1861–February 9, 1865 |
| RADM Charles Henry Davis | April 28, 1865–May 15, 1867 |
| RADM Benjamin F. Sands | May 15, 1867–February 23, 1874 |
| RADM Charles Henry Davis | February 23, 1874–February 18, 1877 |
| RADM John Rodgers | May 1, 1877–May 5, 1882 |
| VADM Stephen C. Rowan | July 1, 1882–May 1, 1883 |
| RADM Robert W. Shufeldt | May 1, 1883–February 21, 1884 |
| RADM Samuel R. Franklin | February 24, 1884–March 31, 1885 |
| Commo George E. Belknap | June 1, 1885–June 7, 1886 |
| Capt Robert L. Phythian | November 15, 1886–June 28, 1890 |
| Capt Frederick V. McNair | June 28, 1890–November 21, 1894 |
| Commo Robert L. Phythian | November 21, 1894–July 19, 1897 |
| Capt Charles H. Davis II | July 19, 1897–April 16, 1898 |
| Commo Robert L. Phythian | May 18, 1898–November 1, 1898 |
| Capt Charles H. Davis II | November 1, 1898–November 1, 1902 |
| RADM Colby M. Chester | November 1, 1902–February 28, 1906 |
| RADM Asa Walker | February 28, 1906–November 12, 1907 |
| RADM William J. Barnette | December 2, 1907–April 19, 1909 |
| Commo T. E. DeWitt Veeder | April 20, 1909–October 16, 1911 |
| Capt Joseph L. Jayne | October 16, 1911–February 11, 1914 |
| Capt John A. Hoogewerff | February 14, 1914–March 31, 1917 |
| RADM Thomas B. Howard | March 31, 1917–March 4, 1919 |
| RADM John A. Hoogewerff | March 4, 1919–June 8, 1921 |
| Capt William MacDougal | August 1, 1921–October 5, 1923 |
| Capt Edwin Pollock | October 5, 1923–September 19, 1927 |
| Capt Charles S. Freeman | September 19, 1927–May 21, 1930 |
| Commo J. F. Hellweg | June 9, 1930–February 18, 1946 |

Capt Ralph S. Wentworth       February 18, 1946–September 1, 1946
Capt Guy W. Clark             September 1, 1946–April 20, 1949
Capt George W. Welker         April 20, 1949–April 27, 1951
Capt Frederic A. Graf         July 23, 1951–June 30, 1956
Capt Charles L. Freeman       October 17, 1956–August 21, 1957
Capt Carl G. Christie         January 8, 1958–June 30, 1959
Capt Byron L. Gurnette        June 30, 1959–June 30, 1961
Capt Alvin W. Slayden         June 30, 1961–June 30, 1962
Capt Thomas S. Baskett        July 1, 1962–June 30, 1965
Capt Joseph M. McDowell       September 10, 1965–June 30, 1968
Capt J. Maury Werth           July 1, 1968–August 31, 1970
Capt John R. Hankey           September 1, 1970–June 30, 1972
Capt Edward A. Davidson       June 30, 1972–September 28, 1973
Capt Sherwin J. Sleeper       September 28, 1973–September 17, 1976
Capt Joseph C. Smith          September 17, 1976–September 24, 1979
Capt Raymond A. Vohden        September 24, 1979–August 27, 1982
Capt Charles K. Roberts       August 27, 1982–September 16, 1986
Capt Richard A. Anawalt       September 16, 1986–July 27, 1988
Capt James B. Hagen           July 27, 1988–August 14, 1991
Capt Winfield Donat III       August 14, 1991–June 29, 1993
Capt Richard E. Blumberg      June 29, 1993–August 11, 1995
Capt Kent W. Foster           August 11, 1995–August 11, 1997
Capt Dennis G. Larsen         August 11, 1997–January 28, 2000
Capt Benjamin Jaramillo       January 28, 2000–

### Scientific Directors of the Naval Observatory

William Harkness       December 21, 1894–December, 1899
                       ("Astronomical Director")
Stimson J. Brown       December, 1899–March, 1901
                       ("Astronomical Director")
(Office vacant         1901–1958)
Gerald Clemence        February 1, 1958–1963
K. A. Strand           September 1, 1963–February 28, 1977
Gart Westerhout        June 20, 1977–October 2, 1993
Kenneth Johnston       October 3, 1993–

### Superintendents of the Nautical Almanac Office

Lt Charles Henry Davis        July 11, 1849 (ordered)–November 23, 1856
Professor Joseph Winlock      November 23, 1856–August 9/10, 1859
Cdr Charles Henry Davis       August 10, 1859–September 18, 1861
Professor Joseph Winlock      September 18, 1861–May 1, 1866
Professor John H. C. Coffin   May 1, 1866–September 15, 1877
Professor Simon Newcomb       September 15, 1877–September 20, 1894

**Directors of the Nautical Almanac Office (title changed September 20, 1894)**

| | |
|---|---|
| Professor Simon Newcomb | September 20, 1894–March 12, 1897 |
| Professor William W. Hendrickson | March 12, 1897–June 30, 1897 |
| Professor William Harkness | June 30, 1897–December 15, 1899 |
| Professor Henry D. Todd | December 15, 1899–August 24, 1900 |
| Professor Stimson J. Brown | August 24, 1900–March 25, 1901 |
| Professor Walter S. Harshman | March 28, 1901–October 1, 1907 |
| Professor Milton Updegraff | October 1, 1907–November 2, 1910 |
| Professor William S. Eichelberger | November 2, 1910–September 18, 1929 |
| James Robertson | September 18, 1929–May 31, 1939 |
| Walter M. Hamilton | May 31, 1939–February 1, 1940 |
| Wallace J. Eckert | February 1, 1940–February 28, 1945 |
| Gerald M. Clemence | February 28, 1945–January 31, 1958 |
| Edgar W. Woolard | January 31, 1958–January 31, 1963 |
| Raynor L. Duncombe | January 31, 1963–1975 |
| P. Kenneth Seidelmann | 1975–1990 |

In September, 1990 the Astronomical Applications Department was created and the Nautical Almanac Office became a small branch of that organizational unit.

Chief, Nautical Almanac Office

| | |
|---|---|
| LeRoy Doggett | 1990–1996 |
| Alan Fiala | 1996–2000 |
| Sethanne Howard | 2000– |

**Directors of Astronomical Applications Department**

| | |
|---|---|
| Paul Janiczek | 1990–1997 |
| John Bangert | December 1997– |

**Director of Orbital Mechanics Department**

| | |
|---|---|
| P. K. Seidelmann | 1990–1994 |

**Directors of Time Service and Their Predecessors** (compiled by Whitney Treseder)

Officer in charge of the Transit Circle (and the Time Service)

| | |
|---|---|
| Professor J. R. Eastman | 1880–1881 |

Head of the Department of Chronometers

| | |
|---|---|
| Lt E. K. Moore | 1882–1883 |

Head of the Department of Chronometers and Time Service

| | |
|---|---|
| Lt E. C. Pendleton | 1884–1885 |
| Lt S. C. Paine | 1886–1887 |
| Ensign A. N. Mayer | 1888 |
| Lt Hiero Taylor | 1889–1890 |

| | |
|---|---|
| Ensign Thomas Snowden | 1891 |
| Professor S. J. Brown | 1892 |
| Lt L. C. Heilner | 1893–1895 |
| Lt William V. Bronaugh | 1896 |
| Assistant Astronomer H. M. Paul | 1897 |
| and Lt Charles E. Fox | |
| Lt A. N. Mayer | 1898 |
| and Professor H. M. Paul | |
| Lt A. N. Mayer, | 1899 |
| and Computer M. E. Porter, and | |
| Lt B. W. Hodges | |
| Lt-Cdr B. W. Hodges | 1900 |
| Lt-Cdr Charles E. Fox | 1901 |
| and Computer M. E. Porter | |
| Lt-Cdr Edward Everett Hayden | 1902–1906 |
| Lt-Cdr E. E. Hayden | 1906 |
| and Assistant E. A. Boeger | |
| Cdr E. E. Hayden | 1907–1910 |

Superintendent of Compasses/Head of Department of Chronometers and Time Service

| | |
|---|---|
| Lt-Cdr J. H. Sypher/Lt-Cdr Edward McCauley Jr | 1911 |

Head of the Department of Compasses, Nautical Instruments, and Time Service

| | |
|---|---|
| Lt-Cdr Edward McCauley Jr | 1912 |
| Lt-Cdr W. R. Gherardi and Cdr Edwin T. Pollock | 1913–1915 |
| Cdr W. D. MacDougall | 1916 |
| Lt-Cdr James P Murdock and Cdr George C. Day | 1917 |

Head of the Division of Nautical Instruments and Time Service

| | |
|---|---|
| Cdr C. T. Jewell and Lt R. S. Patten | 1918 |
| Cdr C. S. Graves | 1919–1920 |
| Lt-Cdr J. G. Stevens | 1921–1923 |
| Lt-Cdr J. F. Crowell | 1924–1925 |
| Cdr A. G. Stirling | 1926–1927 |

Director of Time Service

| | |
|---|---|
| Paul Sollenberger | 1928–1955 |
| William Markowitz | 1955–1966 |
| Gernot Winkler | 1966–1995 |
| Demetrios Matsakis | 1995– |

On June 1, 1994 the Directorate for Time was established, which included the Time Service and Earth Orientation Departments. On that date Dr Gernot Winkler became the first Director of the Directorate. In 1995 Dr Dennis McCarthy became the second Director.

**Directors of Meridian Instruments**

5.3-inch Transit Instrument and 4-inch Mural Circle
M. Yarnall                1852–1877

8.5-inch Transit Circle
Simon Newcomb         November 8, 1865–October 9, 1867
Asaph Hall             October 9, 1867–December 31, 1867
Simon Newcomb         January 1, 1868–December 31, 1869
William Harkness       January 1, 1870–May 29, 1874
J. R. Eastman          June 1, 1874–July 1, 1891
                       (dismounted and converted to 9-inch)

9-inch Transit Circle (Division established September 20, 1911)
A. N. Skinner          1894–1900
F. B. Littell          July 30, 1901–September 3, 1903
W. S. Eichelberger     September 3, 1903–September 30, 1908
F. B. Littell          October 1, 1908–November 11, 1913
H. R. Morgan           November 11, 1913–1944
The 9-inch transit circle was decommissioned at Morgan's retirement in 1944

6-inch Transit Circle (Division established September 20, 1911)
J. C. Hammond          1911–December 31, 1933
C. B. Watts            1934–1959
A. Norwood Adams       1959–1969
Benny L. Klock         1969–1976 (see Transit Circle Division below)

7-inch Transit Circle
Francis P. Scott       1948–June 12, 1970
Jack L. Schombert      1970–1976 (see Transit Circle Division below)
In 1976 the instrument was placed under a single Transit Circle Division. The instrument was located in Argentina 1966–1973, and in New Zealand, 1984–1996.

Transit Circle Division
Jack L. Schombert      1976–1977
James A. Hughes        1977–1982 (see Astrometry Department below)

Prime Vertical
J. S. Hubbard          1862–1863
Simon Newcomb
Asaph Hall                          –1867
Edgar Frisby           May 10, 1893–
George A. Hill         December 9, 1893–1927

In 1893 this instrument was remodeled to a Fauth-type instrument.

Altazimuth
G. A. Hill                  1894–1903
Frank B. Littell          1903–1933

**Directors of Equatorial Telescopes**
9.6-inch
James Ferguson           1848–1867
Asaph Hall               1868–1874
J. R. Eastman            1874–1882
Cdr. W. T. Sampson       1883–1884
Edgar Frisby             April 2, 1885–December 19, 1895

26-inch
Simon Newcomb            November 3, 1873–June 16, 1875
Asaph Hall               1875–June 16, 1891
Stimson J. Brown         1893–December 15, 1899
T. J. J. See             December 18, 1899–September 20, 1902

**Division of Equatorial Instruments**
A. N. Skinner            September 20, 1902–July 10, 1907
W. S. Eichelberger       July 10, 1907–July 28, 1908
Asaph Hall Jr            July 29, 1908–June 30, 1929
H. E. Burton             1929–June 30, 1948
John S. Hall             September 1, 1948 – September 5, 1958

**Astrometry and Astrophysics Division**
K. A. Strand             August 1, 1958–September 1963
Stewart Sharpless        September 29, 1963–August 1964
Otto Franz (Acting)      August 1, 1964–August 1965
Victor Blanco            September 1, 1965–14 July, 1967
Paul Routly              1 September, 1968–1982
In 1982 the Exploratory Development Staff became a part of the Astrometry Department.

**Astrometry Department**
James A. Hughes          1982–January 15, 1992
Clayton Smith            January 1992–March, 1993
F. Stephen Gauss         May 9, 1993–June 2, 2000
Theodore R. Rafferty     October 8, 2000–

On June 1, 1994 the Directorate for Astrometry was established, including the Astrometry Department, the Astronomical Applications Department, and the Flagstaff station. Dr P Kenneth Seidelmann was named the first Director.

**Directors of Flagstaff Station**

| | |
|---|---|
| Arthur Hoag | 1955–1965 |
| Gerald E. Kron | September 1, 1965–September 1973 |
| Harold Ables | 1974–October 2, 1995 |
| Conard Dahn | January 1996– |

**Directors of Naval Observatory Time Service Station, Richmond, Florida**

| | |
|---|---|
| Gerald C. Whittaker | ?–1962 |
| Donald R. Monger | 1962–1985 |
| James Martin | 1985–1987 |
| Alice K. Babcock (Monet) | 1987–1990 |
| Timothy S. Carroll | 1990–1996 |

In 1996 the functions of the Richmond station were transferred to Falcon Air Force Base, Colorado. The alternate Master Clock at Falcon (later renamed Schriever Air Force Base) became the reference for the GPS Master Control Station on September 12, 1996.

**Librarians of the USNO** (compiled by Brenda G. Corbin)

| | |
|---|---|
| J. S. Hubbard, Professor of mathematics | May, 1845–August, 1863 |
| William Harkness, Professor of mathematics | August, 1863–October, 1865 |
| J. E. Nourse, Professor of mathematics | October 1865–February, 1879 |
| E. S. Holden, Professor of mathematics | February, 1879–February,1881 |
| Lt E. F. Qualtrough | February, 1881–June, 1882 |
| Lt G. E. Yardley | June, 1882–July, 1883 |
| Lt J. C. Wilson | July 1883–August, 1885 |
| Lt L. L. Reame | August, 1885–May, 1887 |
| W. D. Horigan (acting) | May, 1887–May, 1889 |
| Assistant Astronomer H. M. Paul | May, 1889–October, 1892 |
| W. D. Horigan (assistant and later librarian) | October, 1892–July, 1927 |
| Prof. Asaph Hall Jr | July, 1927–June, 1929 |
| Grace O. Savage | August, 1929–November, 1951 |
| Marjorie S. Clopine | March, 1952–1962 (?) |
| Lettie Bevis (later Multhauf) | 1962 (?)–1968 (?) |
| Veronica Williams (acting) | 1968–1969 |
| Anne G. Grundstein | March, 1969–April (?), 1973 |
| Brenda G. Corbin | April, 29, 1973– |

Names of Secretaries of the Navy and their dates of tenure can be found at http://www.history.navy.mil/faqs/faq37-1.htm.

# Appendix 3: Selected astronomical instrumentation and standard clocks at the U. S. Naval Observatory, 1844–2000

Equatorial

| 1830 | 1860 | 1890 | 1920 | 1950 | 1980 | 2000 |

9.6-inch Merz & Mahler refractor — 1844

26-inch Clark refractor — 1873

12-inch Clark refractor — 1895

10-inch photo refr. — 1912

1934 — 15-inch refractor — 1996

40-inch Ritchey–Chrétien reflector — 1935

61-inch reflector — 1963

8-inch astrograph — 1970

**Meridian**

4-inch Troughton & Simms mural circle
1844 — 1877

5.3-inch Ertel/Merz & Mahler transit instrument
1844 —

5-inch Pistor & Martins prime vertical
1844 — 1867 — 1893 — 1912 — 1921–25

4.5-inch Ertel & Sons transit circle
1844 — 1862

8.5-inch Pistor & Martins transit circle
1866 — 1891

9-inch transit circle
1893 — 1945

altazimuth
1897 — 1933

6-inch Warner & Swasey transit circle
1897 —

7-inch transit circle
1955 — 1995

1996

**Time Determination**

5.3-inch Ertel/Merz & Mahler transit
1893 —

6-inch
1916 — 1925

Prin transit

1915 photographic (variation of latitude)

1934 zenith tubes (time)

radio telescopes
1984 —

**Timekeeping Standards**

Riefler
1903 —

Shortt
1931 —

Quartz
1946 —

Atomic
1966 —

**Time Dissemination**

Time Ball at USNO
1845 —

Telegraph
1866 —

At Old Executive Office Bldg
1887 —

Radio
1904 —

1912

Arlington Towers + others
1936 — 1941

VLF, LORAN, etc.
1955 —

GPS

# Appendix 4: U. S. Naval Observatory: Six generations of selected personnel

| 1830 | 1860 | 1890 | 1920 | 1950 | 1980 |
|------|------|------|------|------|------|
| Generation 1 | Generation 2 | Generation 3 | Generation 4 | Generation 5 | Generation 6 |

**Administrative**

[Military]

30 —— 44 —— 61 —— 86 —— 98 ——
Goldsborough   Maury   Davis   Phythian   Davis
Wilkes
Gilliss

—— 82
Sands/Rodgers
—— Gilliss

[Civilian]

48 —— T. Harrison

21

94 99 01
Harkness/Brown

27 — 30 —— 46
Freeman   Hellweg

58 —— Clemence
63 —— Strand
77 —— Westerhout
93 —— Johnston

**Equatorial**

61 —— 77
Newcomb (see NAO)

48 —— Ferguson   67

62 —— A. Hall Sr.   91   98

05 A. Hall Jr.   29 Burton   48 Hall   58 Strand 68 Routly  82
09 —— Burton

Peters   31  36 Mikesell   70  69 Harrington

20   Willis   44   60   Worley

**Meridian**

Hubbard 45–63
Eastman 61
Yarnall 52–79
Littell 91–98, 33
Morgan 10–44
Hammond 98–33
Harkness 62–99
Watts 99–11
Skinner 70–07
Adams 31–69
Frisby 68–99
F. P. Scott 29–70
Eichelberger (see NAO) 96–09
Smith/Hughes/Gauss/Corbin 59
Seidelmann 90

**Time**

Gardner 64–98
See 99–30
Sollenberger 14
Markowitz 36–53
Robertson 29–39
Eckert 45–58
Clemence 53
R. G. Hall 53–82
McCarthy
Winkler 66
Matsakis 95
Woolard/Duncombe/Seidelmann 90

**Nautical Almanac**

Davis 49–56
Winlock 66
Coffin 77
Newcomb 97
G. W. Hill 61–92
Lewis 05–51
Eichelberger 10–29

**Flagstaff**

Hoag 50
Kron 74
Ables
Dahn 96

# Appendix 5: Key legislation related to the Naval Observatory and Nautical Almanac Office[a]

| Congress, session | Date | Subject | Appropriation | Statute[b] (volume, page) |
|---|---|---|---|---|
| 27th, second | August 31, 1842 | To authorize construction of a Depot of charts and instruments for the Navy of the United States | $25,000[c] $10,000 | 5, 576 |
| 30th, first | August 3, 1848 | Pay of Superintendent of Naval Observatory; Superintendent a Captain, Commander, or Lieutenant | $3,000 | 9, 266 |
| | Ditto | Professors of Mathematics appointed by President | None | 9, 272 |
| 30th, second | March 3, 1849 | Preparation of Nautical Almanac Construction of a Magnetic Clock | None $10,000 | 9, 375 9, 374 |
| 31st, first | September 28, 1850 | Use of Greenwich and Observatory meridians | None | 9, 515 |
| 38th, second | March 3, 1865 | Rank of Superintendent as Captain, Commander or Lieutenant repealed | None | 13, 533 |
| 41st, second | July 5, 1870 | A refracting telescope of the largest size (appropriations made over several years ending June 6, 1874) | $50,000[c] $10,000 | 16, 334 |
| Transits of Venus | | | | |
| 41st, third | March 3, 1871 | Preparing instruments for the Transit of Venus | $2,000 | 16, 529 |
| 42nd, second | June 10, 1872 | Purchase and preparation of instruments for the Transit of Venus | $50,000 | 17, 367 |
| 42nd, third | March 3, 1873 | Transit-of-Venus expeditions | $100,000 | 17, 514 |
| 43rd, first | June 23, 1874 | Completion of transit-of-Venus work and return of expeditions | $25,000 | 18, 210 |
| 47th, first | August 7, 1882 | To organize parties for 1882 transit of Venus | $75,000 | 22, 323 |
| 49th, first | 26 July, 1886 | Completion of reduction of transit-of-Venus observations | $3,000 | 24, 150 |
| New Naval Observatory | | | | |
| 45th, second | June 20, 1878 | To appoint a Commission to ascertain cost of removing the Naval Observatory | None | 20, 241 |
| 46th, second | February 4, 1880 | To locate and purchase a new site for the U. S. Naval Observatory | $75,000 | 21, 64 |
| 49th, first | July 26, 1886 | For commencing the erection of the new Naval Observatory | $50,000 | 24, 156 |
| 49th, second | March 3, 1887 | For continuing the erection of the new Naval Observatory | $60,000 | 24, 585 |

| Congress, session | Date | Subject | Appropriation | Statute[b] (volume, page) |
|---|---|---|---|---|
| 50th, first | September 7, 1888 | For continuing the erection of the new Naval Observatory | $50,000 | 25, 463 |
| 50th, second | March 2, 1889 | For completing the new Naval Observatory | $240,000 | 25, 814 |
| 51st, second | March 2, 1891 | For bookcases, instruments, buildings, accessories, roads, and residence of Superintendent | $136,689 | 26, 806 |
| 53rd, second | August 1, 1894 | To establish an Observatory Circle | None | 28, 588 |
| 56th, second | March 3, 1901 | Appointment of a Board of Visitors Superintendent a line officer of the Navy, not below rank of Captain | None | 31, 1,122 |
| 64th, first | August 29, 1916 | Providing no further appointments to the Corps of Professors of Mathematics | None | 39, 577 |
| 68th, second | December 5, 1924 | To complete purchase of land lying within Observatory Circle | $4,041 | 43, 689 |
| 70th, second | December 10, 1928 | Superintendent's quarters assigned to Chief of Naval Operations | None | 45, 1,018 |
| 71st, second | June 11, 1930 | To provide for modernization of Naval Observatory | $225,000[c] $0 appropriated | 46, 556 |
| 71st, third | February 28, 1931 | Astrographic and research work, and modernization | $50,000 | 46, 1,453 |
| 72nd, first | June 30, 1932 | Ditto | $110,000 | 47, 445 |
| 81st, second | June 17, 1950 | Relocation of Naval Observatory to Charlottesville, Virginia | $7,000,000 (not expended) | 64, 239 |
| 86th, second | June 8, 1960 | Research, development, and testing facilities at Flagstaff, Arizona | 1,900,000[c] $0 appropriated | 74, 171 |
| 86th, second | July 12, 1960 | Ditto (61-inch reflector) | $1,900,000 | 74, 463 |
| 93rd, second | July 12, 1974 | Official Residence of the Vice President of the United States | None | 88, 341 |

Notes:
[a] Exclusive of routine annual appropriations. Most of the legislation listed here is part of the annual Navy Appropriations Acts. Transit-of-Venus appropriations were in the Sundry Civil Expense Acts. Construction at Flagstaff is part of the Military Construction Act. Increasingly during the Space Age, the Naval Appropriations Acts account for only a portion of the Naval Observatory's budget, which was supplemented by funding from the Office of Naval Research, NASA, the Air Force, and other sources.
[b] *Statutes at Large*.
[c] Authorized, with appropriations delayed or spread over several years.

# Index

Page numbers in italics refer to illustrations.